PRODUCTION TO NEAR NET SHAPE

Source Book

A collection of outstanding articles from the technical literature

PRODUCTION TO NEAR NET SHAPE

Source Book

A collection of outstanding articles from the technical literature

Compiled by
Consulting Editors

C. J. VAN TYNE
Assistant Professor of Metallurgical Engineering
Lafayette College

B. AVITZUR
Director of the Institute for Metal Forming
Lehigh University

American Society for Metals
Metals Park, Ohio 44073

Contributors to This Source Book

STANLEY ABKOWITZ
Dynamet Technology Inc.

N. AKGERMAN
Battelle's Columbus Laboratories

J. M. ALEXANDER
University College, Swansea

TAYLAN ALTAN
Battelle's Columbus Laboratories

R. W. BALLIETT
Phelps Dodge Brass Co.

J. L. BARTOS
General Electric Co.

JOSEPH BENEDYK
American Can Co.

JOHN C. BITTENCE
Senior Editor, *Materials Engineering*

M. J. BLACKBURN
Pratt and Whitney Aircraft Div.

E. F. BRADLEY (ret.)
Pratt and Whitney Aircraft Div.

GOVIND R. CHANANI
Northrop Corp., Aircraft Div.

HARRY E. CHANDLER
Editor, *Metal Progress*

C. C. CHEN
Wyman-Gordon Co.

L. PAUL CLARE
GTE Sylvania

JAMES E. COYNE
Wyman-Gordon Co.

T. A. DEAN
Dept. of Mechanical Engineering
University of Birmingham

J. L. DUNCAN
McMaster University

RICHARD EDMONDSON
National Machinery Co.

H. FISCHMEISTER
Chalmers University of Technology

K. M. FISHER
GKN Group Technological Centre

P. H. FLOYD
National Aeronautical Establishment
National Research Council Canada

J. A. FORSTER
McMaster University

G. H. GESSINGER
Brown Boveri & Co. Ltd.

RANDOLPH GOLD
Editor, *Precision Metal*

ALICE M. GREENE
Technical Editor, *Iron Age*

ROBERT E. HARVEY
Senior Editor, *Iron Age*

R. L. HEWITT
National Aeronautical Establishment
National Research Council Canada

W. T. HIGHBERGER
NAVAIR, Naval Air Systems Command

J-P. A. IMMARIGEON
National Aeronautical Establishment
National Research Council Canada

C. A. KELTO
AFML/LTM Wright Patterson AFB

K. M. KULKARNI
Cabot Corp.

G. D. LAHOTI
Battelle's Columbus Laboratories

TEIZO MAEDA
University of Tokyo

R. B. MICLOT
U.S. Army Weapons Command

MADAO MURAKAWA
The Institute of Physical and
    Chemical Research

JOHN H. MOLL
Colt/Crucible Research Center

TAKEO NAKAGAWA
University of Tokyo

N. M. PARIKH
IIT Research Institute

J. W. H. PRICE
Nuclear Power Company (Risley) Ltd.

S. RAJAGOPAL
IIT Research Institute

RUSSELL H. RHODES
GTE Sylvania

NOTE: Affiliations given were applicable at date of contribution.

GREGORY V. SCARICH
Northrop Corp., Aircraft Div.

S. L. SEMIATIN
Battelle's Columbus Laboratories

ROGER R. SKROCKI
TRW Inc.

R. A. SPRAGUE
Pratt and Whitney Aircraft Div.

D. STAWARZ
La Salle Steel

M. J. STEWART
Dept. of Energy, Mines & Resources,
   Canada

T. L. SUBRAMANIAN
Battelle's Columbus Laboratories

Y. VAN HOENACKER
Dept. of Mechanical Engineering
University of Birmingham

T. WATMOUGH
IIT Research Institute

CHARLES WICK
Managing Editor, *Manufacturing
   Engineering*

G. WILLIAMS
GKN Group Technological Centre

JOHN T. WINSHIP
Associate Editor, *American Machinist*

R. H. WITT
Grumman Aerospace Corp.

PREFACE

Traditionally, metal manufacturers have relied on several basic forming techniques such as casting, forging, and machining to impart the desired geometrical shape to their products. In recent years, because of stiffer industrial competition, the development of new alloys, shortages of certain metals, and the increase in energy costs, these traditional processing methods are being critically analyzed and reevaluated. It is becoming very desirable to produce the final product in fewer processing steps and with as little waste as possible. Several techniques for the manufacture of components to ''net shape'' or to ''near net shape,'' based on the firm foundation of the traditional processes, are being developed to meet these challenges of today and of the future. This Source Book brings together articles that discuss many aspects of these ''net shape'' or ''near net shape'' production processes.

The fundamental incentive for these net shape processes is economic. In spite of the fact that these methods often require extra capital equipment and/or special handling procedures, the over-all cost per product in many cases can be substantially decreased by a net shape forming method. The elimination of trimming, machining, and some final finishing operations, the better use of critical materials and, in some cases, substitution of materials, as well as the possible decrease in total energy consumption provide the justification for these techniques.

The process of producing to final net shape is a desirable goal for many metal products. In fact, casting, a very old and time-proven process, can produce a useful solid object directly from the molten state and thus can be considered as an early version of the new family of net shape processes. The areas this Source Book examines are the metalworking processes that have recently been developed to achieve the desirable shaping to present-day standards, in a minimum number of steps. The types of processes the articles in this book deal with cover the gambit of modern net shape processing techniques. Such

methods as flashless forging, hot-die forging, isothermal forging, hot isostatic pressing, squeeze forming, and superplastic forming are presented. Although these processes are based on seemingly different techniques, all can be used to form metals to net or near net shape.

Flashless Forging

The process of flashless forging has many advantages over the more common forging methods in which a flash is produced. One obvious advantage is that the elimination of the flash reduces the scrap produced. Since no flash is formed, the preform is smaller in size and thus the amount of energy needed to heat the material to the forging temperature is also reduced. A reduction in the number of finishing steps that are often required in the more traditional processes can also occur. In order to use this technique effectively, close and precise control must be exercised over the volume of the blank entering the forging dies. Undersize blanks will prevent complete filling of the die, whereas oversize blanks may damage the dies or jam the press because there is no escape provided for the excess material. This necessitates very close size tolerance on the material which enters the forge. Thus the cutting or cropping of the blanks must be done with precision. In order to achieve these close tolerances, techniques like fine blanking or fine shearing must be used.

Hot-Die and Isothermal Forging

Heating the dies in a forging press is another technique that can be used to form parts to net or near net shape. The high-temperature dies (especially if they are at the same temperature as the metal being formed) allow for easier plastic flow of the material and thus a more complete filling of the dies. This technique can also decrease the excess volume required to ensure a complete filling of the critical cavities within the die. By proper preform design large deformations can be achieved and very precise parts can be made via this technique. The cost, however, is high, and presently the technique is reserved to high-priced components usually made of expensive alloys, with which material conservation is of primary importance.

Hot Isostatic Pressing

Forming complex-shaped objects from powder metals is a method that has been used for many years. Once formed, the part must be sintered (or plastically formed further) in order to cause a strong bonding between the individual powder particles. With the new developments in rapid solidification technology, extremely good properties of the metal powder can be achieved. One of the most recent developments in the processing of these powder metals is hot isostatic pressing (HIP). Many high-temperature alloys can be made to net or near net shape via this technique. Furthermore, solid solution of elements that are desired but cannot be produced by casting from the melt, can nevertheless be produced by the powder technology. Although HIP is a more complicated and costly process than the forging and machining techniques that are used to make precision parts, HIP minimizes waste and can often be a net or near net shape process. Thus, for components of intricate shapes and made of costly alloys, HIP offers a distinct advantage.

Other Processes

Other net shape processes that can be used under certain conditions are hot pressing of scrap material to produce a net shape product; squeeze forming, wherein the metal is molten or semimolten during the mechanical forming steps; and superplastic forming, whereby extremely large deformations can

be achieved in certain alloy systems with low pressure being applied. These techniques are more specialized and can be used only when the metal or alloy has certain unique properties that allow these processing techniques.

In spite of the fact that each production method briefly described above is more expensive or requires more control than the traditional techniques, the savings in raw material and energy consumption and the elimination of final finishing steps often justify the newer net shape methods.

This Source Book attempts to bring together some of the recent articles in this rapidly expanding field. Included among its contents are descriptions of each technique in detail and discussions of both the advantages and limitations of each method.

We hope that the articles in this book are of value to those who contemplate using and wish to learn these methods as well as the people who want to improve upon their present net shaping techniques. The references listed in these articles can also guide the reader toward further, more intimate studies of each selected technique.

*　　*　　*　　*　　*

The American Society for Metals extends grateful acknowledgment to the many authors whose work is presented in this book, and to their publishers.

C. J. VAN TYNE
Assistant Professor of Metallurgical Engineering
Lafayette College

B. AVITZUR
Director of the Institute for Metal Forming
Lehigh University

CONTENTS

SECTION VI: ISOTHERMAL FORGING

SECTION VII: ISOSTATIC PRESSING

SECTION VIII: OTHER NET SHAPE PROCESSES

PRODUCTION TO NEAR NET SHAPE

Source Book

A collection of outstanding articles from the technical literature

SECTION I
Introduction

Edging closer to 'net shape'

By John C. Bittence, Senior Editor

One of the virtues of powder metallurgy is that the process can hold close dimensional tolerances. As a result, little secondary processing is required to finish a PM part.

For years, this virtue has been valued by manufacturers of aerospace and other components made primarily from costly metals that were nearly impossible to machine. However, with rising labor and ma-

This article was a featured presentation by Materials Engineering *at the 1979 Winter Meeting of the Powder Metallurgy Parts Association.*

terials costs, even low-carbon steel chips can be expensive to produce; now manufacturers of production-line, high-volume commercial products are also interested in turning out parts as close as possible to finished — or net — shape.

While PM has been the dominant near-net-shape process for the last two decades, other metalworking processes are now becoming economical for production-line manufacture. Although some of these processes have been lying dormant for up to a decade, recent processing refinements, along with the changing economics of today's marketplace, have brought these near-net-shape processes into a position where they can be competitive with PM.

How near net shape?

Manufacturing processes generally are not interchangeable. Usually you cannot simply switch from one process to another without affecting certain mechanical properties in the finished part.

Likewise, near-net-shape capabilities vary from process to process. Close tolerances can't always be maintained in all directions without the need for secondary machining or processing. While some metalworking processes may be able to hold close-tolerance net shapes in one or two directions, variations in other directions may be ten times larger.

A variety of manufacturing processes are currently regarded as good near-net-shape candidates.

3

What makes fine-blanking so fine?

Fine-blanking is a carefully engineered metal-stamping and forming process in which the tooling is designed to control the amount of elastic deformation in the workpiece precisely. In addition, permanent plastic deformation is limited only to regions where it is needed to separate the part from the blanking stock.

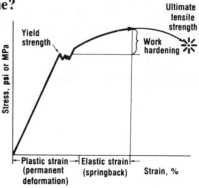

The heavy line, superimposed over the stress-strain curve, shows how a metal deforms during a typical metal-stamping or forming process. Metalworking is elastic plus plastic deformation, which takes place beyond the yield point. Once the applied load or stress is removed only the plastic deformation remains. Elastic deformation disappears through a phenomenon called springback or recovery.

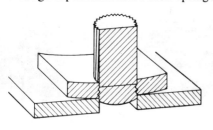

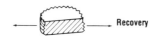

Recovery

Conventional stamping

During conventional stamping, both elastic and plastic strain exceed the ultimate strength of the metal, and the part breaks away from the stock under shear and rupture stresses. The exaggerated drawing of the material as it is about to break shows the excessive deformation that takes place, much of which is recovered in the form of springback as soon as the part breaks away.

The edge of the part is partially smooth and partially rough, from the final rupture. Because of the gross distortion that takes place in the workpiece, close-tolerance features can't be included in the conventional stamped blank.

Fine-blanking, on the other hand, uses two additional pieces of tooling — a counterpunch below that eliminates unnecessary elastic and plastic deformation, and an impingement ring or stripper (usually on top) to help retain the material. The combined effect of these added elements virtually eliminates the rough surface of the workpiece edge. The part separates from the raw material with a "clean" shear surface. Because stresses are concentrated only where needed, there is little over-all gross distortion during the process. Consequently, tiny and precise details can be stamped into the part.

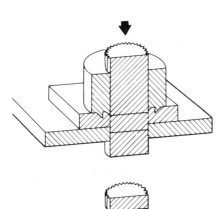

Fine blanking

Punch and die clearance for ordinary stamping is usually about 5% of the dimension across the die. For fine-blanking, this clearance is about 1% — sometimes less than 0.5%.

Tolerances in fine-blanked parts are improved, typically by an order to magnitude, over ordinary stamping. Small holes, with diameters as little as 60% of the stock thickness, can be produced. Webs — that is, metal between features such as holes — can be as little as 60% of the stock thickness. This is two to four times closer than allowed in conventional stamping design.

With larger presses, parts can be fine-blanked from plate as thick as 0.750 in. (19 mm). And you can even form blind holes, claim some producers.

How near net shape?

Manufacturer's comparison of fine blanking against PM for precision ring gears

	Typical dimensional tolerance (in.)	Critical dimensions (in.)	Other tolerances	Secondary processing
Fine blanking	±0.0005	±0.0002	±0.0002 hole-to-hole location Sheet, flat within 0.001 in./in. Coil, flat within 0.002 in./in.	Deburring
PM	±0.002	±0.0005	Flat within 0.002 in./in.	Coining, deburring, lapping, I.D. grinding. Metal removal: 5% weight of finished part.

However, some of these processes, such as hot isostatic pressing and isothermal forging, are highly specialized and not ready for high-volume, low-cost applications. Even injection molding of high-strength polymers has reached a point where some parts can compete with metal equivalents both mechanically and from a near-net-shape standpoint.

Two metalworking processes, however, have emerged recently as true high-production near-net-shape contenders that can rival PM; they are cold or warm forging and fine blanking. Under the right conditions, these processes can turn out parts that equal PM in cost and sometimes require less secondary processing.

In addition, investment casting and die casting have evolved into fair near-net-shape processes, thanks to research programs launched over the last few years. In some cases, today's castings can be almost as near-net-shape as pressed and sintered PM parts — and mechanical properties can be just as good.

Fine blanking comes to life

Fine blanking is now new. It's an old Swiss process that first made headlines about 10 years ago. But the process never was appreciated in the U.S. — probably because economics of "near-net-shape" were not important back then. But by mid-1978, the world's largest fine-blanking press was installed in the U.S. by E. R. Wagner Manufacturing Co. in Milwaukee — a sign that the process is now receiving attention in this country.

Parts now in regular fine-blanking production include ring gears (some over 12 in. diameter, 305 mm, and 1/2 in. thick, 13 mm) and automotive brake and transmission parts. In most cases, the only secondary processing required is deburring.

Why cold forging is getting hot

"Forging," in its broadest sense, is simply displacement or plastic deformation of metal. That definition includes extrusion, heading, upsetting, roll-forming, rotary forging and so on. An important distinction is the temperature at which these processes take place.

Hot forging occurs at temperatures that cause the grain structure to recrystallize. That's about 1800 F (1255 K) for a typical steel. Warm forging is below this point — typically below 1400 F (1030 K) for common steels. Cold forging is essentially at shop temperatures, although a "cold" part may have emerged recently from some thermal treatment.

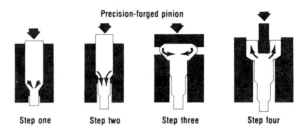

Precision-forged pinion

Step one   Step two   Step three   Step four

Although the sketches here are hypothetical, they approximate steps required to produce a near-net-shape pinion shaft complete with gear teeth.

Step One: the barstock, which has been cut to blank length, is forward-extruded to form the first diameter. Step Two: the second diameter is extruded. Step Three: a heading and upsetting strike that roughs the head and — although this cutaway illustration can't show it — rough forms the teeth on the diagonal face of the head. In the fourth step, an optional backward extrusion might be used to produce a blind hole and size the OD.

Ideally, this part will require deburring, perhaps finish-grinding on one or two diameters, and roll-forming to bring the gear teeth into tolerance. Less than 2% of the weight of the part is removed in secondary machining.

How near net shape? Manufacturer's comparison of precision cold-forging against PM for pinion gears	Typical dimensional tolerance (in.)	Critical dimensions (in.)	Secondary processing
Cold forging	±0.010	±0.002	Roll forming, O.D. grind, metal removal: 2% weight of finished part
PM	±0.003	±0.001	Same

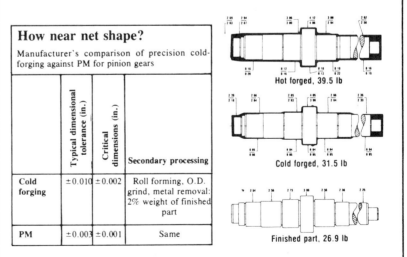

Hot forged, 39.5 lb

Cold forged, 31.5 lb

Finished part, 26.9 lb

An actual example of how cold forging stacks up as a near-net-shape process is this shaft, drawn from data provided by Burns Cold Forge Co., Minerva, Ohio. Stock removal, where needed, is less than half of that required for hot forging. The cold-forged blank yields 8 lb fewer chips in secondary machining than the hot-forged blank. The finished part appears below.

Equivalent conventional stampings may require subsequent machining, ranging from coining to reaming. PM gears may require a second coining or hot-forming step, plus ID grinding.

One automotive automaker is currently experimenting with fine-blanked connecting rods for one of its smaller V-8 engines.

Today, virtually all stampable steels can be fine-blanked. Monel, aluminum, BeCu, bronze, and cold-working stainless steels can be fine-blanked if properly prepared. Sometimes prior annealing is required.

How fast is fine-blanking? The process itself is slower, if measured in press-cycle time, than conventional stamping. Automatically fed, larger, hydraulic fine-blanking presses operate at around 30 strokes per minute to produce thick parts. Smaller parts are produced at up to 60 strokes per minute.

However, over-all production time per part usually is less, since the need for secondary processing is reduced or completely eliminated in fine blanking. One fine blanker claims that a production lot of 1000 conventional stampings required a total of nine hours to complete — including secondary processing. The same number of parts could be fine-blanked in only one hour, thanks to a significant reduction in secondary processing steps.

Forging, cold and warm
More progress has taken place in forging technology over the last few years than in any other field of metalworking — and this trend is expected to continue into the future.

Aside from just plain improvement in tool design and process versatility, process engineers have been toying with forging variables to tighten the part-to-part repeatability of the workpiece as it comes

out of the forging die.

For example, as forging temperature is decreased, more energy is needed to forge the part. On the other hand, part-to-part variations are reduced. Elastic deformation is much smaller and more controllable, and parts can be forged much closer to finished dimensions, often right to final size.

Conventional hot forgings require about 1/8 in. of metal removal per surface to account for part-to-part variations in size as well as for the irregular and "dirty" hot-forged surface.

In contrast, a typical cold-forged part requires only about 0.050-in. stock removal per surface (1.5 mm). Part-to-part variations on a cold-forged diameter are typically within 0.020 in. (0.4 mm).

Over the last few years, cold, warm, and combinations of cold and warm forging have gained substantially in popularity. Gears, bearing races, a number of automotive products, and even wrench sockets are typical products routinely forged by these new techniques today.

Typical production speeds for bearing races indicate that races in the 1-1/2 to 4-in. diameter (38 to 102 mm) range can be warm forged on units equipped with automatic feeding at about 20 parts per minute in a 2100-ton press. The manufacturer found that doubling production speeds reduced tool life by 50%, however.

Cold forging similar parts on a 2700-ton press can yield nearly 35 to 40 parts per minute. Production rates for manual-feed set-ups are about half these values.

At these rates, tooling will last typically between 40,000 to 60,000 cycles — up to 120,000 for some applications. Tool life for cold forging is longer because the tool steel is not subjected to damage from thermal cycling.

Although production figures are difficult to compare, one manufacturer suggests that warm forging is economically practical for volume requirements over 10,000 parts/month. Cold forging may pay off for requirements as low as 1000 parts/month.

News from the casting front
Casting processes — particularly die casting — recently have benefited from great amounts of highly concentrated engineering research. With these "engineered" die castings, overdesign can be systematically eliminated. Die castings that once had 0.060-in.-thick walls (1.5 mm) are now pared down to 0.030-in. (0.8 mm). It's possible to get down to 0.015-in. (0.4 mm) if necessary.

Ferrous die casting is also on the way. One interesting development, and one that will no doubt find a niche in the metalworking arena, is stainless-steel die casting. It's now being done for wrench sockets and valve balls on a production basis. Both Federal Die Casting, Chicago, and Worcester Controls Corp., Worcester, Mass., are among the pioneers in stainless die casting.

Dies for the process are made of molybdenum/tungsten alloy to withstand 2500 F (1640 K) molten-metal temperature. Producers claim that the technology to produce these castings is much like that for aluminum die casting, except of course, for the temperature. An 800-ton machine can knock out from 60 to 300 shots per hour, depending on the number of cavities filled per shot and other die-design variables. Most of the demand for stainless die casting so far has primarily been from customers interested in replacing stainless investment castings, since die casting is nearer net shape than investment casting. ∎

Forging technologies of the twenty-first century

Economic and performance pressures are placing new demands on producers of aerospace forgings. Near net shape techniques are meeting tough production part requirements and are developing into the forging technologies of the future.

By RANDOLPH GOLD, *editor*

The earliest and crudest form of forging was to move metal by heating it and hitting it with a hammer. Parts were simple: horseshoes, basic tools and implements. The blacksmith's forge did not produce parts to tolerance or to meet mechanical specifications, and people were not particularly concerned with economics of production.

Since the production of the first impression die forging, the technology available to produce forgings has expanded to meet increasing property and economic requirements. Forging has always been the process used for many aerospace, nuclear and other high technology components. But stronger, increasingly complex forgings with tighter economics are requiring entirely new technologies.

In the forefront of new forging technologies are three processes, developed partly by and being applied at Wyman-Gordon Co. The essence of these technologies can be described in the words "near net shape." The primary advantage of these technologies is economic, derived from the ability to make more detailed, closer tolerance parts with less starting material and secondary machining. Technical benefits may breed new economies for the forging user in terms of improved product performance.

The three processes for making near net shapes in titanium, nickel base

Editor's note: All photos and illustrations courtesy Wyman-Gordon Co.

The 1800 ton isothermal forging unit is shown discharging a finished gas turbine forging. Off the photo to the left, preforms are loaded into the vacuum chamber. A 3000 ton isothermal forging unit is being added to Wyman-Gordon's isothermal forging capability.

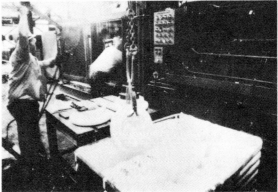

Preforms, produced in an earlier isothermal forging operation with a blocker die, are loaded into the vacuum chamber adjacent to the furnace section of the isothermal forging unit. Billet heating, forging and discharge are all carried out in a vacuum.

and other high-temperature/high-strength superalloys are hot die forging, isothermal forging and hot isostatic pressing (HIPing). The first two processes are closed impression die forging techniques, while HIPing is a powder technique that can be used to either make near net shapes directly, or to make billet or preform material for subsequent processing by hot die or isothermal techniques.

Closed impression near net forging techniques work because of the same principle that operates in conventional hot forging. When metal is heated, its strength decreases, and the forces needed to make it flow plastically decrease. In hot die forging, however, the die is also heated. Die temperatures of 1500-1700F (816-927C) are commonly used in hot die forging, compared to conventional die temperatures of 700-800F (370-427C). Because of the relatively small temperature gradient across the die and forged material in hot die forging, metal flow at any given forging pressure is better, producing a finer detailed, closer tolerance part.

Isothermal forging, as the name implies, occurs when the die is heated to the same temperature as the forging stock, and the benefits are accordingly greater than those obtained from hot die forging. Gatorizing®, developed by Pratt & Whitney and used by Wyman-Gordon under license agreement, is a variation of generic isothermal forging. The primary difference is that in Gatorizing, metal is in the superplastic state.

Isothermal forging, according to Paul Wisniewski, vice-president, sales and marketing, of Wyman-Gordon's Eastern Division, is not new. "Information on isothermal forging has been in the public domain for about 30 years.

"A number of universities and research laboratories have done work on isothermal forging over the years, but it was Pratt & Whitney's development of the superplastic Gatorizing technique that really kicked off isothermal forging for turbine components on a commercial basis."

View of hot die forging operation shows production of a titanium airframe part. Dies are heated by gas-fired infrared burners and electric resistance elements.

"Superplasticity is a phenomenon in metal where the amount of energy needed to deform the metal diminishes greatly by producing the right structure and grain size," Jim Coyne, Wyman-Gordon vice-president, technical director, tells us. "IN100 material for Gatorizing is produced by encapsulating the nickel-base powder and subsequently extruding it to obtain a forging billet (not a preform) with a very fine grain structure.

"Superplastic material flows under a very low stress at forging temperature. We are producing isothermal (Gatorized) IN100 parts on an 1800-ton (16 MN) press that you could not produce conventionally even on an 18,000-ton (160 MN) press.

"The isothermal forging of superplastic material is the closest I believe anybody is getting to net shape."

With the forging die at the same temperature as the forging stock in isothermal forging, a logical question would be what kind of die material do

you use. The answer was found in a molybdenum die material, but, as Wisniewski puts it, "At 2100F (1150C), the dies would vaporize in the atmosphere. Therefore, these molybdenum base dies are used either in an inert atmosphere or a vacuum."

Hot isostatic pressing is a technique whereby a container the shape of the desired part is filled with metal powder, then placed in a pressure vessel which subjects the filled container to a very high pressure (in argon gas) and temperature.

The force exerted on the exterior of the container is equal over the container's entire area (hence the term isostatic), and the container "shrinks" as the particles of metal powder inside consolidate into a homogeneous solid material.

Unlike conventional cold pressed P/M parts, HIPed parts exhibit wrought density. According to Coyne, "You cannot sense any difference in density between a HIPed part and a wrought part. These are 100% dense parts, and if you sectioned a HIPed

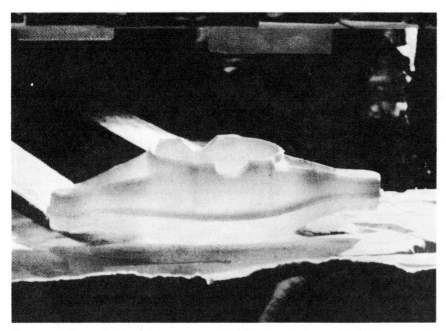

A near net shape titanium airframe component is removed from a 6,000 ton hydraulic press after hot die forging. The die, heated to 1500-1700F, appears white because it is glowing red just like the part it is forging.

part and took a micrograph, you would see no porosity."

Wisniewski and Coyne share the same view regarding the future application of HIPing for forging preforms. Coyne believes that ". . . there will be certain shapes best suited to preform/forge. Other parts will be HIPed with no forging involved."

In Wisniewski's opinion, "Powder techniques may or may not be used with either closed die near net shape processes. I personally believe there is a future for HIPing in the production of forging preforms."

HIPed parts currently in production at Wyman-Gordon are used as-HIPed and as preforms for hot die forging. As-HIPed components are jet engine discs and shafts which receive only heat treating, inspection and rough machining to facilitate testing.

The current limiting factor in near net shape forging is not the forging pro-

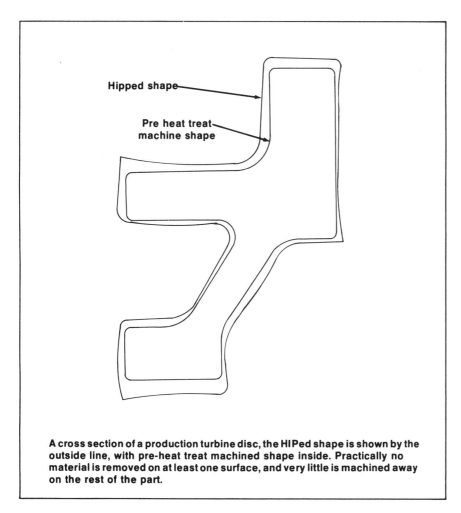

A cross section of a production turbine disc, the HIPed shape is shown by the outside line, with pre-heat treat machined shape inside. Practically no material is removed on at least one surface, and very little is machined away on the rest of the part.

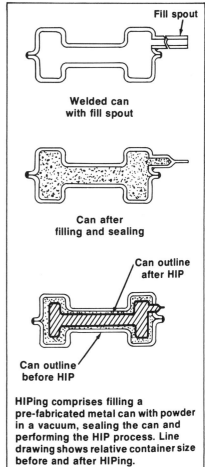

HIPing comprises filling a pre-fabricated metal can with powder in a vacuum, sealing the can and performing the HIP process. Line drawing shows relative container size before and after HIPing.

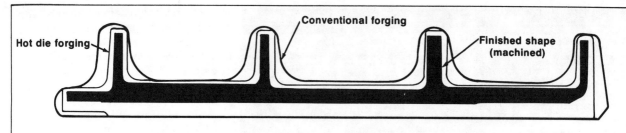

What near net shape is all about. The outermost shape represents conventional closed impression die forging for a 6AL-4V titanium bearing support. The next shape is that produced by hot die forging, followed by the finished machined configuration. Note how close the hot die forging is to final finished dimensions. As cost of material increases, the benefits of near net shape production techniques become dramatic. One could expect even closer to net shape results with isothermal forging.

cess, but the inspection process. Current ultrasonic techniques used to inspect turbine forgings can only give reliable results up to a certain distance from the surface. If, for example, an ultrasonic system can see a defect no closer to the surface than 0.100 in. (2.5 mm), inspection must be performed on a surface at least 0.100 in. (2.5 mm) from the finished part surface to be assured that defects in this uninspectable zone are outside the finished machined product. It is the shape generated by this 0.100 in. (2.5 mm) of extra material (the so-called sonic shape) that is the existing minimum configuration that may be forged to.

According to Coyne, "We are getting a 0.050-0.060 in. (1.3-1.5 mm) cover over the shape we want to inspect on three isothermal parts we are forging. Total envelope is about 0.100 in. (2.5 mm). If the NDT equipment was sophisticated enough, we could forge to even closer tolerances.

"I personally believe that, were it not for the limitations of inspection, we could forge to an accuracy of net on one surface and to within 0.050-0.060 in. (1.3-1.5 mm) on the other.

"Work is being conducted by the jet engine industry on 'smart' sonic inspection equipment that does not require a flat surface to inspect, and on equipment that will narrow the current 'dead band' or range of inspection depth that cannot be read reliably."

The implications for near net shape forgings both now and in the future are staggering.

Wyman-Gordon's hot isostatic press with its 80 metric ton wire-wound steel frame subjects containers filled with metal powder to 15,000 psi of argon gas at temperatures in excess of 2000F. Under pressure, the container shrinks and consolidates the powder into a homogeneous material of near net shape. Wyman-Gordon is HIPing turbine discs from Rene 95 metal powder for General Electric.

"The new technologies that are coming on stream are really exciting," says Coyne. "As far as I am concerned, these techniques open up new avenues for metallurgy and materials engineering.

"Pratt & Whitney brought out its IN100 powder and made it work with the Gatorizing isothermal forging process for the F-100 engine.

"General Electric is bringing out the as-HIPed Rene 95 powder on its F-404 and T-700 engines.

"Twenty years ago, it would have taken us three times as long to bring all this technology into existence. Now, everything is going so fast; we just recently saw the first production F-404 engine delivered to McDonnell Douglas for the F-18 program. With the Boeing 767 and 757 projects, we are looking at new parts for these as well."

Coyne sees new alloys coming along because of the ability of near net shape technologies to economically forge them. "I also see these techniques extending the capabilities of current alloys. That is what is happening with Rene 95, which is producing more reliable parts than it did with cast/wrought technology.

"With these technologies we can do more things with current alloys, do it better, and hopefully do all this at a lower cost." **PM**

How Forging Has Put New Punch into Its Act

The forging industry is undergoing a period of major change. Both internal and external pressures are forcing it to update its practices.

"**Y**ou can't teach an old dog new tricks."

Maybe so, but whoever invented that hackneyed expression could not have had the forging industry in mind.

Forging, as an industry, goes back at least a couple thousand years. Yet, since World War II, more has happened in the industry and to the industry than in all its previous years. And it's still happening. And if we listen to the experts on the shape of things to come, we'll be seeing even greater strides in the future.

It used to be that the hammer was the chief tool of the trade. And it will continue to be an important tool whether it be a power hammer, an air hammer or a board hammer.

But the strong trend is toward presses—mechanical, hydraulic and screw. Today's equipment will do whatever it takes to make a forging efficiently—squeeze the metal, wedge it, impact it, roll it, cross roll it or just upset it. Even the hammers are programmed to controlled cycles.

Some plants make their forgings without heat. Some warm the metal. Most still form the forgings hot. But wherever heat is used, it's done far more efficiently than it has ever been done before.

The change that's taking place in the industry is extensive. "In the forging industry today," says G. W. Weinfurtner, technical director of the Forging Industry Association (FIA), "we're seeing both internal and external pressures on profit, return on investment and on our markets. These pressures are forcing the industry to update its methods and procedures . . . The need to produce high quality forgings at a reduced cost is becoming stronger every day."

Certain trends are having a very significant effect on the industry. "The first trend," says Mr. Weinfurtner, "is the breaking down of the heavy traditions in the industry and replacing them with the science and knowledge involved in metallurgical and mechanical engineering. This applies to all plants, not just the larger ones."

Mr. Weinfurtner continues, "A second trend is the turn to near-net-shape forging . . . I feel very strongly that this is probably the major trend affecting our industry today."

Near-net-shape forging is particularly advantageous where exotic high-cost materials must be used. For this reason, the objective is to start with as little of the costly material as possible and deliver a forging that requires little or no machining.

Three relatively new processes are standouts in this area—hot die forging, isothermal forging and HIPing (short for hot isostatic pressing). Wyman-Gordon Co., Worcester, Mass. has adopted all three of these processes for the hard-to-forge super-alloys used for turbine engines and airframes. It did so as a counter measure against the continually increasing costs of these already expensive materials.

Both hot die forging and isothermal forging are closed impression die forging processes. In both, the alloys are heated until they're plastic, then formed to their highly complex shapes in a forging press.

In the hot die process, Wyman-Gordon normally heats the dies to within a range of 1400° and 1600°F. It heats the forging materials to within the 1750° and 2100°F temperature range, depending on the alloy—considerably

Forging dies at Rockwell Automotive are within 0.0002 to 0.0003 in. so that these truck axle gears (left) require no machining. Crankshafts at right are not only forged, but machined as well at Ellwood City Forge.

11

Isothermal Forging Trims Machining Cost

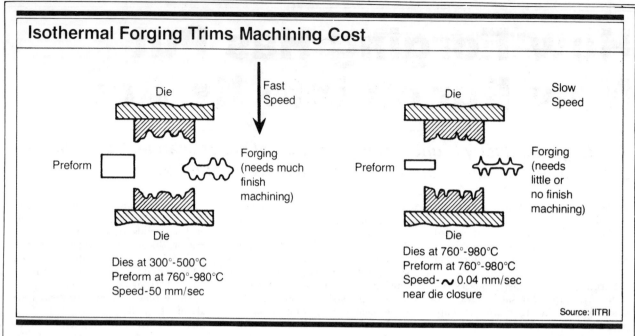

Dies at 300°-500°C
Preform at 760°-980°C
Speed-50 mm/sec

Dies at 760°-980°C
Preform at 760°-980°C
Speed- ∿ 0.04 mm/sec
near die closure

Source: IITRI

above the 600° to 900°F range for conventional forging.

The higher die temperatures improve the flowability of the forging material by reducing heat loss from the material to the dies. Thus, with better material flow, less material is required to produce a forging with good detail and to much closer tolerances.

With the use of higher temperature in the hot die process, the costs for die materials also run higher. One way to offset some of these costs is to use forging alloys that are "tailor made" for the hot-die process. Wyman-Gordon cites as an example the use of beta titanium alloy which can be forged in the 1200° to 1400°F range using less costly die materials.

Isothermal forging differs from the hot-die process in that the die temperature is higher for a given forging material. Dies are heated to the same temperature as that of the forging material. Forging of nickel-base super alloys, for example, requires heating to a range of 2000° to 2100°F. In this temperature range, molybdenum-alloy dies are called for and forging is done in a vacuum or inert gas atmosphere.

As with the hot-die process, isothermal forging at these higher temperatures contributes significantly to the flowability of the material and produces forgings that are much close to the required net shape. "In fact," says James E. Coyne, Wyman-Gordon's vice president-technical director, "isothermal forging is the closest, I believe, that anybody is getting to net shape."

The saving in the weight of the starting material for some turbine engine parts is quite significant. In some cases, the starting weight of the forging material is only about a third of that required for an equivalent conventional forging.

"And," says Mr. Coyne, "we're currently able to produce isothermally-formed parts on an 1800-ton press that might otherwise be made conventionally on an 18,000-ton press."

The isothermal forging process adopted by Wyman-Gordon is one which it licenses from Pratt & Whitney Aircraft. It carries the tradename of Gatorizing.

HIPing, the third of the closer-tolerance forging processes, is also used to produce turbine engine hardware. In this case, however, the starting material is a powder metal which is placed in a metal container. Then, under heat and pressure in an autoclave, the powder is compacted to 100 pct of its theoretical density. The protective container is later stripped either by machining or pickling.

Pressures in the autoclave may range from 10,000 to 30,000 psi. A typical isostatic cycle at Wyman-Gordon may reach a pressure of 15,000 psi, a temperature between 2000° and 2200°F, and may last about 8 hours.

HIPed parts may be used in their as HIPed condition. They may also serve as powder preforms for further forging by conventional means, hot-die forging or isothermal forging. Where HIPing is followed by forging, the parts usually acquire much more detail than that obtained by HIPing alone.

Automation is still another trend that's shaping the future of the forging industry. "Granted," says Mr. Weinfurtner, "this is not a new process or method. However, the number of companies using high-speed automated systems is expanding." Some newer processes and machines are producing forgings at rates that are eight to ten times above those for conventional forging.

In other areas, heating systems are being monitored and controlled much more accurately. Dimensional control and accuracy is much more sophisticated. Numerically-controlled EDM is already accepted practice for die sinking and computerization of the operation is on its way. And the constant shift to higher-strength materials is something the customers demand and the forging industry is learning to live with.

At the Chicago conference of the Forging Industry Education and Research Foundation (FIERF) held a little over a month ago, Dr. Robert P. O'Shea, manager, basic manufacturing process research, and Otto Novelli, manager, metalworking, International Harvester Co., told of their work with precision warm forging of ferrous parts.

They define warm forging "as the bulk deformation of steel in the temperature range of 1000° to 1400°F." Their work over the past 18 months has been focused on precision forging of bevel gears and gear blanks. The parametric study took into account such things as the influence of workpiece temperature, press load, lubrication, die temperature, die construction, preform geometry and other factors on the production of precision flashless forgings. (Cont. on p. 79)

How Conventional and Programmed Forging Compare

Results of forging by "skilled operator." Production—100 pct							Optimizing forging operation with programmed control.* Production—230 pct						
Section, mm In	Out	Stroke	Pene-tration	Mani-bite	Length	Time	Section, mm In	Out	Stroke	Pene-tration	Mani-bite	Length	Time
450x450	360x472	200	90	220	1670	0.4	450x450	360x472	150	90	220	1670	0.3
472x360	382x377	200	90	220	1980	0.9	472x360	382x377	150	90	220	1980	0.7
377x382	297x398	200	80	220	2400	1.4	377x382	297x398	150	80	220	2400	1.1
398x297	318x310	200	80	220	2860	2.0	398x297	318x310	150	80	220	2860	1.6
310x318	250x335	150	60	220	3500	2.6	310x318	240x335	150	70	220	3500	2.1
335x250	265x264	150	70	220	4050	3.5	335x240	265x254	150	70	220	4200	2.8
264x265	254x267	25	10	50	4160	5.2	254x265	248x266	50	7	120	4300	3.7
267x254	257x256	25	10	50	4280	7.0	247x266	250x250	50	16	120	4500	4.6
256x257	248x258	25	8	50	4420	8.8							
258x248	250x250	25	8	50	4500	10.7							

*Using Forgemaster I Programming System developed by Pahnke Engineering.

Metallographic examination clearly shows that the warm-forged gears have a much better macrostructure than the cut gear blanks. But the process has other advantages, including savings in materials, less energy usage than for conventional hot forging, elimination of post-forge normalizing and a sharp reduction in machining, say the researchers. And because warm forging requires fewer process steps, it's conducive to higher productivity.

More and more, the forging operation is being taken out of the hands of the operator, however skilled he may be, and the control turned over to a computer, a microprocessor or other type of control. Forging operations have become too sophisticated, too integrated, too costly and too complex for an individual operator to decide what the forging techniques should be.

Computer control, for example, allows a forging plant to deliver forgings that are far more accurate and uniform from workpiece to workpiece, and especially so with forging of complex shapes.

Computer control also lends consistency to the forging technique, eliminating variations that are bound to exist among individual operators.

Product quality can be maintained more readily by the fact that the computer issues precise instructions at each stage of the forging operation.

Also, the fast response of the computer sharply reduces manipulator time between passes, giving the forging press more time to work while the workpiece is within the prescribed temperature range. This reduces the number of reheats, conserves energy and thereby helps to reduce overall costs.

The storage capability of the computer speeds the assessment of new work and provides reliable data on the essential parameters used for previous forgings. The compilation of forging records over a period of time could serve as the basis to generate new forging schedules.

But computer control isn't the answer for every forge shop's problems. Sven Sonnenberg, chief engineer, Towler Hydraulics, Urbana, Ohio, told the FIERF conference of early attempts to develop a so-called 'Computerized Forging System' after solid-state controls gained acceptance.

And Mr. Sonnenberg adds, "The introduction of computerization into the controls of the forging process has opened a new boundless area of possibilities. But the microprocessor, because of its miniature size and flexibility, made it possible to build a compact, customized unit which can be applied in an average forge shop environment . . . Exploration in this direction has just begun," he concludes, "but the potential is great."

Perhaps no stronger and steadier trend is being set today than that of custom parts production by automatic hot forging. It's a system in which mill-length steel bars are fed in at one end of a machine and pierced, precision parts emerge from the other.

Until not too long ago, primary interest in automatic hot forging was shown chiefly by companies such as Caterpillar Tractor, Ford Motor Co., certain divisions of General Motors, plus others. This was simply because they could afford to purchase the equipment, which is quite costly, for their own needs.

Today, it's a different picture. Any number of job shops now have the capability for automatic hot forging of steel bar stock, several manufacturers now produce equipment for the process, and both the shapes of parts and their weights have been extended considerably.

"To date, we've had 300 installations of Hatebur automatic hot formers worldwide," says K. H. Beseler, vice president, Girard Associates, Cleveland, the North American representative for this line of equipment. "Most were bought for job-shop work although in the U.S., captive installations currently outnumber the independents.

"We estimate that our equipment in the U.S. alone turns out more than 235,000 tons of forged steel parts annually," he continues, "with some machine models producing at speeds of 180 units a minute, or up to 10 times faster than conventional forging presses."

An important reason for the rapid growth of the process is its ability to use the lowest-cost steel bars, i.e., hot rolled and unannealed but only seam-free stock of forging quality. All carbon steel grades, all alloy steels, and even bearing grades such as AISI 52100, can be hot forged automatically. The Hatebur equipment can handle bar diameters from ⅝ in. to 3 in. and lengths from 9 ft to 30 ft with 24 ft being among the most popular.

The term "automatic hot forging" actually stands for two different ways of attaining a forged product. One process is that of automating the conventional forging system whether

These truck gears, forged at Rockwell Automotive Operations, are cleaned on trees by shot blasting, then machined only on their back face.

Also, many parts as used "as-forged" with little or no machining, and little or no cleaning.

To answer its own need for high quality forgings at low cost, the Automotive Operations of Rockwell International allocated $10 million a couple years ago for a precision forging plant. Its goal was to provide near-net shape forging for other Rockwell plants.

Rockwell specified and built one of the cleanest, most highly automated precision forging plants to be found anywhere. And, it's quiet, too—80 db maximum on all equipment. All machines are hydraulic; most have microprocessor control.

Rather than renovate or add onto an existing forge (Rockwell Automotive now operates three), the plant was developed as a separate facility to incorporate the very latest in forging technology and assemble what Rockwell calls "a center of excellence." And among automotive forges, it is unusual in employing strictly state-of-the-art technology.

"We could have done it more cheaply," observes Robert Gurnitz, vice president and general manager of the Supply Division, "but we would be making the trade-off between investment and operating cost, and you get what you pay for.

"If you want to put the money up front, you can have lower operating costs and higher quality and, in fact, the machines in the plant could have had substantially less automation. But what you pay for then is increased operating costs," he observes.

Besides the great degree of automation applied to production of its three current product lines, operating costs at the forge are minimized with the use of low-cost TVA power—one of the main reasons for locating the plant in Morristown, Tenn. About 7000 kva is being used at the 71,000-sq ft plant.

The products coming out of the presses now are precision forged gears supplied to Rockwell's Truck Axle division; S-head brake camshafts for the Brake and Trailer Axle divisions; and drive axle spindles for the Trailer Axle and Truck Axle divisions.

Rockwell Automotive buys from some 30 different forges, but the precision forging plant increases Rockwell's forging capacity, principally to offset what it sees as the inadequate domestic supply base for the near-net shaped parts it needs.

The company specified or designed all but one piece of equipment in the plant—a Hasenclever press. Plant management points out that the envi-

it be by hammer or vertical mechanical forging press. Automation is built into the system by mechanically feeding-in and transferring the steel forging from die to die until it completes its multi-station cycle.

This process, like conventional forging, can produce practically any forgeable shape and size. It produces 325-lb crankshafts and 265-lb front axles for trucks. Also, it uses round-cornered squares up to 7⅛ in. Flash must be trimmed. Draft angles are the normal 3° to 5°. And top speeds are about 20 units per minute.

Equipment of this type is made by such companies as Amforge-Hasenclever of Broadview, Ill., Ajax Mfg. Co. of Cleveland, Chambersburg Engineering Co., Chambersburg, Pa., Erie Press of Erie, Pa., National Machinery, Tiffin, Ohio, and Verson Allsteel Press of Chicago.

The second and higher-speed process is based on the Hatebur machines and equipment built by National Machinery and Industrial Fasteners of Marieville, Quebec. In this process, individual blanks are transferred mechanically through three or four die stations. The equipment ejects a finished precision forging at each stroke.

Both processes use closed-die machines that can range in price from $1 to $2 million for small parts. Machines for larger parts can cost from $6 to $10 million.

Despite the high capital cost, automatic hot forging continues on its sharp growth curve. The economics of parts production dominates the process. In addition to its very high processing speeds, the process saves anywhere from 20 to 30 pct in initial stock by flashless forging. And the machines require a minimum of labor and skill.

Also, because of its process speed, there's little heat loss between the first and last dies. Ejected parts are at a temperature of about 1900°F, then cooled slowly. This often obviates any need for annealing or normalizing the parts before machining.

ronmentally clean plant is the result of investment, again. "What we've wound up doing is installing processes which are inherently clean, such as induction heating and saw cutting, rather than attempt to clean up older equipment," comments Joseph Sniezek, plant manager.

As part of the brake cam line, the Hasenclever operates at rates up to 450 pieces an hour. The automated horizontal forging machine can both gather stock and press forge at the same time. After cut bars are induction heated, the progressive die operation upsets, forges and does the final trim. The formed cams are cold-coined for a nearly scale-free finish.

The largest system in the plant is the spindle extrusion line which includes two fully automated fast-acting Lasco hydraulic presses. The first makes preforms and punches out the centers. The parts are then automatically transferred several feet away to the second press for extrusion. The total operation is done in five stages.

To make accurate axle spindles and avoid machining time—no roughing is required—Rockwell installed a system that automatically weighs and sorts billets by volume.

After shearing, the system puts each billet into one of four categories. The data are fed back to the stock gage for automatic correction. Getting the precise weight the first time precludes having to shear and then cut to size, thereby reducing material waste.

Capacity on the spindle line is about 175 pieces an hour. With one press, 90 pieces would be possible.

Dies for the spindles receive periodic graphite lubrication. The atomized lubricant apparently has been the only airborne dirt in the plant. But the lubrication cycle adds about 50 pieces an hour to production. So, to remedy the graphite overspray, Rockwell has ordered a collector. "That's been the limit of the unexpected during the start-up," notes Mr. Gurnitz.

The third product—pinion gears—are forged and finish formed in the same press. No machining of the teeth is required. The Morristown plant has complete responsibility for the part. No gear has yet been returned by a user plant for any reason, says Mr. Sniezek.

The bars, 8620 carbon steel, are saw cut to control the volume of the slug, then centerless ground for roundness. With its weight at a few ounces over that of the finished gear, the slug is placed manually in three progressive dies set in a 1300-ton Ajax press. The flash weighs only 0.2 lb. The finished gear weighs 2.5 lb. It's smaller, light-er, yet stronger and more wear-resistant than a conventional gear.

After forging, gears are placed on a conveyor where they go through controlled cooling for normalization and scale prevention. The gears are shot cleaned, the centers reamed and back faces machined. They're then 100-pct inspected and packed. The profile of the gear teeth is held to a repeatable 0.0002 to 0.0003 in. Quality sampling is constant.

Behind the accuracy of the gear teeth are the dies. All dies are cut on Cincinnati Milacron EDM equipment and an NC lathe. The plant's toolroom can produce all dies and components the plant requires.

The extent of the die support is mandated by the unusually high production rates in the forge despite the fact that the number of parts produced per die is at least equal to that of an older forge.

The need to produce high-quality forgings at reduced costs is becoming stronger with every day.

It's interesting to note that because of the EDM process to make the dies, the surface finish on the parts is satin rather than glossy. This enables the gears to carry a more efficiently distributed oil film than conventionally-cut gears.

Rockwell Automotive appears to have succeeded in minimizing its costs while improving the quality of the three forged parts. It is getting almost unheard accuracy in forging the near-net shape parts.

Overseeing its highly automated operations are relatively few workers. Total plant employment is 70. And those on the floor are technicians who tend the equipment rather than artisans of forging.

The plant is now about 90 pct launched. A 2500-ton National press will arrive shortly. Once it arrives, Rockwell can consider making universal joint parts and additional sizes of axle shafts and drive pinions—some of which may be for outside customers.

Ellwood City Forge Corp., Ellwood City, Pa., likewise has set its goal at being the epitome of the modern forging plant. But unlike the Rockwell plant, Ellwood City Forge starts forging with ingots up to 80-in. diam and weighing up to 80,000 lb. Its major equipment consists of two forging presses—one of 2500-ton capacity and the other of 1000-ton.

Both presses were modernized by converting them from steam/water to direct-drive, oil-hydraulic, using Towler "D" pumps. The direct-drive system provides a full power stroke of the ram without intensifiers or accumulators and also allows for a second, higher oil pressure capability that's necessary for upsetting. The conversion to oil-hydraulic increased forging speed about 20 to 25 pct and improved productivity accordingly, notes Bob Barensfeld, president of Ellwood.

Electronic controls were also installed on these presses, using a binary digitizer so that the press operators can preselect the desired forging dimensions on the control console and set the precise distance between the open dies.

The smaller of the two presses can handle blooms up to 25 in. square and 10,000 lb in weight. Working in conjunction with the presses are forging manipulators—one a 30-ton rail-bound Alliance and a 10,000-lb capacity mobile nip by Herr Voss.

At the heart of the Ellwood City Forge operations is a new control system—a Honeywell TDC 2000—which provides the ultimate in plant flexibility and furnace and forge management.

The system manages the interaction of flame control, oxygen sensing, pressure control, zone air control and heat transmission. It allows various combinations of each of these elements by simple keyboard entry. It takes less than five minutes to make any process change—something that previously required complete controller changes and extensive relay wiring modifications.

"Faced with rising costs of natural gas is what finally pushed us into this," he says. "We expect to pay for the system in less than two years, have printed records of our entire process and control on more points with tighter control ranges than we have previously achieved."

Investigating and selecting the system for the first six furnaces took three man-years. Installation for the next 11 furnaces means simply plugging into the system.

Local control rooms are interconnected to one master control room where complete furnace operations are started, controlled and cycled automatically by microprocessors and program controllers. □

Castings whose fatigue life can be predicted with reliability high enough for any critical application.

Forgings so near net shape that they can be used "as is" in close-tolerance assemblies.

Both metalworking processes, casting and forging, have made significant technological advances in the last five years. In fact, in many applications today, parts made by either of the two processes could be specified. Often, the tradeoff made when selecting one process over the other is based more on cost and availability than on engineering differences.

Generally, competitiveness between the processes is strongest for smaller parts, where part shape and configurations would be nearly the same for either process. Larger parts are more likely to be tailored to accommodate process characteristics and limitations — press capacity for forgings or gate and riser locations for castings, for example. In such cases, process interchangeability may not be a simple matter.

New data for lightweight castings

Many of the properties of cast metal vary with the cross-sectional area of the part. Consequently, as parts are downsized (either to reduce weight, as is the case in the automotive industry, or to conserve costly materials), properties such as fatigue life and fracture toughness can diminish. "Downsizi[ng] good for forgings," boasts a represe[nta]tive of the forging industry, referri[ng] this characteristic of cast metal.

The Steel Founders' Society [of] America, however, is sponsoring a[n on-]going program to develop accurate [de]sign data for smaller, lightweight parts. Fatigue and fracture-tough[ness] data is being developed that will a[llow] users to design and produce cast[ings] with the minimum amount of m[etal] while accurately predicting strength [and] life expectancy, and keeping costs [at a] competitive level.

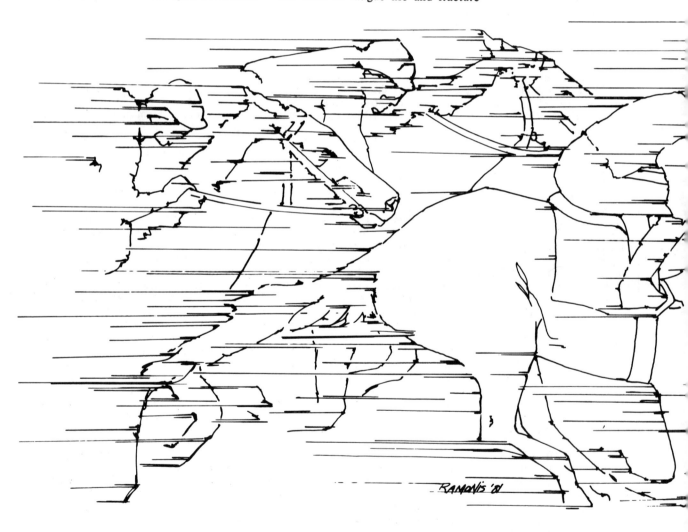

RAMONIS '81

e SFSA research program divides ~~t~~e life into three stages: crack initi-~~a~~, crack propagation and failure. ~~Mode~~ls are being developed for five of ~~m~~ost commonly used cast steels.

~~Th~~e fracture-toughness portion of the ~~resea~~rch is already beginning to gener-~~ate gu~~idelines for specifying castings for ~~low-t~~emperature applications such as ~~arcti~~c pipelines. Research was con-~~ducte~~d using the dynamic tear test, ~~whic~~h is similar to the Charpy V-notch

impact test, but the samples measure 0.625 in. thick, 1.62 in. wide and 7 in. long (15.8 X 41 X 180 mm). Midway along the tension side of the specimen is a 0.375-in.-deep notch (9.5-mm) or brittle-crack starter weld.

The dynamic tear test yields a longer fracture path and a sharper transition region than the Charpy test. for a

CASTING or FORGING?

Today, two age-old metalworking processes are running neck-to-neck

By John C. Bittence, Editor

Forging for near net shape

Today, most forging research has been aimed at producing parts that require "as little as possible" subsequent machining. In many cases, this goal has been achieved. By adjusting forging temperatures and die temperatures, producers have been able to forge to tolerances closer than ±0.003 in. (±0.08

quenched and tempered low-alloy steel, for example, the conventional Charpy test shows a ductile-to-brittle transition that falls within a 120°F temperature range (70°C). The dynamic tear test narrowed the transition down to within 40°F (25°C). In some cases, the dynamic test correlated better in predicting actual service failures than the Charpy test.

mm). Surface finishes can approach 63μ in./in.

Although precision, near-net-shape forging is possible with many metals and alloys, state-of-the-art near net shape forging of aluminum is about 10 to 15 years ahead of ferrous technology, according to the Forging Industry Association. Consequently, much of the current research is being devoted to

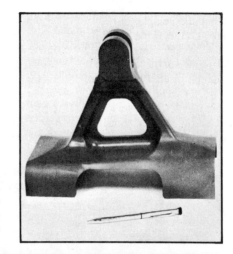

Vacuum-cast damper bracket is reliable enough to hold down vibration in a helicopter rotor blade. This 15-5PH steel casting is produced by Arwood Corp. for Boeing Vertol, and it replaces a machined forging.

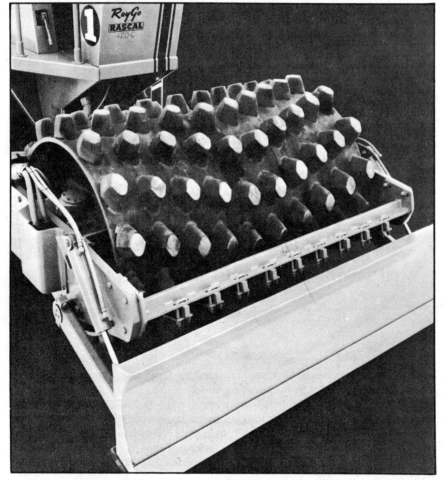

Isotropic strength, uniform in all directions, is a characteristic of cast metal needed for the steel treads of this soil compactor. The cast properties of ASTM A 148 Grade 120-95 are required to resist in-service vibration of over 100 treads attached to the 50-in.-dia drum, according to the Steel Founders' Society of America.

"catching up" in ferrous forging.

At a precision forging seminar co-sponsored by the American Society for Metals and FIA last fall, ten precision techniques — either already or "just-about" in production — were described. Precision-forged metals included superalloys, titanium, aluminum and various steel alloys. Many of the processes were limited to one or two of these materials, however.

Many of the developments in ferrous forging are based on reducing the forging temperature. Generally, four temperature ranges have been defined. So-called cold forging takes place between ambient temperature and 200°F (93°C). Two versions of warm forging exist: one below the material's transformation temperature (which is between 1000 and 1400°F [540 and 760°C] for steel) and the other above the transformation, between 1600 and 1800°F (870 and 980° F). Conventional hot forging for most steel alloys takes place in the 2000 to 2400°F range (1090 to 1320°C).

A new hot forging process, which creates nearly no flash (only 8-1/2% or less of the weight of the initial billet being converted into flash) can produce parts weighing up to 16 lb (7.3 kg) to tolerances better than ±0.010 in. (±0.25 mm).

Cold forging, particularly cold heading and extrusion, has been used successfully to produce precision, near-net-shape parts such as splines and gears. One typical process can turn out parts up to 40 in. long (100 cm) and 5 in. dia. Gear or spline teeth can be held to within 0.0003 in. (0.008 mm) profile error. A 30-in.-long part can be straight within 0.0004 in. (0.012 mm) over its full length.

Various forms of ferrous "flashless" forging have been developed, although they are mostly proprietary. Many of these processes impose size and shape limitations, however. In one warm-forming process, for example, the initial billet must be weighed to an accuracy of ±0.035 oz (±0.001 kg). Underweight billets will produce an incomplete finished part; overweight billets would ruin the forging dies.

Material savings from both flashless as well as other near-net-shape processes can be impressive. In most cases, at least 30% less raw material is required. Machining time, too, is reduced. Most flashless and near-net-shape precision processes reduce machining costs from 30% up to 100%.

Energy can be saved with these new processes. Isothermal forging, for example, needs less energy because fewer forging stages are required than for conventional hot forging. Cold formed or extruded splines require from 50 to 60% less energy than their hot-forged counterparts, while warm-formed bevel gears need 40% less energy.

CAD/CAM, of course

Some of these high-technology forging and casting techniques may be too tricky for an operator to control. As a result, human judgment is being replaced by computer control. In the foundry, the computer is now being used to design the molds, locate critical gates and risers, and even to direct robots in moldmaking and pouring.

In the forge shop, the computer provides part-to-part consistency that human judgment could not. When the computer directs workpiece transfer, for instance, cycle time increases. Also, computer control over forging variables such as temperatures helps conserve energy.

The computer can also design tooling. In foundries, the computer is expected to improve mold and coremaking productivity. Today's manual foundry methods lead to mold and core scrap rates ranging from 25% to as high as 70% for complex or critical parts. The computer is expected to cut this rate by at least half.

Computer design of forging dies is expected to eliminate the traditional cut-and-try aspect of diemaking. In addi-

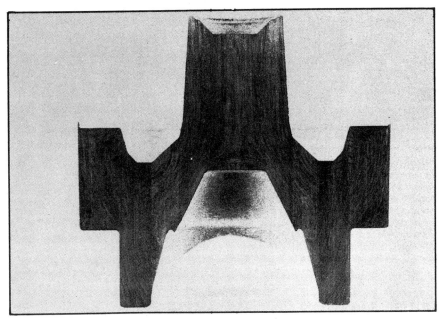

Macroflow in forged metal is frequently cited as a strengthening advantage of forging. This metal flow is clearly seen in the isothermally forged titanium-alloy turbine disc produced by Wyman-Gordon Co.

Cast or forged? Although these Titanium 6Al-4V bulkheads are forged by Wyman-Gordon, castings for similar aircraft applications have been tried in other designs. Unfortunately, the webbed aluminum A357 castings, which had wall thicknesses down to 0.1 in., were never used commercially. This forging has a 3,580-in.² surface area. Ill-fated castings were somewhat larger.

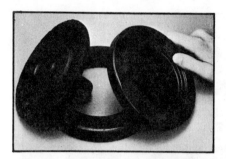

Hot forging can also turn out flashless parts such as these steel gear blanks. National Machinery developed both the hot-forging process and press. Tolerances on these parts can be held to within 0.03 to 0.04 in. on the diameter and 0.04 in. on thickness. The press, said to be the largest of its type in the world, is capable of 45 strokes per minute.

Flashless warm forming, one of several proprietary techniques developed recently to approach "near net shape" production, converts 100% of the billet into the forging, according to MSP Industries. Tolerances on parts such as this medium-carbon steel tank trackwedge can be held to ±0.010 in., but initial billet must be sized to within ±0.035 oz.

Seven-lb adjusting cap for a heavy truck suspension used to be a 12-lb, two-piece casting that required broaching and assembly. Now it is cold forged in one piece in medium-carbon steel. As-forged, internal and external hex flats are radially in line within 0.003 in., according to Molloy Manufacturing Co. Drilling and tapping three holes is the only subsequent machining required.

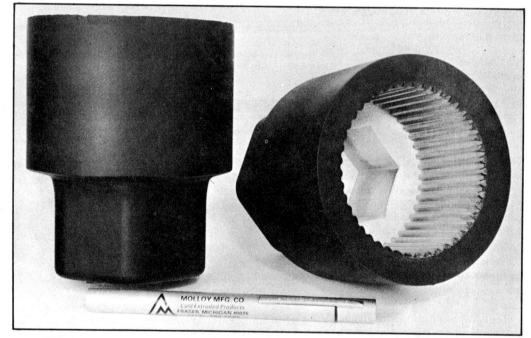

tion, a true CAD/CAM system will be able to translate a conceptualized die design directly into numerical-control machining instructions. The die could then be designed and machined without human interaction.

But can you get 'em?
The supply picture for both castings and forgings is expected to be good through 1981, according to the Forging Industry Association and Steel Founders' Society. Capability to produce steel castings has increased at a slow but steady rate of about 2-1/2% a year over the last five years. Unfortunately, clean-air and other environmental restrictions have held this rate down — although most high-technology foundries have been able to cope with these regulations.

Sophisticated forge shops, capable of producing the high-technology products discussed here, have been operating at around one-half capacity for the last two years, according to the Forging Industry Association. Predictions indicate that this figure will be about the same for 1981, indicating plenty of capacity for near-net-shape and other exotic forgings. ∎

Cutting Metal Loss Tied To Near Net Shapes

By Robert E. Harvey

With the rising demand for titanium and superalloys, researchers are reviewing manufacturing methods with an emphasis on near net shapes.

Manufacturers have always been concerned about the large amounts of metal wasted in metal forming and cutting operations. The concerns, however, have grown more acute in recent years.

Metal prices have risen and the availability of certain metals has become sporadic, while applications for these scarce metals are growing by leaps and bounds.

As the demand for such materials as titanium and superalloys continues to escalate, fabricators and researchers are reviewing manufacturing methods—this time with a closer eye to near net shapes.

Most metal wasted during product manufacturing is lost in the initial forming operation—usually casting or forging. A number of new developments in these two operations gives some promise of dramatically reducing metal loss. Also promising are new ways of working with powdered metals (PM) to reduce material waste.

Many of the new uses of casting, forging and PM involve novel application of old processes or the immediate implementation of technologies that have been gathering dust in research laboratories.

One process that has been around for some time but is finally appearing in commercial applications is squeeze casting. Gould Inc., for example, has begun making pistons using this process. The use of turbocharging has allowed diesel engine manufacturers to provide customers with more power while holding down weight and cost, explains a Gould spokesman.

More power, however, means higher loads and temperatures and Gould found that pistons made by conventional means were inadequate for the job.

Pistons are almost all cast, reports M. F. McGuire, director, product development at Gould. In castings, especially aluminum castings, porosity is the overriding problem, he explains. An efficient foundry is able to avoid or reject castings that display gas po-

rosity or gross shrinkage. Microshrinkage is tolerated, however, because it is so difficult to detect with typical testing methods. Casting also produces a relatively coarse structure because cooling rates are limited, notes Mr. McGuire.

Gould has looked to forging as a means to avoid the problems inherent in casting. But it has found that forging has generally not been cost effective, even when technically feasible.

Squeeze casting is a process "which combines forging and casting in a single operation," explained Joseph C. Benedyk, manager at IIT Research Institute (IITRI), in a paper to the Society of Manufacturing Engineers.

"Because of the pressure applied to solidifying metal, it is possible to

squeeze cast a variety of shapes having improved structure, mechanical properties, surface finish and dimensional control at reduced cost in a variety of ferrous and nonferrous metals and alloys."

"Squeeze casting is one step," emphasizes S. Rajagopal, manager, metalworking technology, IITRI. "You are not taking a porous casting and squeezing it. Squeeze casting is a viable alternative for a variety of applications. Because of direct conversion of molten metal into finished parts, the metal usage is highly efficient. Fine details and complex contours can be produced in a number of alloys, including aluminum alloys of cast and wrought composition, cast iron, high-strength steels and nickel-

Improvement in Mechanical Properties of Cast Aluminum Alloys by HIP at 15,000 Psi

Alloy	Condition	Fatigue Life, Cycles-20,000 PSI	Yield Strength, PSI	Ultimate Tensile Strength, PSI	Elongation at Rupture, Pct
Sand Mold Cast C355	As-Cast-T6	147,500	32,400	34,200	0.8
	HIP-T6	2,905,000	36,200	40,000	1.8
Permanent Mold Cast A 356	As-Cast-T61	452,000	28,900	38,500	7.5
	HIP-T61	4,650,000	30,500	42,200	11.1
Permanent Mold Cast 142	As-Cast-T4	709,000	29,100	32,600	—
	HIP-T4	10,346,000	29,600	36,000	—

Source: TMO report No. 5 Alcoa Technology Marketing Div.

Improvement in Mechanical Properties of Superalloy Castings by HIP

	Temp. F/Stress, PSI	Average Life, Hour	Elongation Pct	Reduction in Area, Pct
IN-738, As-Cast	1800/22,000	19.0	11.8	20.0
IN-738, HIP	1800/22,000	52.5	20.5	20.6
RENE 77, As-Cast	1500/20,000	77.0	10.0	20.0
RENE 77, HIP	1500/20,000	165.0	9.0	18.0
RENE 77, As-Cast	1800/22,000	51.0	19.4	37.0
RENE 77, HIP	1800/22,000	68.0	22.0	55.0
IN-792, As-Cast	1600/45,000	175.0	9.2	6.5
IN-792, HIP	1600/45,000	283.0	12.1	22.0

Source: Proceedings Second International Conference AIME MCIC 72-10

base superalloys.

"Squeeze castings," he continues, "are characterized by a fine as-cast microstructure. Their mechanical properties are isotropic, unlike forgings, and the property levels are higher than typical as-cast properties and approach those of a cast-wrought structure in many applications."

While squeeze casting is a one-step forge/casting process, Hot Isostatic Processing (HIP) is an up-and-coming two-step technique. HIP compacts castings to a quality forging through high heat, and pressure from inert gas.

HIP has been around for some 20 years and has long been used to compress metal and cermet powders into nearly net shapes. Only recently, however, has HIP been used to upgrade the structural properties of castings. This use has considerably broadened the range of applications for these relatively low-cost materials.

Using high temperatures and pressures obtained from isostatic gas press, the process involves the simultaneous application of heat and isotropic gas pressure to a usually complex workpiece, reported D. A. Seifert and H. D. Hanes of Battelle Laboratories in a presentation on "Improving Bronze Castings by Hot Isostatic Processing."

"Healing defects in castings is the most recent application of HIP," the report states. "The initial work in this area, performed by Alcoa, demonstrated significant improvements in fatigue properties of aluminum alloy castings. . . . More recently others have applied HIP to healing of superalloy investment castings with similar results" (see aluminum and superalloy tables). Similar improvements have been achieved in the properties of cast titanium alloys."

The report stated that in the above-mentioned materials the elimination of cast-in defects "generally increased stress-rupture, elongation and low-cycle fatigue life." When the internal defects in a casting were closed, according to the report, there was little or no dimensional distortion, a definite plus when the goal is a near net shape.

Numerous companies and several government agencies, including the Air Force, are showing interest in HIP densification of castings. The Air Force, for example, is sponsoring a program with General Electric as prime contractor and Battelle Columbus Laboratories, Precision Castparts Corp., and Anacast as subcontractors.

The program aims to integrate casting, HIP, heat treating and manufacturing processes into a fully-integrated system. The program would result in a premium quality casting that would be inexpensive, energy-efficient, and materially frugal.

The program is attempting to reduce costs by increasing casting yield and "permitting compromises in casting practices (such as fewer gates etc.) through the application of subsequent hot isostatic pressing."

Another aim is to improve the quality and properties of HIP castings so that forgings can be replaced by castings. Costs can also be reduced by decreasing cycle time or combining HIP with alloy heat treatment. Another byproduct would be the reduction of manufacturing costs such

The price of titanium is just too high to justify such waste.

as chem-milling, welding and scrap loss.

Much of the research that is going into HIP-produced parts is concentrating on titanium applications because: forging wastes too much metal and the price of titanium is just too high to justify such waste.

Currently, titanium alloy impellers are used in high-performance radial engines. Heavy forgings are used to fabricate the part. A 22-lb forging is used to make the T62T-40 impeller weighing 2.6 lbs, reported Arthur G. Metcalfe and Alvin N. Hammer of Solar Turbines International at a Tri-Service Metals Manufacturing Technology Program status review. Again, the alternative which would provide materials and cost savings is casting, but the metal quality of the castings is just not good enough.

Mr. Metcalfe and Mr. Hammer are part of an Army-sponsored HIP program which has already filled the thin blade sections of the impeller and improved surface quality after etching. The project has also developed heat treating to a point where the part had double the fatigue strength of previous castings.

The researchers report that cast titanium impellers are potentially viable, although problems of unsatisfactory casting recovery and weld repair still require solutions, increased capacity for casting recovery and weld repair is needed as well as increased fatigue strength. New production methods of heat treating are

also on the most-wanted list.

HIP, before it was used to densify castings, was used in isostatic diffusion bending and hot-isostatic pressure consolidation of powders. In fact the same processes can be performed on practically the same equipment. And even though HIP of powdered metals has been around commercially for several decades, it is being reexamined for the same reason that scientists are looking at HIP castings—large material savings. Again, the leading edge of the HIP PM technology is aimed at reducing the cost of aircraft parts made of such metals as titanium and aluminum.

Powdered metal HIP is still by far the most popular use of the process. Powder metals were first developed in order to produce near net shapes from costly, difficult-to-fabricate materials. Care was taken to make preforms that would require little machining.

The preforms (shaped and compacted powdered metal) were then sealed into a container, degassed, then submitted to equal pressures from all sides: voilá, a fully dense part. An advantage of the process was full part density achieved at temperatures below those needed for sintering and hot pressing.

According to Hugh D. Hanes of Battelle, this process produced finer grained material than before and enabled more microstructural control to be exercised, "permitted better inspectability, and enhanced the possibilities for the creation of new alloys or new microstructural forms. Thus, it can be seen that the advantages of HIP processing extend beyond the ability to compact powders into complex shapes."

These advantages permit HIP PM to take a leading role in aerospace technology.

A report by R. H. Witt, Grumman Aerospace Corp., Cliff Kelte, Air Force Materials Laboratory, and Ted Highberger, Naval Air Systems Command, titled "Titanium Net Shapes by a New Technology," discusses recent interest in HIP PM.

The inability to produce close-to-fine-dimension titanium shapes, the long lead times for forgings, and overall machining requirements produce delays, slow production and rising costs, explains the report.

As a result, major research efforts have been started "in the fields of HIP, isothermal forging, hot die forging, superplastic/diffusion bonding, titanium casting and others."

The research is examining the Crucible Inc. Ceramic Mold HIP process.

It consists of encapsulating metallic prealloyed titanium powders in a suitably shaped ceramic mold and inserting the mold into a special can which is surrounded by a compressible medium. Then the can is evacuated and sealed. The complete assembly is isostatically pressed at between 1,600° and 1,700° F at 10 to 15 ksi in a gas (usually agron) filled autoclave.

Much work has been done on improving ceramic technology for PM shapes at the Colt Crucible Research Center in Pittsburgh. Scientists at the facility have pioneered the development of the "lost wax" ceramic mold process for near net shape air frame and turbine engine parts. Development has been based on the ability to integrate existing investment casting technology and PM super-alloys, titanium and high-speed steel.

A paper prepared by F. J. Rizzo and V. K. Chanhok of Colt Crucible and R. J. Ondercin of the Air Force Materials Laboratory at Wright-Patterson Air Force Base lists the pros and cons of the ceramic mold.

"There are some obvious economic advantages to producing parts by the ceramic mold process," explains the report. "The investment casting industry has already established the mold making technology...."

However, simply the use of "investment casting and conventional cast and wrought results in large yield losses during processing, most of which are not recoverable. Thus there is the concept of high buy-to-fly ratios. In other words, the ratio of the starting materials weight to the finished machine part weight is very high. This necessitates a large amount of costly machining or other material removal.

"In the HIP and HIP plus forge processes, the input weight is usually at least one-half to one-third the size of that for the more conventional processes."

The paper goes on to report that when comparing the as-HIP near-net to the HIP and forge process, engineers found that "isothermal forging is an extra very costly processing step. Tooling costs are a major factor."

In addition, more metal removal is needed in many cases after forging because shape definition is poor.

There are three promising as-HIP shape making technologies, explains the article: ceramic mold, sheet metal can and hard or soft die. Generally these processes are technology competitive, however, economic differences exist.

"In the ceramic mold process, a

Hot Die Forging Closes in on Near Net Shapes

At **Wyman Gordon Co.'s** isothermal forging unit in Worcester, Mass. near net shape turbine discs are possible on the 1800-ton unit.

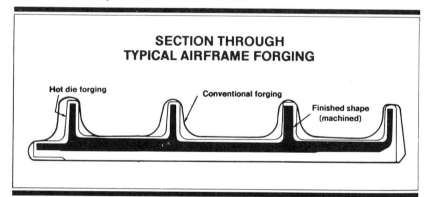

SECTION THROUGH TYPICAL AIRFRAME FORGING

Hot die forging   Conventional forging   Finished shape (machined)

This airframe forging section shows the conventional forging envelope. Note that the hot die forging outline is much closer to the finished machined shape which enables a dramatic savings in material and machining time.

wax pattern can be made inexpensively by machining or injecting an aluminum die. However, shape iterations require changes in starting dimensions and thus wax machining for these parts or for a limited production run can be more cost-effective than wax injection. With sheet-metal cans, steel dies of some sort have to be made even for the first iteration. Each subsequent shape trial will require die rework and a production run may require a total remake of dies for more permanent tooling.

"With the hard or soft die technique, a nonreusable steel die is machined for each part. The larger or more complex the part, the more costly the die material and the more difficult the preparation of steel dies.

Wax injection dies are permanent tooling made of aluminum and these are not required until the pre-HIP dimensions are set and even at this point, it may be less costly to NC machine waxes depending on the total number of parts desired.

"Machining waxes would, in most cases, be less expensive than forming sheet metal can components and would certainly be significantly less expensive than machining a steel die."

Besides combining PM and investment casting, PM has also been teamed up successfully with injection molding. The wedding has been so successful that two awards for design excellence were awarded at the 1979 Powder Metallurgy Part-of-the-Year

Design Competition for products made by injection molding techniques.

Parmatech Corp. won an award for the fabrication of a nickel screw seal used in the wing flap ball screw assembly of Boeing 707 and 727 aircraft. The firm won again for its columbium alloy thrust chamber and injector for a liquid propellant rocket engine.

The company used a metal powder injection molding method called the Wiech process. To produce the parts, metal powders are mixed with a plastisol-like material and molded at approximately 300°F and 1,000 psi. After forming, the organic material is removed and the part is sintered. The geometry and size limitations of the process reportedly are the same as any injection molded part.

Besides being injection molded, investment cast, and HIPed, powdered metals are also being forged. Basically, PM forging involves deforming a conventional sintered PM preform with enough temperature and pressure to produce a completely dense part. What in fact usually happens is that different densities appear. Research has found that much of the variation can be traced to the chilling and friction between the die and the workpiece.

A number of companies including IITRI and TRW Inc. have been working to solve the problem. IITRI combined PM processing with isothermal creep forging to produce high-strength aluminum alloy components.

In isothermal forging the die and the workpiece are at similiar temperatures, reducing chilling problems. And creep forging simply means that since the flow stress of aluminum alloy is sensitive to the speed of deformation, it is better to use slower deformation rates.

"The main advantage of the creep forging technique," says S. Bhattacharyya of IITRI, "is that it enables forming of complex and high-precision components that cannot be formed by other conventional forging techniques. In addition, the forging pressure is greatly reduced—in effect, increasing the capacity of existing presses to make larger forgings." Reduced machining and better material utilization, reportedly, are other advantages of the process.

"Powder metal forgings has been projected to be a manufacturing process of great promise because it combines the cost saving advantages of press and sinter powder technology with the part performance enhancement associated with forging," reported B. L. Ferguson of TRW Materials Technology Laboratory in a paper on PM Forging Components for Army Applications.

"While classified as a PM process, PM forging is truly one type of precision forging, with resulting parts having full density and mechanical properties sufficient for high performance applications," he adds.

Researchers at TRW used three types of forging to produce three different parts. "For a small part with a high surface area to volume ratio, minimum deformation forging was used to succesfully forge 4640 steel powder to full density," explains Mr. Ferguson.

"For axisymmetric parts such as

Forging with powdered metals is not only way to forge near net shapes.

differential gears, preforms requiring gross metal flow were utilized to maximize properties and minimize cost. For a complex shape . . . isothermal forging of steel powder was used."

In all three cases cost savings were projected because of more efficient metal use and reduced machining costs.

Forging with powdered metals is not the only way to forge near net shapes. Wyman-Gordon Co., which, among other activities, fabricates titanium and superalloy parts, is working both with PM and with solid workpieces to reduce metal loss.

"The return on investment is there," explains an official at Wyman-Gordon, "we are just trying to pick it up."

The procedure for reducing the metal waste of a forging is similiar to that of a PM forging: increase the temperature of the die.

"In recent years, the demand for improved cost-effectiveness coupled with better materials utilization in manufacturing processes has brought about recent developments and advancements in the hot-die forgings of titanium alloys to near-net structural shapes," reported C. C. Chen of Wyman-Gordon in a paper titled Nearnet Shapes of Titanium structures by Hot-die Forging. "The major contribution of the hot-die forging is to reduce or to eliminate the influence of die chilling and material strain hardening."

As a result, shapes are refined, material is better used, machining costs are down and the number of forging operations are reduced, he explained.

Wyman-Gordon engineers have worked with two types of "hot die" forgings: isothermal, where die temperatures are the same; and hot die (near isothermal), where die temperatures are slightly below the forging.

"In the case of conventionally forged titanium alloys," explains Mr. Chen, "increased web thicknesses, or large fillet radii, increased rib and flange widths along with decreased depths are design features generally required for adequate filling of the die cavity using unit forge pressures of reasonable magnitudes.

"However, with a die system operating at increased temperatures, the decreased differential between the forging stock temperature and the tool temperature allows a more refined shape of the forged component to be produced in a given operation."

He goes on to explain that with hot die forging "extensive metal flows are possible within one die cavity providing preform shapes have previously distributed the necessary volume of material in localized zones from which the new shape is generated."

There are problems with hot die forging, however. The dies are expensive, lubricants could be more effective, heating times are slow and process difficulties exist.

In fact, one trouble with much of the near net shape technology is that it is expensive. Work is being done in many areas to reduce prices.

Gorham International, Inc., for example, is seeking sponsors for a two-year project that will attempt to reduce the cost of HIP for the manufacturing of precision high alloy parts that would be competitive with forgings and machined parts. Metals to be studied for the process include 316 stainless, Monel, M-2 tool steel, Stellite 6 and 21, titanium, aluminum and vanadium alloys, and 4640 steel.

The project will also attempt to determine specific commercial alloys the technical, economic and market feasibility of replacing castings, forgings, and wrought machined parts by hot isostatically densified centered PM parts. □

ENERGY CONSERVATION ASPECTS OF NEAR NET-SHAPE PROCESSING

by

S. RAJAGOPAL
Materials and Manufacturing Technology Division
IIT Research Institute
Chicago, Illinois 60616

ABSTRACT. The energy consumed in manufacturing a finished part depends on the material used, the type of manufacturing process, the material yield in the process and, in many cases, the extent to which process scrap is recycled. Process flow diagrams and energy models are described herein for casting, forging, and powder metallurgy (P/M) processes. Material yield is seen to play a dominant role in determining the energy use in manufacturing steel parts, while scrap recycling takes precedence in the case of aluminum. The most energy-efficient processes are those that can use 100% scrap as raw material and convert it, without remelting, into near net-shape parts. The feasibility of this form of processing was shown, on an experimental basis, by hot pressing scrap turnings of aluminum into strong, dense, near net-shape parts.

NOMENCLATURE

e_1 Energy requirement for primary ingot production, in MBtu per ton of ingot

e_2 Energy requirement for remelting and casting/atomizing, in MBtu per ton of ingot

e_3 Energy requirement for pressworking, in MBtu per ton of workpiece material

e_4 Energy requirement for metal removal, in MBtu per ton of scrap generated

e_5 Energy requirement for sintering P/M compacts, in MBtu per ton of sintered compacts

e_6 Energy requirement for cleaning and pulverizing process scrap, in MBtu per ton of scrap

E Energy consumed in manufacturing each finished part, in MBtu

W Weight of each finished part (net-shape), in tons

x Fraction of total scrap which is recycled

y_1 Material yield fraction in casting

y_2 Material yield fraction in upset forging

y_3 Material yield fraction in finish forging

y_4 Material yield fraction in P/M

y_5 Material yield fraction in scrap hot pressing

INTRODUCTION.

The escalating cost of energy makes its conservation in the manufacturing industry an issue of growing importance. The first step in a conservation scheme is to audit the energy usage in manufacturing processes, recognize the factors which influence energy consumption, and determine the extent of this influence. Energy-intensive operations can thus be identified, and new or modified processes developed to avoid the dependence on such operations.

Near net-shape processes convert raw material into shapes approaching the required final shape and thus involve little or no secondary machining. Near net-shape alternatives currently exist within the broad categories of metal casting, metal forming, and powder metallurgy (P/M). Die casting and low-pressure permanent mold casting are examples of near net-shape casting; isothermal forging exemplifies near net-shape

forming; while virtually all P/M processes produce parts at or near net-shape. In all cases, the processing is completed in closed metal dies for good surface finish and dimensional precision, with minimum allowance necessary for machining. Near net-shape processes are efficient in their use of raw materials and, consequently, in energy utilization as well.

In the present study, energy models were developed for manufacturing processes to analyze and compare the energy content of finished parts produced by casting, forging, and P/M processes. Flow diagrams were used to follow the material from ore to net shape, through primary processing, metalworking operations, and machining. Weight balances and specific energy requirements were considered for each operation within a given process to arrive at the energy content of the finished product.

The main findings of the study were that (a) a high degree of material yield (net shape capability) helps conserve energy, (b) where net-shape capability is lacking, similar benefits may be derived by effective recycling of scrap, especially for aluminum alloys, and (c) to minimize energy utilization, the use of primary metal and remelting operations must be discouraged.

Based in part on the last finding, an experimental study was initiated to consolidate mechanically processed scrap into finished shapes without remelting or sintering. It was found for aluminum alloys that machined chips (swarf) could be readily hot pressed in tool steel dies at about 850°-950°F (450°-510°C) under a pressure of 12-20 tsi (165-275 MPa) maintained for 2-5 s. In addition to consuming only 10% of the energy required for aluminum P/M, and only 1% of the energy consumed in casting or forging, the parts were strong, fully dense, and could be made reproducibly with good tolerances and surface quality.

ENERGY MODELS.

Flow diagrams (Fig. 1) were used to describe material flow and energy requirements at various

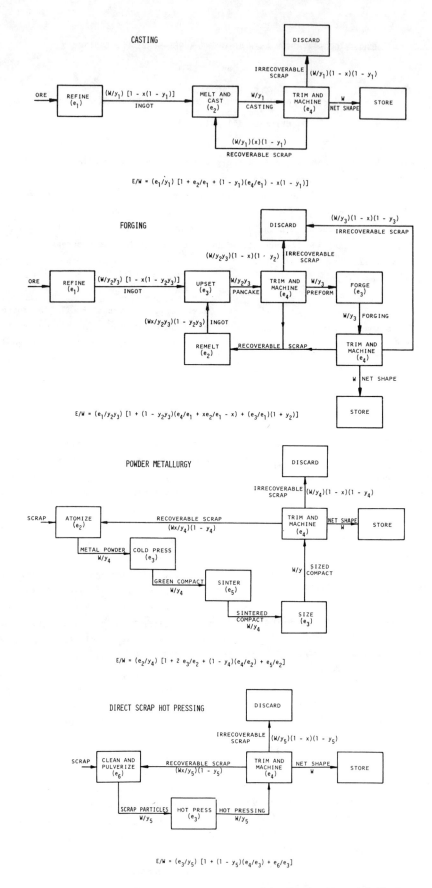

Figure 1. Manufacturing Process Energy Models Showing Material Flow, Weight Balance, and Energy Requirements in Casting, Forging, P/M, and Scrap Hot Pressing.

stages in casting, forging, P/M, and scrap hot pressing. In each process, the objective was to produce finished parts of weight W using either primary metal or scrap metal or both, as dictated by the nature of the process. The energy consumed in each process to do this, as seen by the equations for E/W in Fig. 1, was dependent on the specific energy requirement (e_1, e_2, etc.) for each operation in the process, on the yield obtained (y_1, y_2, etc.) and, for casting and forging, on the extent of scrap recycling (x). In the case of P/M and scrap hot pressing, the starting material was assumed to be 100% scrap (no refining of primary ingot from ore) and scrap recycling within the process was thus inconsequential. In all cases, scrap was assigned an energy value of zero.

In a typical casting operation, primary ingot derived from ore is melted, together with any recycled scrap, and cast into a shape approximating the desired end configuration of the product. The casting is then trimmed of risers and gates, if any, and machined to net shape. Of the resulting scrap, a good portion is usually collected and remelted while a smaller fraction is lost as irrecoverable scrap. Typical values of the specific energy components are as follows:

Refining: 244 MBtu per ton of aluminum ingot[1]
(e_1)   24 MBtu per ton of steel ingot[1]

Remelting: 10 MBtu per ton of cast aluminum
(e_2)   7 MBtu per ton of cast steel

Machining: 2 MBtu per ton of machined aluminum
(e_4)   and steel scrap.

In the case of forging, ingots processed from ore and recycled scrap are first upset or otherwise worked to refine the cast structure prior to imparting shape detail. (A material yield factor of $y_2 = 0.8$ was assumed in the first-stage deformation.) The resulting preform is forged to near net-shape and then machined to net shape. The energy constants are:

Refining: 244 MBtu per ton of aluminum ingot[1]
(e_1)   24 MBtu per ton of steel ingot[1]

Remelting: 10 MBtu per ton of cast aluminum
(e_2)   7 MBtu per ton of cast steel

Pressworking: 1 MBtu per ton of forged aluminum
(e_3)   and steel

Machining: 2 MBtu per ton of machined aluminum
(e_4)   and steel scrap.

In powder metallurgy, scrap metal is first melted and atomized into powder. The powder is compacted in dies, sintered, re-pressed to size, and then machined (usually minimally) to finished shape. The energy consumed in each stage of the process was assumed to be:

Atomizing: 25 MBtu per ton of aluminum powder
(e_2)   19 MBtu per ton of steel powder[2]

Pressworking: 0.5 MBtu per ton of compacted aluminum
(e_3)   and steel

Machining: 2 MBtu per ton of machined aluminum
(e_4)   and steel scrap

Sintering: 15 MBtu per ton of sintered aluminum
(e_5)   17 MBtu per ton of sintered steel

In direct scrap hot pressing, a process whose technical feasibility is explored subsequently in this paper, machined chips are degreased and mechanically downsized, hot pressed isothermally in heated dies and, finally, machined to net shape. In this case, the constants are:

Pressworking: 2 MBtu per ton of hot-pressed aluminum
(e_3)   4 MBtu per ton of hot-pressed steel

Machining: 2 MBtu per ton of machined aluminum
(e_4)   and steel scrap

Cleaning/
Pulverizing: 2 MBtu per ton of aluminum and
(e_6)   steel scrap.

DISCUSSION OF RESULTS. By substituting the above energy constants into the mathematical models described in Fig. 1, it was possible to determine the influence of material yield and scrap recovery on the overall energy consumption (Fig. 2). The significant features of the curves shown in Fig. 2 are:

1) For casting and forging, energy consumption decreases as the scrap recycling percentage increases. The influence of scrap recycling is much more significant for aluminum than for steel, due to the very high energy content of primary aluminum ingots.

2) Energy consumption decreases with increasing material yield. In casting and forging, material yield becomes more important when scrap recovery is poor. Conversely, with higher yield, there is less scrap generated and so its recovery is not as important as in a low-yield process. In the case of aluminum casting and forging, the effect of yield on energy consumption is very marginal when there is 100% scrap recycling, because of the relatively low energy requirement for remelting.

3) For steel, the energy savings stemming from scrap recycling are relatively small; consequently, material yield is the main consideration. With aluminum, the reverse is true: scrap recycling takes precedence over yield.

4) P/M reduces energy consumption for aluminum quite significantly. In the case of steel, P/M actually requires more energy than casting or forging at similar levels of yield. However, the yield in P/M is usually much better than in either casting or forging and, for this reason, the process consumes 20-50% less energy.

Source: North American Metalworking Research Conference Proceedings, 1981, 215-220

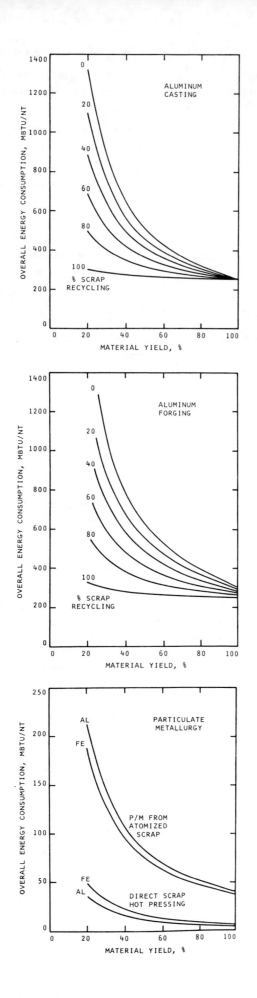

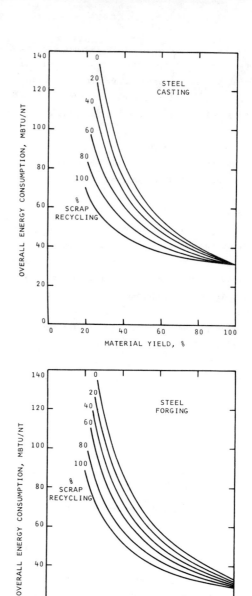

Figure 2. Influence of Material Yield and Scrap Recycling on Energy Consumption Per Net Ton of Finished Products in Casting, Forging, P/M, and Scrap Hot Pressing of Aluminum and Steel.

Note: In the case of forging, indicated material yield is for finish forging. First-stage (upset) forging was assumed to produce 80% yield.

Table 1. Raw Material Requirement
Per Net Ton of Finished Product.

PROCESS	YIELD, %	RECYCLING, %	TONS OF RAW MATERIAL PRIMARY	SCRAP	TOTAL
CONVENTIONAL CASTING	50	50	1.50	0.50	2.00
		80	1.20	0.80	2.00
GATELESS CASTING	80	50	0.94	0.31	1.25
		80	1.05	0.20	1.25
CONVENTIONAL FORGING	40*	50	1.75	0.75	2.50
		80	1.30	1.20	2.50
PRECISION FORGING	64**	50	1.28	0.28	1.56
		80	1.11	0.45	1.56
POWDER METALLURGY	90	0-100	0	1.11	1.11
DIRECT SCRAP HOT PRESSING	90	0-100	0	1.11	1.11

*80% YIELD IN UPSET FORGING AND 50% YIELD IN FINISH FORGING

**80% YIELD IN UPSET FORGING AND 80% YIELD IN FINISH FORGING

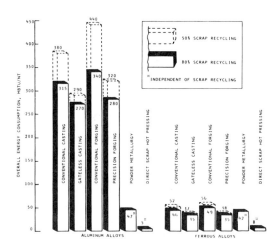

Figure 3. Typical Energy Consumption Per Net Ton of Finished Product in Casting, Forging, P/M, and Scrap Hot Pressing for Aluminum and Steel.

5) The most dramatic energy savings of all results from direct scrap hot pressing. Ore refinement and scrap remelting--the most energy-intensive components of conventional manufacturing processes--are eliminated in scrap hot pressing. This lowers the energy consumption by one to two orders of magnitude in comparison with casting, forging, and P/M.

Table 1, based on the weight balance in Fig. 1, lists the quantities of primary metal and scrap required to manufacture one ton of finished product by casting, forging, P/M, and direct scrap hot pressing. Increased recycling reduces the quantity of primary metal required, as does increased material yield. As noted earlier, P/M and scrap hot pressing were assumed to rely entirely on process scrap, and with a 90% yield of finished product from raw material.

For the material yields given in Table 1, Fig. 3 depicts the expected energy utilization per net ton of finished aluminum and steel products manufactured by casting, forging, P/M, and scrap hot pressing. In general, aluminum is seen to be nearly 10 times as energy intensive as steel when primary metal is used (casting and forging), but the two are nearly equal in scrap-based processes (P/M and scrap hot pressing).

Again, the drastic reduction in energy content of finished products due to scrap hot pressing is clearly noted.

FEASIBILITY OF DIRECT SCRAP HOT PRESSING. The most energy-intensive operations in typical cast/wrought processes are refining of primary ingot and remelting of scrap. In P/M processes, powder production and sintering consume most of the total energy. Development of a

new process which avoids these high-energy operations requires:

1) Use of scrap, without remelting, with mechanical downsizing (by crushing) for die fill purposes, and

2) Direct consolidation into finished parts, without the need for sintering.

Preliminary experiments on solid slugs hot pressed from scrap turnings of aluminum showed the feasibility of this approach. The scrap chips were degreased, transferred into a die at 950°F (500°C), and compacted under a pressure of 12 tsi (165 MPa) for 2 s. The slugs were completely free of porosity and exceeded conventional P/M expectations of strength and density.

Further work was directed at producing a stepped collar using the die arrangement shown schematically in Fig. 4. The raw material was 2014 aluminum alloy swarf produced in a milling operation. These chips were degreased, charged into the hot die at 850°F (450°C), and consolidated with a single-action punch at 20 tsi (275 MPa) pressure for 5 s. The collars were ejected from the die and then solutionized and aged to the T6 condition. In comparison with wrought 2014-T6 alloy, the density and strength of the hot-pressed parts were 100% and 80%, respectively, of wrought values. The failure of these parts was somewhat more brittle than a wrought part and similar, in this respect, to conventional powder metallurgy. Figure 5 shows the microstructure of the hot-pressed collar to be clean, fully dense and with flow lines within each individual chip from the cold work of machining.

Direct hot pressing of scrap thus appears to be one possible means of reducing energy utilization in manufacturing.

Source: North American Metalworking Research Conference Proceedings, 1981, 215-220

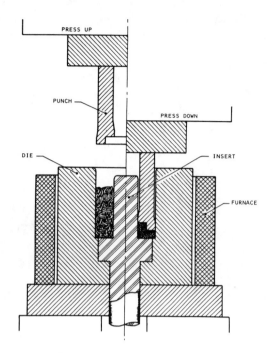

Figure 4. Experimental Setup for Direct Hot Pressing of 2014 Al Scrap Turnings into Stepped Collars.

Figure 5. Optical Micrograph of 2014-T6 Al Stepped Collar

CONCLUSIONS. Energy models were developed to measure the energy content of finished parts produced by casting, forging, and P/M processes. This study showed that

1) A high degree of scrap recycling is effective in reducing the energy content of cast and forged parts, especially for aluminum alloys.

2) A high degree of material yield reduces the overall energy consumption in all processes, particularly in the case of ferrous alloys.

3) Energy savings in powder metallurgy stems largely from the near net-shape capability of P/M processes.

4) Drastic reductions in energy utilization require eliminating the use of primary ingot and remelting operations.

Direct hot pressing of scrap particles into finished parts was shown feasible in terms of the resulting strength and density. The energy requirement for this process is smaller, by one to two orders of magnitude, in comparison with casting, forging, and powder metallurgy.

REFERENCES

(1) "Energy Use Patterns in Metallurgical and Nonmetallic Mineral Processing," Battelle-Columbus Laboratories, U.S. Department of Commerce Report No. PB-245-759, p. 5 (1975).

(2) S. M. Kaufman, "Energy Consumption in the Manufacture of Precision Metal Parts from Iron Powder," The International Journal of Powder Metallurgy and Powder Technology, Vol. 15, No. 1, pp. 9-19 (1979).

SECTION II
Flashless Forging

Flashless forging is here

Careful control of preforms, tooling, and equipment ensures just the right amount of stock to fill the dies—no more, no less

FORGING FLASH is on its way out! The metal that extrudes from closed-impression dies is targeted for extinction at MSP Industries Corp (Center Line, Mich), a W.R. Grace subsidiary manufacturing special fasteners.

MSP contends that its patented Warmflow process is a cost-competitive alternative to conventional hot forging with impression dies, powder-metal processing, and cutting from wrought stock on such equipment as screw machines. And, says MSP, the process is particularly suited for high-volume production.

Benefits realized from the no-flash breakthrough are substantial. Gone are several undesirable forging byproducts:

■ Material waste. Records show that, in conventional closed-die forging operations, flash averages 20-40% of the original workpiece. Either it's salvaged and sold at a fraction of the metal's original cost, or it's thrown away. Flashless forging uses 100% of the stock.

■ Energy waste. Heating metal that will become flash, transporting it, and then deforming it in such machines as presses and parts formers expends energy for no useful purpose.

■ Production waste. Flash formation creates undesirable internal structures in forgings. Grain flow is often at 90° to the direction it should be, degrading the integrity of the part. Flash also must be trimmed, which is an additional operation requiring a press—or at least a snag-grinding station. In turn, this means more labor, part-handling, and equipment. Overall, the result is slower and costlier progress through the shop.

■ Equipment waste. The press equipment used for flashless forging can be smaller because there's not as much metal to deform. Tonnage ratings can be somewhat lower, and so can equipment investment.

In general, as MSP points out, these savings and advantages combine to make flashless forging less costly than the conventional method—and more profitable.

The weapon used to kill forging flash is precise volume control, achieved through tight regulation of forging equipment, preforms (billets), and particularly tooling. Its trademarked designation: Warmflow. A longer but more descriptive name is MSP Flashless Warm Forming. (At MSP, as in other forging environments, *forming* is the slightly more-generic equivalent for *forging*.)

Two years ago, after several spent on research, W.R. Grace demonstrated its confidence in the process by constructing a $2.5-million facility for the production of parts for automotive and other original-equipment manufacturers. The 12,000-sq-ft shop, designated Plant 5, is a subsidiary building near MSP's main plant. It currently houses four forging lines, each with heating unit and press.

The name Warmflow is a key clue to the process. Workpieces are heated only to the level at which the metal flows just enough, under forging pressure, to fill the die but not so freely as to extrude from the closing die halves before the cavities are filled. This factor combines with precise control of the other process variables to produce the desired result.

Currently, the parts that can be made by the process must fit within a 6- x 3- x 3-in. rectangular envelope, and their top cross-sectional area can be no larger than 9 sq in. Weight limitations depend on the material being forged: for steel, brass, or copper, the range is 0.3-5 lb; for aluminum, 0.1-3 lb. Wall thickness can be a minimum of 0.100 in., depending on the configuration of the part. Many materials can be formed: carbon, alloy, boron and stainless steels, titanium, aluminum, copper, and brass.

For the ferrous stock currently used in flashless production, the temperature range of metal as it enters the dies is 1000-1800F. Composition of the workpiece material—mainly the carbon content—and the nature of the part to be forged determine which of two temperature ranges is required. If the part's

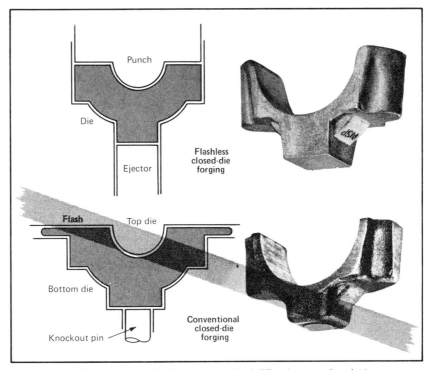

Connecting-rod cap made by flashless forging (top) differs in more than just appearance from same part forged conventionally: strength and life span are greater

By John T. Winship, associate editor

Reprinted with permission from American Machinist, June 1981, 138-141, © 1981 McGraw-Hill Inc.

metallographic structure is to be kept body-cubic (BCC), the temperature will be held between 1000F and 1400F. If a face-center-cubic (FCC) structure is required, the level will be established at some point between 1600F and 1800F.

The transformation temperature is between these ranges, in the vicinity of 1500F. It's the point at which rapid expansion—as much as 7%—occurs, and MSP technicians compensate for it in their efforts to eliminate flash formation during the forging stage.

Although specialty fasteners are MSP's chief product, the first part made on its flashless forging lines was a connecting-rod cap. The part is roughly 3.5 in. wide and 1 in. thick in the plane normal to forging-die motion; MSP calls the plane the "punch planar area." Other parts produced by the flashless process are components for automotive transmissions, tracks for military vehicles, and special pistons for racing cars.

The preform, or billet, that will become a con-rod cap looks like a small cigar. It's sheared from uncoiled wire and shaped on a 1.00-in. National Machinery Co cold header.

Delivered in tote bins to the flashless-forging facility, the billets are taken to one of the forging lines and dumped into an elevating conveyor that orients the parts lengthwise, then advances them by means of a pusher mechanism into the induction-heating unit.

The electric induction-heating units, built by Induction Heating International, are each rated 200 kN, 10 kHz. Their coils are interchangeable; the largest can accommodate 2-in.-dia stock. MSP also uses stock with the rounded-square cross-section.

Key: tension-knuckle press

A 660-ton (US) Komatsu tension-knuckle press is the key component in each of the forging lines now in operation. Key features are a die-transfer mechanism and a digital tonnage monitor, both supplied by Komatsu.

The parts produced on these lines are dimensionally precise to ±0.010 in.; for some, it's as close as ±0.005 in. These are much tighter tolerances than the average ±0.030 in. obtained by conventional impression-die forging, and they're held consistent, part after part. But the improvements don't stop there.

Draft angle is now a term for forging-history books; flashless forging produces parts with vertical surfaces—if required—and, likewise, right-angle corners sharper than any that can be made by conventional forging. The as-forged surfaces are smooth, too—within 125 μin. or better in many cases.

Flashless-forged parts also have better

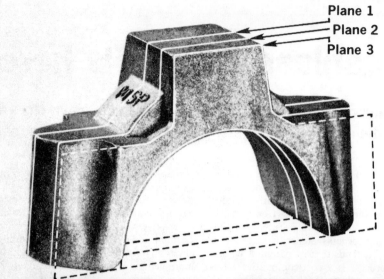

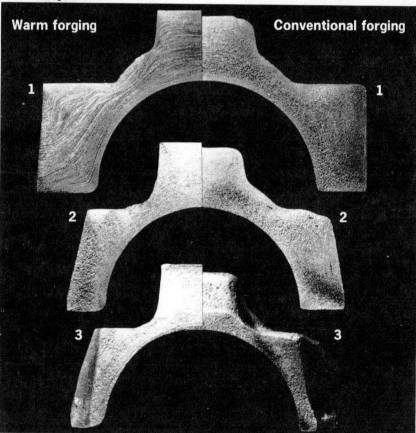

Warm forming—no hotter than 1800F—produces grain flow parallel with the part contour (sections on left). This is another benefit of the flashless method

integrity and strength than the other kind because internal grain structure follows the flow of metal within the part. In parts forged with flash, the grain pattern flows into it, and subsequent removal, by machining, press-trimming, or grinding, severs the metal fibers in the flash zone. This interruption in grain orientation weakens the part in this area and reduces its ability to endure high stresses.

Four years ago, these accomplishments were in the planning stage, objectives considered attainable and worth pursuing by MSP and its parent. A test program begun at Battelle's Columbus (Ohio) Laboratories first produced more worn-out tools than satisfactory parts. Fine-tuning during six or seven die trials eventually yielded 5000 satisfactory test parts from one set of tools. Research was

Flashless warm-forming stations at MSP ingest precisely cut billets from another location, heat them by induction, then feed them to tension-knuckle presses

Three steps to a finished part, a trackwedge for a military tank: cold forming of billet (left), flashless warm forming (note 90° corners), and machining

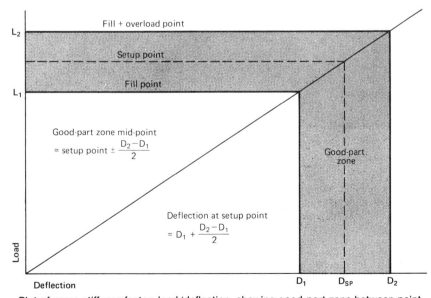

Plot of press stiffness factor, load/deflection, showing good-part zone between point where die is completely filled and point where press becomes overloaded

expedited by application of CAD/CAM principles and guided by Dr Richard P. McDermott, MSP's director of operations at the new facility.

Ideally—and MSP comes very close—the process requirements for flashless forging are as follows:

■ The preforms, or blanks, must be equal in weight.

■ Each preform must be heated to the same temperature.

■ Tooling must be kept at a constant temperature.

■ No tool wear can be tolerated.

■ Punch/die clearance must be close to zero.

Collectively, these requirements are intended to limit the amount of stock confined within the dies to an acceptable tolerance. McDermott's approach to the attainment of this goal focuses first on the allowable weight (which relates to volume) variation of the part as it is completed within the press dies. This is actually a function of (1) the design of the part to be forged and the weight variance that can be tolerated without adversely affecting the part's function and (2) the characteristics of the forging machine.

Once these are quantified, the factors that contribute to a properly formed, flashless forging can be evaluated with the objective of making their algebraic sum equal to or less than the allowable variations in part volume.

Maximize weight tolerance

The first condition is the simplest to explain: set a weight tolerance that is as large as possible while preserving critical dimensions and function.

The second condition is the selection of a forging press with a low stiffness factor (defined as load/deflection). MSP's tension-knuckle-joint presses meet this condition.

These conditions combine into an equation that MSP depends on to ensure the success of its flashless-forging operation: $V_t + V_c + V_o \leqq \Delta V$.

To determine the factors in the equation, McDermott plots the stiffness factor as a diagonal line on a graph (see left) where deflection is the abscissa and load the ordinate. Load L_1 and resulting deflection D_1 determine the fill point, when the stock has just filled the die cavity. L_2 and D_2 are the practical upper limits for the press (before damage begins). Between them is what he calls the "good-part zone," the center of which is D_{sp}, the setup point.

Thus, the permissible variation of die-cavity volume, ΔV, resulting from the action of the forging press, is equal to the projected area of the part being forged times the press-deflection value of

Source: American Machinist, June 1981, 138-141

35

$(D_2 - D_1)/2$, which is, in effect, the allowable part-height tolerance.

For example, for MSP's con-rod cap, which has a design weight of 200g, ΔV is ± 0.011 cu. in., or 1.4g; this variation of 0.7% is within the part's permissible weight tolerance.

On the other side of the equation, opposite ΔV, are the key conditions that must be precisely controlled. If not, flash will almost certainly appear. Preform volume is the first one.

If the part's design weight is less than 1 lb, the stock is coiled rod. (Wire larger than a certain diameter can't be coiled.) Preparation for flashless forging may include annealing, pickling, coating and drawing, cutoff, and processing in a cold-heading machine. Weight and, thus, volume are precisely determined at the header's cutoff station.

Preforms required to weigh 2 lb or more—for parts of the same weight—begin as bar stock that's cold-sized, cut on a saw, then sent through a weight-classifier scale that separates the on-spec and the off-spec billets.

For either of these preform steps, MSP claims a maximum tolerance of ± 1 g or $\pm 0.5\%$ of specified weight, whichever is largest, and designates this factor V_c.

The thermal effect on forging-stock volume is another factor plugged into the flashless equation, V_t. A simple example shows how McDermott determines it for every 50F increment; this value also happens to be the extreme temperature variation permitted as the stock exits from the induction heater. Accurate monitoring at this point ensures consistently heated stock within this range.

The example: steel's coefficient of cubic expansion equals 18×10^{-6} in.^{3}/in.^{3}/deg F; a temperature rise or drop of 50F, therefore, produces a percent volume change of $(18 \times 10^{-6}) \times 50 \times 100$, or 0.09%.

MSP's third factor, called V_o, is an amalgam of several other volume effects. One of them is the build-up of lubricant on the surfaces of the punch and die cavity. This condition is kept within practical limits by use of proper compounds and correct application methods.

Die-temperature variation is another contributor to volume changes in the forging process. So dies are held within close temperature limits by a closed-loop, controlled-flow lubrication system.

Die wear is crucial

Die wear is considered to be the most crucial factor, however, and is recognized as an inevitable source of flash. For all intents and purposes, says McDermott, it should not be allowed to occur. But it does, and so MSP tries to keep dies in acceptable, non-flash-producing condition for as long as possible.

For example, punch/die clearance is maintained at nearly zero; if it's too large, the result is an underfilled die cavity and flash. If clearance is too small, the impact of punch against die produces the heat and consequent expansion that spell tooling failure. The company believes that one way to maintain a low punch/die clearance is to use presses designed for accurate stroke alignment; MSP engineers are convinced that the Komatsu presses feature this characteristic.

Die wear can also be minimized by use of the proper die material, says McDermott. He points out that the forging operation, particularly the flashless kind, creates severe mechanical and thermal stresses. The con-rod cap, for example, requires a setup load $[L_1 + (L_2 - L_1)/2]$ of 250 tons. Exerted over the part's projected area of 3.5 sq in., the average stress is thus 150,000 psi. And, 40 times each minute, the die undergoes expansion and contraction stresses induced by alternating exposure to billets at 1700F and ambient air at 70F. ∎

Cold-heading shift spurs MSP's growth in fasteners

Few, if any, shops can match MSP's cold-heading operations: 40 cold headers range along four aisles in a 120,000-sq-ft facility that also houses equipment for uncoiling and wire drawing, a quality-assurance department, and machinery used in weigh-counting and packaging the fasteners.

Located in MSP's main plant, this large operation, still the mainstay of the firm's business, is a far cry from the small battery of six- and eight-spindle automatic screw machines that began turning out conventional fasteners for Michigan Screw Products Co in 1945.

In a phased transition that started four years later, the screw machines were replaced by cold headers. And the firm's standard catalog line of screw-machine parts was converted to products that started out as cold-headed blanks and were then finished by such secondary operations as threading and plating.

The result was higher output, better quality, lower piece costs, and the growth of MSP Industries Corp into a leading US producer of special fasteners, as well as cold-headed and cold-extruded products made of carbon and alloy steel. These poured into the OEM channels for use by such industries as automotive, agricultural equipment, and off-highway and over-the-road vehicles.

Incoming coils of stock are prepared for cold heading in MSP's Plant 2. The so-called "wire" runs a conditioning gamut that can include annealing, pickling, and coating with either phosphate or lime. Wire normally ranges from 5/16 to 1 in., although stock as large as 1½ in. dia is processed occasionally.

Destination of the conditioned coiled wire is one of the 40 cold headers at the main plant. Each of these machines is equipped with still another conditioning machine, an Ajax-Hogue wire-drawing unit that precisely sizes the wire before it feeds into the header. The products made on these machines can range in size from ⅛ in. dia x ¼ in. long to 1½ in. dia x 8 in. long. After discharge from the cold headers, they are packaged for shipment or are further processed by any of nearly 100 different machines in the MSP shop; these include thread rollers and trimmers.

Coils of wire await processing by the 40 cold headers in MSP's main plant

The feasibility of flashless forging

by **T. A. Dean**, Department of Mechanical Engineering, University of Birmingham

This paper discusses the use of conventional forging equipment for the manufacture of components without flash. The likely range of billet volume variation, the major problem associated with flashless forging, is enumerated for practical situations. An analysis is presented for the estimation of forging loads and die stresses in completely closed cavity dies, when over-sized billets are encountered. It is indicated that maintaining die stresses within working levels is more practical on a percussive machine with good energy control, than on a mechanical press. Various forms of die design for overload protection are illustrated and discussed.

Relative to total industrial output, the market for drop forgings is diminishing. Increasingly, engineering components which were traditionally drop forged are being replaced by other products, such as castings, sintered powder parts, fabrications and at a rapid pace, plastics. The main reason for this, is that these other products have better physical properties, compared with drop forgings, which are inaccurate and expensive when the necessary post-forged machining operations are included. The mechanical properties of forgings have become less outstanding, with improvements in the alternative processes and designers are tending to compensate for the slightly inferior dynamic properties of other components, in order to be able to exploit the precision available.

Considering the limitations inherent in the conventional forging process, Brookes[1] has questioned the existence of the long-term future of the forging industry. He said, that if the industry is to survive, it had to become less labour and material intensive and that the forgers 'Holy Grail' must be the pursuit of flashless or near-flashless forging, together with closer dimensional precision.

Some 30% of stock material entering a forge leaves it as waste. Hot forging is material intensive, 50% of the total component cost being due to material cost. Consequently small improvements in material utilisation result in large savings. If these can be achieved together with greater component accuracy related to the finished product, the competitiveness of forgings will be significantly increased.

Most of the materials wasted is as flash. To determine how this waste may be reduced or eliminated it is important to understand the nature of flash.

FLASH FORMATION

The formation of a peripheral band of flash is inevitable, when the parting line of the dies is transverse to the line of ram motion. For this reason alone, it is necessary to put more metal in the die than is required for the basic component shape. However, forgers have collaborated with the inevitable by designing into dies, lands and gutters, to accommodate the flash. Having done this, it is possible to use excess metal purposefully, to compensate for variations in billet volume and die cavity growth due to wear. The design of preform shapes and flash land and gutter

Presented at the Drop Forging Research Association Conference, Leicester, 20-21 May, 1977.

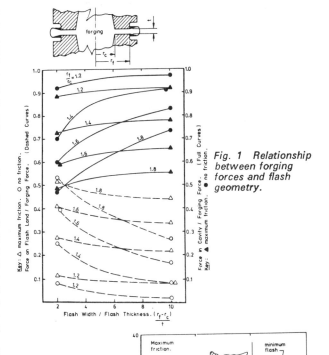

Fig. 1 Relationship between forging forces and flash geometry.

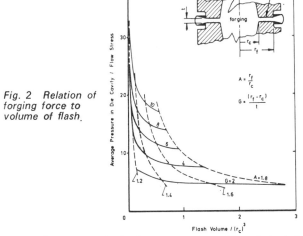

Fig. 2 Relation of forging force to volume of flash.

geometries, to permit the least amount possible of stock to be used, is a skill of high order. Unfortunately, bad design practice can often be obscured by the use of billet volumes greatly in excess of the optimum, resulting in unprofitable production.

Force requirements

A study of the mechanics of flash formation for axi-circular shaped forgings has already been made and references 2 and 3 are recommended for detailed information on the subject. Relevant data are summarised in figures 1 and 2. Figure 1 illustrates how the forces on dies relate to the flash geometry and the surface conditions on the flash land. The possible extremes of the latter are represented by no friction and maximum friction curves. It is likely that, in normal forging practice, conditions approaching maximum friction will obtain. It is seen that the force on the flash land can be expected to be, at best, about 10% of the total forging force and in some cases is likely to approach half of the overall forging effort. To obtain high proportions of the applied load within the cavity, where the load is needed for die filling purposes, narrow flash lands (small ratios or r_f/r_c, in figure 1) and thin flash (high values of flash width/thickness ratio) are required. The limits to which these requirements can be met are generally set by the exingencies of strength and durability of the flash lands. Values of flash width to thickness ratio in the range 4 to 8 are common and it can be seen in figure 1, that the proportion of the total load within the die cavity, increases little for higher values, at realistic levels of friction.

Judging from these curves, it may be expected that the load necessary to forge a shape can be reduced dramatically by eliminating flash lands. The exact amount by which the load can be diminished will depend on the shape of the cavity section and the manner of the material flow in filling it. This matter is currently the subject of our investigations.

Material wastage

For good economy it is necessary to produce the minimum quantity of flash, whilst concurrently obtaining sufficient pressure in the body of the forging, to fill the die cavity. Figure 2 shows how the average forging pressure in a die cavity is related to the volume of flash between the lands. This represents the absolute minimum amount of material necessary to achieve the cavity pressure and it is unlikely, in practice, that such small quantities could be realised. The curves in figure 2 show that higher cavity pressures can be achieved, by reducing the width of flash lands (lowering values of r_f/r_c) and by increasing the ratio of flash width to thickness (raising values of r_f-r_c/t). These changes are accompanied by a reduction of the required volume of flash material. They also permit a greater proportion of the total effort to be applied in the cavity (see figure 1) and therefore promote a higher load 'efficiency'. A range of flash geometries most likely to be used in practice, is marked on figure 2. It is seen that, within this range a flash volume of 0.2, compared with the cube of the major die cavity radius is easily reached. This could constitute a large proportion of metal in a common type of forging. For instance, in a cylindrical forging with a diameter equal to twice the height, the flash volume would be about 50%.

FLASH ELIMINATION

The concept of eliminating flash from the forging operation implies the use of 'completely closed die cavities'.

A 'completely closed cavity' obviously cannot be used with 'off-the-bar' techniques, so forging from billets only can be considered.

Lack of a normal flash line introduces two immediate difficulties to production:

(a) excess billet volume will cause overloading of machines and overstressing of dies;

(b) the forged height will depend on billet volume.

These points, which will be dealt with in more detail later, are both a result of variation in billet volume.

BILLET VOLUME VARIATION
Billet production

The bar sizes considered will be in the range 1 in (25 mm) to 3 in (75 mm) diameter. These may be regarded as including the most popular dimensions for forging stock.

The accuracy to which billets may be prepared depends on, the variation of cross-section and the variation in cut-off length. The current guaranteed limits of standard circular

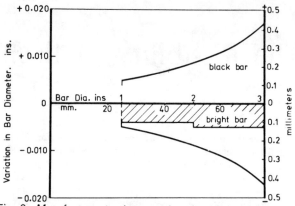

Fig. 3 Manufacturers' tolerances for circular bar stock.

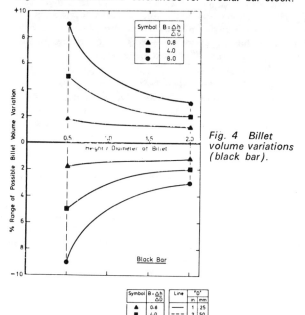

Fig. 4 Billet volume variations (black bar).

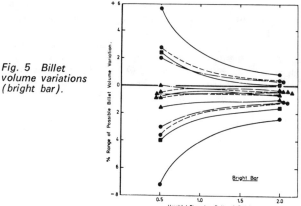

Fig. 5 Billet volume variations (bright bar).

black bar and bright bar sections, are shown in figure 3. Within the range of nominal diameters considered, the tolerance on black bar varies between $\pm0.42\%$ and $\pm0.53\%$ of diameter. The bright bar tolerance is negative, being 0.004 in (0.1 mm) on bars between 1 in (25 mm) and 2 in (50 mm) diameter and 0.005 in (0.125 mm) on bars having diameters from 2 in (50 mm) to 3 in (75 mm).

The information obtainable regarding the accuracy with which billets may be produced, is less exact. Two methods are available for separating billets: sawing and shearing. Because no material is wasted in the process, shearing is

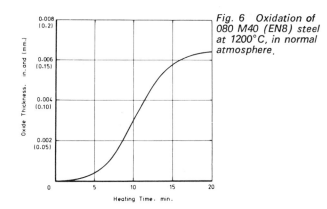

Fig. 6 Oxidation of 080 M40 (EN8) steel at 1200°C, in normal atmosphere.

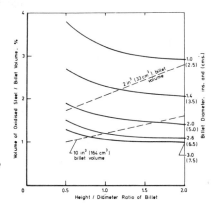

Fig. 7 Dependence on billet geometry of volume of steel oxidised.

preferred. The fact that end-faces are less perfect when produced by this method, is of little importance in drop forging, consistency of profile being more significant. The highest accuracy of sheared length encountered by me has been with the Petro-crop[4] which has sheared bright bar stock to within 0.003 in (0.075 mm) of nominal length. Peddinghaus claims that an accuracy of ±0.004 in (0.1 mm) on nominal length, can be obtained with black bar using the 'Caddy' bar stock shear[5]. The total variation of billet volume, due to diameter and length variations, can easily be determined, as demonstrated in Appendix 1. Here it is shown that if the deviations of height and diameter of the billets are independent of the billet size, then the greatest accuracy of volume is obtained when the ratio of the height to the diameter of the billet equals the ratio of the variation in height to the variation in diameter. Figure 4 shows the possible range of billet volume variation for black bar. It is seen that the accuracy obtainable increases with the height-to-diameter ratio of the billet and with the precision with which billets are separated from the bar (B represents the ratio of possible variation in cut-off length to the variation of bar diameter). Volume variation is independent of the nominal bar diameter. The curves associated with a B value of 0.8 represent volume accuracies obtainable with the best billet separation equipment and it can be seen that in this case, volumes within ±2% of nominal should be obtainable. With inferior billet separation techniques, inaccuracies of about ±6% may have to be tolerated.

Volume variations in billets cut from bright bar are seen, in figure 5, to decrease with increasing height-to-diameter ratio, in a similar fashion to black billets. Contrary to that of the former, the volume accuracy of bright billets is dependent on bar diameter; each mutually increasing. Because of the minus tolerances on bright bar the tendency is for the error in billet volume to be negative. For the largest bar diameter considered, the volumetric accuracy obtainable with precision cut-off equipment may be maintained within 0 to −0.6% for all billet geometries. However, a practical range at best, might be expected to be within 0 to −1.5%, and under production conditions +2% to −3%.

Billet oxidation

Unless the environment is controlled, oxidation of the billet surface is inevitable when it is brought to forging temperature. Figure 6 shows some experimental results obtained for the thickness of oxide formed on 080 M40 (En8) bright billets, heated to 1200°C in an electric resistance muffle furnace. It can be seen that for moderate-to-long heating times, a scale thickness of about 0.16 mm (0.006 in) is to be expected. The volume of steel associated with the oxide formed, has previously been estimated[6] and the resulting proportion of the billet lost in oxidation, for a range of billet shapes and sizes is shown in figure 7. The loss of

stock due to oxidisation is seen to be more significant on billets of small volume with small diameters. However, oxidisation is controllable and should not constitute a serious addition to the factors causing unpredictable changes in billet volume.

DIE CAVITY VOLUME

Billet volume accuracy is important only when related to the volume of the die cavity. This will increase during the course of a forging run, due to erosion. In normal forging practice, the part of the die surface suffering most wear is usually the flash lands. Having eliminated these in completely closed cavity dies, the regions likely to wear most significantly are the edge radii. The die cavity volume associated with a particular size of edge radius may readily be calculated, as shown in Appendix 2. A range of results for a representative component shape, is given in figure 8, where the axis is plotted as the ratio of cavity intersection radius R, to corner r, and the abscissa represents the volume associated with the corner radius, divided by the cube of the cavity intersection radius. This figure clearly shows that the relative cavity volume increases rapidly with a reducing ratio of cavity radius to corner radius. The effect of draft angle although significant is secondary. Assuming the profiles of edge radii to remain circular, even when worn, the information contained in figure 8 allows the effect of cavity erosion on cavity volume to be determined.

British Standard 4114 allows a variation of +50% −25% on edge radii up to 10 mm. This means that if a die is machined with a radius on the lower limit, an increase of 75% can be tolerated. Figure 9 shows how the cavity volume would vary if the corner radius, in three sizes of the forged shape illustrated, should increase by 75% due to

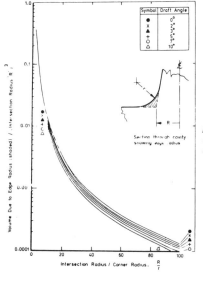

Fig. 8 Contribution of edge radius to die cavity volume.

wear. It is seen that smaller volumetric changes are occasioned in large forgings having small transition radii.

BILLET VOLUME PRODUCTION REQUIREMENTS

To ensure that the die cavity is always filled completely, the billet volume, at the lowest value of the tolerance range, must be at least equal to the cavity volume. Study of figure 4 for black bar, shows that for moderate cut-off accuracy, B=4 say, the nominal billet volume should be an additional 5% of the cavity volume for squat billets and an additional 2% for slender billets (height/diameter 2). For bright bar of 25 mm (1 in) diameter cut with an accuracy B=4, a nominal billet volume of 4% in excess of the cavity volume is required for a height-to-diameter ratio of 0.5 and for a height-to-diameter ratio of 2, an excess volume of about 1.6% is needed. For larger diameters of bright bar, a nominal excess volume of less than 1% is required.

From these estimates, it may be judged that billets will be produced, up to 10% in excess of the die cavity volume, using moderately accurate cut-off methods and up to 4% excess using precision billet-separation methods. It is of interest to note that if bar section tolerances could be maintained always positive, the required excess billet volumes would be much reduced, compared with the above figures.

How cavity wear affects billet volume, depends on production methods. If the cavity volume could be monitored continually and the billets cut to suit, then the required excess volumes need to be no more than those given above. However, if conventional production methods are pursued and billets made in one batch for a complete forging run, the nominal volume of the billets will have to be increased. The amount to be added will be determined by considerations similar to those from which the data in figures 8 and 9 were derived. To allow for cut-off inaccuracy, the nominal addition will be double the calculated wear volumes. For the forging shown in figure 9, for example, an additional 3% of volume could be required.

FORGING LOAD AND DIE STRESS

The effect of oversized billets on forging loads and die stresses, will depend largely on the shape of the die cavity and the type of forging machine used, for a given stock material and preheat temperature. To obtain a general picture of the principal features of the situation, a cylindrical forging, as shown in figure 10, is considered. For analytical ease, it is assumed that the circular container is suspended in such a manner that there is effectively no friction between it and the vertical surface of the billet. Friction between punch and anvil and billet can easily be allowed for. The calculation of required data is explained in Appendix 3.

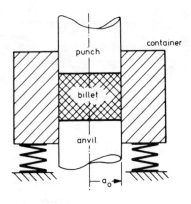

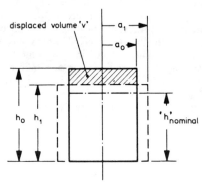

Fig. 10 Cylindrical forging in closed die cavity.

(a) Scheme of Cylindrical Forging. (Billet just fills container.)

(b) Deformation of Oversized Billet

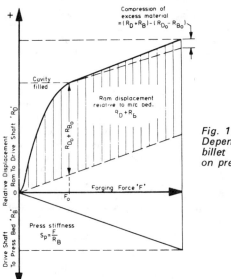

Fig. 11 Dependence of billet compression on press stiffness.

Forging with a mechanical press

With a mechanical press the displaced volume v shown in Figure 10b, does not represent the whole of the excess stock, as during the height reduction from h_o to h some billet volume will be accommodated in machine distortion. This point is illustrated in figure 11 where it can be seen that the closing of the dies is reduced by the elastic distortion of the press. When the die cavity is filled, little further distortion of the billet will occur if the press has appreciable flexibility. The forging load-excess billet volume relation existing for a particular situation may be obtained as indicated in Appendix 3. The excess material which can be accommodated, for a given forging load, is dependent on the press stiffness. The greater this is, the higher is the forging load for a particular addition to the billet volume. The way

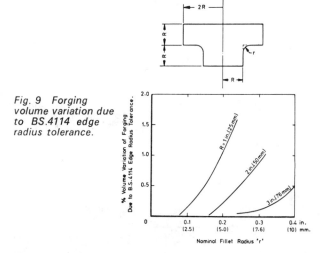

Fig. 9 Forging volume variation due to BS.4114 edge radius tolerance.

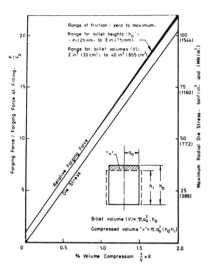

Fig. 12 Relation of forging force to volume compression of billet.

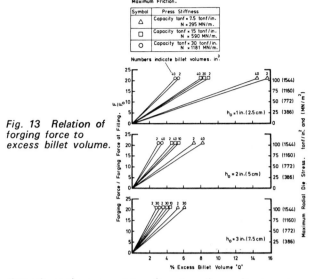

Fig. 13 Relation of forging force to excess billet volume.

data may be obtained from this analysis depends on the specific approach to a production problem. One of the major considerations is the accuracy with which billets can be produced as upon this depends the size of press required. But the size of the press also depends on its stiffness, since a flexible machine will be able to deal with greater excesses of billet volume than a stiff machine.

In figure 12 are plotted the values of relative forging load and radial die stress for a range of cylindrical forgings with a material flow stress of 5T/in² (77 MN/m²) compressed

Table 1
The relations between forging load and excess billet volume (maximum friction)—mechanical press forging.

HEIGHT OF FORGING (cm/in)	VOLUME OF FORGING (cm³/in³)	FORGING FORCE X=0 (MN/Tons)	FORGING FORCE X=0.02 (MN/Tons)	STANDARD PRESS CAPACITY (MN/Tons)	PRESS STIFFNESS (MN/m / T/in)	REL. EXCESS BILLET VOL. Q (EQN.8) (%)
2.5 / 1	33 / 2	0.13 / 13.1	2.86 / 287	3 / 300	117685 / 300000	5.4
					58842 / 150000	8.9
					29420 / 75000	15.7
	330 / 20	1.96 / 197	43.5 / 4368	45 / 4500	26480 / 67500	8.3
	660 / 40	4.73 / 475	105 / 10547	100 / 10000	3530 / 9000	5.1
					1765 / 4500	8.2
					883 / 2250	14.5
5.0 / 2	33 / 2	0.06 / 5.5	1.20 / 121	1.49 / 150	1765 / 4500	3.3
					883 / 2250	4.6
					441 / 1125	7.2
	165 / 10	0.31 / 31	6.79 / 682	7.97 / 800	4707 / 12000	4.8
	660 / 40	1.48 / 149	32.64 / 3276	35 / 3500	41190 / 105000	3.5
					20595 / 52500	5.0
					10297 / 26250	8.1
7.5 / 3	33 / 2	0.035 / 3.53	0.769 / 77.2	0.996 / 100	1177 / 3000	2.8
					588 / 1500	3.7
					294 / 750	5.3
	165 / 10	0.188 / 18.87	4.12 / 413	4.98 / 500	2942 / 7500	3.8
	495 / 30	0.612 / 61.45	13.44 / 1349	13.95 / 1400	16476 / 42000	3.0
					8238 / 21000	4.1
					4119 / 10500	6.2

through various amounts after the die cavity has been completely filled. If an infinitely stiff press were available, the volume compression v, would be equal to the excess billet volume. It is seen that for the range of billet shapes and sizes considered, the relation between the relative forging force and volume compression is virtually linear and all results fall within a single band with differences of less than 1%. The die stress for all conditions is a single linear function of the compressed volume. It should be noted that the values obtained in figure 12 are specific to the container geometry chosen in Appendix 3. Considerations of die life would probably limit the radial stress to about 60 T/in² (427 MN/m²) for monobloc container and even with a duplex design it is unlikely that a volume compression of greater than 2% would be allowed.

If a practical situation is considered, in which allowance is made for press stretch, a different situation arises. In the data presented, three press stiffnesses have been considered:
1. Stiffness (T/in), (MN/m) = Press capacity (T),(N) x (7.5),(295).
2. Stiffness (T/in), (MN/m) = Press capacity (T),(N) x (15),(590).
3. Stiffness (T/in), (MN/m) = Press capacity (T),(N) x (30),(1181).

Table 1 lists the quantities used to compile the curves of figure 13. As relations are virtually linear, over the range of variables considered, it is necessary to calculate a result for one value of excess billet volume only, to obtain the plotted line. In Table 1 a value of proportional volume compression X=0.02 has been chosen. Obvious differences arising from considerations of finite press stiffnesses (c.f. figure 13 with figure 12) are:
1. a single relationship no longer exists for different forging considerations;
2. far greater excess billet volumes can be accommodated for a given increase in forging load;
3. the relation between increase in forging load and excess billet volume is not greatly dependent on billet volume but is markedly affected by the height of a forging. Taller forgings can cause greater overloads;
4. press overloading is less severe when its stiffness is low.

Forging with a percussion machine
In this case it is assumed that the blow energy of the machine can be preset. The energy level must be chosen so that full filling of the die occurs with a billet which is the maximum oversize. Two important situations can then arise:
1. the billet volume is the maximum oversize and the

Source: Metallurgia and Metal Forming, November 1977, 489-498

Symbol	Billet volume		Billet height at die filling	
	in.³	cm.³	in.	cm.
●	2	33	1	2.5
○			3	7.5
■	40	655	1	2.5
□			3	7.5

Maximum Friction. $R = \dfrac{h_s}{h_o}$ (appendix 4)

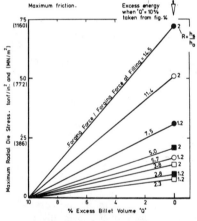

Fig. 14 Relationship between energy requirement and excess billet volume.

Blow energy levels taken from figure 14 at 10% excess billet volume for each forging condition.

Symbol	Billet volume		Billet height at die filling	
	in.³	cm.³	in.	cm.
●	2	33	1	2.5
○			3	7.5
■	40	655	1	2.5
□			3	7.5

Maximum friction. Excess energy when 'Q'= 10% taken from fig. 14

Fig. 15 Dependence of die stress and forging load on excess billet volume with preset blow energy.

height of either 1 in (2.5 cm) or 3 in (7.5 cm) are imagined. Billet deformation, before die filling, is represented by height reductions ($R = h_s/h_o$, Appendix 4) of 1.2 and 2.

Figure 14 shows that the proportional excess energy is greater for small forgings than for large (for a given proportional excess billet volume). The situation also holds for greater billet reductions prior to die filling, although the effect is very small for the larger forgings. Assuming that monitoring of individual billets is not possible, the blow energy of the machine must be set to the value required for the largest billet size. If a billet excess of up to 10% of nominal volume arises, the energy setting would be, from figure 14, 11.8% for the small volume forging (1 in (2.5 cm)) high, deformed through a height ratio of 2, whilst 1.2% is sufficient for the range of large forgings.

Figure 15 shows how the die stress and forging force vary with excess billet volume if the excess energy is set to that required for 10% excess billet volume. It is shown again, that the smaller and shorter the forging the more the probable overload.

If figures 15 and 13 are compared, it will be seen that generally, equivalent excess billet volumes cause less overload on percussive machines than on mechanical presses. No account has been taken of the variation of energy about the nominal on the percussive machine which would cause higher overloads, but then the elasticity of machines such as screw presses would reduce the estimates shown in figure 15, which are based on the assumption of completely rigid, perfectly efficient machines.

These factors can easily be incorporated into the suggested analyses.

It is important to note, that if a facility is available for changing the blow energy rapidly, it is worthwhile to monitor billet volume when using a percussive machine. The exercise is of no value if a fixed stroke press is used.

LOAD LIMITING METHODS

Forging with completely closed cavity dies is feasible, provided that sufficient control is maintained over process conditions. Limited examples of the practice are available. Probably the most obvious are the Hatebur press and the Lasco hammer press. However, although the former relies on frame stretch to withstand overload, the slowly changing variables, such as cavity volume and stock section are compensated manually through operator vigilance.

The techniques employed by the users of high energy rate forging machines were largely those associated with

Fig. 16 Gears hammer-forged in completely closed hammer die.

die is just filled as the ram stops;
2. the billet volume is less than the maximum oversize and a proportion of the deformation energy is used in lateral die distortion. In this case, the different characteristics of machines with a load-limiting device, such as screw presses and those without, such as hammers, are important. Using the former, some of the excess energy apparent when billets of less than the maximum volume are used, can be dissipated in a slipping clutch, thus reducing forging and die loads. A means of reducing die stresses (but increasing maximum forging load) on both types of machine, is to use clash faces. It is assumed that the consequent vertical loading of the die is harmless.

It is important to know the ensuing forging loads and die stresses when excess energy is used. A general picture can be obtained by considering a cylindrical forging once more. An approximation to the relation between deformation energy and small billet volume variations, can be readily calculated, as shown in Appendix 4. In the results, shown in figure 14, two conditions of forging, covering a wide range of practical situations, are introduced. Two forging volumes, 2 in³ (22 cm³) and 40 in³ (655 mm³) each having a

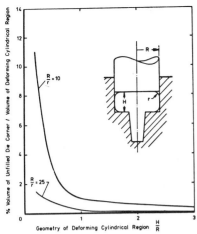

Fig. 17 Volume of unfilled die corners related to geometry of deforming region.

flashless forging. Figure 16 shows some spur gears which have been made in a completely closed-die cavity on a Petro-Forge machine[T]. No overloading of the die cavity occurred, but billet accuracy was ensured by machining overall and preheating in an inert atmosphere and the blow energy of the machine was accurately controlled.

Accommodating space in die cavity

As a means of reducing overloads, space can be allowed within the die cavity to accommodate excess billet material. The region in which this space is situated will depend, to some extent, on the component shape. In general, it should be positioned at the place which is the last to be filled, otherwise underfilling of the die can occur. For a simple forging, deviating little from cylindrical form, the edge radii are probably the last features to be filled and it is possible to deliberately leave these unfilled to allow for billet volume variations. Figure 17 shows that the advantage to be had from this concept is greater for squat forgings and that the unfilled die volume (proportional to the volume of the forging) rapidly reduces above a forging height-to-diameter ratio of 0.5. I have used this technique for the preform stage of a two-stage forging illustrated in figure 18. The final impression utilised conventional flash lands.

Fig. 18 Forging sequence with completely closed preform die (corners for preform remain unfilled).

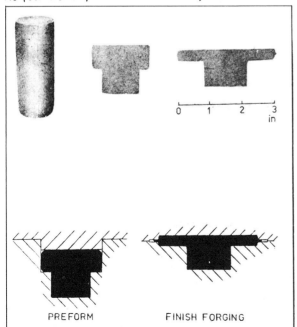

PREFORM   FINISH FORGING

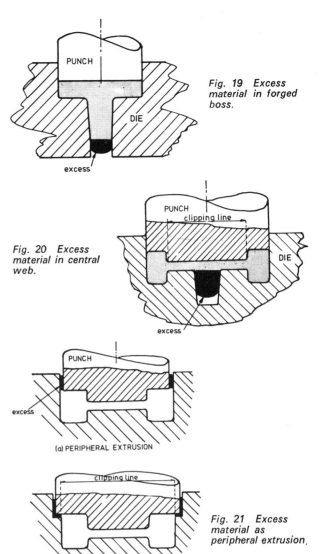

Fig. 19 Excess material in forged boss.

Fig. 20 Excess material in central web.

(a) PERIPHERAL EXTRUSION

(b) STEPPED PERIPHERAL EXTRUSION

Fig. 21 Excess material as peripheral extrusion.

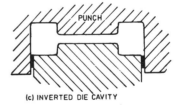

(c) INVERTED DIE CAVITY

If the shape of the forging is such that the central region is the last to fill, accommodation for excess billet material must be made here. In the case of a flange and boss for instance, the bottom of the recess can be left 'open', providing the ratio of boss diameter to flange diameter is small enough for the outer corners of the flange to fill first, as shown in figure 19. The disadvantage of this method is that relatively small volumes of material can be contained in practical lengths of extruded boss.

Dies for components having a thin central web which is punched out to form a hollow, can be designed to accommodate excess billet material in this region, as sketched in figure 20. The advantage of this configuration is that the waste material can be removed in a conventional punching operation when the bore is formed. A drawback is the limited space available for the excess, as too big a cavity

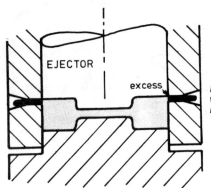

Fig. 22 Die set with built-in trimming facility.

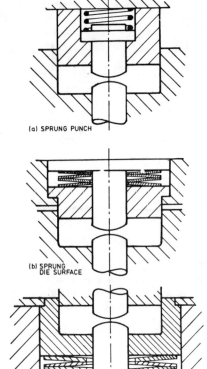

Fig. 23 Multi-part dies with springing.

(a) SPRUNG PUNCH

(b) SPRUNG DIE SURFACE

(c) SPRUNG INSERT

may result in underfilling other regions of the forging, higher punch pressures being present on the axis of the die.

As a punch and container design is the obvious one for a completely closed cavity, a natural expedient is to allow excess material to extrude up the side of the punch as shown in figure 21(a). The punch and container may be tapered to provide an extrusion gap which decreases in thickness as forging proceeds. It has been suggested[8] that if the extruded ring is stepped as shown in Figure 21(b), the waste is more easily removed by normal clipping methods compared with the former case when turning would have to be employed. A disadvantage arising from this design is that the horizontal section formed will increase the forging load required. For easier access to the forging, an inverted configuration such as shown in figure 21(c) may be used.

Figure 22 illustrates a proposal[9] for accommodating and clipping flash in the same die. The waste material flows into four peripheral slots during forging and is sheared from the component on ejection. The discard is cleared from the die during the subsequent forging operation. It appears that the success of this idea under production conditions, would depend on the maintenance of sharp corners at the entrance to the lateral slots. If this is not achieved, distortion and sticking of the component is likely to occur.

Multi-part dies

If the accuracy of billet volume is sufficient it may be possible to obtain components that are always within tolerance, without provision in the die for excess material. This situation is more likely to occur with squat forgings than slender ones, as the variation in thickness, for a certain variation in billet volume, is less. However, it is likely, particularly when mechanical presses are used, that overload protection will be necessary. If conventional forging machinery is to be used, the load-limiting device must be within the tooling. Several methods of springing parts of a forging die are possible. Considering a die for the manufacture of a component with a central web, one or both of the punches can be sprung, as shown in figure 23(a). In this way billet variations are absorbed by the web, which assumes various thicknesses and which is subsequently removed from the forging. The design of figure 23(b) allows relative movement of the main surface of the die and a greater excess of material can be accommodated for a given spring movement than in the previous design. However, in this case an important component dimension is varying significantly. A major drawback of both these tooling configurations is the presence of sliding surfaces in the region of hot deforming oxidised metal and it is doubtful that required fits could be maintained for long. A design which overcomes this problem is shown in figure 23(c); the whole of the bottom die insert is sprung and the excess stock volume would be spread evenly over the surface of the forging.

A major disadvantage of metal springs is the probability that they will lose their temper in the working environment.

To reduce this possibility they should be located as far as possible from the hot metal and cooling applied. The type of spring used will have a significant effect on the performance of the tool set (figure 24). The ideal spring would be infinitely stiff until the die filling force was reached. It would then deform with no increase in load. With practical springs, overload is inevitable. The amount can be reduced with disc springs having decreasing spring rate/load characteristics, but with coil springs sufficiently stiff to cause little relative movement during the earlier parts of deformation, considerable overloads can be introduced. The consequences are that larger forging machines have to be employed and the probability of forming a fin of material between punch and container, is increased. Fin formation is to be avoided as it can cause sticking to the punch and can hasten the erosion of the punch nose.

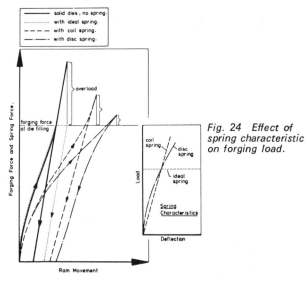

Fig. 24 Effect of spring characteristic on forging load.

The feasibility of flashless forging

by **T. A. Dean,** Department of Mechanical Engineering, University of Birmingham

Part 2: Conclusion and Appendices

The elimination of conventional flash is practically possible using conventional forging equipment. For best results, accurate billet volumes are required. Analysis indicates that if this is achieved, it is possible to forge without overload protection and that lower die stresses can be maintained on a percussive machine with precise energy control than on a mechanical press. However, production considerations will probably dictate the use of the latter machine when it will be advantageous to use a method of load limiting. Vertically split dies, as used in the brass stamping industry, have been employed for cored steel components, but it is doubtful that these types of dies will become widely used, considering the types of component generally made in the drop forging industry.

Acknowledgement

The information contained in this paper is the outcome of research being supported by the Science Research Council.

References

1 Brookes, R.P.: Opening address at the Seminar on Drop Forging N.A.D.F.S. I.Prod.E., Nottingham University, May, 1970.
2 Dean, T.A.: The Mechanics of Flash in Drop Forging— Temperature and Speed Effects. *Proc.E.Mech.E.*, 190 33/76, pp. 457-466.
3 Dean, T.A.: Load and 'Rise'; Their Dependence on Flash Geometry and Machine Characteristics in Drop Forging. *Proc. 17th Int.M.T.D.R. Conf.*, 1976, pp. 363-369.
4 Personal communication from Professor S.A. Tobias.
5 Private communication. Also see 1976 Peddinghaus catalogue.
6 Kellow, M.A., Dean, T.A., and Bannister, F.K.: The Oxidation of Steel at High Temperature and its Effect on Die Surface Temperatures in Hot Forging. *Proc. 17th Int. M.T.D.R. Conf.*, 1976, pp. 355-361.
7 Abdel-Rahman, A.R.O. and Dean, T.A.: Production Considerations for the High Speed Forging of Spur Gear Forms. *Proc. 14th Int.M.T.D.R. Conf.*, 1973, pp. 807-813.
8 Anon: Sanksmidning utan skägg (Flashless die forging). IVF—Resultat 74602, Jan. 1974.
9 Akaro, I.L. and others: Warmumformung von im Planrunden Schmiedestücken mit geringen Verlusten. *Kuznecno-stampovocnoe proizvodstvo*, 13 (1971) 3.

APPENDIX 1

Billet volume variation

Let the nominal diameter and height of a billet of circular cross-section be D and H respectively.

Then the billet volume is: $V = \dfrac{\pi.D^2.H}{4}$

If the variation in diameter and height is ΔD and ΔH respectively. The corresponding variation in volume is:

$$\Delta V = \frac{\pi.D.H}{2}.\Delta D + \frac{\pi.D^2.\Delta h}{4} \qquad (1)$$

Substitute $H = \dfrac{4.V}{\pi.D^2}$ in (1)

Gives $\Delta V = \dfrac{2.V.\Delta D}{D} + \dfrac{\pi.D^2.\Delta H}{4}$

If ΔD and ΔH are constant values:

$$\frac{d(\Delta V)}{dD} = -\frac{2.V.\Delta D}{D^2} + \frac{\pi.D.\Delta H}{2}$$

For minimum volume variation $\dfrac{d(\Delta V)}{dD} = 0.$

$$\frac{h}{D} = \frac{\Delta h}{\Delta D}$$

from (1) $\dfrac{\Delta V}{V} = \dfrac{2\Delta D}{D} + \dfrac{\Delta h}{h}$ If $\dfrac{h}{D} = A$ and $\dfrac{\Delta h}{\Delta D} = B$

$$\frac{\Delta V}{V} = \frac{\Delta D}{D}\left(2 + \frac{B}{A}\right)$$

B depends on the length accuracy.

$*\dfrac{\Delta D}{D}$ is approximately ± 0.005 for back bar

$\dfrac{0.004}{D}$ for bright bar 1 in (25 mm)

~2 in (50 mm).

and $-\dfrac{0.005}{D}$ for bright bar 2 in (50 mm)

~3 in (75 mm).

*See figure 3 (last month).

APPENDIX 2

Die cavity volume due to an edge radius

The accompanying sketch represents the section of a die cavity containing two regions connected by a radiused corner r. The longitudinal region of the cavity has a draft

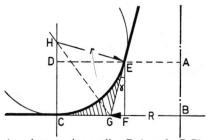

angle α and an intersection radius R (equals G B) as shown.

Section through cavity at corner radius
The volume associated with the shaded section ECG in the sketch may be calculated by determining the volumes swept when various sections are rotated about the axis. If

$$K = \frac{R}{r}$$

the volume due to:

1. Rectangle ABCD is: $\pi r^3 (1-\sin\alpha)(\tan\phi + K)^2$

2. Sector HEC is: $2\phi\pi r^3 \left(\tan\phi + K - \frac{2\sin^2\phi}{3\phi}\right)$

3. Triangle HDE is: $\pi r^3 \sin\alpha \cos\alpha \left(K - (1-\sin\alpha)\tan\alpha + \frac{2}{3}\cos\alpha\right)$

4. Triangle EFG is: $\pi r^3 (1-\sin\alpha)^2 \tan\alpha \left(K - \frac{1}{3}(1-\sin\alpha)\tan\alpha\right)$

5. Rectangle ABFE is: $\pi r^3 (1-\sin\alpha)[K-(1-\sin\alpha)\tan\alpha]^2$

The volume ECG due to the radius is then:
$1 + 3 - (2 + 4 + 5)$

APENDIX 3
Forging load and die stress due to an oversized billet

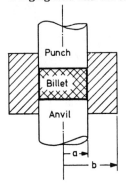

FORGED BILLET

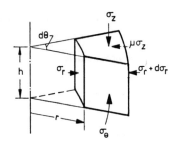

STATE OF STRESS ON AN ELEMENT

Nomenclature
σ = normal stress
h = current billet height
a = current billet radius
b = outside radius of container
r = radius
z, r, θ = axial, radial, circumferential directions respectively
μ = coefficient of friction at billet end faces
ν = Poisson's ratio of container
Y = flow stress of billet
σ_a = radial stress between billet and container

Adopting the usual assumptions of plastic stress analysis together with the exclusion of friction between billet and container; also homogeneous deformation of the forging and uniform stressing of the container, the same stress system as for simple upsetting may be adopted, as shown above. The differential equation of equilibrium is then:

$$\frac{d\sigma_z}{\sigma_z} = 2\mu \frac{dr}{h}$$

which on integration becomes

$$1n\sigma_z = \frac{2\mu r}{h} + C$$

The boundary condition is that
where $\qquad r = a, \ \sigma_r = \sigma_a$

Hence $\qquad \sigma_z = (Y + \sigma_a) \exp \frac{2\mu}{h} \overline{r-a}$

Giving a total load of

$$F = \pi a^2 (Y + \sigma_a)\left(1 + \frac{2}{3}\mu\frac{a}{h}\right)$$

(3)

Considering the container, for a radial displacement U of the bore:

$$\sigma_a = \frac{UE}{a} \left/ \left(\frac{a^2 + b^2}{b^2 - a^2} + \nu\right)\right.$$

where E is Young's modulus of the die steel.

If b/a is 10, about the maximum useful ratio, then for a steel container

$$\sigma_a = \frac{UE}{1.32a}$$

Substituting this in equation (1) gives:

$$F = \pi a^2 \left(Y + \frac{UE}{1.32a}\right)\left(1 + \frac{2}{3}\mu\frac{a}{h}\right)$$

(4)

Suppose that, when the billet just fills the container, it has a height h_o (compared to h_N the nominal height) and radius a_o and at full compression the height is reduced to h_1. The radius a_1 can be calculated using the constancy of billet volume, i.e.

$$\pi a_o^2 h_o = \pi a_1^2 h_1$$

The radial die displacement is then:

$$U = a_1 - a_o = a_o \left(\sqrt{\frac{h_o}{h_1}} - 1\right)$$

If the volume of the billet displaced after complete die filling (shaded region in figure 10b) is
$$v = \pi a_o^2 (h_o - h_1)$$
and $x = v/V$ where V is the total billet volume
Then

$$F = \frac{V}{h_o(1-X)}\left(\gamma + \frac{E}{1.32}(1-(1-X)^{\frac{1}{2}})\right)\left(1 + \frac{2\mu}{3\pi}\frac{V^{\frac{1}{2}}}{h^{\frac{3}{2}}(1-X)^{\frac{3}{2}}}\right)$$

(5)

The associated radial expansion of the die is:

$$U = \left(\frac{V}{\pi h_o}\right)^{\frac{1}{2}}\left[\left(\frac{1}{1-X}\right)^{\frac{1}{2}} - 1\right]$$

(6)

and the maximum radial die stress:

$$\sigma_s = \frac{E}{1.32}\left[\left(\frac{1}{1-X}\right)^{\frac{1}{2}} - 1\right]$$

(7)

Press forging

If the forging force when the billet just fills the die (and is h_o high) is F_o and the forging force after a reduction in height to h_1 is F, then the press spring during displacement of volume

$$v = XV \text{ is: } (F - F_o)/S_p$$

where S_p is the overall press stiffness.

The amount of excess billet volume accommodated in this press distortion is:

$$A_1 . (F_1 - F_o)/S_p$$

where A_1 is the area of the top surface of the billet

The total excess volume requiring a forging force F_1 is:

$$XV + A_1(F - F_o)/S_p$$

The relative excess volume is:

$$Q = X + \frac{A_1}{V}(F - F_o)/S_p$$

or $$Q = X + \frac{(F - F_o)}{h_o(1-X)S_p}$$

(8)

APPENDIX 4
Relationship between deformation energy and billet volume variation in simple upsetting

For a billet of current height h and current radius a the required forging force is shown, in Appendix 1 to be:

$$F = \pi a^2 Y \left(1 + \frac{2}{3}\mu\frac{a}{h}\right)$$

where Y is the flow stress and μ the coefficient of friction.

If the original billet height is h_s and the finished forging height h_o, the required deformation energy is:

$$E = \int_{h_o}^{h_s}\left[\pi a^2 Y\left(1 + \frac{2}{3}\mu\frac{a}{h}\right)dh\right]$$

which becomes

$$E = VY\left[\ln R - \frac{4}{9}\mu\frac{1}{\sqrt{\pi}}\frac{V^{\frac{1}{2}}}{h_o^{3/2}}(1 - R^{3/2})\right]$$

where V is billet volume and $R = \dfrac{h_s}{h_1}$

(9)

For a small change in billet volume ΔV the change in energy is seen to be

$$\Delta E = Y\left[\ln R - \frac{2}{3}\frac{\mu}{\tau}\frac{V^{\frac{1}{2}}}{h_o^{3/2}}(1 - R^{3/2})\right]\Delta V$$

(10)

SOME METAL FLOW PHENOMENA ARISING IN AXISYMMETRIC FLASHLESS FORGING

Y. Van Hoenacker* and T. A. Dean*

(Received 1 November 1979)

Abstract This paper describes aspects of metal flow in flashless forging dies which are either undesirable in manufacture or lead to defective components. Three different forms of flow were identified; asymmetric filling of top and bottom cavities when shapes with peripheral flanges are made and friction is high, splitting of the body of similar shapes of this type, when friction is low and fold formation in shapes with a central boss, under certain geometrical conditions. Velocity fields have been proposed which describe these happenings and enable estimates to be made of the range of conditions under which they will occur.

INTRODUCTION

THE ELIMINATION of flash from drop forgings has been the subject of several recent papers by these authors [1–3]. It is now well established that useful material savings can be made and in addition lower maximum loads and reduced energies may be required.

Of necessity, a die cavity for flashless forging comprises basically a container and a punch, in contrast to the two opposed, open faced cavities normally used. This construction can cause differences in material flow which have a significant effect on both the process and the product quality.

This paper describes and analyses peculiarities of flow that have been observed in the formation of axisymmetric shapes which typify two ranges of drop forged components.

EXPERIMENTAL DETAILS

The shape of the two types of forging made and an outline of the die cavity constructions used are shown in Fig. 1. The forgings were all axicircular, and die inserts were used to vary the diameter of the central boss of the type 1 forging and the thickness of the peripheral flange

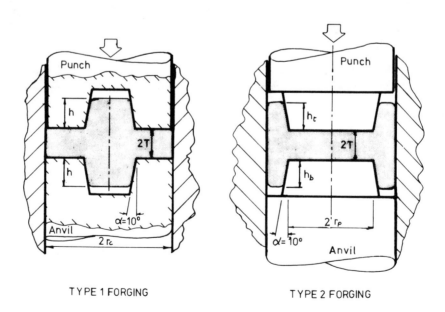

TYPE 1 FORGING   TYPE 2 FORGING

FIG. 1. Forged shapes and die cavities.

* Department of Mechanical Engineering, University of Birmingham, England.

Reprinted with permission from International Journal of Machine Tool Design and Research, Vol. 20, 1980, 45-53, © 1980 Pergamon Press Ltd.

TABLE 1. GEOMETRICAL DATA – TYPE 2 FORGINGS

Test number	Web diameter (2 rpm m)	Web thickness (2T mm)	Billet height (Ho mm)
19	25.40	7.62	23.75
20	30.48	7.62	20.02
21	35.56	7.62	15.65

of the type 2 forging. Both types of cavity were symmetrical about their midheight. The maximum diameter of the forgings remained unchanged at 45.7 mm.

Two stock materials were used; H30 aluminium alloy, forged at room temperature and M40 080 medium carbon steel, preheated to 1200°C.

Room temperature forgings were made, either with clean dry dies, or with dies and billets liberally coated with lanolin as a lubricant. Clean dry dies only were used for the hot forgings. Cylindrical billets were used in all cases. For the type 1 forgings, billets of various heights and diameters were used, but for the type 2 shapes, only the billet height was varied, the diameter being kept the same as that of the cavity, which was 45.7 mm for both forging types. Geometrical details of the three unlubricated, type 2 forging tests referred to later, are given in Table 1.

ASYMMETRICAL FLOW

Zalesski and Tyurin [4] have previously established for shapes similar to those described here that due to friction at the walls, differences in flow into top and bottom die cavities occur. Symmetry can be established by using opposing punches or a "floating" container, but simplicity and ruggedness of design make the arrangement shown in Fig. 1 the most likely to be encountered.

With the thin webs formed in the type 1 forgings made in these tests ($T/c = 0.22$) and the use of billets with diameters less than that of the cavity resulting in contact with the die walls late in the process, the surface areas over which wall friction was effective were relatively small. As a consequence, no asymmetry was recorded, for any test conditions.

Metal flow into top and bottom dies was noticeably asymmetric when type 2 forgings were made, using poorly lubricated dies.

Velocity field

An Upper Bound solution, using the velocity field shown in Fig. 2, was used to determine

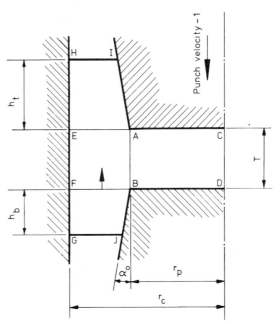

FIG. 2. Velocity field for Type 2 forgings.

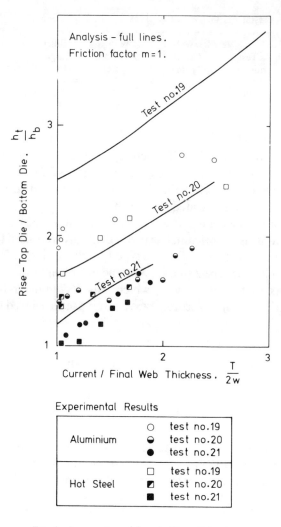

Analysis – full lines.
Friction factor m = 1.

Test no. 19

Test no. 20

Test no. 21

$\frac{h_t}{h_b}$

Rise – Top Die / Bottom Die.

3

2

1

1   2   3

Current / Final Web Thickness. $\frac{T}{2w}$

Experimental Results

Aluminium	○	test no. 19
	◐	test no. 20
	●	test no. 21
Hot Steel	□	test no. 19
	◪	test no. 20
	■	test no. 21

FIG. 3. Asymmetry of flow in Type 2 forgings.

theoretical overall flow patterns in type 2 forgings.

In region ABFE and ACDB parallel velocity fields are assumed to exist. In the regions EAIH and FBJG, conical streamlines directed towards the intersection of the tapering walls of the die were used.

Optimisation for minimum power dissipation was obtained using the vertical velocity of a particle on BF, as a parameter. A constant friction factor has been assumed over all the forging.

Figure 3 shows a comparison of experimental and theoretical results. It is seen that the theoretical analysis is able to predict well the decreasing trend of the ratio of rise top die/bottom die, with deformation. However, the value of this ratio is always over-estimated, particularly for the test in which a small diameter web was formed. It is to be noticed that in all three cases, the height of rise into the top die is greater than that into the bottom, but that the difference decreases with increasing punch diameter. No definable difference between experimental results for hot and cold forgings can be seen, indicating that neither material properties nor temperature significantly affect flow. As a result of the asymmetric flow, very little displacement of the material entering the top cavity occurs. This is particularly evident in the case of the small diameter punch, when the process is akin to a piercing operation.

The results of the analysis for a wide range of geometries displayed in Fig. 4, show that the ratio rise in top die/rise in bottom die is increased with the increase of the final web thickness of the forging but that the starting height of the billet has no effect on the initial value of the ratio. Increasing flange thickness (reducing punch radius r_p) is again seen to increase the flange height in the top die compared with that in the bottom.

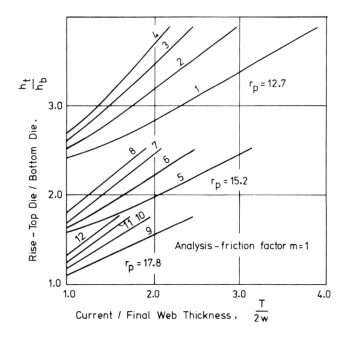

No.	r_p mm.	H_o mm.	$2w$ mm.
1		21.2	5.1
2	12.7	23.7	7.6
3		26.3	10.2
4		28.8	12.7
5		17.5	5.1
6	15.2	20.0	7.6
7		22.6	10.2
8		25.1	12.7
9		13.1	5.1
10	17.8	15.6	7.6
11		18.2	10.2
12		20.7	12.7

FIG. 4. Theoretical effects of geometry on flow—Type 2 forgings.

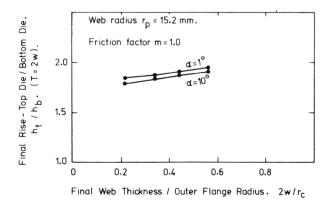

FIG. 5. Effect of draft angle on asymmetry of flow—Type 2 forgings.

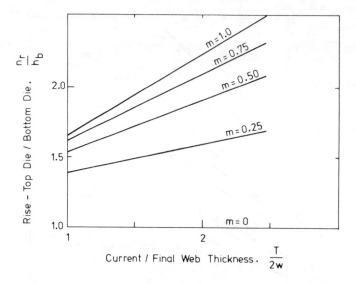

FIG. 6. Effect of friction of asymmetry of flow—Type 2 forgings.

The theoretical curves of Fig. 5 shows that the value of draft angle, within the range likely to be encountered in forging practice, has little influence on asymmetry. Friction however, significantly increases asymmetry, as seen in Fig. 6. If friction is eliminated, equal flow into the top and bottom die cavity is achieved.

PERIPHERAL DEFECTS

Type 1 forgings

Central folds. One way to reduce the energy required for flashless forging is to delay contact between stock and the cavity wall. This can be achieved by increasing the height to diameter ratio of the billet. However, using billets with a large aspect under lubricated forging conditions may result in a bollard shape during the early stages of deformation, due to inhomogeneity of metal flow.

Investigation of the incidence and extent of bollarding was carried out using cylindrical billets of H30 aluminium alloy. Five different billet diameters, ranging between 12.7 and 31.7 mm and four different central boss diameters of 8.1, 11.4, 13.7 and 18.3 mm were used. Lanolin was applied generously on both die surfaces and end surfaces of the billets. Deformations corresponding to billet height reductions of 14% and 50% were imposed and the extent of bollarding was measured by the index E, described in Fig. 7.

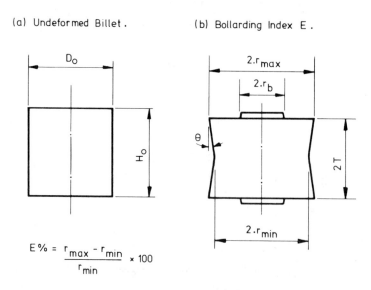

FIG. 7. Bollarding of billets—Type 1 forgings.

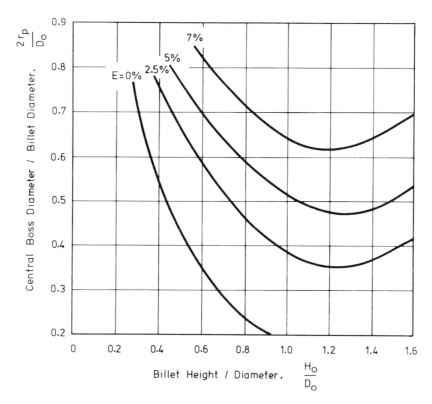

FIG. 8. Extent of bollarding at 14% reduction of billet height.

Experimental values of the bollarding index at a height reduction of 14% are shown in Fig. 8. It is seen that E is a function of the initial aspect ratio and also of the ratio of the diameter of the central boss to the initial billet diameter. On the left hand side of the line corresponding to zero E, no bollarding occurred and slight tendencies to barrelling were detected. Generally the bollard type distortion increases with increasing geometric parameters. The tendency to a reversal of this trend, for larger values of the billet aspect ratio, can be attributed to the limitations of the index chosen. It is probable that the angle θ, in Fig. 8, would be a more representative index, but this is not easily measured with accuracy. When the initial billet height was reduced by 50%, it was found that the distortion of billets with a previous E value of less than 3%, had regressed and in some cases had disappeared, but that billets formerly having an E value larger than 5% developed a fold. The geometric limits leading to these conditions are shown in Fig. 9. Here it is seen that the graphical region indicating no defect has enlarged, compared with that associated with the smaller height reduction, particularly in the area corresponding to slender billets and small boss diameters. This indicates that the ratio of boss diameter/billet diameter predominates in the formation of a surface defect in the finished forging. Figure 10 is a photograph of a partially deformed billet and a section of a completed forging, showing the peripheral defect.

Type 2 forgings

As discussed earlier, low friction allows symmetrical shapes to be made, but in some cases, during the forging of well lubricated aluminium alloy billets, defects appeared at the periphery of the body of the forging.

Depending on the degree of deformation, this defect was found to take one of two different forms. The first consisted of superficial cracks. These generally originated from machining marks remaining on the original billet surface, but even when the surface roughness was reduced to 800 μm, the cracks still formed, as shown in Fig. 11. In these cases the cracks did not correspond to the original machining marks and were seen to develop as deformation progressed. Johnson [5] has observed the formation of defects on smooth surfaces in components of shapes similar to those described here. They were referred to as a "sucking in"

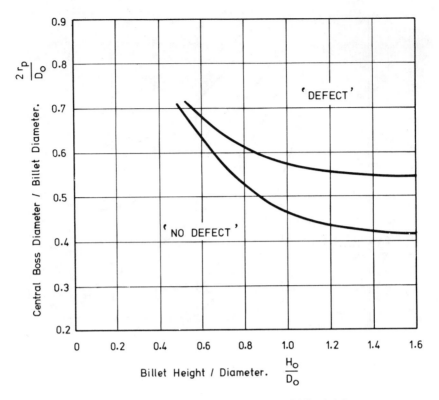

FIG. 9. Peripheral defects at 50% reduction of billet height.

of metal and a plane strain analogue, consisting of a velocity field in which the surface of the deforming region moved inwards from the wall of the cavity, was proposed. This situation was not recorded in these experiments. It is thought that the existence of a draft angle and the early opening of cracks suppressed such tendencies.

Analysis of crack formation

The axisymmetric velocity field shown in Fig. 12 was constructed as an analogue describing the onset of cracking. It consists of two conical surfaces of velocity discontinuity AK and BK and two rigid blocks AEK and BFK. The point K, normally at the periphery of the forging is allowed to move inwards to K'. Neglecting the power required to form new crack surfaces, the total applied power can be expressed as a function of three parameters; velocity of movement into the flanges, radius to the crack tip r_k and height of the crack

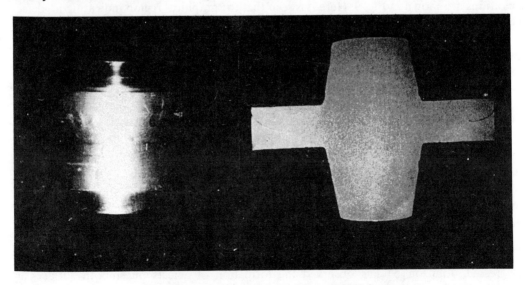

FIG. 10. Sections showing peripheral defect Type 1 forgings.

FIG. 11. Peripheral cracking of Type 2 forging.

opening z_k. The onset of cracking is defined when the minimisation of the applied power yields a value of r_k less than r_c. The results for various forging conditions are shown in Fig. 13. It is seen that the critical web thickness, at which a crack is likely to form, is a function of both container and flange dimensions and decreases both with increasing friction and increasing flange height. A comparison of the frictionless axisymmetric results, with that from Johnson's plane strain analogue, indicates that the crack is likely to occur much earlier in the process than the "sucking in" type of defect, with the die shape utilised in these tests. Examination of

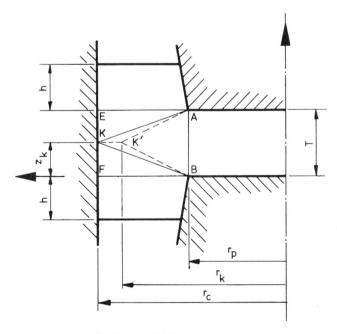

FIG. 12. Velocity field incorporating crack.

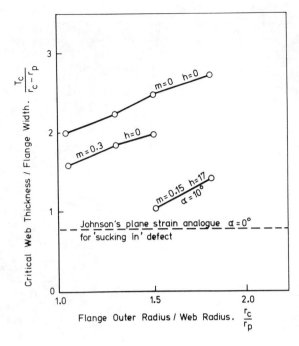

FIG. 13. Theoretical prediction of crack formation—Type 2 forgings.

forged shapes enabled cracks to be seen when the ratio of web to flange thicknesses was about two. No attempt to establish the exact stage, at which cracking started was made, because of the difficulty of establishing the exact moment at which a feature is no longer an aspect of surface roughness and constitutes a crack.

CONCLUSIONS

From the investigations described in this work, several features of metal flow of concern in the bulk forming of flashless components have been discovered. For forgings with peripheral flanges asymmetrical flow into top and bottom dies can occur. The degree of asymmetry is increased by friction, but is also dependent on the forged shape. Thick webs and thick flanges both increase flow into the top die compared with that into the bottom one. An axisymmetric velocity field has been proposed, which enables predictions of flow to be made for various process conditions.

Surface defects can arise in both central boss and flange forgings and those with webs and peripheral flanges. In the former, the defect is due to the folding in of the side of the billet, when it forms a bollard shape early in the deformation. Low friction, slender billets and a large central boss diameter all contribute to the formation of this defect. In forgings with a peripheral flange, defects can occur due to the opposing directions of metal flow at mid-height. For conditions of very low friction, cracks can be formed. An axisymmetric velocity field, proposed as a model for this situation, allows predictions of the critical web thickness at which cracking will occur for various forging situations. It is shown that critical thickness depends on both container and flange dimensions.

Acknowledgement – The work reported here is part of the outcome of a programme of research sponsored by S.R.C. Y. Van Hoenacker worked as a National Research Council of Canada Scholar.

REFERENCES

[1] T. A. DEAN, *Metallurgia and Metal Forming* **44**, 488–498, 542–544 (1977).
[2] Y. VAN HOENACKER and T. A. DEAN, *Int. J. Mach. Tool Des. Res.* **18**, 81–93 (1978).
[3] T. A. DEAN, *I. Mech. E. Proc. Instn mech. Engrs* Vol. 193.
[4] V. I. ZALESSKI and N. I. TYURIN. *Kuznechno-shtamp. Proizv.* **1**, 4–8 (1959).
[5] W. JOHNSON, *Appl. scient. Res. A.* **8**, 52–60 (1959).

56

THE APPLICATION OF UPPER BOUND ANALYSES
TO REAL SITUATIONS—A COMPARISON
OF FLASH AND FLASHLESS FORGING

Y. Van Hoenacker and T. A. Dean*

(*Received* 5 *January* 1978)

Abstract – This paper describes the application of a heuristic method in which calculations using upper bound analyses are combined with experimental results to produce an indication of frictional behaviour and the effects of work hardening and strain-rate sensitivity in practical forging situations. It also demonstrates that considerable load reductions are achievable by eliminating flash from two types of forging considered.

INTRODUCTION

THE USE of upper bound analyses for load determination in metal forming processes is now a well established practice. Early work on steady-state plane-strain situations, where simple geometric or algebraic manipulations were adequate, has been extended to triaxial deformations and the necessary calculations are often laborious unless computerised. Problems in which patterns of flow vary with time, due to changes in shape, are gaining increased attention from metal forming engineers. Outstanding for complexity in this category, is drop forging, in which a combination of stock geometries with a range of die cavity shapes, provides a virtually infinite variation of deformation patterns.

An early approach to the forging problem was suggested by Johnson [1], who proposed a plain-strain analogue for the flash formation stage. Kudo [2], in tackling problems of the type met in cold forging, introduced the concept of a unit deforming region, to facilitate calculations. Within these units, both parallel and triangular velocity fields were constructed, those requiring the least dissipation energy being accepted as the better solution. McDermott and Bramley [3], have extended the range of elemental regions to include shapes approximating to those encountered in commercial drop forging dies. The resulting analyses are too complex to be handled manually, in the manner of Kudo and computer programmes have been employed.

In the above work, situations have been idealised by eliminating the space and time variations of flow stress due to work hardening and strain-rate variations. If the characteristics of real materials and the practicalities of actual forming situations are accepted, problems arise in the choice of material properties and frictional restraints to be used in upper bound analyses. One objective of this paper is to elucidate the effects of work hardening, interface friction and strain-rate in the forging process, through an interaction of analysis and experiment.

The reduction of waste material from the conventional forging process is of major importance, as has recently been emphasised [4]. Maximum savings can be made by forging without flash in a completely closed cavity die. If this is done, it is likely that the load requirement will differ from that of the conventional flash forming process.

The other objective of this paper is to obtain a comparison of the forging loads required for two shapes (typifying ranges of components) forged both with and without flash.

EXPERIMENTAL METHOD

A die set was constructed to make two types of axicircular shape, as shown in Fig. 1. By means of inserts the diameter of the central boss (type 1 forging) and the thickness of the

---

*Department of Mechanical Engineering, University of Birmingham.

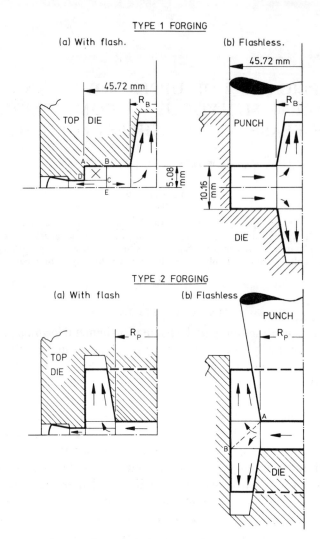

TYPE 1 FORGING

(a) With flash.   (b) Flashless.

TYPE 2 FORGING

(a) With flash   (b) Flashless

FIG. 1. Chosen velocity fields – final forging stage.

peripheral flange (type 2 forging) could be altered and forgings of the same size made, with or without flash. The height of the central boss and the peripheral flange were varied by varying the billet volume. For type 1 forgings, billets of various diameters were used. For type 2 forgings, billets of diameter equal to the maximum diameter of the die cavity were employed.

Cylindrical billets of two stock materials were used;

(a) H 30 aluminium alloy, fully annealed, was forged at room temperature
(b) 080 M40 (En8) a medium carbon steel, was heated in an electric muffle furnace to 1200° C before forging.

Stress strain data for the H 30, which work hardens but is strain-rate intensive [5] and 080 M40 which is highly strain-rate sensitive [6] are shown in Fig. 2. Both lubricated and unlubricated forging conditions were used. For the unlubricated tests, the dies were cleaned with emery paper and acetone. Lanolin was applied to both billet and dies as a room temperature lubricant, whilst colloidal graphite in spirit was brushed on the dies for the lubricated hot forging tests. Forgings were made from the billets in one operation, on a mechanical press.

To enable equivalent forged shapes to be made, in both types of die cavity, shapes with flash were first forged. When these had been measured and the volume of flash calculated, billets of the same geometry as those used for the flash forgings and with volumes appropriate to the equivalent flashless forging were made. Table 1 shows the range of shapes produced. Forging loads were recorded with the now familiar strain gauged load cell situated above the top die.

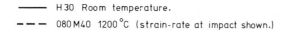

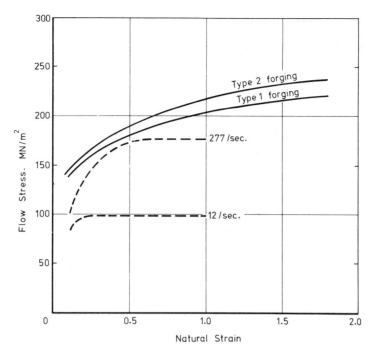

Fig. 2. Stress–strain relations of stock materials.

TABLE 1. MAJOR PARAMETERS OF FORGING TESTS

Type 1 forging				Type 2 forging			
		Total forging height mm			Total forging height mm		
Test No.	R_B mm	Cold	Hot	R_p mm	Cold	Hot	Test condition
1	5.7	38.6	43.8	15.24	30.9	36.8	LUBRICATED
2	5.7	38.4	39.2	15.24	30.4	27.2	
3	5.7	31.7	31.7	15.24	34.9	30.3	
4	9.15	43.8	41.0	12.70	38.5	35.2	
5	9.15	39.0	34.9	15.24	37.9	33.2	
6	9.15	34.1	30.9	17.73	37.8	29.7	
7	5.7	31.8	46.1	15.24	27.4	29.1	UNLUBRICATED
8	5.7	28.8	35.4	15.24	27.2	29.3	
9	5.7	31.1	40.5	15.24	30.0	31.5	
10	5.7	30.4	35.0	12.70	33.5	36.1	
11	9.15	35.9	44.3	15.24	32.0	33.8	
12	9.15	32.2	36.9	17.73	30.5	31.4	
13	9.15	37.5	42.9				
14	9.15	34.9	38.8				

UPPER BOUND ANALYSES

The simple velocity fields, shown in Fig. 1, related to the observed patterns of flow were adopted. They are seen to be based on the unit deforming regions of Kudo. For the forging with flash, type 1a, a quarter section only, of the shape was considered, due to the symmetry of flow. Region ABCD was a dead metal zone and power dissipation was minimised by varying the radius of the neutral surface BE. Conical flow into the boss, was assumed, whilst the flash which has reached the gutter (free flash) is allowed to deform in the radial and circumferential directions only. A maximum energy consumption of 3 % of the total was recorded in this region.

Due to the asymmetry of friction about the mid height, both top and bottom halves of the type 1b flashless forgings were considered. It is noteworthy that the frictional power dissipation in this "parallel" velocity field, where the vertical velocity decreases linearly to nil, from top to bottom of the flange, is twice that occasioned by symmetrical flow in which velocity decreases linearly from top and bottom of the flange to nil at its mid-height. The "rise" into the top and bottom cavities was assumed equal and experimental evidence on the whole, supported this. It was found that rise into the top die was slightly greater than that into the bottom, for the hot forgings. This was due to heat losses, prior to forging, from the region of the billet in contact with the die.

Conical flow into the flanges of the type 2 forgings, was assumed also. The treatment of the flash in the type 2 shapes, was the same as that for type 1. For these shapes, a maximum of 11 % of the total energy was found to be dissipated in the free flash. Asymmetry of sidewall friction again required that both top and bottom halves of the type 2b forgings be considered. For these forgings two velocity fields were offered. One with orthogonal lines of discontinuity and one with a conical surface of discontinuity AB. The field requiring minimum power for a particular geometry, was chosen. In all the upper bound computations, the material was assumed to be rigid-plastic, yielding according to Von Mise's criterion.

Equivalent flow stress

If the experimentally recorded forging load is L and the computed forging load for an identical shape forged under nominally the same conditions of surface friction is denoted $\sigma L'$, then the equivalent flow stress for the stock material may be obtained by dividing L by L'. The equivalent flow stress represents a value which may be used, assuming homogeneous material properties, to calculate deformation loads.

Interface friction

Frictional restraint was taken to correspond to a shear factor m, of the flow stress. Two approaches were made to the adoption of values of friction factor. In the first, m values, uniform over the whole interface were used, for both lubricated and unlubricated conditions. However, due to variations in surface flow, it is more reasonable to assume that friction will vary over the die for a particular nominal condition of the surface. In the second approach therefore, different shear factors were ascribed to different regions of the die. The initial choice of friction factors was based on both current and past experimental evidence, as explained below for each case, and their appositeness to a particular forging situation was judged from a reaction with experimental results, the argument being as follows:

As the chosen velocity fields for both lubricated and unlubricated flashless forgings are the same and experimental evidence showed no significant differences in flow for the two conditions of forging, the equivalent flow stresses should be the same. The overall level of friction chosen, should therefore, lead to equal values of equivalent flow stress being calculated for these two cases. Moreover, if the general disposition of the friction values is correct, calculated ratios of loads, flash forging/flashless forging should agree with experimental results. Thus two checks are available on the validity of the chosen state of friction.

DISCUSSION

Type 1 forging

H30 cold

Figure 3 shows the calculated values of equivalent flow stress for the various forged shapes and two types of tests. The results for unlubricated conditions are identical within the significance of experimentation. It can be inferred from this, that the assumption of a uniform flow stress does not introduce significant inaccuracies, even in the work hardening material,

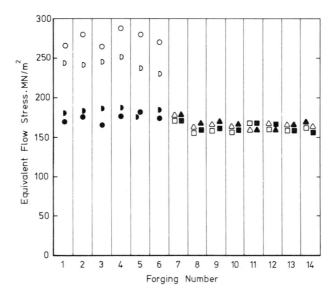

		With flash		Flashless	
		Non-uniform friction	Uniform friction	Non-uniform friction	Uniform friction
Lubricated		●	○	◗	D
Unlubricated		■	□	▲	△

FIG. 3. Equivalent flow stresses – type 1 forgings, cold.

when large total strains are imposed, despite the fact that strain histories and final strains are markedly different between flash and flashless forgings. That the flow stress was substantially uniform in the final forgings, was confirmed by microhardness tests on a sectioned shape. As similar flow is occasioned in both lubricated and unlubricated forgings, the equivalent flow stress extracted from these tests should be the same for both conditions. The fact that this is not so, for the uniform friction approach, is readily apparent in Fig. 3. The relatively high values of the results for lubricated forgings, indicate that the chosen levels of friction factor are inappropriate. It was evident from the experiments that high friction existed over all surfaces, in the unlubricated forgings, so there is little justification in reducing the levels of shear factors already chosen. However, study of the lubricated forgings indicated that lubrication was impaired in the boss of the cavity, due to the scraping action of the extruding material and worse on the flash lands, due to the greater severity of flow. The lubricant remaining in these regions after forging, was either greatly diminished, or completely eliminated. Choice of the friction factors shown in Table 1, produced the results designated "non-uniform" in Fig. 3 with these values of friction, all values of equivalent flow stress are within $\pm 10\%$ of 170 MN/m². This is lower than the value to be expected from inspection of the stress–strain curve in Fig. 2 and reflects the nature of an upper bound solution.

Calculated and experimental values of flash/flashless forging load ratios are compared in Fig. 4. The disposition of results reinforces the above argument concerning friction. The analysis employing the nonuniform concept, producing results very close to the ideal line, whilst the use of uniform friction, for lubricated tests, leads to theoretical results which are low, compared with experiment. The ratio of flash/flashless forging load is apparently dependent on the forging geometry. Forgings 4, 5, 6, 11 and 12 having a large boss and 1, 2, 3, 7, 8, 9 and 10 having small diameter bosses. In the forgings with large bosses, the rate of power dissipated in the boss is low, relative to that in the flash, whereas in the forgings with small bosses, the difference is not as great. It would appear then, that the proportion of the power consumed in the flash is a dominant factor influencing flash forging loads, relative to the flashless counterparts. Exceptions to the above groupings are forgings 13 and 14, which have large bosses. This is because both of these forgings had very small volumes of flash (Fig. 6)

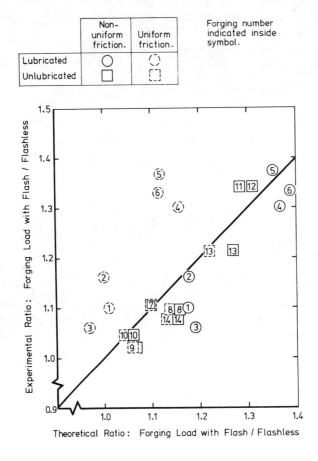

	Non-uniform friction.	Uniform friction.
Lubricated	○	( )
Unlubricated	□	[ ]

Forging number indicated inside symbol.

FIG. 4. Comparison between experimental and theoretical load ratios type 1 forgings, cold.

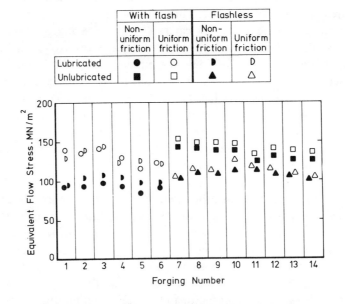

	With flash		Flashless	
	Non-uniform friction	Uniform friction	Non-uniform friction	Uniform friction
Lubricated	●	○	◗	◖
Unlubricated	■	□	▲	△

FIG. 5. Equivalent flow stresses – type 1 forgings, 1200° C.

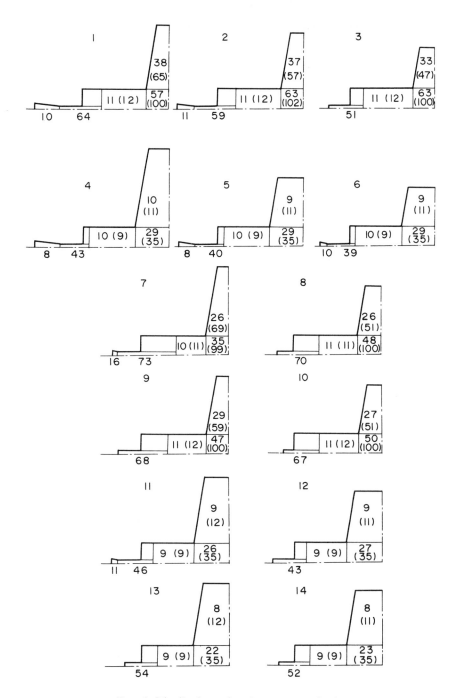

FIG. 6. Distributions of strain-rate type 1 forgings.

which reduced the load compared with the flashless shape. It is of interest to note that the forging load at the instant flash starts to form, is less than the load for the equivalent flashless shape, because the former is constrained less. Forgings 7 and 8 correspond to higher values than 9 and 10 because they were made from billets with lower height/diameter and hence more flash resulted.

*M*40 *hot*

Calculated equivalent flow stresses are shown in Fig. 5. The uniform friction factors were chosen as 0.2 and 0.85 respectively, for lubricated and unlubricated conditions. Although variations in strain-rate and temperature will influence forging loads, using a similar argument to that above, the equivalent flow stresses for both lubricated and unlubricated flashless forgings should be the same. The fact that they vary when a uniform flow stress is

TABLE 2. NON UNIFORM FRICTION FACTORS – TYPE 1 FORGINGS

| Material | Test condition | Friction factor | | |
		Flash land	Central boss	Other surfaces
H30	Lubricated	0.60	0.40	0.15
	Unlubricated	1.00	1.00	0.95
080M40	Lubricated	0.70	0.60	0.40
	Unlubricated	1.00	0.950	0.95

used indicates once more, inaccuracy of friction factors. The work of Keung and others [7] using ring tests, indicates that for M40 steel at 1200° C friction factors between 0.75 and 1.0 for no lubrication and 0.35 and 0.6 for dry graphite lubrication can be expected, according to the conditions of oxidation of the hot steel. It is thought that the reduced oxide on the flash, due to the severity of deformation, and the extended time of deformation, compared with the ring test, causing greater decomposition of the lubricant, will result in higher friction in the forging tests. Using similar reasoning as for the cold forgings, the friction factors shown in Table 2 were chosen to obtain the most consistent non uniform results in Fig. 5. The results for the lubricated and unlubricated forgings are now more nearly equal. The slightly greater values associated with the unlubricated shapes can be accounted for by a small fin which was extruded up the side of the punch under these conditions. However, noticeable differences between other test results remain. It is possible to estimate the influence of strain-rate, in the hot forgings by examination of Fig. 6, which shows a breakdown of values. They are the average values derived for each region from the values of the internal power of deformation and of the power dissipated on the discontinuities, provided by the upper bound analysis. The power dissipated on the discontinuities has been apportioned equally to the two adjacent regions except where a dead metal region exists. In such cases all the energy has been credited to the single contiguous region. The values for flashless forgings are bracketed, in this figure. If the strain-rates corresponding to the central part of the web are disregarded, the general levels of strain-rate in the narrow regions which include the flash, are 3–4 times higher than elsewhere, for forgings 11, 12, 13 and 14. With a strain-rate exponent of 0.213, these higher strain-rates correspond to an increased flow stress of 26–34%, which acting in the appropriate volume of material, results in an increase in load (or equivalent flow stress) of 8–13%. This could explain in part, the difference in equivalent flow stresses between the unlubricated flash and flashless forgings. The complexity of strain-rate distribution is too great to allow this explanation to be pursued for the lubricated test results. However, it is noteworthy that, for lubricated forgings, the strain-rate in the flash is less dominant. That considerations of strain-rate cannot fully eliminate the discrepancies in the calculated equivalent flow stresses for the steel, indicates that the unaccounted non-uniform temperature distribution has a significant influence for the velocity fields together with corrected values of friction factor have been shown adequate to deal with extreme conditions of deformation for cold forgings.

The facts that work hardening is unimportant in the cold forgings and strain-rate sensitivity is significant in the hot forgings, can be used to explain why greater "rise" is obtained with steel than with aluminium, using clean dry dies. For the flow stress is virtually homogeneous in the cold forgings, but in the hot forgings an increase in flow stress in the flash commensurate with the high strain-rates, causes greater flow into the cavity.

Experimental and calculated load ratios, for the steel forgings, are shown in Fig. 7. The disposition of the points is similar to that for the cold forging tests, with the exception that departure from the ideal line of results for unlubricated tests, is much greater, due to the reasons discussed above.

Type 2 forging

H30 cold

Values of equivalent flow stress are shown in Fig. 8. Those derived on the basis of uniform

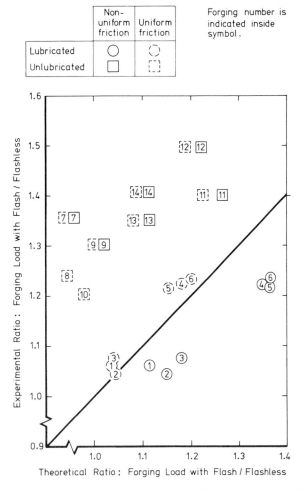

FIG. 7. Comparison between experimental and theoretical load ratios – type 1 forgings, 1200° C.

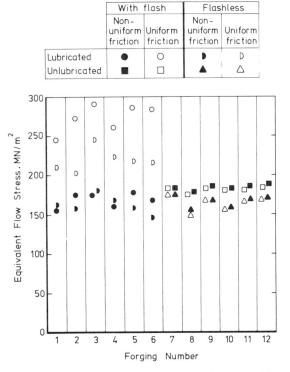

FIG. 8. Equivalent flow stresses – type 2 forgings, cold.

TABLE 3. NONUNIFORM FRICTION FACTORS – TYPE 2 FORGINGS

Material	Test condition	Flash land	Friction factor Peripheral flange	Central webb
H30	Lubricated	0.60	0.30	0.15
	Unlubricated	1.00	0.95	0.95
080M40	Lubricated	0.70	0.40	0.30
	Unlubricated	1.00	0.95	0.90

friction vary significantly. Several computations leading to the choice of friction factors shown in Table 3, have enabled the flow stress of the flashless lubricated and unlubricated forgings to be nearly equalised, in line with the argument proposed above. It will be noted that the friction factors are similar to those arrived at for the type 1 shapes, slightly lower values obtaining in the peripheral flange, compared with the boss and the average value of flow stresses is virtually the same as that derived for the type 1 forging. However greater variation of flow stress remains for this forged shape than for type 1. Two factors are proposed to account for this;

(1) Micro hardness measurements on a sectioned forging indicated that only low strains were reached in the flanges, and the material in this region was not fully work hardened. The experimental forging load would be sensitive to changes in average deformation such as are likely to be encountered in these tests.

(2) Inevitable inconsistencies of lubrication and hence surface friction, would have a significant effect on loads for these forgings, which have large surface areas in contact with the die.

Forging load ratios are compared in Fig. 9. The trend of the relation of experimental to theoretical results follows the ideal line, but the theoretical values are low. This can be said to

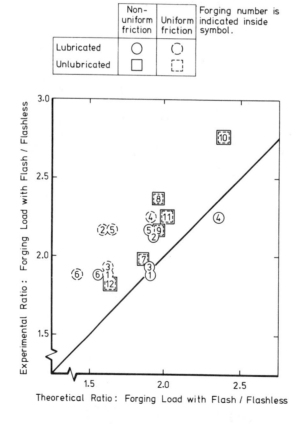

FIG. 9. Comparison between experimental and theoretical load ratios – type 2 forgings, cold.

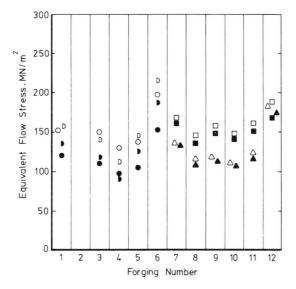

FIG. 10. Equivalent flow stresses–type 2 forgings, 1200° C.

be due to the low flow stresses in the flanges (due to low deformations) compared with those in the flashes.

*M*40 *hot*

Figure 10 shows values of equivalent flow stress derived assuming both uniform friction ($m = 0.2$ lubricated, $m = 0.85$ unlubricated) and non uniform friction, the values of which (Table 3) are based upon previous argument. Although the average levels of the lubricated and unlubricated related points are similar a large variation exists between individual results. Figure 11 illustrates the computed pattern of strain-rates existing in various forgings at a time near to the end of deformation, the bracketed values being for the flashless shapes. (No significant difference between the strain-rates in the bodies of flash and flashless forgings was observed, so in these regions bracketed figures are omitted.) The increasing values of strain-rate in the flanges as their thickness are decreased (forgings 4, 5, 6 and 10, 11 and 12) are reflected in the equivalent flow stresses for these forgings shown in Fig. 10. Other differences can as yet, only be explained by reference to temperature variations which are unknown.

Figure 12 shows the forging load ratio relations. The experimental results for clean dry dies are greater than the corresponding theoretical ones. The largest difference (forging 10) corresponds to the largest difference in strain-rate between body, and flash, whereas the smallest difference (forging 12) is related to the smallest strain-rate difference. The lubricated forging results are close to the ideal line, although a similar trend is evident. The relative disposition of points is identical to that for the H 30 forgings. The thickness of the peripheral flange is dominant in determining load ratios, as can be seen by noting the relation of results for forgings 1, 2, 3, 5, (7, 8, 9, 11) to 4, 6, (10, 12) respectively.

CONCLUSIONS

The equalisation of equivalent flow stress and comparison of calculated and experimental loads for forgings made under different conditions, may be used to investigate the distribution of interface friction and the effects of nonuniform flow stress in real forging operations.

Results indicate that friction forces vary over the surface of the forging dies, for a nominally

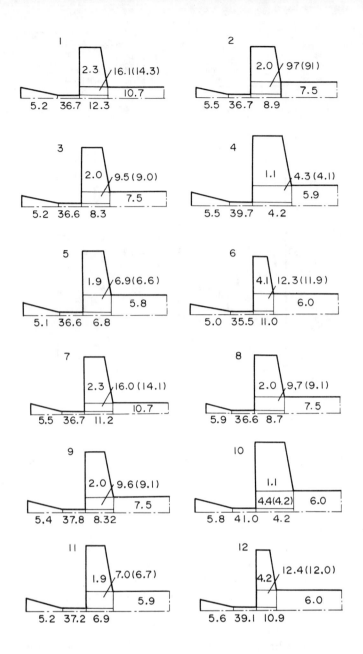

FIG. 11. Distributions of strain-rate – type 2 forgings.

uniform condition of lubrication. Higher restraints are apparent where surface flow is greater. For a severely deformed strain hardening material, load ratios flash/flashless, can be adequately calculated assuming a uniform flow stress. For a hot (strain-rate sensitive) material the load ratios are sometimes underestimated for high interfacial friction conditions.

For the two types of forging chosen, the required load may be reduced by up to 100%, by eliminating flash. In general, greater reductions are obtainable for forgings having relatively low deformations and deformation rates, in the main body, compared with those in the flash.

Acknowledgment – The work reported here is the outcome of a research programme supported by S.R.C. Y. Van Hoenacker is a National Research Council of Canada scholar.

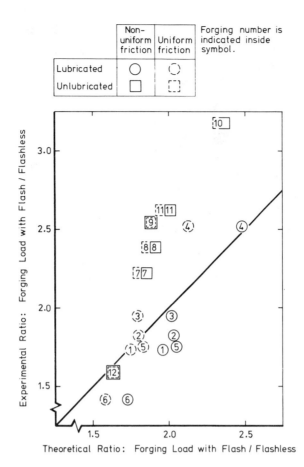

	Non-uniform friction	Uniform friction	Forging number is indicated inside symbol.
Lubricated	◯	◌	
Unlubricated	▢	⸬	

Fig. 12. Comparison between experimental and theoretical load ratios–type 2 forgings, 1200° C.

REFERENCES

[1] W. Johnson, *Proc. 3rd U.S. Cong. App. Mech.* p. 571 (1958).
[2] H. Kudo, *Int. J. mech. Sci.* **2,** 102–127 (1960).
[3] R. P. McDermott and A. N. Bramley, *Proc. 15th Int. M.T.D.R. Conf.* (1974).
[4] T. A. Dean, *Metallurgia & Metal Forming,* Nov. & Dec. (1977).
[5] T. A. Dean, *Metallurgia & Metal Forming* **42,** 4–8 (1975).
[6] T. A. Dean and C. E. N. Sturgess, *Proc. I.Mech.E.* **187,** 523–533 (1973).
[7] W. C. Keung, T. A. Dean and L. F. Jesch, *Proc. 3rd N.A.M.R.C.* (1975).

High-speed Flashless Forging

Close tolerance forgings can be progressively formed automatically at rates up to 150 per minute on horizontal hot formers. Elimination of flash provides substantial material savings

RICHARD EDMONDSON
Vice President, Engineering, National Machinery Co.

AUTOMATIC HOT FORGING of close tolerance parts at rates up to 150 per minute is possible with hot formers made by National Machinery Co. These horizontal, closed die machines are self-contained feeding, shearing, and forging units. Bar stock heated to forging temperature is progressively converted into finished parts automatically with little or no material waste.

Advantages. Faster production with less equipment, floor space, and manpower — which results in lower cost per part — is probably the single most important advantage of using hot formers. For example, one size gear blank is being automatically produced on a Model 6-4 hot former at 70 per minute, which is six times the rate possible on a hand fed, vertical forging press, and 3 1/2 times that from a 1600-ton (14 234-kN) automatic Maxipres.

Material savings is another major advantage of the process. Parts are automatically forged without flash, thus reducing material requirements and eliminating the need for trimming. The only waste is when parts require piercing of a hole, which can be done in the last operation on the hot former. Little or no draft is required, and this combined with the elimination of flash can provide material savings of up to 20% over conventional press forgings. Also, closer tolerances are maintained and smoother surface finishes produced, saving additional material removed in subsequent processing of press forgings, as well as the cost of such operations.

Another advantage is that tool life is longer than with conventional forging. This is because contact time of the hot material with the dies and punches is less with this higher speed process. Also, the hot formers have flood coolant systems which keep the tools cooler.

Compared to Cold Forming. Where parts can be produced by cold forming, it may be a more economical process

*1. TYPICAL PARTS
produced on hot formers.
Round, short, and symmetrical parts
are ideally suited for the process.*

*2. SINGLE FORGING
provides an outer and inner
bearing race, plus a center web.*

4"
100 mm

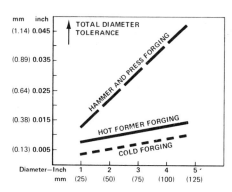

mm   inch
(1.14) 0.045
(0.89) 0.035
(0.64) 0.025
(0.38) 0.015
(0.13) 0.005
Diameter—Inch
mm

3. COMPARATIVE TOLERANCES that can be maintained on diameters with different forging processes.

because closer tolerances and smoother surface finishes can usually be obtained, and little or no subsequent machining is required. However, there are cases where the parts and/or materials cannot be formed cold, and others where hot formers are more economical.

One reason for considering the hot former is that it is about twice as fast as a cold former, which means that less equipment and manpower are required for the same output. Also, in some instances, there can be substantial savings in raw material costs. There are also some cases where the closer tolerances and smoother surface finishes possible with cold forming are not required for the function of the part.

Volume Requirements. The relatively high cost of hot formers makes large volume requirements essential for economic justification. However, while long runs are the most efficient, it is economically feasible to produce shorter runs of several different parts because of fast and easy changeover features. One purchaser of a large 6-4 hot former was using it to produce 11 different parts during the first six months of operation.

It is generally desirable to have a minimum lot size that would require operating the hot former at least one full shift. On a small hot former this could mean 72,000 parts, and on a large machine, about 20,000 parts in 8 hours. Actual minimum lot size per run can only be established from economic studies, taking into account design of the part, the material from which it is made, possible savings compared to the present production method, and future repetitive requirements for the part.

Applications. Parts ideally suited for hot forming are typically round, symmetrical, and short, with lengths usually less than their diameters. The maximum diameter of such parts is about 4 3/4" (120 mm), with a maximum weight of about 7 lb (3.2 kg), and many have through holes. Shapes shown in *Figure* 1 give some idea of the process versatility. More complex and even unsymmetrical parts are possible by combining upsetting with forward and backward extruding operations.

Practically any material now being hot forged, with the possible exception of some heat resistant materials and superalloys, can be formed in this way. Straight hot-rolled steel bars of random length and with no special mill requirements are most commonly used. Some experimental work has been done with coil stock, but uncoiling and straightening for passage through the induction heating coils present problems.

Hot forming is being used in the fastener industry to produce nut blanks, and in the bearing industry for races. Automotive applications include gear blanks and a wide variety of hubbed and flanged parts. Cluster gear blanks can be made by friction welding two or more hot formed parts.

Some configurations permit producing more than one part from a single forging, as is the case with the bearing races seen in *Figure* 2. The completed forging consists of an outer race, an inner race, and a center web. Some manufacturers perform secondary machining operations before separating the two races, while others want all three separated before machining. In the latter case, special tooling at the final station punches out the center web, separates the inner and outer races, and discharges all three into separate chutes.

Tolerances Maintained. Smooth surface finishes and close tolerances maintained in hot forming often eliminate the need for subsequent machining on some surfaces, and minimize stock removal on others. Machining allowance

on hot formed parts is only 0.020 to 0.040" (0.51 to 1.02 mm) per side, depending on forging size. Diameter tolerances, *Figure* 3, are closer than hammer and press forgings, but not quite as close as cold formed parts of comparable size. Typical length or thickness tolerances are ±0.012" (0.30 mm), and radii for corners and fillets are usually made about 1/8" (3.2 mm).

Machines Available. Hot nut formers, available for years from National Machinery, are widely used for making hex nuts as large as 2 3/8" (60 mm) across the flats at rates up to 3300 per hour. Because of their design, however, these machines offer little opportunity to form other parts. Hot formers now available are complete systems with much more versatility. An automatic transfer unit, seen in the heading illustration, moves the forging between each of three or four dies. Operations performed depend on the shape of the part, but often include upsetting, rough and/or finish forging with controlled metal displacement, and punching a through hole.

The line of hot formers made by National Machinery presently includes four sizes, some with three dies and others with four. On the smallest 2-3 machine, parts weighing less than 0.8 lb (0.36 kg) can be automatically hot formed at rates up to 150 per minute. Cutoff capacity of this machine is 1 1/4" (32 mm) diam x 2 1/2" (63 mm) long, and it is equipped with a 50-hp (37-kW) motor. The largest 6-4 machine, *Figure* 4, can produce parts weighing up to about 7 lb (3.2 kg). Production rates are up to 70 per minute, depending on the part. This machine, powered by a 400-hp (298-kW) variable speed d-c drive motor, has a cutoff capacity of 2 3/8" (60 mm) x 5 1/2" (140 mm) long.

Heating, Feeding, and Shearing. Induction heating is recommended for bringing the bar stock to the forging temperature — in the 2250°F (1230°C) range — because it is rapid and minimizes scale. Four rolls, each operated separately, give good control of bar ends when feeding straight bars. In-

4. HOT FORMER at S.A.F.E. (Renault Group) in France forging a 4" diameter gear blank at up to 70 per minute.

5. *FOUR ROLLS keep ends of heated bars together when feeding, and are water cooled internally.*

ternal channels are provided for water cooling the feed rolls and their shafts, *Figure 5.*

Heated bars stop against a stock gage that precisely measures cutoff length, and that can be adjusted while the machine is running to provide accurate control of blank volume. To eliminate the need for an operator to gage random bar lengths and manually drop short ends, the machine can be equipped with an automatic bar end detector. This unit, interlocked with the machine controls, senses the bar ends, measures linear movement of the bar into the machine, and automatically drops short sheared ends but allows required length blanks to be transferred.

Before shearing, a gripper firmly holds the bar against the stationary cutter to minimize distortion during shearing. After shearing, the gripper opens to permit drag-free feeding of the bar. Cutoff motion is from complementary cams that give positive drive in both directions without the need for large springs or air cylinders. Force is applied through a safety bolt that prevents shearing a cold bar.

Transfer Unit. The transfer unit moves horizontally to carry parts between dies, and opens vertically to clear the tooling. A guided vertical movement allows the punch holders to move close to the dies. This permits making the punches short and rigid, which helps lengthen their life. Simple

6. *TOOLING LAYOUT on a four-die hot former for automatically producing gear blanks.*

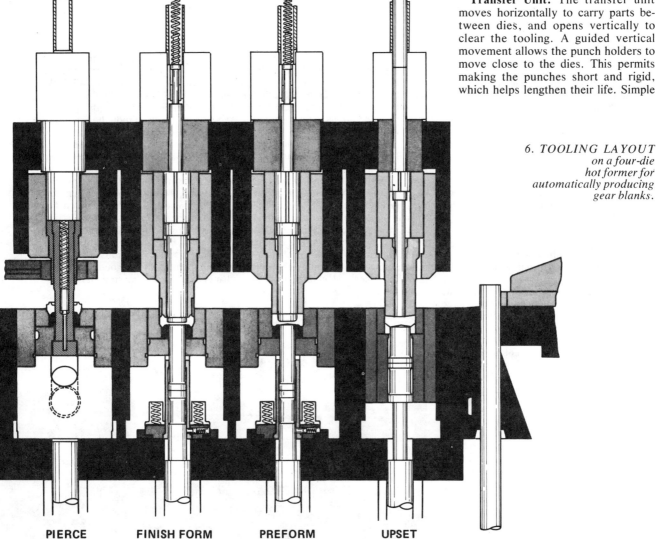

PIERCE   FINISH FORM   PREFORM   UPSET   SHEAR

blank-gripping fingers are easy to change for producing a different part. Transfer drive is through a rack-and-gear and spring loaded clutch.

Die Design. Dies are mounted in a stationary block with each die having its own cam controlled, adjustable kickout mechanism to push forgings out of the die and into the transfer fingers. Punches are mounted on the machine slide, which reciprocates continuously during production, and each punch can be adjusted horizontally and vertically in its own holder. All punch holders are mounted on a common plate removable from the slide as a unit.

Some part shapes are formed inside the punches. Timed knockouts can be provided on the machine slide to push such parts free of the punches as they move away from the dies, thus keeping the parts in the dies for ejection into the fingers of the transfer unit. Use of these timed knockouts simplifies tool design, and extends the range of shapes that can be formed.

Hot work tool steels are used, with die inserts made from AISI Types H12, 13, or 26, and punches and knockout pins from H21. The hot forming tools are made harder than normal — about R_C50, compared to the usual R_C36 to 44 for press forging tools. This increased hardness reduces tool wear rate and lengthens their life.

Typical Operations. *Figure* 6 illustrates a tooling layout, as viewed from the top of the machine, showing four punches on top and four dies at the bottom for progressively forming gear blanks. Blanks are sheared from the heated bar at the right. The first die upsets the hot sheared blank, shortening it by more than half, increasing its diameter, and making it concentric. A pin at the bottom of this die, about the same diameter as the sheared blank, makes a conical indentation in the upset that matches a pin in the second die.

Preforming is done in the second die with controlled metal displacement. Punch penetration forces the metal to flow outward from a hole into the die cavity, and forms a depression in the bottom of the part. Since blank thickness is nearly the same before and after displacement, little extrusion occurs. The third die and punch forms the part to final shape and dimension, with the web still in place. Web thickness should be as small as possible to minimize material waste, but thinner webs increase pressure requirements. To improve tool life, a sleeve is pressed on the end of the third punch. Proper venting is important for the second and third dies to let air, water, and steam escape.

Punchout of the web to produce the required through hole is done by the fourth punch, with the die supporting the part to prevent distortion. Inside the punch is a spring loaded pin to dislodge any punchouts that may stick to the end of a roughened punch. A stripper removes completed forgings from the piercing punch, and the gear blanks fall into a discharge chute. For parts that do not require through holes, the last die can be used for finish forming or coining, and a kickout provided to clear the die.

The second and third tools of the layout shown have two special mechanisms, one to assist in placing parts in the dies, and the other for kickout required because of the concave bottoms on the forgings. A spring assembly in each punch holds the sliding punch pins forward, and friction devices in the dies keep the kickout pins extended. Workpieces are gripped between these pins while the transfer fingers move out of the way of the larger punches. The pins then guide the workpieces into the dies, and hold them

the hot former runs another. An extra transfer unit can also be set up on another fixture outside the machine, ready for the next job.

A hydraulic clamping system on the machine slide can release all the punch holders, which lift out as a unit. Motorized draw screws and clamps are used to release the die block, which also lifts out as a unit with all dies in place. Then the preset punch holder assembly and die block can be clamped in place quickly and easily. Kickout rod adjustments for each die are also motorized and secured in place with hydraulic locknuts.

Cooling System provided flushes heat away from the tooling and machine. External flooding cools the punches, cutoff gripper, and guide, while coolant circulated through internal channels cools the dies, feed rolls, and their shafts. Spray heads mounted on a manifold are automatically lowered as the machine slide retracts to shower coolant on the punch ends, and are then raised to clear the tools for the forming

7. ALIGNMENT FIXTURE that holds an extra die block and punch holder assembly outside the hot former.

for the forming blows. The spring-return kickout pins eject the formed parts beyond the faces of the dies, and then retract to clear the workpiece.

Quick-change Features. Optional equipment is available that permits complete changeover of a hot former from the production of one part to another in four hours or less. This equipment can be divided into two groups: the tooling itself, and quick clamp and release mechanisms on the machine. The tooling includes an extra die block and punch holder assembly that fits onto an alignment fixture, *Figure* 7. Here, tooling can be preassembled and prealigned for one job while

blow. This mechanism also serves as a safety device, interrupting the stroke and stopping the machine through electrical interlocks should a formed part stick on a punch.

Plain water is used as the coolant, with a water soluble wax sometimes added for corrosion protection. A high volume flow, up to 36,000 gallons (136 000 L) per hour, is used. This copious flow on the tools reduces wear, prevents deep penetration of heat, minimizes failure from thermal shock, and permits higher production rates. The tools are kept cool enough to permit immediate removal by hand when the machine is stopped. ∎

Precision Forging of Spiral Bevel Gears

ROGER R. SKROCKI

ADAPTATION of present-day precision forging technology to gear manufacturing should minimize or eliminate many time consuming, costly tooth cutting operations by directly preforming the teeth. This process can produce a metallurgically superior gear, semifinished to a point where it is ready for heat treatment and finish machining. Furthermore, this approach to gear manufacturing is less dependent upon the availability of specialized machine tools and the skilled labor required to set up and operate them.

Various types of gear forms are currently being manufactured by precision forging methods, and these gears have demonstrated improved cost effectiveness compared with gears manufactured by conventional gear cutting methods. To date, however, the spiral bevel gear form has not been extensively produced by precision forging because this tooth design presents special problems during forging. In particular, the spiral bevel die cavity configuration is more difficult to fill because metal flow during forging must follow a spiral path into a blind cavity. Also, after forging, the part must be released from the die through a spiral path to avoid damage to the forged teeth. Other problems are 1) lack of a practical method for measuring tooth form in the die cavity, and 2) the necessity of controlled cooling methods to prevent distortion of the tooth form after forging.

This report summarizes results of a development program for precision forging integrally formed gear teeth in a spiral bevel gear and pinion design. Objectives of the development program were:

•Demonstrate suitability of a mechanical crank

ROGER R. SKROCKI is associated with TRW Inc., Cleveland, Ohio.

press for precision forging a specific gear with integrally formed teeth.

•Determine economics of this forging approach as compared with conventional processing.

•Compare fatigue properties of forged teeth to those of conventionally cut teeth.

•Develop and document procedures for tool design, tool fabrication, and forging for production use.

DESCRIPTION OF GEAR SET, PROCESS, AND TOOLING

The gear set chosen for this development and test program was the CH-47 helicopter main transmission spiral bevel gear and pinion. These gears, produced from AMS 6265 steel, transmit engine power at high speed to the reducing gear train to drive the main rotor. The gear and pinion set, in both the as-forged and finished condition, is shown in Fig. 1.

Specification for Gear and Pinion

Gear form data, materials, tolerances, finishes, and other technical requirements of the gear set have been incorporated in the precision forged gear design. Thus, evaluation of processing costs and gear performance can be directly compared to the conventionally fabricated gear. Basic gear and pinion drawing dimensions are given in Table I.

The input pinion operates at 14,720 rpm, transmitting 3750 H. P. and developing a pitch line torque of 16,056 in.-lbs. It is expected to sustain 1100 h of operating time between overhauls. The calculated design maximum tooth bending stress is 29,000 psi

Source: Metals Engineering Quarterly, February 1976, 58-64, © 1976 American Society for Metals

with a tooth contact stress of 205,000 psi. The design load per tooth is 4523 pounds.

Process Design

The technique for precision forging of integrally formed gear teeth is based on the use of a modified crank press. The press is a conventional design single-action type which has been used for precision production forging for over 40 yrs. Widely available in industry, it is capable of high-speed production of

Table I. Basic Gear and Pinion Dimensions

	Pinion	Gear
Number of teeth	35	43
Pitch	4.930	4.930
Pressure angle	22°30'	22°30'
Spiral angle (mean)	25°0″ L.H.	25°0° R.H.
Pitch diameter	7.099	8.722
Pitch angle (basic)	39°9″	50°5'
Root angle (basic)	37°1'	48°11'
Face angle	41°49'	52°59'
Circular tooth thickness	0.344 to 0.337	0.291 to 0.294
Addendum	0.199	0.146
Dedendum	0.184	0.237
Normal chordal thickness at P.D.	0.287	0.241
Load side of tooth	Concave	Convex

Fig. 1—Spiral bevel gear set. Top: as-forged gear, left; finished gear. Bottom: as-forged pinion, left; finished pinion.

forgings. Production tolerances on selected dimensions can be held to a few thousandths of an inch. A typical forged part is produced in one or more preform blows, followed by a coining blow. Several operations may be accomplished simultaneously at each cycle of the press. This multiblow method is usually chosen, even though the press may have sufficient energy to forge the part in one blow, to allow use of small diameter stock. Relatively small diameter starting material can be more accurately and economically cut into billet lengths, which can then be upset to produce preformed forgings of larger diameter prior to a final coining operation. For this program, a 2000 ton capacity press was used.

Tooling and forging sequence for the gear and pinion are shown in Figs. 2 and 3 respectively. For both parts, deformation is accomplished by a hot preform blow, followed by a hot coining blow which forms tooth details.

In the gear preforming operation, the starting billet is converted from a 3 1/2 in. diam × 5 1/2 in. long cylinder to a flat conical disc approximately 9 in. in diam and 1 3/4 in. long. The starting billet is manually positioned in a recessed locating seat for preforming to achieve centrality, and to assure even distribution of material in the preform. The preform die cavity is machined with the conical surface in a downward direction. The parting line is near the top of the part so that any unevenness in filling, which may still occur, will be at the extreme rim of the blank. Later the material in the rim moves into the flash area, which is external to the location of the gear teeth. Having the conical surface facing downward also facilitates forming a small conical indentation in the upper surface to serve as a centrality locator for positioning the preform in the hot coining station. A smooth flow of material radially outward is achieved as the starting billet is compressed in an axial direction.

Draft angles were selected so that the preform would be retained in the lower die cavity on the upstroke of the press. The preform is lifted from the die by a pneumatically actuated ejector pin, facilitating removal and transfer to the hot coining station by the operator. It is then turned over, and placed on the locator in the hot coining die. The lower die block is of completely open design to permit rapid placement and centering of the preformed part, thereby minimizing heat loss during transfer. (This is an important processing feature because accomplishing both preforming and coining operations during the same heating cycle minimizes exposure for metallurgical deterioration of the surface material.)

The profile of the lower hot coining die was made to match the profile of the preform punch when the hot coining die is in a fully closed position. When the coining die is in the open position, there is an intentional mismatch due to protrusion of the central locating pin. This mismatch accomplishes two objectives. First, it permits the larger conical surface to function on a "rough" locator while the pin itself serves as a final locator. Secondly, it supports the hot preform a short distance above the die block, preventing excessive heat loss via this sink. After coining is completed, the dies separate, instantaneously actuating the pin by means of a mechanical spring. A fast response would always be desirable, and is essential for a spiral bevel gear or pinion forging when a negative draft condition exists in the tooth portion of the die. Failure to achieve immediate release of the forging from the lower die block as the press ram starts on the upstroke could strip the teeth from either the die or the forging, probably the latter.

A second spring-actuated pin is used to force the forging from the upper die (punch) an instant after the forging is released from the lower die block. This step requires slight rotation of the forging, which can only be accomplished after the holding friction on the lower surface has been overcome. Therefore, the balance between the spring forces of the opposing stripper pins is critical.

It was estimated that a billet temperature in excess of 1900°F would be necessary to provide sufficient plasticity to achieve acceptable fill in the tooth portion of the forging. Prolonged contact between the hot billet and the die would have a deleterious effect on die life. To minimize contact time for the most

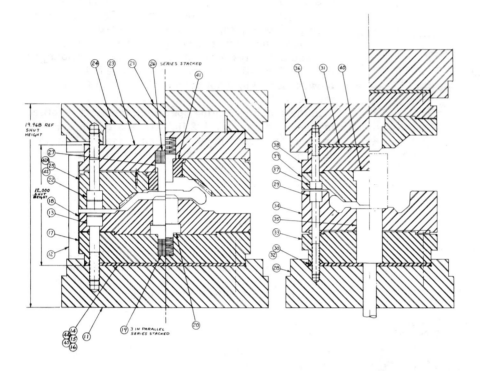

Fig. 2—Gear forging tooling.

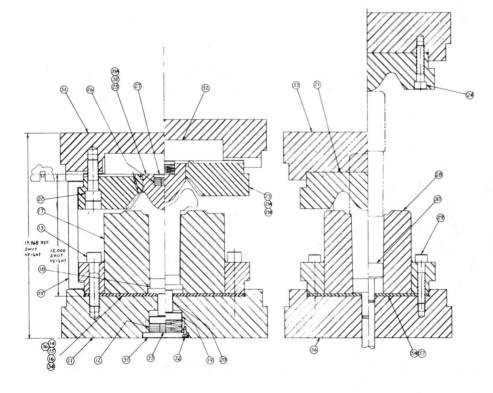

Fig. 3—Pinion forging tooling.

critical elements of the tooling, the tooth form was designed into the punch portion of the die set. Thus, any delay which might be encountered after placing the preformed billet in the hot coining station would not result in a transfer of heat to the tooth portion of the cavity. Furthermore, on the upstroke of the press, gravity and inertia of the forging, in addition to the force imparted by the spring stripper, help to free the forging from the cavity.

Design of the forging tooling for the spiral bevel pinion was similar to that for the gear. Pinion configuration is more complicated due to the integral shaft. This, in effect, increases the number of forging operations required to transform the material from a starting billet shape into the final pinion forging. It was estimated that a minimum of three forging operations would have been necessary to do this. One of these operations was eliminated in the development program by machining the first preform shape from the bar material, appreciably reducing the forging tooling cost and time required to develop the forging. For the small number of pieces involved in this program, the extra upset forge tooling could not be justified economically.

In the preforming operation, the portion of the billet protruding above the lower die is converted from a 3 1/2 in. diam × 5 1/2 in. long cylinder to a flat conical disc approximately 7 in. in diam by 2 in. thick. Positioning of the billet is easy because the stem on the billet is shorter than the depth of the die cavity and has a matching taper. In this way, positioning of the billet is as accurate as the centrality of the stem on the billet. The locating cavity in the coining station also matches the tapered stem of the billet, and will thus again assure centrality.

Practically all of the heat loss to the dies will be from the billet stem. Since no planned deformation occurs in the stem, a heat loss here does not appear to create a problem. Upper dies at both the preform station and the coining station contact the hot forging for a very short time, limiting heat transfer to the critical part of the tooling containing the gear tooth configuration.

Due to the combination of geometric factors such as pitch, cone angle, spiral angle, and depth of gear tooth, there is negative draft angle in the axial direction on the end portions of the pinion tooth. This negative angle prevents the pinion from being withdrawn from the die in a purely axial direction without an accompanying rotational motion. Thus, unless the pinion is free to rotate, upward motion of the punch would deform or otherwise damage the gear teeth. To allow the forging to rotate, the release in the lower die must act immediately as the ram starts to move past bottom dead center. The heavy-duty spring-activated stripper pin provided in the lower die for this purpose must have sufficient force to overcome frictional resistance of the slightly tapered die wall plus the weight of the forging plus the force of the shedder pin in the upper die. Stripper and shedder pins are designed to have a very small movement since only a slight displacement of the forging away from the tapered die wall is necessary to break the frictional holding force, allowing the forging to rotate. Therefore, spring movement is kept small so that a large preload can be used to

achieve a high release force almost instantaneously as the ram starts its upstroke.

FORGING DEVELOPMENT

The forging development phase consisted of normal tool design and process iterations to achieve the proper flow and die fill for producing the desired gear shapes. These iterations consisted of selection of optimum billet size, adjustment of preform tooling dimensions for optimum material distribution, adjustment of tooth profiles in hot coining dies, and development of optimum billet heating techniques. Four developments for the gear and three for the pinion were required to achieve a workable process. A summary of the final process details, not including tool information, is listed in Table II. The adequacy of the forging development phase for obtaining the desired tooth configuration is indicated by the resultant material envelope available for cleanup by finish machining. In Table III, the range of finish stock removal for conventionally semifinished versus forged gears and pinions are compared. The amount of stock removal for the forged gear was comparable, for all practical purposes, to conventionally semifinished gears which are cut from blanks.

MECHANICAL TESTS AND METALLURGICAL EVALUATION

Mechanical Tests

A program of testing was performed to evaluate the mechanical properties of forged gears, and to compare these properties with those obtained in conventional cut gears of the same configuration, material, and heat treatment. The main variable in this comparison, therefore, was the forging versus machining method of obtaining the tooth configuration. Precautions were taken to minimize influence of uncontrolled variables in test gears. Both types of

Table II. Gear Forging Data

Material—AMS 6265 CVM steel, hot roll finished bar
Billet Size—3½ in. diam × 5½ in. long (gear)
　　　　　3½ in. diam × 11⁷⁄₈ in. long (pinion)
Heating temperature—1950°F ±25°F—45 min
Billet heating atmosphere—exogas, 7:1 ratio
Die temperature—400°F
Transfer time furnace to preform die—10 s, average
Transfer time preform die to coin die—12 s, average
Coining load—3.5 × 10⁶ pounds

Table III. Range of Stock Removal

	Cost Side	Drive Side	Root
Cut pinion (semifinished)	0.007 to 0.010 in.	0.006 to 0.010 in.	0.005 to 0.009 in.
Forged pinion (as-forged)	0.008 to 0.013 in.	0.005 to 0.008 in.	0.005 to 0.014 in.
Cut gear (semifinished)	0.009 to 0.011 in.		
Forged gear (as-forged)	0.008 to 0.011 in.		

test gears were made from the same AMS 6265 steel, and were heat treated and finish machined together as a group according to the manufacturer's specification.

Mechanical testing consisted of a comparative single tooth fatigue evaluation under unidirectional loading conditions. The loading arrangement utilized a standard conventionally cut pinion to provide the load to the gear teeth, Fig. 4.

All fatigue evaluations were conducted at $R = 0$ (minimum load = 0). The cyclic frequency for each test depended upon the maximum applied load, and varied between 10 and 20 Hz. Fatigue test results are presented graphically in Fig. 5. It should be noted that loads shown on the graph are the applied actuator loads; tooth loads are approximately 33 pct greater than the actuator loads.

In reviewing the data shown in Fig. 5, it appears

Fig. 4—Single tooth fatigue test fixture.

that the endurance limit of the forged gears is significantly higher than that of the cut gears, as indicated by the results at the 15,000 pound load level. Three of the four forged gear tests at this level did not fail above 4.6 million cycles, whereas the three cut gear tests all failed below 300,000 cycles. Results at the higher load levels did not indicate a significant difference between forged and cut gears; if anything, the cut gears fared somewhat better. The greater importance of the results at the load levels near the design point becomes apparent when considering the magnitude of the design torque relative to the torques applied at the various test levels. The design torque per tooth is approximately 16,000 in. pounds at maximum horsepower, whereas the torque applied at the lowest test load level is 60,000 in. pounds per tooth.

The large amount of scatter, which is typical of gear tooth fatigue tests, prevented precise endurance limit determination with the limited number of tests allotted. Future tests to supplement the data in Fig. 5 will be concentrated at the lower load levels, decreasing from the 15,000 to 20,000 pound load level. To quantitatively assess the difference in fatique properties between cut and forged gears, it would be most informative to determine the difference in the endurance limit between them.

Metallurgical Structure

A primary technical objective of this program was to obtain a favorable forging flow line pattern in the gear teeth. It is generally acknowledged that fatigue endurance and tensile ductility are enhanced in the direction of forging flow. Thus it was desirable to obtain a flow line pattern parallel to the tooth surface as the principal loading stresses are applied there when the gear is in normal service.

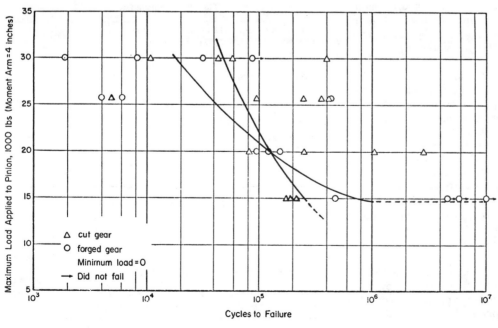

Fig. 5—Results of single tooth fatigue tests.

The forging flow line pattern obtained within a gear tooth in the forged test gears was determined by sectioning to obtain a tooth profile cross section, and then macroetching with a concentrated hydrochloric acid solution. Fig. 6 indicates that a favorable forging flow line pattern, consisting of lines parallel to the surface of the tooth, is present. This contrasts favorably with the pattern generally obtained in cut gears, in which flow lines originating in the blank forging operation intersect the tooth surface at most locations.

The investigation of metallurgical structures also included comparative evaluations of carburized case depth, core structure, and carbide distribution for cut versus forged gears. Core structure and carbide distribution were found to be essentially identical for both types of gears. Carburized case depth, however, was consistently lower for forged gears. This difference could be a result of the modified grain flow pattern near the tooth surface and/or possible nickel alloying.

ECONOMIC ANALYSIS

Cost information compiled in the performance of this program is the basis for an analysis of the economics for the precision forging gear teeth and finishing the forging to the drawing requirements for production. Since the conclusions drawn from this analysis are extrapolations of available information, some assumptions and precepts were necessary in the evaluation. These consisted of 1) the use of standard hours instead of dollars, 2) a forging lot size of 500 pieces, 3) a machining lot size of 100 pieces, 4) materials and purchased services given in actual dollars, 5) analysis applied only to operations before heat treating, and 6) quality control and inspection, which are assumed to be comparable.

The resultant comparative cost summary is shown in Table IV on a cost per piece basis for conventional cut and forged gears. Results show that the forging of gear teeth results in an economic advantage over a conventionally manufactured gear.

CONCLUSIONS AND RECOMMENDATIONS

Economics:

1) Precision forging of gears is a feasible cost-reduction approach in the production of certain spiral bevel gear designs.

2) Dimensional and surface quality of forgings are reproducible, and the yield for the precision forged gear should be good.

3) The lead time to process quantities of gears through the operations up to heat treatment is reduced due to the elimination of the Gleason gear cutting operation, which tends to bottleneck the flow of parts.

4) An allowance must be made for some development for each new gear design.

Table IV. Comparative Cost Summary

Item	Conventional Process Standard Hours Piece	Conventional Process $ Cost/Piece	Precision Forging Process Standard Hours Piece	Precision Forging Process $ Cost/Piece
Material		29.88		10.40
Forging			0.235	
Set-up			0.030	
Machining labor	2.072		1.045	
Set-up	0.20		0.190	
Labor			0.014	
Mat. ODC				1.60
Total	2.272	29.88	1.514	12.00

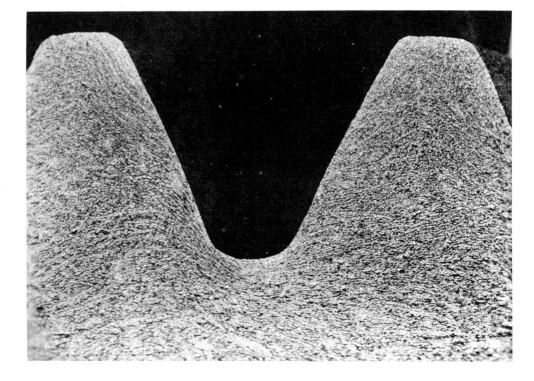

Fig. 6—Gear tooth profile showing forging flow lines (HCl etch).

Technical Factors:

1) Forged teeth have a higher fatigue limit as compared to cut teeth. Additional fatigue testing and folling contact testing are required.

2) Metrology requires some refinement for more economical production, but approaches developed on this program were proven sound in concept.

3) Surface quality of the forged teeth was adequately preserved by nickel plating the billet. Other methods of eliminating scaling should be developed to further improve the surface and reduce costs.

4) It is recommended that finish lapping of forged tooth gear sets be investigated. The value of the precision finish grinding the gear teeth for interchangeability is not realized if gears are replaced as sets.

Production Potential:

1) Processes developed and presented on this program are practical and readily applicable to production. However, adoption by industry, especially the aircraft industry, will be limited by considerations other than performance and economics. Qualifying the process to standard regulations, is required before production will result.

2) Forging forces required on the crank press are relatively low to achieve good definition. Existing presses could form gears up to 30 in. diam if the die sinking electrode and EDM equipment were available.

ACKNOWLEDGMENT

This gear manufacturing development program, performed by TRW Equipment, Materials Technology, was sponsored by U. S. Army Aviation Systems Command, St. Louis, Missouri.

SECTION III
Fine Blanking

Forming high precision parts

Fine blanking has proven to be the number one cost reduction agent when it is applied to the right part. Savings can run up to 500 pct.

If you can manufacture a part which meets your requirements by conventional means, you don't need fine blanking.

But, if secondary operations are necessary to give the proper surface finish; if tolerances are such that conventional blanking cannot produce the part in one operation; if flatness and consistent uniformity are an issue; if heavy materials are involved; or if hole diameters and wall thickness are less than material thickness—fine blanking may be the answer.

"Fine blanking has been proven to be the number one cost reduction agent when applied to the right part," states E. H. Scheitlin, president of Hyflow Blanking Press Corp Tarrytown, N.Y. Hyflow is the U.S.A.—Canada representative of Hydrel AG, Romanshorn, Switzerland.

"Up to 500 pct savings have been achieved over conventional methods," adds Mr. Scheitlin.

Fine blanking, of course, is a process developed in Switzerland for the forming of high precision components without any die break whatsoever on the sheared surfaces in one operation, employing one tool.

Conventional tools and presses used for conventional stampings produce a stamping with a granular fracture zone which is created as soon as the elastic limit of the metal to be blanked is exceeded and tensile fracture results. This is called die-break.

Fine blanking completely eliminates die-break and therefore also eliminates secondary operations such as shaving, broaching, grinding, milling and reaming. It therefore becomes a real cost-saving process.

And if it is considered in the design state of the part, it can more than justify the cost of the press which will by many times the cost of a conventional press.

Hydrel is the second largest importer of fine blanking presses to the U.S. (IA, Aug. 9, 1973, pp. 39-41). In fact, this company pioneered the use of fine-blanking presses in the U.S.

The major application area at that time was office machinery—adding machines, calculators, cash registers and so forth. As electronics have replaced many of the parts necessary for mechanically operated machines, this market has decreased.

But, in its wake, came the automotive market which is now the largest user of the process.

And as more people become aware of the technology and what it can offer in terms of tolerances, clean sheared edges on all contours, superior flatness, repeatability, reduced machining, the process is finding its way into other areas.

An interesting new application of the process is the fine blanking of ½-in. thick solid silver for The Franklin Mint.

The Franklin Mint, the world's largest private mint which is located in Franklin Center, Pa., recently initiated a program of using 1000 grain and 5000 grain sterling silver ingots for several new limited edition medallic art programs.

Fine blanking is about the only way to produce such parts economically. Before the company purchased their own fine blanking press from Hydrel the operation was farmed out.

The cost of handling and transporting sterling silver in quantity would have been enough of an argument for buying the more than $100,000 worth of equipment required to feed the press, fine blank the ingots and shear the scrap.

In addition, however, noticeable improvement in quality has been achieved. Before, four secondary operations were required. Now, this number has been reduced to one.

G. D. Stroud, manager, Project Engineering, The Franklin Mint estimates a four-year payback on the press, with most of the savings due to transportation, including insurance, armored trucks and so forth.

For this application, the Hydrel Model HFP-200 with three independent and individually adjustable hydraulic actions is being used.

The 1000 grain ingots are blanked two-up and have a thickness of 0.257 in. The material is sterling silver with an approximate R_b reading of 68-70. Production from strips is about 70 pieces per minute or 35 spm.

The 5000 grain ingots are blanked one-up and have a thickness of 0.456 in. Again, sterling silver is used with an approximate R_b reading of 60-62. Production from strips is about 25 pieces per minute.

Mr. Scheitlin describes the modifications to the press needed to handle silver.

"Since for the subsequent coining, surfaces have to be kept absolutely scratch free, our push-pull feeding units Model PV-280 had to be altered by adding special nylon pads on guides and grippers.

"For the ejection of the mini-ingot (1000 grain), a specially designed pneumatic cylinder ejection system, again with nylon padding, had to be developed with nylon pads leading from the die plate to a special chute in which parts ride on an air cushion to a conveyor.

"The chute features a tight hole

equires fine blanking

pattern to create an air cushion in order for the part never to touch the surface.

"A special tool design was developed in order to reduce the roll-over or die-roll on edges and corners to an absolute minimum thus maintaining full surface even on corners without washout. This facilitates subsequent coining and provides full registration all around.

"A special lubricant is being used for the blanking of these silver ingots. It is applied automatically with a lubricating device located between infeed and die," Mr. Scheitlin concludes.

The 200-ton Hydrel fine blanking press used actually delivers an overall tonnage of 225 tons (all three actions included) and the 5000 grain ingot calls for an overall tonnage of almost 200 tons.

In any discussion of fine blanking, one must keep in mind that the forces inherent in fine blanking a component are anywhere from 1½ to 2½ times higher than in conventional blanking. Fine-blanking tools therefore must be designed and constructed far more solidly.

In a typical fine blanking tool, the material is held firmly around the contour of the component to be blanked (in the scrap) by means of an impingement or stinger pad. Its main purpose is to hold the material from "running away" from the blanking punch.

Almost simultaneously, the blanking punch will blank or extrude, the part into the die cavity against the third action exerting counter pressure from the bottom up.

For this purpose all fine blanking presses have to have three, independently adjustable forces: Stinger pressure; actual blanking force; counter and ejector pressure.

Alice M. Greene

What materials can be fine-blanked?

Basically, any material with adequate ductility is suitable. All materials with cold-forming qualities lend themselves to fine-blanking. Here are listed some of the most common materials and in what temper they can be fine-blanked:

Steel:

1008—1024   soft, 1/4—1/2 and full hard
1030—1095   only in fully annealed, spherodized condition
4130/4140/8620/8630/8640   same as above

Stainless Steel:

302/4/5 and 405   All rated A (not heat treatable)—very good in fully annealed condition
301/3/9/10/16/47   All rated B—good to bad, only fully annealed
403/10/42/46   same as above

Bainite:

Sandvik # 11 or the US equivalent in a Rockwell C range of 32-36 can be fine-blanked up to 0.080" thickness with reasonable tool life, the latter, however, far below the figures given heretofore.

Aluminum:

All H hardness—cold rolled alloys such as 1100—3003 and 5052 series with excellent results.
All T hardness—heat treatable alloys such as 2014—2024 6061 and 7075 only in T-0 temper with the exception of 2024 which is not fine-blankable.
In thinner gages, 6061 and 7075 can be fine-blanked in T-3 or T-4 condition.

Brass:

Only Brass without lead content is suitable.
CA-260   70/30
CA-268   65/35   very good in soft, 1/4 or 1/2
CA-274   63/37   hard condition—even full hard in thin gages.

Be-Copper:

Only in soft condition or roll-hardened, not in heat treated condition.

Copper; Soft Bronzes; Monel; Nickel-Silver; Fine-Silver; Sterling Silver:

Excellent Results

 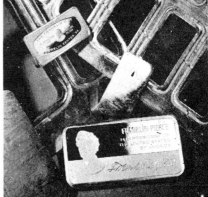

Fine blanking at The Franklin Mint produces ingots just about ready for coining.

Fine-Edge Blanking and Piercing

FINE-EDGE BLANKING (also known as fine blanking, smooth-edge blanking, or fine-flow blanking) produces precise blanks in a single operation without the fractured edges characteristically produced in conventional blanking and piercing. In fine-edge blanking, a V-shape impingement ring (Fig. 1) is forced into the stock to lock it tightly against the die and to force the work metal to flow toward the punch, so that the part can be extruded out of the strip without fracture or die break. Die clearance is extremely small, and punch speed much slower than in conventional blanking.

Fine-edge piercing can be done either separately or at the same time as fine-edge blanking. In piercing small holes, an impingement ring may not be needed.

No further finishing or machining operations are necessary to obtain blank or hole edges comparable to machined edges, or to those that are conventionally blanked or pierced and then shaved. A quick touchup on an abrasive belt or a short treatment in a vibratory finisher may be used to remove the small burr on the blank.

Specially designed single-operation or compound blanking and piercing dies are generally used for the process.

Process Capabilities

Holes can be pierced in low-carbon steel with a diameter as small as 50% of stock thickness. In high-carbon steel, the smallest hole diameter is about 75% of stock thickness. Holes can be spaced as close to each other, or to the edge of the blank, as 50 to 70% of stock thickness. Total tolerances obtainable are: 0.0005 in. on hole diameter and for accuracy of blank outline; 0.001 in. on hole location with respect to a datum surface, and 0.001 in. on flatness.

No die break shows on the sheared surface of the hole. Blank edges may be rough for a few thousandths of an inch

of thickness on the burr side of the part when the width of the part is about twice the stock thickness or less. Finish on the sheared edge is governed by the condition of the die edge and the land within the die. Parts fine-edge blanked from stainless steel will have a surface finish of 32 micro-in. or better. Smooth edges also are produced on spheroidize-annealed steel parts.

Burr formation increases rapidly during a run, necessitating frequent grinding of the cutting elements.

Chamfers can be coined around holes and on edges. Forming near the cut edge, or forming offset parts with a bend angle up to 30°, is possible under restricted conditions.

Metals up to 0.125 in. thick having a tensile strength of 85,000 to 115,000 psi are easily blanked. Parts up to ½ in. thick can be blanked if press capacity is available. Material thicker than 0.125 in., especially steel having a carbon content of 0.25% or more, requires an impingement ring on the die so that the corners on the part will not break down. The edges of parts made of 1018 steel work harden as much as 7 to 12 points Rockwell C during blanking.

In tests on 0.60% carbon spring steel with a hardness of Rockwell C 37 to 40, the surface finish on the sheared edges was 32 micro-in. or better, but punch life was only 6000 pieces.

The cutting speed for fine-edge blanking is 0.3 to 0.6 in. per sec.

Work Metals

Low-carbon and medium-carbon steels (1008 to 1035), annealed or half-hard, give good blanked edges and normal tool wear. High-carbon steels in the spheroidize-annealed condition can be blanked easily; blanking of steel with 0.35% carbon or higher is recommended only when it is spheroidize-annealed. Steels quenched and tempered to about Rockwell C 30 are well suited to fine-

edge blanking, because they do not require subsequent heat treatment, which could result in deformation.

High-carbon steels and alloy steels such as 4130, 4140, 8620 and 8630 cause considerably higher tool wear than low-carbon plain carbon steels, but surface finish is smoother. Leaded steels are not suitable for fine-edge blanking, because of their low deformability.

Parts made of stainless steels of types 301, 302, 303, 304, 316, 416 and 430, in the form of bright rolled fully annealed strip, have good blanked edges, but cause higher tool wear than steels of low and medium carbon content.

Good results have been experienced with aluminum alloys 1100 (all tempers), 5052-O to 5052-H38, 6061-O to 6061-T6, and others having similar yield strength and elongation. Blanked edges on parts made of aluminum alloy 2024 generally are rougher than edges on other aluminum alloys. Brasses containing more than 64% copper are especially suitable. Nickel alloys, nickel silver, beryllium copper, and gold and silver also are easily fine-edge blanked.

Blank Design

Limitations on blank size depend on stock thickness, tensile strength and hardness of the work metal, and available press capacity. For example, perimeters of approximately 25 in. can be blanked in 0.125-in.-thick low-carbon steel (1008 or 1010). It is possible to blank smaller parts from low-carbon or medium-carbon steel about ½ in. thick.

Sharp corner and fillet radii should be avoided when possible. A radius of 10 to 20% of stock thickness is preferred, particularly on parts over 0.125 in. thick or those made of alloy steel. External angles should be at least 90°. The radius should be increased on sharper corners or on hard materials.

Parts with tiny holes or narrow slots to be pierced, or with narrow teeth or

Source: Metals Handbook, Eighth Edition, Vol. 4, 1969, 56-59, © 1969 American Society for Metals

projections to be blanked, may be unsuited to fine-edge blanking. The ratio of hole diameter, slot width, or projection width to metal thickness should be at least 0.7 for reasonably efficient blanking, although a ratio as small as 0.5 has been successful with some parts. The spacing, between holes or between a hole and the edge of the blank should not be less than 0.5 to 0.7 times metal thickness, in order to maintain the quality of hole-wall and blank-edge surfaces, and to avoid distortion.

These limitations have been exceeded. For instance, a ⅝-in.-diam hole was pierced in each end of a 1018 steel link 1 in. wide and ⁵⁄₁₆ in. thick. Since the part had a ½-in. radius on each end, the wall thickness was ³⁄₁₆ in. The part was offset 0.100 in. in the same die. In a part made of 0.156-in.-thick aluminum alloy 5052-H34, 0.125-in.-diam holes were pierced leaving a wall thickness of 0.040 in. A 0.062-in.-diam hole was pierced in the same part.

The sheared faces of holes pierced during fine-edge blanking are usually vertical, smooth and free from die break, provided the maximum hole dimensions are not more than a few times the stock thickness. As in conventional piercing, there is a slight radius around the punch side of the hole, but there are no torn edges on the die side of the blank. A rough sheared surface on the blank may be caused by too great a punch-to-die clearance, or improper location and height of the impingement ring for the material being blanked. On parts blanked to a small width-to-thickness ratio, a small rough surface may be noticeable, but may not be detrimental (see Example 45).

Presses

A triple-action hydraulic press or a combination hydraulic and mechanical press is used for fine-edge blanking. The action is similar to that of a double-action press working against a die cushion. An outer slide holds the stock firmly against the die ring and forces a V-shape impingement ring into the metal surrounding the outline of the part. The stock is stripped from the punch during the upstroke of the inner and outer slides. An inner slide carries the blanking punch. A lower slide furnishes the counteraction to hold the blank flat and securely against the punch. This slide also ejects the blank.

The stripping and ejection actions are delayed until after the die has opened at least to twice the stock thickness, to prevent the blank from being forced into the strip, or slugs from being forced into the blank. Because loads are high, and clearance between punch and die is extremely small, the clearance between the gibs and press slides must be so close that they are separated by only an oil film.

Force requirements for fine-edge blanking presses are influenced not only by the work metal and the part dimensions, but also by the special design of the dies and pressure pads used for fine-edge blanking. Depending on part size and shape, a 100-ton press can blank stock up to 0.315 in. thick; a 250-ton press, up to 0.470 in. thick; and a 400-ton press, up to 0.500 in. thick.

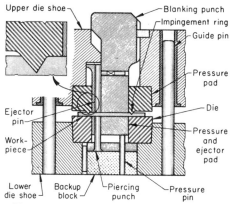

Fig. 1. Typical tooling setup for fine-edge blanking a simple shape

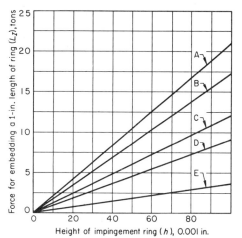

A — Stainless steel
B — Prehardened alloy steel
C — Mild steel; half-hard brass; hard copper; series 6000 aluminum alloys, hard
D — Soft copper; series 6000 aluminum alloys, half hard
E — Commercially pure aluminum, hard

Fig. 2. Force required for embedding impingement rings of various heights into several different work metals

The total load on the press in fine-edge blanking is the sum of three components: the cutting force (L_C); the lower blankholder force (L_{LB}), or counterforce; and the clamping force on the impingement ring (L_{IR}) on the pressure pad. The first two components comprise the total force on the inner slide, and the third component is the force on the outer slide.

The cutting force, in pounds, is calculated from the equation:

$$L_C = 0.8\ Sl_Bt$$

where 0.8 is an experimentally determined constant; S is the tensile strength of the work metal, psi; l_B is the total length of cut (sum of perimeters of blank and holes pierced in blank), in.; and t is the thickness of the work metal, in.

The counterforce, or lower blankholder force, in pounds, is calculated from the equation:

$$L_{LB} = P_C A$$

where P_C is the counterpressure on the lower side of the blank, psi; and A is the area of the blank, sq in. The counterpressure usually is about 10% of the tensile strength of the work metal.

The clamping force on the impingement ring on the pressure pad, in pounds, can be obtained from:

$$L_{IR} = L_I l_{IR}$$

where L_I is the force to embed a 1-in. length of the impingement ring into the work metal, lb; and l_{IR} is the total length of the impingement ring, in. The force L_I for different work metals, as determined by experience in fine-edge blanking, is given in Fig. 2.

When impingement rings are used on both the pressure pad and the die, the calculation of force is still based only on the pressure-pad impingement ring. The reduced height of impingement rings when used in pairs allows the use of a lower clamping force, and thereby reduces the over-all load on the press. This is because the lower impingement ring is impressed into the workpiece by the reaction force.

If coining, embossing or other forming is done during the blanking, the additional force required for those operations must be added to the force requirements as calculated above.

Tools

The design of tools for fine-edge blanking is based on the shape of the part, the method of making the die, the required load, and the extremely small punch-to-die clearance. The considerable loading and intended accuracy require that the press tools be sturdy and well supported to prevent deflection. The small clearance presupposes precise alignment of the punch and die.

Design. A basic tool comprises three functional components: the die, the punch, and back-pressure components. To produce good-quality blanks, the punch-to-die clearance must be uniform along the entire profile and must be suitable for the thickness and strength of the work metal. The clearance varies between 0.0002 and 0.0004 in.

The components of a typical tooling setup for fine-edge blanking of a part of simple shape are shown in Fig. 1. The profile part of the blanking punch is guided by the pressure pad. A round punch is prevented from rotating by a key fastened to the upper die shoe. The hardened pressure pad is centered by a slightly conical seat in the upper die shoe; this pad contains the V-shape impingement ring.

Some diemakers put a small radius on the cutting edge of the die. This causes a slight bell-mouth condition, which produces a burnishing action as the blank is pushed into the die, improving the edge finish.

If holes are to be pierced in the part, the blanking punch contains the piercing die. The slug is ejected by ejector pins, or through holes in the punch.

The die is centered in the lower die shoe by a slightly conical seat, as is the upper pressure pad. Both the die and the upper pressure pad are preloaded to minimize movement caused by compression. The pressure and ejector pad is guided by the die profile, and is supported by pressure pins and the lower slide. The backup block for the piercing punch also guides the pressure pins.

The die components are mounted in a precision die set with precision guide pins and bushings. Some designers prefer pressing the guide pins into the upper shoe.

Materials and Life. Because of the high loads, close tolerances, and small clearances involved in fine-edge blanking, the die elements are made of high-carbon high-chromium tool steels, such as D2 or D3, or of A2 tool steel, heat treated to about Rockwell C 62.

Punch and die life vary with tool material and hardness, punch-to-die clearance, type of work metal, and workpiece dimensional and surface-finish tolerances.

For most work metals under the usual operating conditions, punch life for fine-edge blanking of ⅛-in.-thick stock is 10,000 to 15,000 blanks between regrinds — assuming that the blanks are of simple shape and that punch wear is such that only 0.002 to 0.005 in. of metal need be removed to restore the punch to its original condition.

The effect of work material on punch and die life can be illustrated by the following data.

In one application, after blanking 33,000 pieces made of 1010 cold rolled steel (No. 2 temper), 0.009 in. was ground from the punch and 0.006 in. from the die. Production rate was 35 pieces per minute. When blanking 8617 and 8620 steel, it was necessary to grind 0.009 in. from the punch after 12,000 pieces and 0.007 in. from the die after 23,000 pieces. The production rate was 27 to 30 pieces per minute.

In another instance, 15,000 to 30,000 pieces per punch grind were produced when blanking annealed 1040 and 1050 steel; and 25,000 to 50,000 pieces for 1010 steel (No. 3 and 4 temper). Punch life for blanking fine-tooth gears made of annealed high-carbon steel was 10,000 to 15,000 pieces, and for steel with a hardness of Rockwell C 32 to 34 was 5,000 to 15,000 pieces. The reason for grinding the punch was to remove the small radius on the edge of the punch, which must be kept sharp and flat to obtain a good edge on the part.

Total die life may be 200,000 to 300,000 blanks per tool. The die is usually sharpened once for each two or three punch sharpenings. It may be necessary to remove from the die an amount of metal up to half the work-metal thickness to restore the die to its original condition.

In some production applications of blanking simple shapes from 0.100-in.-thick 1010 steel, life between regrinds was about 40,000 blanks for punches, and about 80,000 blanks for dies, when punch and die wear of 0.005 to 0.007 in. was allowable and the surface finish of the cut edge was 63 micro-in. or better.

Pressure-Pad Impingement Rings

The most important consideration in the design of a pressure pad for fine-edge blanking is the special construction required to lock the workpiece tightly against the die and force the metal to flow against the punch. A V-shape impingement ring is provided on the pressure pad surrounding the outline of the part (see Fig. 1). The lip of the pressure pad between the ring and the shear line has a difference in elevation of 0.002 to 0.005 in. from the outer

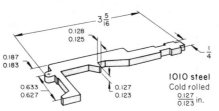

For any ring height (*h*), distance (*s*) should be in the shaded area of the graph. Note the 0.005/0.002-in. difference in elevation between the blankholding surfaces inside and outside the impingement ring (inset in graph).

Fig. 3. Relation of height of impingement ring to distance from edge of die opening

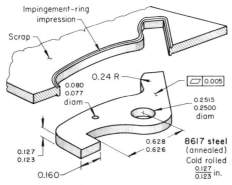

Fig. 4. Lock lever that was fine-edge blanked to close tolerances (Example 45)

When this part was produced by conventional blanking, the small hole had to be drilled, the large hole pierced undersize and reamed, and the periphery shaved.

Fig. 5. Clutch dog that was fine-edge blanked and pierced with 100% land on edges and in holes (Example 46)

surface of the pressure pad (see inset in Fig. 3). The ring (which penetrates to its full depth into the scrap metal outside the shear line) and the lip of the pressure pad hinder metal flow at the shear line during blanking.

The outline of an impingement ring is a closed shape conforming to that of the blank to be produced. Figure 3 shows the minimum and maximum distance of the ring from the edge of the die opening, for rings of various heights. An impingement ring with a 60° angle, instead of the 45° angle shown in Fig. 3, has been used, but more force was required to imbed it into the work metal.

Height of impingement ring depends on the thickness and ductility of the work metal. The height (penetration) of the V-shape is 20% of stock thickness for materials of low ductility. The

more ductile materials require a penetration of 32 to 35% of stock thickness.

If rings are used on both sides of the stock, the height of each ring should be half the total penetration required for the metal. Thus, if the penetration required were 0.042 in., the height of each ring would be 0.021 in. The distance to the edge of the die opening (*s*, in Fig. 3) would be reduced, and so would the length of the impingement ring.

The selection of the ring height and of the exact location within the range defined in Fig. 3 is based on experience. A shallow ring near the shear line has about the same effect as a deeper one farther from the shear line. If the ring is too near the die opening, a large portion of the metal in the zone will flow into the shear or edge radius and impair the efficiency of the ring. When a larger ring is located a greater distance from the die opening, a larger amount of stock is used, and more force is required to impress the ring into the work metal. Improper location and size of the impingement ring can cause rough sheared edges on the blank.

Effect of Stock Thickness. Stock up to 0.156 in. thick usually requires a ring on the pressure pad only. Stock up to 0.188 in. thick may need a partial ring on the die in addition to a full ring on the pressure pad. Full rings on the pressure pad and the die may be necessary for stock over 0.188 in. thick.

Although an impingement ring on the die reduces the edge radius on the blank more than does a similar ring on the pressure pad, its use is avoided when possible, because it makes re-sharpening of the die difficult.

The need for a full or partial ring on the die, to supplement the ring on the pressure pad, can be reduced by properly orienting the blank design on the strip. More precise and intricate cutting can be done on the side of the blank adjacent to the incoming strip, where ample stock is available to restrain metal flow, than along surfaces near the edges of the strip or along the narrow portion of the web where blanks have already been removed. When a blank cannot be oriented on the strip so that an ample width of stock is adjacent to all critical sections of the shear line, it is usually more economical to use partial rings or straight knife-edge projections on the die than to provide large edge and web widths.

Effect of Part Shape. The impingement ring ordinarily follows the contour of the part at a distance depending on ring height (see Fig. 3), but it cannot follow narrow slots in the part.

Impingement rings are not necessary around holes pierced in blanked parts, particularly holes with dimensions that are only a few times the metal thickness. However, the blank must be securely clamped between the punch and the pressure pad.

In Example 48 a circular ring was used, because it was not feasible to follow the contour of a fine-tooth gear. A gear with larger teeth can be made with a ring having a scalloped outline. An impingement ring can extend into wide notches, but not into narrow ones. In Examples 45, 46 and 47, rings were used that closely paralleled the outline of the workpieces.

Lubrication

The work metal must have a film of oil on both sides to lubricate the punch and die during fine-edge blanking. The lack of a lubricant on either side can reduce punch or die life between sharpenings as much as 50%. Oils used for conventional blanking usually are satisfactory. In severe applications, a wax lubricant may be used. In Examples 46, 47 and 48, a sulfur-free oil was used to lubricate the strip. An extreme-pressure chlorinated oil was used for the part in Example 45.

Examples of Application

The lock lever in the following example was made of low-carbon steel and had a low width-to-thickness ratio. Although a conforming impingement ring was used, a rough surface appeared on the cut surface adjacent to the upper (punch) side of the blank.

Example 45. Blanking of a Long Slender Lock Lever to Close Tolerances (Fig. 4)

A lever for a pushbutton lock (Fig. 4) was fine-edge blanked to a minimum total tolerance of 0.003 in. The maximum total dimensional tolerance was 0.010 in., except on fractional dimensions, which had a tolerance of 1/32 in.

The lever was blanked from cold rolled, commercial quality 1010 steel, 0.127/0.123 in. thick and 2¾ in. wide. The coil stock had a No. 4 temper (soft), No. 3 edge (slit), and a No. 2 finish (bright). The blank design was positioned at an angle on the strip, with a progression of 0.875 in. An impingement ring 0.040 in. high was used on the pressure pad.

The edge of the blank was smooth and perpendicular to the top and bottom surfaces. On the die side of the blank, edge radius (rollover) was noticeable, particularly at the outside corners. (This effect, typical of low-carbon steel in fine-edge blanking, is less pronounced on high-carbon and alloy steels.)

The die was made of D2 tool steel, hardened to Rockwell C 57 to 60, and ground to a fine finish. During blanking, the die was lubricated with an EP chlorinated oil. Die life was 30,000 pieces per grind. The die was mounted in a special 40-ton triple-action hydraulic press operating at 40 to 50 strokes per minute.

The two examples that follow describe applications in which fine-edge blanking replaced conventional blanking, thereby eliminating the need for subsequent drilling, reaming and shaving. In the first example, the smaller of the two holes pierced during fine-edge blanking had a diameter only 62% of stock thickness. In the second example, the distance between the edge of a large hole and the edge of the part was only 50% of stock thickness.

Examples 46 and 47. Fine-Edge Blanking and Piercing to Final Size and Finish, Which Could Not Be Done by Conventional Blanking

Example 46 (Fig. 5). The clutch dog shown in Fig. 5 was fine-edge blanked from annealed, cold rolled, commercial quality 8617 steel, 0.127/0.123 in. thick and 1½ in. wide. The two holes were pierced at the same time the outline was blanked. The periphery of the part and the holes had a 100% land. There was no edge radius on the die side, and the burr on the punch side was small and easy to remove.

The sequence of operations used to make the part by conventional methods was: blank outline and pierce a 3/16-in.-diam hole; drill the 0.080/0.077-in.-diam hole; ream the 3/16-in. hole to 0.2515/0.2500 in.; and shave periphery of part to print requirements.

In fine-edge blanking, impingement rings 0.035 to 0.040 in. high were used on both sides of the blank. On the punch side, the metal

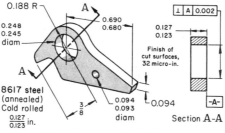

When this part was produced by conventional blanking, holes had to be drilled and reamed, two surfaces shaved, and the part deburred.

Fig. 6. Positive clutch detent that was produced by fine-edge blanking and piercing (Example 47)

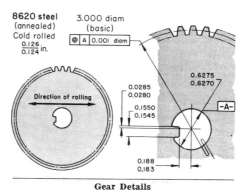

Gear Details

Quality class, precision	I
Number of teeth	48
Diametral pitch	16
Pressure angle	20°
Pitch diameter	3.000 in.
Tooth form	American standard full-depth involute
Circular thickness on pitch circle, max	0.0967 in.
Testing radius (tight mesh with standard master)	1.4995/1.4975 in.
Composite radial variation, max	0.001 in.
Tooth-to-tooth radial variation, max	0.0004 in.
Outside diameter	3.116/3.111 in.
Backlash class	B

Fig. 7. A 48-tooth spur gear that was made in one press stroke by fine-edge blanking (Example 48)

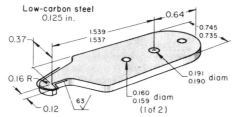

Fig. 8. Latch part that was produced more economically by fine-edge blanking and piercing than by conventional blanking and machining (Example 49)

between the punch and the ring was deformed, indicating draw-in. The same area on the die side of the stock had a sharp outline, but it was about 0.010 in. below the surface.

The blanking dies were made of D2 tool steel, hardened to Rockwell C 60 to 61, and ground to a finish of 16 micro-in. The life between grinds was 12,000 parts. The die was mounted in a special 110-ton triple-action hydraulic press operating at 20 to 30 strokes per minute. Die setup time was 30 min.

The clutch dog was made in lots of 10,000 pieces; yearly production was 50,000 to 70,000.

Example 47 (Fig. 6). The positive clutch detent shown in Fig. 6 was fine-edge blanked from annealed, cold rolled, commercial quality 8617 steel, 0.127/0.123 in. thick. The two holes were pierced and the blank was severed from the stock in one press stroke. There was no distortion where the edge distance was less than work-metal thickness.

The conventional method of making this part was: blank, drill and ream the two holes, shave two surfaces, and deburr.

The part had a small edge-radius along the die-side surface. This radius was more pronounced at the outside corners. The dimensional tolerances were 0.010 in. total. Surface finish in two areas along the cut edge was 32 micro-in.

The die was made of D2 tool steel, hardened to Rockwell C 60 to 61, and ground to a finish of 16 micro-in. Die life was 12,000 pieces per grind. The impingement ring was 0.035 to 0.040 in. high and in the pressure pad only. The die was set up in a special 110-ton hydraulic press operating at 20 to 30 strokes per minute.

A sulfur-free oil was used as lubricant for both methods. Production rate for fine-edge blanking was 26 pieces per minute; lot size was 10,000 pieces, for a total production of 50,000 pieces per year.

The blanking of a 16-pitch gear is described in the following example. The impingement ring was circular instead of scalloped because of the small teeth.

Example 48. Blanking a Spur Gear With a Smooth Edge (Fig. 7)

The teeth, center hole, and key of the 16-pitch, 48-tooth spur gear shown in Fig. 7 were fine-edge blanked in one press stroke from annealed, cold rolled, commercial quality 8620 steel strip, 0.126/0.124 in. thick and 3¾ in. wide. The edges of the hole and teeth were smooth, having no fractured area. On the die side of the gear, the corners at the tip of each tooth had a noticeable edge radius, which decreased along the tooth flank until it was zero in the root. On the punch side, the corners were sharp but had a small burr, which was easily removed by vibratory finishing or belt lapping.

The blanking die was made of D2 tool steel, hardened to Rockwell C 60 to 61, and ground to a finish of 16 micro-in. Impingement rings, 0.035 to 0.040 in. high and 3.218 in. in diameter, were used on both the pressure pad and the die. The feed length (strip progression per stroke) was 3.280 in. The die was set up in a special 110-ton hydraulic press with three slides. The press operated at 10 to 15 strokes per minute. Die life was 12,000 pieces per grind. Lubricant was a sulfur-free oil, applied to both sides of the stock.

The gears were made in 15,000-piece lots at a yearly production of 105,000 pieces. Die setup time was 30 min.

The conventional method for making the gear would have required three or four operations: blank and pierce in a compound die, shave hole and outside diameter, and hob the teeth. Two shave operations might have been required, because the tolerance on the hole diameter and key width was 0.0005 in.

In the following example, it was more economical to form the part by fine-edge blanking than by conventional blanking and machining.

Example 49. Fine-Edge Blanking vs Blanking and Machining (Fig. 8)

The latch part shown in Fig. 8 was made of 0.125-in.-thick low-carbon steel having a No. 1 temper and a No. 2 finish.

Originally, the part was made by blanking and shaving, and then was deburred by vibratory finishing. The three holes were drilled and reamed, and the part was disk ground to the required flatness.

The method was changed to fine-edge blanking. The complete part was made in one press stroke, except that a small burr was removed by fine-belt sanding.

Fine-edge blanking resulted in savings of $5000 for annual production of 35,000 pieces.

Safety

Blanking and piercing involve potential hazards. The articles on Presses and Auxiliary Equipment ((page 1) and on Blanking of Low-Carbon Steel (page 31) contain information and literature references on safe operation.

Fine Blanking ½" Coil Stock

Coil stock ½" thick is not common, but fine blanking of such steel is even more unusual. Yet it is a routine application at Burkland, where 9¼ x 7" plates are being blanked and pierced simultaneously at the rate of 750 per hour

CHARLES WICK
Managing Editor

FINE BLANKING and piercing is a process developed in Switzerland for producing blanks and holes with smoother edges and closer tolerances than possible with conventional stamping practice. This is done in a single-station die with one stroke on a triple-action press. While this process is used more extensively in Europe, it is being applied increasingly in the U.S.

Among the firms specializing in fine blanking in this country is Burkland, Inc., Goodrich, MI. They have complemented the process with equipment for any necessary secondary operations, thus providing a single source responsibility for producing finished parts ready for assembly, as well as subassemblies. Capabilities of the firm were recently increased with the installation of the 880-ton (800 t) — 7828 kN, hydraulic fine blanking press seen in

Figure 1, built in Switzerland and sold by Schmid Corp. of America, also of Goodrich.

Process Advantages. Cost and time savings are important advantages of the fine blanking process for many applications. Smooth edges and close dimensional accuracies of the parts produced eliminate or minimize subsequent operations required for functional surfaces. With conventional blanking and piercing, edges are cleanly sheared through only about one-third of the material thickness, with the remainder roughly broken or torn out (die break). Depending on functional requirements, the surfaces may require milling, broaching, reaming, shaving, or grinding. Fine blanking produces completely sheared edges without any die break.

Accuracy attained in fine blanking depends on the thickness and tensile strength of the material being fine blanked, and the configuration of the part being produced. Tolerances of ±0.001" (0.03 mm) can be consistently held on thin parts made from low carbon steel, and ±0.003" (0.08 mm) on

thicker parts. The smooth sheared edges can be held perpendicular to the material surface within 0.001" per in. (0.001 mm per mm), and surface finishes of 32 μin. (0.81 μm) or less are possible. Flatness is maintained the same as that of the material fed into the die.

Another major advantage of fine blanking is that hole diameters, slot widths, and wall thicknesses can be made smaller than with conventional stampings — as small as 50% of material thickness in low carbon steels. Also, other operations such as bending, embossing, and coining can be combined in the same operation, and the press can be used for conventional stamping dies when not required for fine blanking.

Materials Handled. Many different materials are being fine blanked. These include ferrous materials such as low and medium carbon steels, and some alloy and stainless steels, as well as nonferrous materials such as brass, copper, and aluminum alloys. In fact, any material suitable for cold forming can be fine blanked.

The best nonalloyed steels for fine blanking are those with low carbon contents, with a maximum generally allowable of about 0.70% C. However, thin parts have been fine blanked from plain steels containing up to 0.95% C. Alloy steels should have a low yield strength and high elasticity — the more brittle the material, the poorer the results. For economical production and long die life, the tensile strength of the material should not exceed 85 ksi (588 MPa), except for thin materials (up to about 0.12" — 3 mm thick), where materials with tensile strengths up to 114 ksi (785 MPa) have been fine blanked successfully.

1. FINE BLANKING PRESS is of the triple-action type, with full hydraulic operation. It is rated at 880 ton capacity.

2. TYPICAL PARTS produced by fine blanking are a clutch plate, sprocket, rack, end cap, and seal plate.

Limitations. Fine blanking is certainly no panacea for all stampings. It is best for applications where manufacturing costs can be reduced by minimizing or eliminating secondary operations in producing smooth, accurate functional edges, or for improving product quality. Where the edges produced in conventional stamping are satisfactory for a specific application, fine blanking can seldom be justified economically.

A fine blanking press can cost about twice that of a conventional press with equal capacity, and tooling costs may be up to 70% more than a standard compound die. The process is also slower than conventional stamping. Despite these handicaps, substantial savings are being realized in many applications.

With respect to the design of the fine blanked part, size of the part and material thickness is limited only by the die area and tonnage available on the press. The process is generally limited to materials with a minimum thickness of about 0.03" (0.7 mm), and a maximum of about ⅝" (15.9 mm). Parts made from materials up to about ⁵/₃₂" (4 mm) thick are most common, but applications for precision blanking of thicker materials, which often make the process competitive with machining, are increasing.

Sharp corners cannot be produced with fine blanking, since this could cause tearing of the material and chipping of the tooling. The process does produce edge rollover, usually from 10 to 25% of the material thickness. Also, small burrs, generally having a maximum height of about 0.010" (0.25 mm), are formed, but these can be easily removed by tumbling or a light grinding operation.

Applications. Most early applications of fine blanking involved the production of instrument, watch, and office machine components such as levers, fingers, tooth segments, gear wheels, cams, and similar parts. Now, the process is also being increasingly applied to a wider variety of materials and thicker stock in many industries, including automotive, textile machinery, farm equipment, ordnance, machine tools, printing machines, and household appliances. Tooth forms (gears, racks, sprockets, ratchets, and splines) are easily produced, and are common applications.

Several of the many different parts being fine blanked by Burkland are shown in *Figure* 2. The automotive clutch plate seen at the upper left, about 8" (203 mm) in diameter, is fine blanked and pierced from ³/₁₆" (4.8 mm) thick, SAE 1070 steel. Even with two subsequent operations — straightening and deburring — this processing provides a 35% savings over the previous method, which required shaving of conventional stampings. The industrial sprocket (upper right), about 6" (152 mm) in diameter, is fine blanked from ⁵/₁₆" (7.9 mm) thick, SAE 1020 steel, with a cost reduction of 26% over the previously stamped and shaved parts. Fine blanking of the 14" (356 mm) long gear rack (shown in the center) from ¼" (6.4 mm) thick SAE 1020 steel provided a cost reduction of 58% over the previously hobbed parts.

A transmission end cap seen at the lower left, which is about 7" (178 mm) long x 5" (127 mm) wide, is fine blanked from ⁷/₃₂" (5.6 mm) thick, SAE 1020 steel. In addition to simultaneous blanking and piercing, this part is also semi-pierced near its center to provide the required clearances. The compres-

3. ADAPTER PLATE fine blanked and pierced from ½" thick coil stock on the 880-ton hydraulic press seen in Figure 1.

sor seal plate (lower right) for an automotive air conditioner, about 3" (76 mm) in diameter x ⁵/₃₂" (4 mm) thick, is simultaneously blanked, pierced, and coined, and the O-ring groove is formed in the same operation. Previously, this part was made from an iron casting ⁵/₁₆" (7.9 mm) thick, which required considerable machining.

An adapter plate, *Figure* 3, for a transmission power take-off unit is the part that Burkland is fine blanking and piercing from ½" (12.7 mm) thick coil stock on the new 880-ton press at the rate of 750 parts per hour. The plates are 9.27" (235.5 mm) long x 7.02" (178.3 mm) wide, have 10 holes 0.468" (11.89 mm) in diameter, a contoured center opening about 3⅝" (92 mm) long x 2¾" (70 mm) wide, and weigh 6.2 lb (2.8 kg). The center distance between two dowel holes is held to 0.002" (0.05 mm), and the size of all 10 holes is maintained within ±0.005" (0.13 mm). Web thick-

4. LARGE OPENINGS in press frame permit handling large dies, and provide good accessibility to the die area.

ness (distance from the edges of the holes to the outer edge of the plate) is only 0.276″ (7.01 mm) — an important advantage of fine blanking, which cannot be done with conventional piercing. Flatness and parallelism of the upper and lower plate surfaces is 0.015″ (0.38 mm) after fine blanking, and this is reduced to 0.010″ (0.25 mm) after subsequent double-disc grinding. Grinding reduces the plate thickness to 0.460″ (11.68 mm), removes the slight burrs formed, and provides a smoother surface for the gaskets.

Raw material for the adapter plates is SAE 1010, pickled and oiled, hot rolled steel having a maximum hardness of R_B65 (No. 4 temper). It is purchased in coils, each weighing 6500 lb (2948 kg), because it costs less than strip stock and is safer to handle. The coil stock is 0.484″ (12.29 mm) thick, ±0.014″ (0.36 mm), and 10¼″ (260 mm) wide, which allows about ½″ (12.7 mm) for scrap on each side of the blanks.

These adapter plates were originally made from castings. Then, for cost reduction, processing was changed to a laminated sandwich, made by brazing two ¼″ (6.35-mm) plates together. Now, by fine blanking and piercing from ½″ (12.7-mm) stock, an additional cost reduction of 30% has been attained.

Press Design. The Schmid 880-ton (7828-kN), hydraulic fine blanking press can produce parts at rates up to 2000 per hour, and blank steel stock up to ⅝″ (15.9 mm) thick. It is a triple-action press with each force independently preselected and controlled electronically. Force of the V-ring plate for clamping and stripping is adjustable up to 440 tons (3914 kN), and is automatically reduced during blanking to increase blanking pressure available. Force of the counterpressure plate is adjustable up to 176 tons (1566 kN). In

producing the ½″ thick adapter plates, force of the V-ring plate is preset at 147 tons (1308 kN) and reduced to 73.5 tons (654 kN) during blanking; the counterpressure force is adjusted to 102 tons (907 kN); and blanking and piercing force is set for 683 tons (6076 kN) — resulting in a total force requirement of 858.5 tons (7637 kN).

Press speed, with ram drive through a variable-speed hydraulic motor, is also adjustable and electronically controlled. For fine blanking the adapter plates, fast approach is preset for 283.5 ipm (120 mm/s). Prior to blanking, the speed is reduced to 56.7 ipm (24 mm/s) for sensing — a safety check to automatically stop the press if previously produced slugs or blanks are left on the die, or if the punch or ejector are broken or deformed. During blanking and piercing, speed is maintained constant at 30.7 ipm (13 mm/s), but shearing pressure is simultaneously monitored and the speed is automatically adjusted when increased or decreased cutting resistance is encountered. The fast return speed is 472.4 ipm (200 mm/s).

This fully hydraulic, fine blanking press is of rigid, four-column construction, with exceptional stiffness under heavy loading, and a high degree of accuracy for movement of the cutting elements. The welded steel press frame has large openings for accessibility, *Figure* 4, and the work space holds dies up to 43.3″ square x 19.7″ high (1100 mm square x 500 mm high) to accommodate parts up to 20″ (508 mm) square. Maximum ram travel is 11.8″ (300 mm), ram adjustment travel is 3.94″ (100 mm), and the working stroke is 1.18 to 7.87″ (30 to 200 mm). Stroke length can be controlled within a maximum deviation of 0.001″ (0.03 mm) in more than 1000 strokes. The press can be operated in inching, single stroke, or automatic modes, with selection by a key oper-

ated switch, and can also be used for other forming processes such as normal blanking, coining, embossing, and drawing to a depth of 1.57″ (40 mm). Quiet operation is another feature, with a sound level under 85 dB.

Press Accessories. The press is equipped with both a strip feeder and a decoiler/straightener unit for coil stock. A coil of ½″ thick stock is shown in *Figure* 5, ready to be placed in the hydraulically operated decoiler. Stock is advanced from the coil to feed units on the press between driven pinch and straightening (leveling) rolls. Both push and pull type, hydraulically actuated feeders are provided, one on each side of the press. Positive feed of the stock by double grippers for both in-feed and out-feed provides additional guidance and rigidity, and allows material to be used up to the end of the coil. The feed increment for the ½″ stock used in making the adapter plates is 7⅝″ (194 mm).

Coil stock is automatically coated with a synthetic oil base lubricant on both top and bottom surfaces with felt rolls built onto the in-feed unit. In addition to the automatic safety sensing previously described, the press is equipped with a photoelectric guard (light curtain), seen in *Figure* 4, that stops the press if the operator places any portion of his body into the die area while the press is in operation. The heavy adapter plates are currently being removed from the die manually, but a gripper type unloader (iron hand) is being installed to remove them automatically. Lighter parts are automatically blown from the dies by blasts of compressed air. A scrap chopper at the left-hand end of the press (*Figure* 1) cuts the remaining stock into convenient size pieces which fall into a gondola.

The Fine Blanking Process. Fine

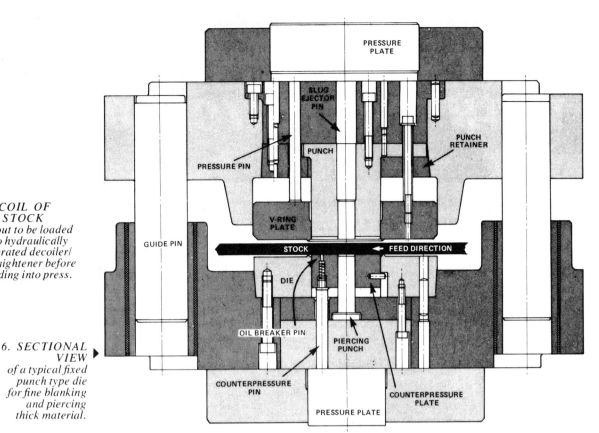

PRESSURE PLATE

SLUG EJECTOR PIN

PUNCH

PUNCH RETAINER

PRESSURE PIN

V-RING PLATE

GUIDE PIN

STOCK ◄ FEED DIRECTION

DIE

OIL BREAKER PIN

PIERCING PUNCH

COUNTERPRESSURE PIN

COUNTERPRESSURE PLATE

PRESSURE PLATE

5. COIL OF
½" STOCK
*about to be loaded
into hydraulically
operated decoiler/
straightener before
feeding into press.*

6. SECTIONAL
VIEW ►
*of a typical fixed
punch type die
for fine blanking
and piercing
thick material.*

blanking and piercing is more of a cold extruding operation than a cutting operation. It is comparable to conventional blanking with a compound die, but with several important differences. First, one or two raised vee-shaped rings or wedges (sometimes called stingers) — located adjacent to and outside the cutting line — restrict the flow of material away from the cutting edge, and press the material to be cut against the punch.

Prior to blanking, the V-ring (or rings) are pressed into the material to emboss an annular groove(s) into the material. Simultaneously, the stock is clamped between the flat undersides of the V-ring plate and punch, and the top surfaces of the die plate and a counterpressure plate. This combination creates high clamping forces both inside and outside the cutting line, and the wedge action of the V-ring(s) creates transverse compression to force material to flow toward the punch before and during blanking. For blanking and piercing thinner materials, only one annular wedge ring is provided — machined on the lower face of the V-ring clamping and stripping plate. For thicker materials, a second wedge ring is machined on the upper face of the die plate for increased clamping and compressive forces.

Another important difference of the fine blanking and piercing process is independent adjustment and accurate control of the three hydraulic pressures (clamping, counterpressure, and blanking) on the triple-action press. A further difference is that only a small clearance is provided between the punch and die — generally about ½ to 1% of the stock thickness, plus corresponding closer tolerances on the punch and die. Also, the punch only enters the die a slight amount — equal to the radius on the die, which is generally about 5% of the material thickness. All of these features combine to prevent fracture cracking of the material, resulting in parts with smooth edges and holes fully sheared throughout the thickness of the material.

Die Design. There are two systems of tooling used for fine blanking and piercing: the sliding (moving) punch system in which the punch is guided by the die plate, and the fixed punch system in which the punch is secured to and reciprocates with an upper die shoe. Schmid recommends and Burkland exclusively uses the fixed punch system, which is similar to the design of compound dies except that the punch is mounted on top.

A cross-sectional view of a typical fixed punch type die for fine blanking and piercing thick material is shown in *Figure* 6. Four long, large diameter, and precision pins and bonded bushings (bronze to steel) in a rigid cast iron die set prevent lateral movement, and assure parallel movement of both halves of the tool. This provides close tolerance parts and long die life. Also, this design permits standardization of tools for a wide range of material thicknesses, and the production of parts having large areas and many holes.

Punches for such dies are generally made from AISI Type M2 tool steel hardened to R_C57 to 61, or D2 with a hardness of R_C60 to 61, and the punch retainer from H13 hardened to R_C53 to 55. Pressure, counterpressure, slug ejector, and oil breaker pins, as well as the pressure plates, are made from Ol tool steel with a hardness of R_C58 to 60.

The V-ring plate encircling the punch for clamping and stripping actions is made from D2 or D6 tool steel with a hardness of R_C58 to 60. The die is produced from the same material, but is made slightly harder — R_C60 to 62. Size of the wedge rings machined on the bottom surface of the plate and top face of the die varies with the material being blanked and its thickness. Height of the vee (usually having an included angle of 80°) generally varies from about 0.008 to 0.118″ (0.2 to 3 mm), and it is located from 0.020 to 0.276″ (0.5 to 7 mm) from the cutting edge — with the height and distance increasing with the material thickness.

Both the counterpressure plate and the piercing punch are generally made from D2 or M2 steels, with the plate hardened to R_C57 to 59 and the punch to R_C60 to 62. Fine blanking and piercing tooling costs more than conventional dies, primarily because of the requirements for closer tolerances and the V-rings. However, they can usually produce 25,000 or more parts before punch regrinding is necessary, and up to 100,000 parts before resurfacing other components. ■

INVESTIGATION ON BURR-FREE SHEARING IN JAPAN

by

MADAO MURAKAWA, The Institute of Physical and Chemical Research
TEIZO MAEDA, Faculty of Engineering, University of Tokyo
TAKEO NAKAGAWA, Institue of Industrial Science, University of Tokyo

ABSTRACT. The authors describe briefly the development of burr-free shearing of sheet metal conducted mainly in Japan. Included among them are such as Reciprocating Blanking, Opposed Dies Blanking in the field of blanking, as well as Counter Cutting and Roll Slitting in the field of slitting. The mechanism of burr formation and the fundamental principle for the prevention of shearing burr are also discussed.

INTRODUCTION. The importance of the burr formation in shearing process and the deburring problem for the product obtained by such process has been unduly neglected since the size of such burr is relatively small in comparison with that resulting from other mechanical workings such as cutting and grinding. However, by considering the fact that the life of shearing tool is actually estimated by the burr height of sheared products, the harmfulness of such burr might be readily appreciated. In fact, such burrs, though relatively small, would turn out to be of a sharp knife-like edge shape with a certain hardness due to work hardening so that they may be harmful or disadvantageous in two ways. On one hand they themselves can be harmful; for example, they will hurt the surface of adjacent products in contact therewith, or they may possibly hurt the fingers of the operators and users, or they will cause the irregularity in the overall thickness of the products in stacked state, or they can give harmful accidents when falling off. On the other hand they may be harmful in the subsequent forming processes; for example, they will reduce the stretch deformability resulting in poor formability, i.e., easy occurance of cracking as for bending, stretch flanging and upsetting. Moreover, in forging they may remain as surface scratches of the forged product, or they can be a cause of failure of lubricating oil film.

Thus it has long been one of the concerns in the shearing industries how to prevent such burrs from occurring since otherwise useful "fine shearing" methods such as fine blanking, finish blanking, shaving and high speed blanking are not effective in suppressing such burrs. Nowadays researchers of burr-free shearing have been rapidly progressing, particularly in Japan, and some of the research have been put into practical use. This paper deals with the brief survey of these research works.

BURR FORMATION IN SHEARING OPERATION. The factors influencing upon the burr formation in shearing can be estimated by considering the mechanism of burr formation. It is obvious that burr will never be generated if cracks occur both from the exact edge points of a pair of shearing tools, punch and die. However, strictly speaking, the cracks inevitably occur at the ponit located slightly apart from the cutting edge because there prevails lower hydrostatic pressure, this being the cause of burr formation. Moreover, in the case of ductile materials, if a sheared or burnished surface is still formed even after the crack origination, then the burr height be further increased. Furthermore, usually crack formation will not take place from both cutting edges of punch and die at the same time, that is to say the crack will occur preferentially either from one or the other of the cutting edges, thus in the extreme case a crack occurring from one cutting edge will grow toward the flank of the other cutting edge in lower hydrostatic pressure

state. In this case the burr will be considerably larger. Generally speaking, it might be reasonably concluded that the burr formation, apart from its size, can not be avoided in the conventional shearing process.

Three conditions, i.e., the geometry of the cutting edge, the clamping condition of the material and the material property, can be pointed as the factors influencing upon the burr size. As the first condition the tool clearance and the dullness of cutting edges are important. Extemely large or small clearance will cause larger burrs, so that 5-15% clearance will usually be adopted in order to minimize the burr height. Moreover, a dull cutting edge will cause the burr height proportional to its dullness, in another word, if cutting edges become dull and have a certain rounded portion then the crack will break out at the terminal point of the portion so that the rounded portion will be transfered as the burr of the sheared blank. The condition of material clamping and restriction can also vary the hydrostatic pressure state thus shifting the crack break out point which results in the change of burr height to some extent. Generally speaking, such clamping condition as will cause lower hydrostatic pressure state may be useful for smaller burr height.

Lastly, in view of the material property hard and/or brittle material will be advantageous for smaller burr level since the crack break out point will come near the cutting edge, however in most cases the wear of cutting edge will also increase the height of burr. While in connection with the material condition, the working temperature and velocity also have the possibility of changing burr height level, higher ductility of the material will cause in most cases bigger burr level.

BURR-FREE SHEARING BY MEANS OF BOTH SIDE SHEAR DROOPS FORMATION.[1] T. Maeda, one of the authors, proposed in 1958 a burr-free blanking method patented and named as "Reciprocating Blanking." The principle of this process exists in the formation of shear droops on both edges of the sheared surface so that no burr will occur. This method has historical importance in the burr-free shearing field since it has become the "core" of ideas of other burr-free shearing methods developed later.

Figure 1 shows an embodiment of this Reciprocating Blanking which comprises two steps. In the first step the material is partially penetrated by the upper punch to a predetermined depth yet not in separated state. The second step is obtained from arranging the die and punch as for the material in opposite direction to the first step, and then by causing the half sheared material to an inverted shearing so as to completely separate the partially sheared portion.

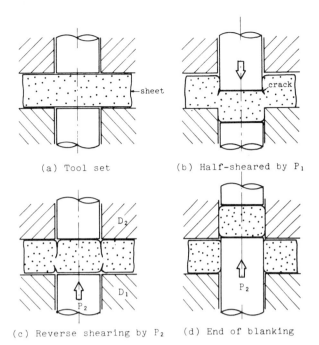

(a) Tool set   (b) Half-sheared by P₁

(c) Reverse shearing by P₂   (d) End of blanking

Fig.1 Process of reciprocating blanking(T. Maeda)[1)]

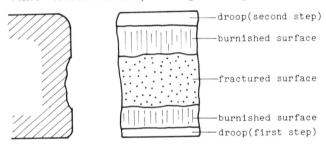

Fig.2 Blanked surface by reciprocating blanking

Figure 2 shows the sheared surface and cross section, obtained by this method where there are formed a first droop by the first step and the second droop by the second step, thus, no burrs being formed on both sheared edges.

The key point of this method is how to maintain or keep the first droop until the second step is finished. For this purpose, the first step must be stopped at least beyond the maximum shearing force in the punch load-stroke curve as shown in Figure 3. If the punch penetration advances too much in the first step, however, a crack will break out from the cutting edge of its punch in the first step thus resulting in the sheared surface with burr. In the most desirable blanking condition, the second step should be performed after there is formed a crack occurring from the die cutting edge in the first step yet not growing completely towards the punch cutting edge. For this purpose it is preferred that a certain degree of roundness is provided on the punch cutting edge of the first step.

Because of the complicated tool arrangement which inevitably leads to a special blanking press, this burr-free blanking method it seems, has very few embodiments which have been put to practical use. However, there seems to be no complete burr-free shearing method up to now other than this one, in which shear droops

are formed on both edges of sheared surface. Thus, this method seems to have something very important in its basic idea. In fact almost all the burr-free shearing methods that have been developed later on owe their principles to this method.

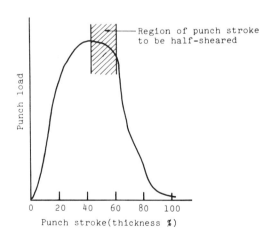

Fig.3 Region of punch stroke to be half-sheared in first step

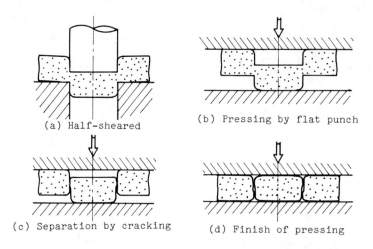

(a) Half-sheared   (b) Pressing by flat punch

(c) Separation by cracking   (d) Finish of pressing

Fig.4 Push back blanking(Makino)[2]

BURR-FREE BLANKING OF THE SHEET METAL

Push Back Blanking.[2] In Figure 4 is shown the so-called "Push back blanking" which was developed by I. Makino (Tokai Press Co., Ltd.). This method seems to be aimed at simplifying the complicated tool arrangement of Reciprocating Blanking in such a way that this can be adapted to a conventional press and tooling. This is characterized in that no blanking punch and die are used in the second step and that instead of them a pair of flat plates is used as the means of pushing back the half sheared blank to obtain the same burr-free product as one by Reciprocating Blanking.

It seems that the operative feature of this method has close relation to so called "push back" feeding system used for transferring sheared blanks. The only difference exists in that while the blanks are completely sheared off in the first step according to the ordinary push back system, the blanks in this case are half sheared in the first step so that a relatively large push back load in the second step is applied. Of course, a knock out operation is necessary in the further step. According to this method it is not only possible to simplify the tool arrangement and the press but also to prevent the additionally occurring burr formation which could be

happened in the second step of Reciprocating Blanking if a poorly aligned blanking tool is used. Owing to the improved burr-free shearing feature this simplified reciprocating blanking method has been successfully put into practical use for blanking such material as aluminum, mild steel and stainless steel of 1.5-3 mm thickness.

Opposed Dies Blanking.[3] This method developed by K. Kondo (Shizuoka Univ.) can be grouped into one of the precision shearing processes which are used to obtain a smoothly sheared surface. As illustrated in Figure 5 which shows the process of this method, its main characteristic is present in that a smooth sheared surface is obtained mainly by means of cutting process. As compared with other precision blanking processes, this can be applied to the blanking of thick and brittle material and the shear droop of blank produced by this process is quite small. This process also has a burr-free feature in view of its shearing since shear droops are formed on both edges of blanked surface before the blank is knocked out and separated from an adjacent scrap chip.

Figure 6 shows the factors effecting the occurance of burr. To get the burr-free blanked parts by this process, deeper penetration a of protruding die, smaller clearance c

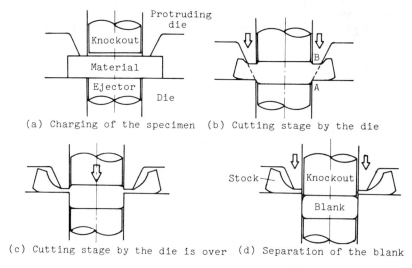

(a) Charging of the specimen (b) Cutting stage by the die

(c) Cutting stage by the die is over (d) Separation of the blank

Fig.5 Opposed dies blanking(Kondo)[3]

between knockout punch and protruding die and positive difference in dimension between die and protruding die are desireable. This process has been successfully applied into the practical use for the burr-free blanking of 0.8 mm thick silver plated stator plate of electric parts.

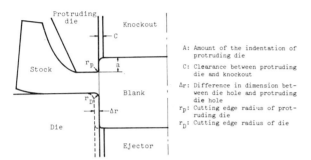

Fig.6 Factors effecting the burr in opposed dies blanking

A: Amount of the indentation of protruding die

C: Clearance between protruding die and knockout

Δr: Difference in dimension between die hole and protruding die hole

r_p: Cutting edge radius of protruding die

r_D: Cutting edge radius of die

BURR-FREE SLITTING OF THE COILED SHEET METAL.

Deburring Problems in Slitting. The shearing of narrow strips and the edge trimming with slitting machine is an important process, because it is an economic and precise method. On the other hand, by the conventional slitting, the burr formation of slit edge is inevitable. To eliminate this drawback on conventionally slit metal strips relatively expensive secondary treatments such as finish milling, mash rolling, and grinding have to be used.

Mash rolling is often used for most kinds of slit products, but this deburring method is unsatisfactory because of easier edge fracture in subsequent cold rolling as well as poorer recoiling capability. Finish milling may be needed when slit strips are subjected to further welding process. Deburring by grinding is necessary for the insulated strips, particularly oriented silicon steel strips for the electrical industry.

Counter Cutting Method.[4] This burr-free slitting was developed in 1968 in West Germany. Mode of action of Counter Cutting is schematically shown in Figure 7. With the initial cutting operation, the cutters penetrate into the material to some extent. In the second step, the circular cutters are arranged in the staggered position opposite to the first pair of cutters, so that separation can be performed from the opposite side. Here the circular cutters are engaged as far as is required for the complete separation of the material thus producing burr-free slit strips.

As will be appreciated from the forgoing description, it might reasonably be suggested

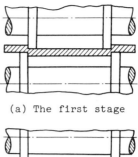

(a) The first stage

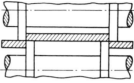

(b) The second stage

Fig.7 Counter cutting method [4]

that this method was originated as to its working mechanism from the Reciprocating Blanking mentioned before. Compared with the conventional slitting where the material is cut through only by the first step cutter, the number of cutter discs in Counter Cutting is naturally doubled, and consequently there would be involved substantially more work in maintenance such as exchanging, regrinding, and realignment of the cutter discs. Moreover, strictly speaking, for the success of such an arrangement of Counter Cutting, it is essential to keep the alignment of the axes of the two pairs of the cutter discs extremely precise, and also it is required that the guide devices for the rolled material should be maintained to operate very precisely. In consideration of such specific arrangement, it is essential to meet a very high severity or precision in the design and manufacture of the slitting stand and all the components pertinent thereto, as well as in the daily operating and maintaining services. Consequently, it is inevitable that the production cost of the slitting machine of such a specific arrangement becomes twice or three times as much as of the conventional arrangement having a similar production capacity.

Roll Slit Method.[5] In consideration of the disadvantages inherent to the Counter Cutting, it would be advantageous if an improvement of such approaches in an attempt to solve the problem of burr formation during the continuous slitting of the rolled material or coil stock be realized and made available without any substantial increase in the initial investment, or without addition of difficulties in the operation and maintenance of the slitting machine.

Roll Slit method which is being developed by two of the authors, T. Maeda and M. Murakawa, is essentially directed to meet such requirements as mentioned above. This method comprises two steps as shown in Figure 8, and the first step is essentially the same one as that of Counter Cutting in which the flat material is partially sheared or penetrated yet not parted from each adjacent part of the material (Figure 8a). The second step is obtained from the insertion of thus partially sheared material into the spacing between an opposing pair of rolls spaced apart substantially to the material thickness so that there occurs flattening by opposing pair of rolls over the both surfaces of the material when engaged (Figure 8b,c). As shown in Figure 9, the sheared edge

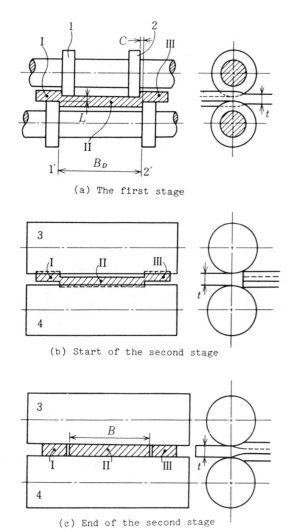

(a) The first stage

(b) Start of the second stage

(c) End of the second stage

Fig.8 Roll slitting method(Maeda, Murakawa)[5]

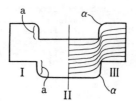

(a) The first stage tip

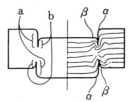

(b) The second stage

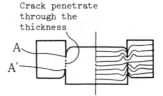

Crack penetrate through the thickness

(c) Just before separation

(d) A cross section of Roll-slit product

Fig.9 Separating procedure in roll slitting

face a obtained from the first step still remains at the initial stage of the second step and meanwhile a burnished surface b is newly formed adjacent to the said sheared face a. In addition, shear droop α formed on the strip edge is reduced gradually, as the strip goes ahead between the nip of flattening rolls, and simultaneously a new shear droop β is formed gradually, becoming greater on (Figure 9a,b). Next, as shown in Figure 9c, the plastic deformability of the half sheared material will come to its limit during the press-down operation of the rolls, and then cracking occurs from the ends (A,A'), whereupon the strip thus partially sheared is now parted into the final slit products I, II, and III as viewed in Figure 8c. Figure 9d exemplifies the cut surface of a completely severed strip (Mild steel strip, t=3.2 mm) where there are formed a small survival droop α' and a fracture surface generated between the sheared face a and the burnished face b, yet having no burrs formed at all.

In order to perform burr-free slitting, how to select the cutters overlapping L and the cutter clearance C as shown in Figure 8a in the first step, is a key point for successful slitting work by Roll Slitting. If the value of L is preset at the value very access to

positive, the strip material will be sheared and severed completely in the first step, thus forming burrs inevitably. On the contrary, too large negative L will cause the half sheared strip to be only pushed back and not be separated completely in the second step. For selection of the appropriate value of C, the experimental test has verified that, when the clearance C varies in direction to positive, the complete separation of a strip material particularly without burrs is not obtained in the second step, regardless of any suitable overlap value L in the first step, and that, when C setting varies reversely in direction to negative the cross-section perpendicularity of slitted strip is not achieved as desired. From the above facts, the second key-point for successful Roll Slitting is to set the clearance at zero or at the negative value approximate to zero. The test results introduced above covers only annealed soft steel. Moreover, the same test was conducted using other different materials such as aluminum, aluminum alloy, stainless steel, silicon steel, pickled hot rolled strip, copper, etc. The results have revealed that the said two key-points can also apply to such materials for satisfactory slitting work.

REFERENCES

1. Maeda T.; Reciprocating blanking method, Science of Machine, Vol. 10, No. 1 (1958), p. 140
2. Makino I.; Burr-free shearing process, Press Technique, Vol. 13, No. 5 (1975), p. 93

3. Kondo K. and others; Application of opposed dies shearing process, Proc. Int'l Conf. on Production Engineering (Tokyo, 1974), Part I, p. 251.

4. West Germany Patent Specification, No. 45-1806305

5. Maeda T. and Murakawa M.; The development of burr-free slitting, Jour. Japan Soc. for Technology of Plasticity, Vol. 18, No. 193 (1977). p. 114

SECTION IV
Net Shape Processes

Three forging production methods favored by aerospace industry

The aerospace industry's growing acceptance of "net shape" forging and P/M parts may force forging companies to embrace newer production methods to meet the demands of the space age.

Spurred by economic concerns, the aerospace industry is currently showing more than a passing interest in newer forging and P/M production methods. Hot die forging, isothermal forging and hot isostatic pressing (HIP) are three methods favored for the production of critical structural and engine components. Preference of these techniques may well push forging companies to change their views on conventional process and invest in new and different types of equipment, die materials, heating systems and lubricants, if they wish to remain competitive.

Economic difficulties are not the only reason for the change in aerospace industry thinking. Tighter performance standards, rising superalloy prices in the past three to five years and mechanical property requirements have contributed to the shift.

As Joseph R. Carter, Wyman-Gordon Co. president says, "We are required to forge essentially the same finished shapes that we have for many years, primarily round wheels and rings for jet engines and rib and web structural components for air frames, but with harder-to-forge materials and to tighter configurations."

The aerospace industry is trying to move toward starting production with as small an amount of superalloy as possible and making a part to near net shape, ready for assembly with a minimal amount of machining required.

It is in this approach that hot die forging, isothermal forging and HIP will have the most effect.

Hot die forging calls for the dies to be heated within 200F (93.4C) or 300F (149C) of the stock temperature to facilitate flow in the die. One example of the process is a turbine engine part made by Wyman-Gordon which, conventionally forged, weighs 185 lbs. (83.9 kg) but using the hot die system, 61.6 lbs. (27.9 kg).

W-G recently used the process to experimentally forge titanium airframe components of 3.4 ft.² (0.32 m²) in plan view area for the F-15. After a period of adjustment, the company believes it will be able to use the hot die process to forge structurals of 5.2 ft.² (0.48 m²) to 6.9 ft.² (0.64 m²) plan view areas.

Such a changeover will necessitate larger capital expenditures. "Obviously, the hot die process requires new die materials, new heating systems and new lubricants and, as the die temperature increases, so does the cost of production," says Robert B. Sparks, manager of W-G's laboratory services. "It appears," he concludes, "that the degree of sophistication obtained in the final forging is proportional to the cost of the die system."

Costs to produce true net shapes, compared to near net shapes, may be prohibitive, especially in forgings larger than 4.2 ft.² (0.39 m²) in plan view area. Although considerable success has been reached in forging large components with this process, no large parts are currently in production.

From the alloy side, both titanium and nickel base alloys are used in producing jet engine net shape components, but more research is underway with the nickel-base superalloys such as René 95, Astroloy and IN-100.

According to Sparks, these alloys, because of their high costs and poor machinability, are candidates

The F-15 keel splice fitting in the center was made by the HIP process and weighs 1 lb. (0.454 kg). After machining, it looks like the part on the right and weighs 0.4 lb. (0.182 kg). The forged fitting on the left weighs 4.67 lbs. (2.12 kg) and requires extensive machining.

Reprinted with permission from Precision Metal, April 1977, 35-36, © 1977 Penton/IPC Inc.

for hot die forging since the potential savings in materials cost can justify the cost of the die system.

A turbine wheel of René 95 is a possible example of such savings. The wheel, conventionally forged, weighs 930 lbs. (422 kg). The same wheel, redesigned for the hot die process, weighs 537 lbs. (243.7 kg), a difference of almost 400 lbs. (181 kg).

"At the price of René 95 and the cost of machining off 400 lbs., you can see how an expensive die system can be justified," says Sparks.

In isothermal forging, dies are heated to the same temperature as the alloy stock. Pratt & Whitney Aircraft Div.'s Gatorizing® is an isothermal process where the entire forging platen, dies and stock are enclosed in a vacuum or inert gas atmosphere with the dies heated to 2000F (1094 C).

W-G is currently using this special process to forge F-100 engine turbine discs from IN-100 alloy and is evaluating isothermal forging for production of titanium components. The firm is using a small Gatorizing unit for turbine hardware parts and has plans to build a larger unit, perhaps with a 6000-ton (53.3 MN) press, to produce much larger isothermal forged parts.

According to Carter, the company expects to make larger and larger parts as experience is gained. He also believes that these processes can be speeded up if aerospace designers utilize some of the new beta titanium alloy types. These new alloys will permit dies to be heated to the 1400F (761C) to 1500F (816C) range, resulting in tooling cost reductions and less wasted energy, plus parts produced to closer tolerances.

Powder metal products are also receiving closer study by aerospace experts. If an ingot is converted into powder, that powder may be blended to form a more uniform product after consolidation. This is especially important, since superalloys are notorious for chemical segregation during solidification; that segregation is not eliminated by subsequent working and thermal treatments.

Through powder metallurgy, powder can be directly forged or extruded to make a billet for forging, or it can be pressed into preform shapes.

One method of making powder preforms currently receiving attention is the hot isostatic pressing (HIP) process. Its biggest advantage is in material utilization . . . the preform shape can be close to the final configuration, and usually only a single die operation is required for finishing.

In the HIP process, powder metal is placed in a glass, metal or ceramic container and heated in an autoclave to a superplastic condition. At this point, the powder is squeezed from all sides by compressed gas at pressures ranging from 10,000 psi (68.9 MPa) to 30,000 psi (206.8 MPa). Heat and pressure combine to consolidate the loose powder to 100% of its theoretical density, as compared with a maximum of 90% to 96% for other P/M techniques.

HIPed parts may be used in their as-produced condition, without subsequent forging operations. W-G has been using a small autoclave to test the process and has recently ordered a large hot isostatic press with a 48-in. dia. (121.9 cm), 60-in. (152.4 cm) long chamber to produce parts at near net shape.

According to Larry Clark, Manufacturing Technology Div., Air Force Materials Laboratory (AFML), the process is so controlled that tolerances currently can be held to ±1% of the required dimensions.

"Mechanical property evaluation studies of HIP parts look so good, we think many parts currently being forged could be made this way," he said. Clark continued:

"The process allows you to begin with less material, then HIP to near net shape. Conventional forgings, on the other hand, are oversized and have an average of seven times the materials required by the final part. This excess must be removed from the forging by expensive, time-consuming machining.

"Also, the HIP process uses 20% less energy than multiple forging operations, a big plus in today's energy-conscious society."

Two parts being manufactured on programs sponsored by AFML are a titanium (Ti6AL-4V) keel splice fitting in the F-15 aircraft and a third stage superalloy disc for the F-102-LD-100 fan engiener.

It would seem that in order to meet the future requirements of the aerospace industry, suppliers will have to commit millions of dollars in equipment and technical know-how. **PM**

The disc and shaft (left) were forged separately by conventional techniques and weigh 103 lbs. (46.8 kg); the part on the right was made integrally by HIP to near net shape and weighs 33 lbs. (14.9 kg). In the center is the machined disc-shaft in which the shaft was attached by friction welding; it weighs only 26 lbs. (11.8 kg.).

Hot-forming gains at Wyman-Gordon

Efficient facilities for hot isostatic pressing, modified isothermal forging, and hot-die forging produce high-strength, reliable, near-net-shape parts for aircraft applications

Three advanced hot-forming processes are now universally known and accepted in the US aerospace industry: (1) hot isostatic pressing, (2) a type of isothermal forging called Gatorizing, patented by Pratt & Whitney Aircraft, and (3) hot-die forging. These processes represent a new era for the forging industry, producing reliable, high-strength turbine-engine hardware and other components for today's sophisticated aircraft. They will most certainly benefit nonaerospace manufacturing in the future.

Wyman-Gordon Co, a leading US producer of turbine hardware for the aerospace industry, currently produces parts by all three of these methods at its plants in North Grafton, Millbury, and Worcester, Mass. The company is also a leader in near-net-shape technology, which is constantly changing and which promises to be a major factor in the future of the forging industry. Listed below are some of the reasons why these new processes represent the state-of-the-art in metals engineering, and why Wyman-Gordon has invested heavily in them.

■ Hot isostatic pressing (HIP) is the newest of the three production techniques. In this process, powder metal is placed in a shaped container, heated to 2000-2200F, and then compressed isostatically by an inert gas at pressures reaching 15,000 psi. Heat and pressure combine to consolidate the powder to 100% of its theoretical density—a decided advantage for jet-engine and other high-performance applications.

In some applications, HIPed parts can be used "as is," with minimal machining after heat treating. Or they can be subsequently finish-forged by conventional, hot-die, or isothermal methods into parts that are either net shape or very near to it.

By John T. Winship, associate editor

The HIP process is able to produce more-complex shapes than would be possible with traditional forging methods.

Both HIP and HIP-plus-forging consume less material than conventional forging uses. Raw-material input is reduced, and less stock is converted to chips during finish-machining.

These advantages are compelling enough, but the fundamental justification for HIP in today's aerospace market is its ability to produce highly defined, sophisticated shapes from metal alloys that do not forge well by conventional methods. They're the high-temperature superalloys that rank high on the aerospace industry's list of best performers in critical applications.

GE's René 95 is one such material. In the HIP process, this hard-to-forge alloy can be made into parts with mechanical properties unmatched in desirability and consistency by those of conventional wrought stock and of materials processed by other hot-forming techniques.

■ Isothermal forging, in which dies and stock are heated to the same temperature to facilitate metal flow during the forming process, is an established technique that has recently undergone sophisticated changes. Steel and aluminum parts were first forged isothermally in laboratories about 30 years ago; titanium parts, more recently.

The original concept of isothermal forging has been modernized by the enclosure of the entire forging system in a vacuum or in an inert-gas atmosphere. This is necessary because the high forging temperatures for some alloys require the use of molybdenum-alloy dies, which vaporize at those temperatures in a normal-air atmosphere.

Several years ago, a modified isothermal-forging method was perfected by Pratt & Whitney Aircraft and dubbed "Gatorizing." In this process, the entire platen, dies, and forging stock are enclosed in either a vacuum or an inert-gas

René 95 superalloy powder sealed in these stainless-steel cans will be hot-isostatically pressed into turbine disks

Stripped of their can enclosures, then heat treated, machined, and inspected, these HIPed turbine disks are ready to ship

Reprinted with permission from American Machinist, January 1980, 124-127, © 1979 McGraw-Hill Inc.

Wyman-Gordon's hot isostatic press

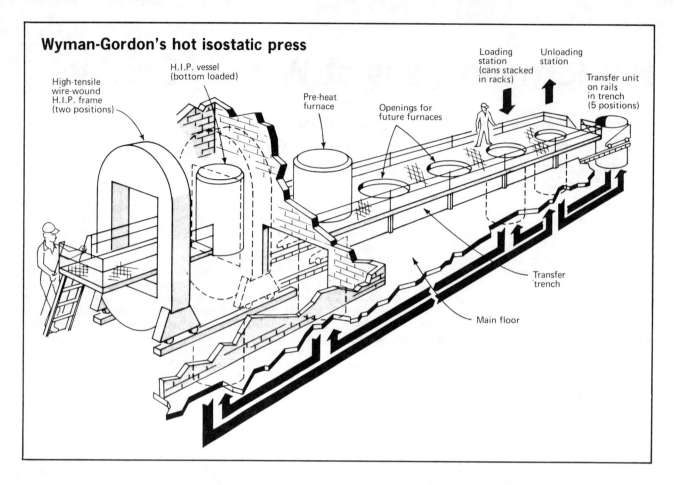

High-tensile wire-wound H.I.P. frame (two positions)

H.I.P. vessel (bottom loaded)

Pre-heat furnace

Openings for future furnaces

Loading station (cans stacked in racks)

Unloading station

Transfer unit on rails in trench (5 positions)

Transfer trench

Main floor

atmosphere, and the dies and stock are heated to about 2000F. Specially consolidated IN-100 powder (iron-nickel-base superalloy) is made superplastic prior to being isothermally forged into near net shapes.

The isothermal process significantly reduces the preforging weight of parts for some major engine programs and permits greater shape sophistication. For such superalloys as IN-100, it is the only feasible forging method. In conventional forging, the steep temperature gradient between dies (at about 800F) and the workpiece (at 2000F) requires a rapid deformation rate to minimize heat loss and associated cracking. The faster this rate, the more difficult shape sophistication becomes.

This difficulty is eliminated in isothermal forging. The effect is improved metal flow, permitting greater shape complexity, less raw material input, and production of parts that are close to net shape. Furthermore, in isothermal forging, the amount of force required to perform a given forging job is considerably less. For instance, Wyman-Gordon uses an 1800-ton press to produce forgings that might otherwise have been conventionally produced on an 18,000-ton press.

■ Hot-die forging, sometimes referred to as near-isothermal forging, is a process in which dies are heated to within 200-300F° of the forging stock in order to reduce heat loss from the stock to the die system. Unlike isothermal forging and Gatorizing, which are performed within an enclosure, hot-die forging is done in the open. The process can be adapted to a wide range of hydraulic presses—from 6000 to 35,000 tons—and the process has been used to produce titanium structural forgings up to 500-600 sq in. in plan-view area.

The major components of Wyman-Gordon's HIP unit (see drawing above) were supplied by ASEA's Quintus Div, which has its US office in White Plains, NY, and its factory in Vasteras, Sweden. Overall length of the new HIP facility is

about 150 ft. Along with its auxiliaries, it occupies several walled-off rooms in Wyman-Gordon's Millbury plant. The working dimensions of its cylindrical pressure chamber are 48 in. ID x 60 in. long. A master console enables one person to control the operation except for loading and unloading.

Currently, the HIP unit is producing 18- to 20-in.-dia René 95 disks for turbine and compressor sections of the F-404 engines that power the US F-18 fighter plane. According to Norman Gustafson, W-G's project manager for the HIP facility, the disks are not the largest the facility has made. A special "spool," which is actually an experimental hub for GE and has as many as ten turbine stages integral in one piece, has been HIPed.

Stainless-steel cans, made by spinning or hydroforming, carry the René 95 metal powder for the HIP process. They're generally supplied as two halves and are, like the powder, rigorously cleaned, inspected, and tested prior to use. The part to be HIPed is designed around the final part, and so the can provides an envelope of the required disk shape, plus an allowance for shrinkage during HIP compaction.

W-G cleans the can components, then welds them together to form a closed container (much like two giant pie plates joined rim to rim), attaches a closable fill spout, and leak tests the assembly prior to powder loading. The welding operation is closely controlled and is designed to prevent contamination of the can's interior. For somewhat the same reason, the powder-filling operation is performed in a "clean-room" atmosphere.

At the HIP unit's charging station, the filled cans—with spouts sealed—are stacked on a charging plate. The stacking arrangement provides proper support so that the cans remain flat when they're subjected to the prescribed heat and pressure in the HIP chamber. For parts the size of the René 95 turbine disks, about 25 cans make up one charge.

The HIP unit's manipulator, or transfer car, is then positioned at the charging station. An electric-powered elevator lowers the cans into the manipulator's cylindrical shell. The cans are then carried to a preheat furnace, where their temperature is raised to about 1500F. As shown in the drawing, the HIP unit has space for two more preheat furnaces, providing for a future increase in production.

After the cans are preheated, they are transferred by the manipulator to the HIP unit's pressure chamber. The chamber's bottom hatch is closed, and the 21-ft-tall wire-wound frame, or yoke, rolls on its own tracks to surround the chamber. Its purpose: to hold the top and bottom hatches of the chamber firmly in place. The cycle is then triggered at the HIP control console and lasts for about eight hours.

Resistance-type electric heaters in the unit gradually heat the cans to about 2100F. Thermocouples placed adjacent to the cans ensure that the temperature throughout the chamber is maintained within 25F° of that desired during the HIP cycle.

Argon gas is pumped into the chamber from liquid-storage accumulator bottles outside the HIP building. Heat and pressure combine to compact the René 95 powder into disk shapes that are close to end-use configuration and weight. Once the HIP cycle has been completed, the argon gas is evacuated, and the temperature is gradually diminished.

When the pressure chamber's bottom hatch is swung open, the manipulator lowers the stack of cans and carries them to the unloading station. From there, they're taken to another

It's not quite conventional forging

The action on Wyman-Gordon's largest forging press is conventional enough, but the machine's capacity is certainly unusual: 50,000 (US) tons of hydraulically powered metal-deforming might. The press, located at W-G's N Grafton, Mass, plant, towers several stories to the plant roof and extends a like distance below ground level. To reach its lowest point, you can take an elevator. This giant, and another of the same rating at the Cleveland plant of Alcoa,

are the US's largest forging presses. In the photo sequence below, W-G's press puts the final twist in an Inconel 625 naval propeller blade (AM—Oct'79,p45). When it's finish-machined for subsequent assembly, along with three others, to make a four-blade controllable-pitch propeller, the blade will weigh 2462 lb. As-forged thickness averages 3 in; width at its mid-section is 60 in. It's one of the largest parts of its kind over forged.

Three separate forging operations on the big press have produced the almost-finished blade. It's shown here in the dies that will give it the final forged shape

Press ram descends slowly enough to permit attendant to make a final check of blade position. Die blocks are made, and sunk by computer, by A Finkl & Sons (Chicago)

When press ram retracts, attendants prepare blade for removal by fork-lift truck. Blade's properties, size, and weight push this job to limit of conventional forging

Press attendant examines blade after it gets its "potato-chip" contour and is removed from dies. The hub and hydrofoil contour will be completely machined

Source: American Machinist, January 1980, 124-127

Isothermally forged turbine parts are removed from air lock of 1800-ton Gatorizing unit. Procedure is done in a vacuum

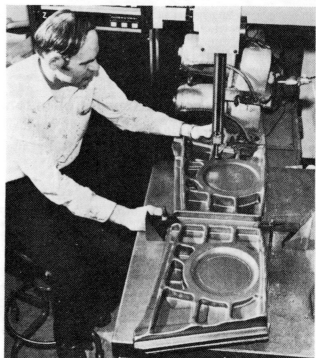

Hot-die "Siamese" forging, 550 sq in. and 92 lb, yields two titanium jet-fighter parts, receiving dimensional check here

room in the plant, where the cans are stripped to expose the finished part. Then the compacted René 95 turbine disks are heat treated and ultrasonically inspected before final machining operations.

Gatorizing IN-100 disks

In one bay of W-G's Worcester Plant, an 1800-ton Verson hydraulic press towers at the apex of a lower, massively-built, L-shaped horizontal structure. One leg of the structure is an electric-resistance-type furnace, its bed fitted with an indexing conveyor that advances workpieces toward the press. The other leg houses a cooling chamber and a removal conveyor. An automatic tranfer mechanism carries each heated workpiece to the proper position between the press dies, then lifts the part away after the forging operation and deposits it on the removal conveyor.

According to Paul Wisniewski, W-G's Eastern Div vice president for sales and marketing, the system's handling mechanisms are among its most ingenious—and costly—components. But the most important characteristic of this continous-flow installation, which W-G designed and built, is that both legs and the press's forging station are kept under vacuum. Transfer locks, to prevent breaking the vacuum, permit loading and unloading.

Wisniewski says that argon gas, such as is used for pressurizing the W-G HIP unit, could be used as the forging atmosphere, but, for reasons best known to W-G, a vacuum is preferable for this kind of forging work instead of an inert atmosphere.

Turbine disks of IN-100 superalloy are forged on this unit, shaped by dies of molybdenum alloy (TZM). This is the best material available for forging IN-100 or any of the principal superalloys. But, at a temperature of 2100F, molybedenum literally vaporizes if exposed to air; hence, the vacuum. W-G has commercially forged IN-100 and titanium alloys in this unit and has successfully forged other alloys on an experimental basis.

Typically in this process, lubricant-coated preforms are loaded through the furnace lock in groups, the number depend-

ing on their size, which is about the same as the disks processed by the HIP unit. In single file, they're indexed into the press, forged, then ejected to the conveyor, and removed through the lock.

After the forged parts are removed from the unit, they are allowed to air cool, then are taken to heat treating and ultrasonic inspection stations. These postforming operations are material-, not process-, dependent; that is, they are not required just because the parts are Gatorized. As is the case with the hot-die forgings, minimal machining is the last major step prior to shipment.

Hot-die forging titanium

One part succesfully forged by the hot-die method at W-G is a titanium-alloy bearing-support housing for a jet fighter. The part is noteworthy for several reasons. First, it is fracture-critical and so must be of the highest quality. It is also large, as most hot-die forgings go, covering an area of 550 sq in. as it is forged, two at a time, "Siamese" style. It is also "net shape" in one area, a pocket that needs no machining.

W-G forges the titanium bearing-support housing at a temperature of 1750F, with dies at 1600-1625F. After trial runs on a 6000-ton hydraulic press failed to produce the correct thickness (the part starts as a round wrought titanium billet), W-G moved the operation to an 18,000-ton press and discovered that the part needed only 8000 tons of force—much less than would be necessary in a conventional forging operation.

Throughout the hot-die-forging process, all the critical variables are precisely monitored and controlled. Among them: average velocity of the upper die, the speed at which workpiece deformation occurs, the strain rate, and the dwell period, during which the workpiece material is allowed to flow to the farthest recesses of the die cavity. Instrumentation is plentiful, including a displacement transducer to indicate press condition and a strip-chart recorder to show such conditions as stroke, displacement, and force; combined mathematically, they determine the amount of energy used to make a specific part. ■

Near Net Shape Processing for Gas Turbine Components

Increased emphasis on more efficient utilization of materials because of the need to conserve materials and energy has led to the concept of near net shape processing. Cost reduction is another powerful factor favoring any initial shaping process which results in a configuration closer to final dimensions. Billions of dollars are spent annually making chips in machining to final part shapes. Manufacturers of aircraft turbine engines are dedicated to advancing technology to minimize this cost via near net shape processing.

Let us consider this product—the gas turbine engine, a large consumer of sophisticated and expensive materials. The principal materials used in the gas turbine engine are steel, aluminum, titanium, nickel, and cobalt. Aluminum, steel, and titanium are used in the cooler fan and compressor sections while nickel, cobalt, and stainless and heat resistant steels make up the hot burner and turbine sections. As turbine engines have become more complex to meet demands for improved performance levels, the weight of raw material which must be purchased and processed to produce one pound of finished component, that is the Buy/Fly ratio, has increased proportionately as shown in Table I. The Buy/Fly ratio has increased from 4.4:1 for the early JT3D engine to 6.2:1 for the more recent JT9D engine. The higher Buy/Fly ratio translates into increased raw material, fabricating costs and energy. It is obvious that the greater the Buy/Fly ratio for a component, the greater the amount of material which must be bought

and then removed by machining to produce the finished component (see Table II). The material costs for chips and the processing costs to produce chips are very large for turbine engine components and represent an enormous area for cost reduction possibilities. Consequently, effort is being made to reverse the trend of escalating Buy/Fly ratios as high performance engines become more complex and sophisticated. This effort is being concentrated in the area of near net shape processing and involves forging, extrusion, casting, and powder metallurgy. Before any process change is considered, certain facts must be established. Among these are 1) the original component mechanical and chemical properties must not be adversely affected, 2) there is a cost reduction potential, 3) the part must remain inspectable to confirm quality and finally the process is usually verified by engine test.

E.F. BRADLEY, Chief Materials Engineer (Retired), Pratt & Whitney Aircraft, East Hartford, CT.

Table I. Material Buy/Fly Ratio

	Buy Weight, lbs	Fly Weight, lbs	Ratio
JT3D-7	18,935	4300	4.4:1
GG4-A7	27,031	5307	5.1:1
JT12	2495	468	5.3:1
TF30-P-408	17,385	3282	5.3:1
JT8D-7	17,014	3155	5.4:1
J52-P-8A	12,042	2118	5.7:1
JT9D-3	52,242	8470	6.2:1
JT9D-7	17,855	8358	5.7:1

Source: Journal of Applied Metalworking, July 1979, 73-79, © 1979 American Society for Metals

Table II. Engine Buy/Fly Material Breakdown

	Buy, lbs	Fly, lbs	Chips, lbs	Material cost Fly	Material cost Chips	Rough Machining Cost
Aluminum	2407	584	1823	$982	$3,080	$2,930
Stainless steel	14,498	1537	12,961	$1,960	$16,480	$44,500
Titanium	11,718	2055	9663	$9,800	$46,200	$89,700
Nickel	17,811	2966	14,845	$17,000	$85,500	$156,000
Possibility for savings				$29,742	$151,260	$293,130

ISOTHERMAL FORGING

Much has already been accomplished. In the area of hot disks, for example, more efficient utilization of material via near net shape processing has been realized in a number of production engine parts. Hot disk development programs have been directed at two targets, improved mechanical properties and reduced costs. One significant result to date is the development of the isothermal forging process (see Fig. 1). In this process, the alloy to be forged is placed in a temporary condition of low strength and high ductility (superplastic) at the forging temperature. Forging occurs isothermally as both the dies and forging stock are heated to the established forging temperature and maintained at that temperature during forging. The process presently utilizes extrusion to consolidate powder into a fine-grain bar which is cut into billet size for forging. The extrusion process introduces sufficient stress in the billet for superplastic condition at the forging temperature. The first application of the isothermal forging process to volume production of parts has been for the high thrust to weight F 100 engine. In the early stages of the development of this engine, the selection of materials was predicated on the highest strength to weight ratio at engine operating temperatures that it was possible to attain. An alloy was needed for the latter stage compressor disk and the turbine disks that was stronger than Astroloy, then the highest strength conventionally forged alloy operating in production jet engines. IN 100, a nickel base superalloy developed for cast turbine blades, showed the desired strength to weight ratio at the operating temperature. However, this highly segregation prone alloy was unusable for large engine disk forgings using conventional state-of-the-art melting and forging practice. By utilizing an all inert powder metallurgy approach in conjunction with the isothermal forging process, compressor and turbine

SUPERPLASTIC BEHAVIOR

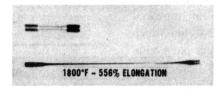

1800°F – 556% ELONGATION

Fig. 1—Isothermal forging of turbine disks.

APPLICATION

Isothermally Forged

disks and spacers were produced for the F 100 engine with a high level and uniformity of mechanical properties. It is interesting to note that isothermal forging was first developed and used for property reasons, that is, it was the only process by which IN 100 could be successfully forged. Isothermal forging allowed this high strength to weight material to be used in the forged condition. However, isothermal forging, as a near net shape process, has enabled the F 100 parts to be made with a reduction of billet input weight of approximately 50 pct compared to the input weight required to forge identical configurations from weaker superalloys. In addition, parts of complex geometry can be readily produced in light capacity forging presses. Fig. 2 shows the most complex forging produced for the F 100 engine. This

part is isothermally forged from the powder metallurgy billet shown in a one step operation. Figure 3 shows several of the F 100 configurations made by the isothermal forging process. In the F 100 engine nine IN 100 parts are produced via powder metallurgy and isothermal forging. It is significant that these parts are the first powder metallurgy disks ever operated in a jet engine.

Up until recently, the full potential of the isothermal forging process to forge a configuration to just over the finish part size was inhibited by the requirement for a square cut outline with a minimum envelope of material to facilitate ultrasonic inspection of the forged part. Improvements in ultrasonic techniques now have been made which make it feasible to inspect a near net shape configuration. As a result, through the use of multiple forging operations, near net shape forgings of IN 100 can be produced for the F 100 engine with a total billet weight of 500 instead of 990 lbs.

Figure 3 illustrates the cost reduction possibilities of the isothermal forging near net shape process. The compressor disk for the initial F 100 development engine was machined from a 250 lb Astroloy forging (see Fig. 4). The current engine disk is machined from a 126 lb IN 100 isothermal forging to make the 15 lb finished part. Thus, the isothermal forging process provides a saving of 124 lbs of chips. It is projected that further improvements will be made (see Fig. 5). The near net shape forging is expected to weigh 50 lbs providing an additional savings of 76 lbs of chips.

HOT ISOSTATIC PRESSING

The use of metal powders to obtain chemically and metallurgically uniform structures for critical rotating

Fig. 2—Isothermally forged IN 100 advanced engine disk.

Fig. 3-Isothermally forged IN 100 disks.

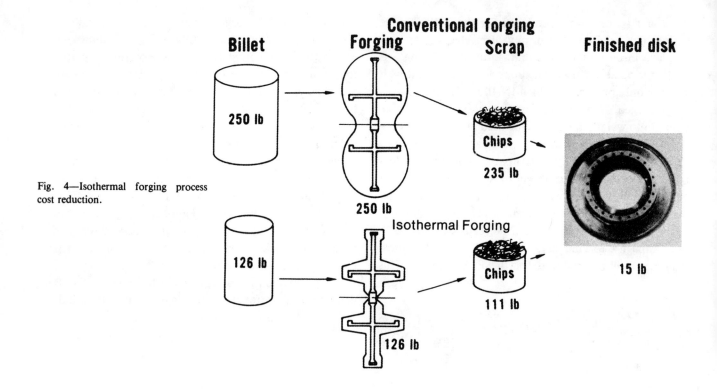

Fig. 4—Isothermal forging process cost reduction.

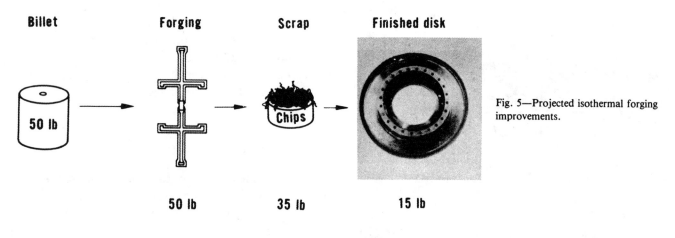

Fig. 5—Projected isothermal forging improvements.

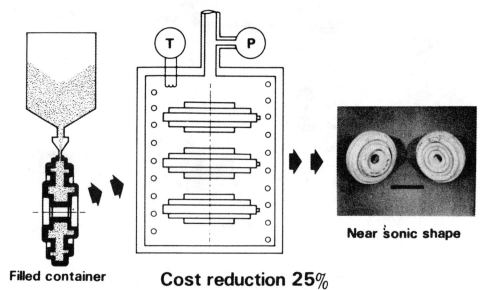

Fig. 6—Hot isostatic pressing.

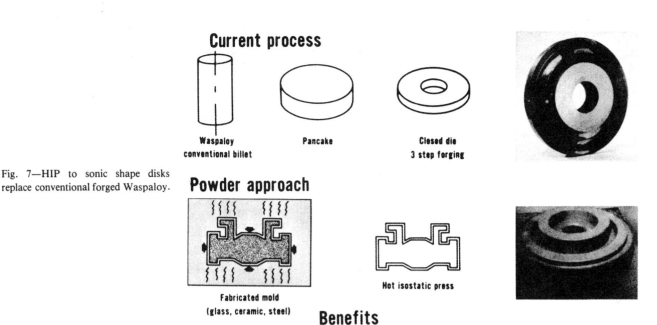

Current process

Waspaloy
conventional billet

Pancake

Closed die
3 step forging

Fig. 7—HIP to sonic shape disks replace conventional forged Waspaloy.

Powder approach

Fabricated mold
(glass, ceramic, steel)

Hot isostatic press

Benefits

- 25% less $ than current Waspaloy

HIP sequence

- Produce powder
- Seal in containers
- Hot isostatic press
- Leach can
- Sonic inspect

Conventional forging

- Melt ingot
- Forge to billet
- Prepare for shape forging
- Upset forge to pancake
- Reheat
- Forge in closed dies
- Reheat
- Forge in closed dies
- Reheat
- Forge in closed dies
- Machine to sonic shape
- Sonic inspect

Fig. 8—HIP reduces cost through process simplification.

gas turbine hardware has been gaining acceptance with the advent of high purity, prealloyed superalloy powders, and the development of suitable shaping techniques such as the isothermal forging process. Another consolidation process of interest in this regard is hot isostatic pressing, known as HIP (see Fig. 6). HIP is a densification process combining simultaneous application of high temperature and isostatic gas pressure. This process, which is being developed for several engine applications, incorporates powder metallurgy techniques to produce disk blank preform shapes very close to the required final configuration, with substantial material, forming, and final machining cost savings. In addition, HIP has the potential of producing superior strength disks because of the uniformity of alloy constituent dispersion and stable substructure which results from the process.

The cost reduction promise of HIP is further highlighted when compared to current conventional forging techniques (see Fig. 7). Large cost reductions are expected through minimizing the amount of chips produced. Additional cost reductions result from the process simplification of HIP as compared to conventional forging (Fig. 8). Comparison of the current conventional forging practice for a typical disk with projected HIP processing shows dramatic improvement (see Fig. 9). The current process involves 360 input lbs to make the forging. The first generation HIP process will reduce input powder weight to 220 lbs. The powder will be pressed to a closer shape preform and then forged in one final die. The second generation HIP process will require only 175 input powder lbs to form the sonic shape, and the long range plan is to HIP to near net shape starting with 110 lbs of powder. The only current production application of HIP in P&WA engines is the first-stage turbine disk for the TF30−P−100 engine. This part utilizes the HIP process to consolidate a preform from Astroloy powder which is then forged conventionally to shape. HIP and isothermal forging can be effectively combined, particularly for IN 100 alloy where HIP may be used to consolidate the billet from powder and isothermal forging used to shape the part. This processing represents a potential cost reduction based on the savings realized from forming the billet by HIP as compared to extrusion. The combination has also resulted in tensile strength and ductility improvements over parts that were only isothermally forged.

Conventional process

• Cast and forge 360 lbs input

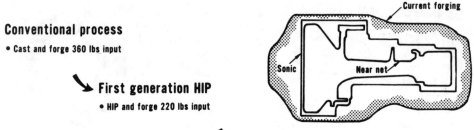

↘ **First generation HIP**

• HIP and forge 220 lbs input

↘ **Second generation HIP**

• HIP to sonic
  175 lbs input
• Container
  technology
• HIP
• Heat treat
  parameters

↘ **Long range program**

• HIP to near net
  110 lbs input
• Container refinement
• Distortion control
• Improved NDE

Fig. 9—HIP of superalloy disks summary.

Benefits

Program:	HIP and forge	HIP to sonic	HIP to near net
Cost savings	15%	35%	55%
Raw material savings	140 lbs	185 lbs	250 lbs

OTHER PROCESSES

A number of other near net shape processes, in addition to isothermal forging and HIP, are either being used or developed for aircraft turbine engines. Extruded, rolled and welded engine cases are now being used rather than conventional forged cases with substantial reduction in input material (see Fig. 10). In one application, elimination of 107 lbs of chips reduced costs by 46 pct.

A rotary forging process developed in Austria to produce high quality large caliber gun barrels has been adapted to produce compressor drive shafts for advanced engines. In this process, a mandrel is placed through a hollow tube heated to the forging temperature and held at one end by a chuck. The chuck spirals the tube as hammers shape the outer configuration and the mandrel shapes the inside. Rotary forging reduces the

Forged and machined ring

• Forged ring   135 lbs
• Finished ring   20 lbs
Chips 115 lbs

Extruded, rolled and welded ring

• Extruded ring   28 lbs
• Finished ring   20 lbs
Chips 8 lbs

Cost reduction 46%

Fig. 10—Extruded and rolled ring.

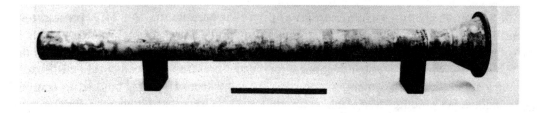

CONVENTIONALLY FORGED
913 lbs

ROTARY FORGED
472 lbs

COST REDUCTION 22 %

Fig. 11—Rotary forged JT9D low pressure turbine shaft.

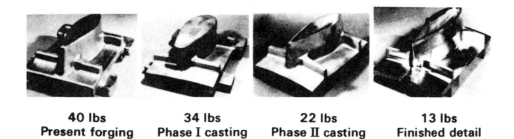

| 40 lbs | 34 lbs | 22 lbs | 13 lbs |
| Present forging | Phase I casting | Phase II casting | Finished detail |

Fig. 12—Centrifugally cast titanium case application.

amount of raw material needed to produce the shaft by nearly 50 pct (see Fig. 11).

The cost of pump gears, although produced in only moderate quantities, may be significantly reduced by taking advantage of automotive powder metallurgy techniques. Complex tooth and spline profiles can be pressed directly out of powder metal leaving only the easily turned bearings and journals to be machined. In this instance, these lightly loaded powder metal parts are pressed, sintered, and coined. The coining gives added density (about 94 pct) and close shape. The powder metal processes discussed earlier provide 100 pct dense products.

In addition, to these various consolidation processes, casting directly to near net shapes offers economic advantages. Precision casting has long been used in the manufacture of turbine engine airfoils. Die casting is another process with near net shape advantages over conventional forging or casting. Far more of the final configuration details of small parts such as pump covers can be obtained by this casting process than by forging. Final machining costs and raw material input weight are reduced. Centrifugal casting is also being developed to produce thin cross section members, such as titanium fan exit cases, with expected raw material saving up to 50 pct (see Fig. 12).

A continuing effort is being made to create, adapt, and develop processes which will reduce engine production costs. A few of these processes have been discussed, some of which are already in production and others are still in the development state. It is clear, however, that these efforts have reversed the trend of the ever increasing Buy/Fly ratio and these cost reduction programs are beginning to yield significant results. The JT9D engine Buy/Fly ratio has been decreased from 6.2:1 to 5.7:1. Results of processes now under development should far exceed this gain. In summary near net shape processing offers substantial cost reduction through minimizing chips and reducing input weight. Some progress has already been made in this regard but many opportunities still exist. HIP processing to near net shape offers great potential for critical large rotating engine parts. The aircraft engine industry will continue to search out methods to cut costs through minimizing chips.

Technology is Key to Production of Superalloy Disc Forgings

By James E. Coyne

ABOUT FOUR YEARS AGO, a new generation of aircraft turbine engines was born — advanced technology engines which burn hotter and produce more thrust than their older counterparts.

The new engines utilize a variety of nickel base alloys with superior high temperature strength to meet the performance criteria established for them. This high temperature strength, however, also made them extremely difficult to forge in a conventional manner. This forging bottleneck came at a time when engine builders were also expecting significantly improved shape sophistication and materials utilization.

This provides the setting for the development of the near net shape technologies which have today reached the production stage in the forging industry. They include isothermal forging/Gatorizing and hot isostatic pressing.

History — The seeds for this new era in aerospace forging technology were planted with the development of the SR-71 aircraft — a milestone in aircraft design because of the materials employed in both airframe and turbine hardware.

The SR-71's J-58 engine featured Waspaloy and Astroloy hot section components. Waspaloy was the first of the superalloys in which superior properties could be achieved by thermomechanical processing. In shop terms, the alloy was finished forged in the lower part of its forging temperature range, which in turn produced the desired metallurgical effect in the postforge heat treat cycle.

The price paid for this metallurgical advance was substantial: a severe reduction in forging range, increased forge shop temperature surveillance, a larger forging envelope to protect the final shape, lower productivity, and an increased tendency toward cracking.

Astroloy, an even stronger high temperature alloy, richer in alloying elements, had an additional problem: improved vacuum melting techniques were needed to help minimize segregation. This resulted in the direct forging of small diameter ingot into turbine wheels using elaborate insulating techniques. The conventional route — forging billet converted from large diameter ingot — could not be taken.

In the early 1970's, General Electric introduced René 95. Production techniques similar to those employed for Astroloy worked effectively for early René 95 turbine hardware. However, there were still drawbacks: poor materials utilization, excessive in-process conditioning, and increased processing time.

Gatorizing Marks Dawn of Near Net Shape Age

Wyman-Gordon Co. has been experimenting with hot die and isothermal forging for almost 15 years. What gave this program impetus was development work for the Pratt & Whitney Aircraft (P&WA) F-100 engine, the powerplant for the F-15 and F-16 fighters.

The F-100 was one of the first advanced technology engines — it required IN-100 turbine wheels made from extruded powder billets subsequently forged using P&WA's patented isothermal process, Gatorizing.

Top, the different envelopes associated with isothermal forging/Gatorizing and conventional forging. Note the substantial improvement in materials utilization possible with the isothermal process. The part is a turbine disc for Pratt & Whitney Aircraft's F-100 engine. It's made of IN-100. Bottom, a hot isostatically pressed, René 95 turbine wheel for General Electric's F-404 engine.

Source: Metal Progress, November 1980, 34-37, © 1980 American Society for Metals

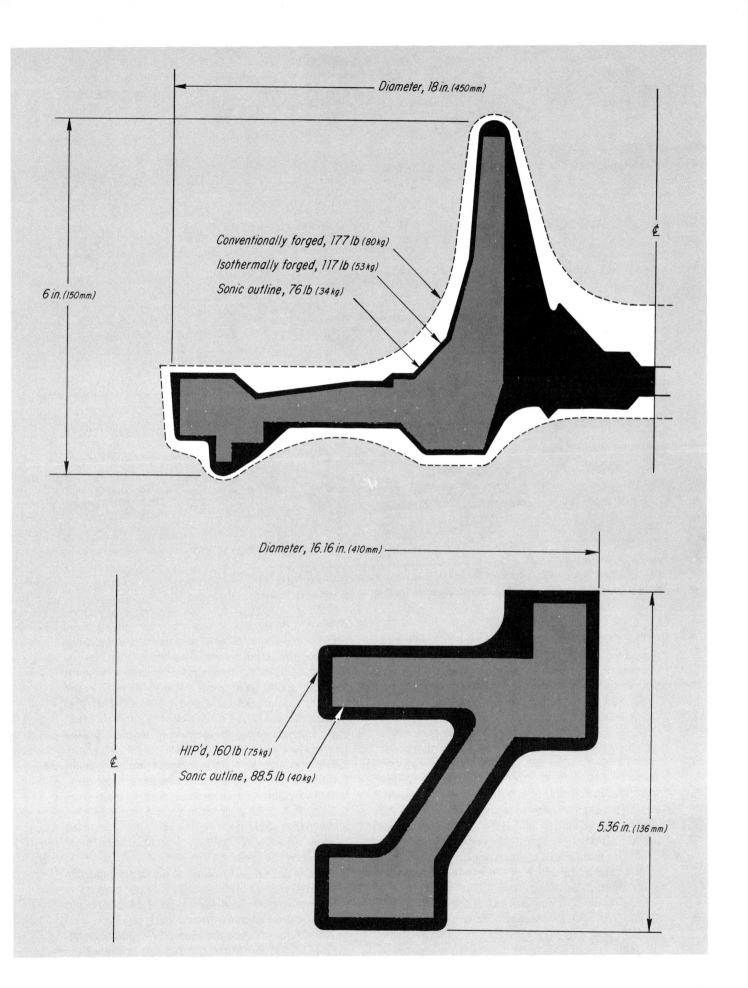

Diameter, 18 in. (450mm)

6 in. (150mm)

Conventionally forged, 177 lb (80kg)
Isothermally forged, 117 lb (53kg)
Sonic outline, 76 lb (34 kg)

℄

Diameter, 16.16 in. (410mm)

HIP'd, 160 lb (75kg)
Sonic outline, 88.5 lb (40kg)

℄

5.36 in. (136 mm)

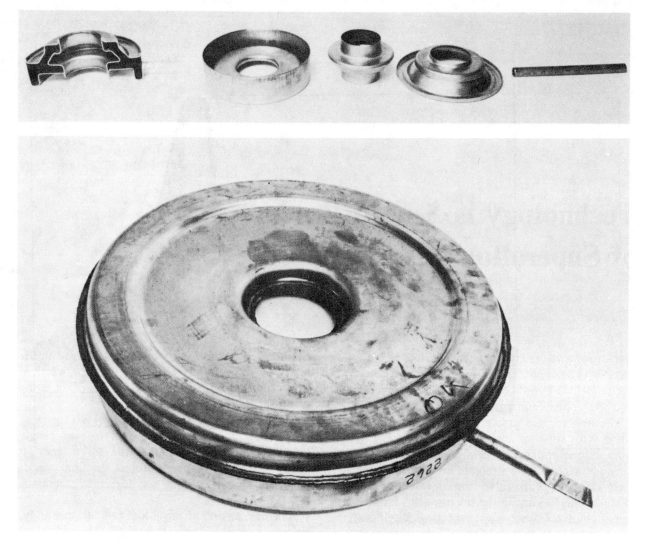

Fig. 1 — Top, can components used in the hot isostatic pressing of René 95 turbine wheels. Bottom, sealed and powder filled can with crimped spout prior to HIP'ing.

The development of Gatorizing introduced a new dimension to the forging of superalloy turbine components. Significant improvements were realized in both shape sophistication and materials utilization.

To qualify as a supplier to P&WA, Wyman-Gordon became licensed to forge IN-100 by their process. The parts would be Gatorized in a newly installed, 1800 ton (16 MN) Verson isothermal press.

Challenges — Our task was to construct a system that would heat and transfer to the press production quantities of extruded powder billets. We also had to establish a quality control program to assure that the process was repetitive and that each part would respond properly to the postforge heat treatment.

Major mechanical and metallurgical problems had to be solved in the design of this production isothermal forging/Gatorizing system. For example, the process involves heating stock to about 2050 F (1120 C) and forging at a very controlled strain rate in dies heated to the same temperature. The best available die material is TZM molybdenum, an alloy that is

very strong at the required forging temperature, but also one that oxidizes rapidly when exposed to air. Consequently, our first unit required a vacuum chamber around the press platen, a superalloy die stack to support the TZM molybdenum die set, and incorporation of temperature controls to assure that the material being forged would remain at the temperature where it is superplastic.

Other system features included a special vacuum furnace to bring the stock to forging temperature, transfer mechanisms for moving the forge mults from the furnace to the dies, and press controls that allowed the programming of the very slow, 0.25 to 0.5 in./min (0.1 to 0.2 mm/s) crosshead movement required by the process.

State of the Art — Forgings of the size produced in Wyman-Gordon's 1800 ton (16 MN) isothermal press were formerly made in large hammers and hydraulic presses, but with not nearly the same level of sophistication. And the superalloys used were not as difficult to forge as IN-100. In some instances, the new shapes are made in a single forging operation where three conventional operations were formerly required. Other shapes are made using multiple isothermal operations, an approach generally pursued when superior materials utilization is possible and can be justified.

The IN-100 forgings for the F-100 engine are true near net shape parts: cover is approximately 0.050 in. (1 to 1.5 mm). Isothermally forged parts are also being made from other nickel base alloys, such as René 95, and from Ti-6A1-4V. These alloys, however, are not made superplastic, so they're not "Gatorized" as described in the P&WA patent. Instead, each is isothermally forged in the upper portion of its forging temperature range where the material's flow stress is very low.

More than 3000 turbine components have been produced using these processes; nondestructive inspection has resulted in the rejection of just three.

We are confident of the future: a second, 3000 ton (27 MN) isothermal press of improved design has recently gone on stream, and engineering for an even larger unit is in progress.

Beta Forged Titanium — In the mid 1960's, the concept of beta forging of titanium was introduced. Although the process, which utilizes higher than normal forging temperatures, is not universally accepted by the metallurgical community, it has found significant use in applications requiring high creep resistance and improved fracture toughness.

Wyman-Gordon has developed a near isothermal beta forging process for Ti-17, a titanium alloy used by General Electric in its CFM-56 and F-404 engines. The components are forged in special presses to near net tolerances which results in an improvement in materials utilization of about 50% and superior mechanical properties.

HIP'ing Handles René 95 Turbine Wheels

Shortly after addressing the isothermal forging/Gatorizing challenge, Wyman-Gordon took on an even greater one in terms of near net shape technology: making nickel base alloy turbine hardware, without forging, via hot isostatic pressing (HIP). Prompting this venture was General Electric's decree that hot isostatically pressed, powder metal René 95 turbine wheels would be used in its new T-700 and F-404 engines.

In September 1978, we accepted and put in operation a 47 by 60 in. (1190 by 1520 mm), ASEA hot isostatic press. (The press is shown on the cover of this issue of *Metal Progress*.) Our only HIP'ing experience to date had been two years with a 10 in. (250 mm) unit.

Now we had to learn not only the time-temperature-pressure parameters needed to consolidate René 95 powder, but also the powder handling and canning techniques needed to produce the desired shape.

Cleanliness is Critical — The key to making satisfactory René 95 turbine hardware by HIP'ing proved to be powder cleanliness.

All low cycle fatigue failures originate at foreign particles. These particles are present in cast and wrought superalloys as well as in powder alloys. In fine mesh powder, however, they're significantly smaller — so small that they are not detectable by today's most sophisticated nondestructive testing procedures.

For example, check the close-up of the fracture surface of a low cycle fatigue test bar in Fig. 2. By any previous standard, the structure is metallurgically clean. Failure, however, is shown to have originated at a very small, 0.001 in. (0.025 mm) in diameter ceramic inclusion.

Solution — It became evident that low cycle fatigue requirements could only be met if powder of improved cleanliness became available. This need was addressed by using finer mesh powder and screening out foreign particles produced in the melting and atomizing processes. Also, stringent cleanliness requirements for incoming powder have been instituted and control samples are periodically taken at various stages in the powder handling process. ◈

For More Information: You are invited to contact the author directly by letter or telephone. Mr. Coyne, FASM, is vice president-technical director, Eastern Div., Wyman-Gordon Co., 105 Madison St., Worcester, Mass. 01613; Tel: 617/756-5111. The article is based on a presentation given by the author at the 10th International Drop Forging Convention in London, England on 26 June 1980.

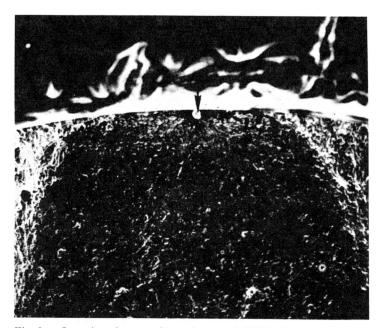

Fig. 2 — Scanning electron photomicrograph (55X) of a low cycle fatigue failure in HIP'd René 95 powder. Arrow points to origin of failure — a ceramic inclusion measuring just 0.001 in. (0.025 mm) in diameter.

ADVANCED TITANIUM METALLIC MATERIALS AND PROCESSES FOR APPLICATION TO NAVAL AIRCRAFT STRUCTURES

W.T. (Ted) Highberger
NAVAIR, Naval Air Systems Command, Washington, D.C.

and

Govind R. Chanani and Gregory V. Scarich
Northrop Corporation, Aircraft Division,
Hawthorne, California

This paper describes three major efforts to reduce the cost or improve the performance of titanium components on naval aircraft: 1) isothermal rolling, 2) hot isostatic pressing (HIP) of titanium powder, and 3) superplastic forming/diffusion bonding (SPF/DB) of Ti-6Al-4V and CORONA-5.

Because of the high cost savings potential for producing long constant cross-section parts, the Navy is sponsoring isothermal rolling process development in two areas: custom Tees and square bend. To demonstrate the custom Tee process, the manufacturing technology necessary to produce the fuel floor support for the F-18 is being developed. The square bend process will be demonstrated by producing an F-14 wing beam.

Three programs to produce near-net shape parts by HIP of titanium powder for two Navy aircraft, the F-14 and the F-18, will be reviewed. The first program demonstrated the feasibility of manufacturing a HIP titanium powder component, an F-14 fuselage fitting, and established its flight worthiness, reproducibility, economics, and NDI criteria. Success with the small F-14 fuselage fitting (1.7 pounds) led to scale-up effort on the F-18 aircraft arrestor hook support fitting. This part has a flying weight of 28.4 pounds. In the third program an F-14 nacelle frame with flying weight of 53 pounds will be fabricated by electron beam welding four HIPed powder subcomponents into an oval-shaped frame section.

Two projects in superplastic forming/diffusion bonding (SPF/DB) of titanium have been initiated, one will make a triangular glove vane for the F-14 and the other, with Air Force funding, will make a typical three sheet SPF/DB sandwich section of CORONA-5 titanium alloy.

1. INTRODUCTION

High strength titanium alloys are used in advanced aircraft systems both as airframe structural members and engine components because of their high strength-to-density ratios, combined with other favorable properties such as elevated temperature creep strength, corrosion resistance, fracture toughness, and thermal expansion compatibility for joining with graphite/epoxy composites. Despite these advantages, titanium alloys are not used extensively because of the relatively high fly-away cost of titanium components. This is due to the high cost of fabricating titanium components and the extremely low economic value of the scrap. If the "buy-to-fly" ratio can be improved, less raw material will be needed and the amount and cost of machining will be reduced. Several different approaches are being evaluated to manufacture net or near-net shapes with acceptable properties. The three approaches described in this paper are isothermal rolling, hot isostatic pressing (HIP) of prealloyed titanium powder, and superplastic forming and diffusion bonding.

Isothermal rolling is particularly suitable for long, slender shapes. This process uses direct resistance heating of the metal being worked, and can be used effectively to roll special shapes from various alloys including titanium.

The HIP of prealloyed titanium powder produces dense products with mechanical properties equivalent to wrought products. In this process, a mold is filled with prealloyed powder, the mold is sealed in an evacuated can, and then isostatically pressed at a high temperature and pressure to produce a fully dense near-net shape.

Superplastic forming and diffusion bonding (SPF/DB) is a relatively well-developed approach that is at present in its early stages of application. With this method, a differential pressure is applied across a titanium sheet at a preselected temperature to produce complex, sheet metal, titanium parts that could not be fabricated by conventional forming processes. The next three sections describe the Navy-sponsored programs and results in these areas.

2. ISOTHERMAL ROLLING

The basic principle behind isothermal rolling is illustrated in Figure 1. Molybdenum tooling, in this case rolls, are used to form the part and also as electrodes to heat the part. Electric current passing from one roll to the other through the workpiece heats the workpiece and enough of the rolls to form a traveling hot zone. Shaping occurs under the combined action of the squeeze force and feed force. The feed force is essential for large reductions that may exceed 95 percent reduction in a single pass.

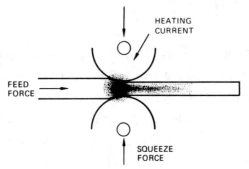

HEATING
CURRENT

FEED
FORCE

SQUEEZE
FORCE

Figure 1. Isothermal Rolling-
Schematic

The basic process can be modified to use one roll operating against fixed tooling, shaped rolls, split rolls, roll-forge dies, or combinations of these[1]. The goal of isothermal rolling is to reduce the cost of titanium components by reducing the amount of material required for each component and by reducing the amount of machining. Two programs presently underway to apply the cost saving benefit to Naval aircraft structures will be described: the custom Tee process for F-18 structural components, and square bend of an F-14 front wing beam.

2.1 ISOTHERMAL ROLLING OF F-18 STRUCTURAL COMPONENTS

Northrop's portion of the F-18 has many titanium components that would benefit by the economies of isothermal rolling. These components are listed in Table 1. Based on the capacity of the presently available isothermal rolling equipment and the potential of implementation, the fuel bay floor support was selected for development and potential application to the F-18. The Heintz Division of Kelsey-Hayes is developing a production isothermal rolling capability which can handle larger components.

The fuel bay floor supports, right and left pair, are shown in Figure 2. The supports are 40 inches long and weigh 1.3 pounds each. Presently the supports are machined from extrusions weighing 7.5 pounds each (buy-to-fly ratio of 5.6). The isothermally rolled support is expected to have a buy-to-fly ratio of less than two, thus saving 10 pounds of titanium per pair.

The first phase of this program is complete. This phase consisted of producing and evaluating a Tee section which includes all of the isothermal rolling steps involved in making the support. The other steps are creep forming to final contour and machining thin and cut-out areas. The evaluation of isothermally rolled parts will use the present F-18 component as a baseline. A minimum of five isothermally rolled pieces are to be evaluated. To accelerate the program the testing was initiated on the basic isothermally rolled shape used to make the full component – a T-shape (Figure 3). This testing is complete and is reported herein. In the next phase, which is underway, the fully formed and machined components will be evaluated. A brief description of the evaluation planned for the Tee and fuel bay floor support is presented in the following paragraphs, followed by the results for the Tee.

Non-destructive testing of all material will include dimensional, radiographic, penetrant, ultrasonic, and residual stress determinations. Me-

TABLE 1. F-18A CANDIDATES FOR ISOTHERMAL SHAPE ROLLING

Component Description	Buy-to-Fly[a] Ratio	Quantity per S.S.[b]	Finish Wt. per S.S. (Lbs)	Rough Wt. per S.S. (Lbs)	Length (inch)
Rudder Spar	14.9	2	6.80	101.4	66
Lower Center Longeron	9.3	2	35.38	327.6	120
Lower Inboard Longeron	9.9	2	22.10	219.8	82
Former Assembly	2.2	2	11.82	26.5	44
Fuel Bay Floor Stringer	8.1	2	2.92	23.7	37
Fuel Bay Floor Support	5.6	2	2.66	15.0	40
Stiffener Keel Beam	12.5	1	0.47	5.9	24
Stiffener Keel Beam	3.2	2	0.80	2.5	13
Stiffener Keel Beam	30.9	1	0.19	5.9	24
Stiffener Keel Beam	30.9	1	0.19	5.9	24
Duct Plate	6.3	2	1.64	10.4	30
Forward Door Hinge Half	6.5	2	1.10	7.1	4
Aft Door Hinge Half	6.4	2	1.50	9.7	4
Arresting Hook Longeron, Segment	4.3	2	0.80	3.4	18
Upper Aft Longeron	9.4	2	7.52	70.9	67
Lower Aft Longeron	6.1	2	24.80	151.3	94
Longeron Splice Plate	5.8	2	1.60	9.2	14
Lower Keel Beam Cap	4.2	1	2.10	8.7	27
Upper Keel Beam Cap	5.8	1	1.10	6.4	20

[a] Raw weight/finished weight

[b] Shipset

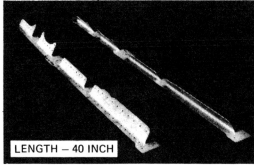

LENGTH — 40 INCH

Figure 2. F-18 Fuel Bay Floor Supports

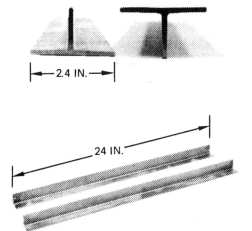

2.4 IN.

24 IN.

Figure 3. Isothermally Rolled Tees

chanical property evaluation will include tension, sharp-notch tensile, bearing strength, S/N fatigue, and fatigue crack growth. All mechanical properties will be compared directly to the existing beta extrusions and where possible to alpha-beta titanium. In the following paragraphs, results are reported for the four Tees evaluated and for the extrusion presently used. Metallography and fractography will be performed to determine microstructure and fracture characteristics.

2.1.1 Tensile Properties

The results are presented in Table 2. Specimens were tested from the legs of the Tees and from the intersection of the legs, which is worked and heated in a different sequence than

TABLE 2. TENSILE PROPERTY COMPARISONS

Location or Thickness	UTS (ksi)	YS (ksi)	% e	% RA
Isothermally Rolled Tee				
Legs[a]	146	135	13	-
Intersection[b]	146	135	12	46
Extrusions Presented Used				
Legs[a]	148	130	12	-
Intersection[a]	145	135	15	28
Design Mechanical Properties Annealed Ti-6Al-4V, MIL-HDBK-5C				
Sheet, Strip, and Plate[c]				
<0.062-in.	134	126	8	-
0.0632-0.1875-in.	134	126	10	-
0.1875-4.000-in.	130	120	10	-
Extrusion[d]				
<4.00-in.	130	120	10	-

[a] Average of 12 tests in longitudinal and transverse orientation. Flat tensile specimens (0.07-inch thick by 0.25-inch wide with 1.0-inch gage length per ASTM E8) were tested from the legs of the parts.

[b] Average of 4 tests in longitudinal orientation. Round tensile specimens (0.16-inch diameter with an 0.65-inch gage length per ASTM E8) were tested from the intersection of the legs.

[c] MIL-T-9046, "Titanium and Titanium Alloy Sheet, Strip and Plate."

[d] MIL-T-81556, "Titanium and Titanium Alloys, Bars, Rods, and Special Shapes, Extruded."

the legs. No important differences in tensile properties between the extrusion and isothermally rolled Tees is seen. All tensile properties exceeded minimum requirements.

2.1.2 Bearing Test

Bearing tests are used to determine the resistance to hole pullout. The results are presented in Table 3. Due to an error in machining the specimens from the extrusions, the edge distance ratio was 1.9 instead of the intended 2.0. An edge distance ratio of 1.9 is a more severe condition than 2.0, nevertheless these results still exceeded MIL-HDBK-5C A-basis values for MIL-T-81556 material for an edge distance ratio of 2.0.

For the isothermally rolled Tees, all values are well above the minimum requirements. The results are similar to those for the extrusion; however,

TABLE 3. BEARING TEST RESULTS

Tested per ANSI/ASTM E 238-68, "Pin-Type Bearing Test of Metallic Materials."

Longitudinal orientation

Specimen width, 1.0 inch; hole diameter, 0.188 inch; Pin diameter, D, 0.1875 inch

Product	Thickness (inch)	Edge Distance Ratio, e/D	Bearing Yield Strength (ksi)	Bearing Strength (ksi)
Isothermally Rolled Tee	0.07	2.0	229	300
Extrusion Presently Used	0.07	1.9	235	286
Extrusion[a]	<2.0	2.0	216	268
	2.01-3.00	2.0	212	268
Sheet and Plate[b]	<0.1875	2.0	208	252
	0.1875 - 2.000	2.0	198	245

[a] MIL-T-81556, "Titanium and Titanium Alloys, Bars, Rods and Special Shaped Sections, Extruded," Type III, Comp. A, Annealed; MIL-HDBK-5C A-basis values

[b] MIL-T-9046, "Titanium and Titanium Alloy, Sheet, Strip and Plate," Type III, Comp. C, Annealed; MIL-HDBK-5C A-basis values

the difference in edge distance ratios makes a direct comparison difficult.

2.1.3 Sharp-Notch Tensile

ASTM E 338 procedures were followed except the specimen was narrower than required due to the size of the fuel bay floor support. These data are presented in Table 4. Both the sharp-notch strength (SNS) and the sharp-notch strength to yield strength ratios (SNS/YS) for the isothermally rolled Tees, which are indicative of the resistance to unstable fracture from sharp cracks such as fatigue cracks, are below those for the extrusion. This is probably due to the extrusion being beta formed and the isothermally rolled Tee being alpha-beta formed.

2.1.4 S/N Fatigue – Notched

The test results are presented in Figure 4. The notched fatigue behavior of the ISR Tees is as good or slightly better than that of the extrusion.

2.1.5 Fatigue Crack Growth

Data for both TL and LT orientations are shown in Figure 5, along with data for the extrusion. In comparing individual results, no differences were seen between TL and LT. In Figure 5, it can be seen that particularly at lower stress intensities the fatigue crack rate for the isothermally rolled Tee is faster than that for the extrusions. This is probably due to the extrusions being beta formed while the ISR Tees are alpha-beta formed.

TABLE 4. SHARP-NOTCH TENSILE RESULTS FOR ISOTHERMALLY ROLLED TEES

Test Based on ASTM E 338-68[a]

Max. Precrack Load = 2,000 lbs, R = 0.1, Load Rate = 100 ksi/min.

Material	Typical Test Parameters				Average of 4 Tests	
	Thickness, h (in.)	Width, W (in.)	Crack Length, 2a (in.)	Max Load,P (kips)	Sharp-Notch Strength, SNS[b] (ksi)	$\frac{SNS}{YS}$
Isothermally Rolled Tee	0.070	1.00	0.35	5.63	123	0.91
Extrusion Presently Used	0.070	1.00	0.35	6.26	135	1.02

[a] Specimen is narrower than required in ASTM E 338.

[b] Sharp-notch strength, $SNS = P/h(W - 2a)$.

[c] Pre-cracked per ASTM E 338 Appendix A2; except crack length, 2a, 0.35 inch.

2.1.6 Non-Destructive Inspection

Non-destructive testing of the isothermally rolled Tees did not reveal any defects, except for an unacceptable out of flatness, which will probably be solved by minor process variation or in the final creep forming step.

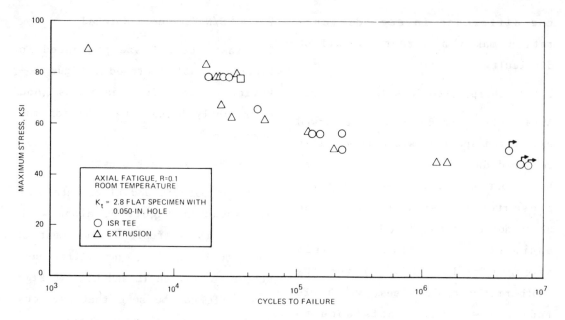

Figure 4. S/N Fatigue Results for Isothermally Rolled Tees

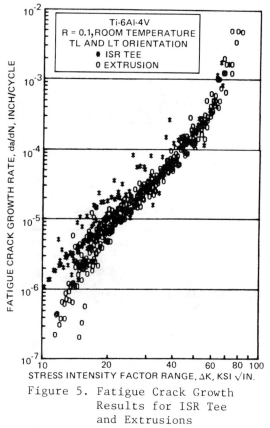

Figure 5. Fatigue Crack Growth
Results for ISR Tee
and Extrusions

2.1.7 Comparison of the Isothermally Rolled Tees and Presently Used Extrusions

Tensile, notched fatigue, and bearing properties of the isothermally rolled Tees and the extrusion were compara-

ble. The fatigue crack growth and sharp-notch tensile properties of the isothermally rolled material were somewhat below those for the extrusion. These two better extrusion properties probably result from the extrusion being formed in the beta temperature range. Beta processed Ti-6Al-4V products generally have better fatigue crack growth resistance and fracture toughness than alpha-beta products.

Although the properties for alpha-beta processed material are generally acceptable for the F-18, most of the parts being considered for manufacture by isothermal rolling were qualified as beta-processed extrusions. The general Navy requirements for material substitutions require property equivalence. However, the component under evaluation is designed by tensile properties and, since these properties are equivalent the isothermally rolled part is expected to qualify for F-18 application.

2.2 FRONT WING BEAM

One of the programs, to demonstrate the isothermal rolling process, is to make a front wing beam for the F-14. This is a titanium (Ti-6Al-6V-2Sn) tapered channel 9-feet long by 6-inches wide by 2.5-inches deep with 0.091-inch metal thickness. The bends are at an angle which changes along the length. The F-14 has a total of eight such wing beams of varying dimensions. The early production parts were hogged out of plate with a buy-to-fly ratio of 11. A slight improvement in the buy-to-fly ratio was realized by starting from a forging. The part is now made by electron beam welding two rough machined extrusions followed by finish machining with a buy-to-fly ratio of about 5.

The wing beam produced by the square bend process will have a buy-to-fly ratio of 1.8. In the square bend process, the part is first brake formed. Then it is rolled or ironed down by molybdenum rolls with localized resistance heating to perform the local deformation. The Heintz Division of Kelsey-Hayes is prepared to supply production parts when the process is qualified by Solar and Grumman.

3. HOT ISOSTATIC PRESSING (HIP) OF TITANIUM POWDER

Recent government-sponsored work has shown that titanium components can be produced at lower costs by HIP of spherical titanium alloy powders than by conventional means. At present one powder production technique and two HIP compaction techniques (the Ceramic Mold and the Fluid Die processes) have the potential for production application. To take advantage of this process and gain acceptance by the industry, both the Navy and Air Force have major efforts underway to scale-up the process, demonstrate the consistent properties in titanium P/M products, and develop a design data base. In this section of the paper, the process will be briefly described followed by a discussion of the Navy-sponsored effort.

3.1 PROCESS DESCRIPTION

The Crucible Ceramic Mold process described elsewhere[2,3] is being used for the Navy programs. Figure 6 shows schematically the principal steps in the process. Since it is difficult to predict consolidation characteristics of complex shapes, such as in this program, several iterations of the steps shown in Figure 6 have to be made to design the tooling to produce an accurate near-net final shape. All of the titanium alloy powders (Ti-6Al-4V ELI and Ti-6Al-6V-2Sn) used to produce parts have been made by the rotating electrode process. The HIPed parts will undergo dimensional analysis, NDI, mechanical testing, and microstructural examination. Furthermore, the parts will also be structurally tested to evaluate the durability and damage tolerance of all three parts. Completion of all the structural tests and coupon testing coupled with cost analysis on these Navy parts is expected

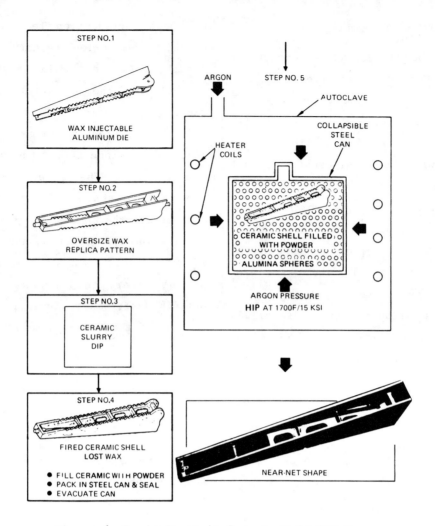

Figure 6. Basic Steps in Ceramic Mold HIP Process

to demonstrate the structural integrity of complex P/M titanium shapes, establish cost-effectiveness, and provide data for a design data base.

3.2 RESULTS AND DISCUSSION

The work on the three parts is interrelated and is at a different stage of progress for each part. Table 5 lists the parts and alloys and shows their buy-to-fly ratios and material utilization factors. The work on the fuselage fitting, the smallest of the three parts, is complete and has been reported in detail[4,5]. Only a brief summary of these results will be presented. The successful production of

this part led to the initiation of work on the arrestor hook support fitting. This work started in late 1978 as part of AFML/Crucible Contract No. F33615-77-C-5005, and will be completed in 1982. The first year's progress was described in detail[6] and will be summarized here together with the new results. The third program which is just starting is the most ambitious of all. It derives from all the experience obtained in the titanium P/M area and will utilize electron beam welding to fabricate the 53 pound F-14A nacelle frame from four HIPed P/M subcomponents.

Part	F-14A Fuselage Brace Support Fitting	F-18A Arrestor Hook Support Fitting	F-14A Nacelle Frame
Alloy	Ti-6Al-6V-2Sn	Ti-6Al-4V ELI	Ti-6Al-6V-2Sn
Flying Weight, lb	1.7	28.4	53.0
B/F Ratio[a]			
o Forging	3.6	9.0	6.6
o HIP	1.2	1.9	1.3
MUF, percent[b]			
o Forging	28	11	14
o HIP	83	52	77

[a] Buy-to-fly ratio = Raw weight/machined weight

[b] MUF = Materials Utilization Factor = (Machined weight/raw weight) X 100

3.2.1 F-14A Fuselage Fitting

The feasibility of manufacturing the F-14A fuselage fitting was demonstrated by establishing flight-worthiness, reproducibility, economics and NDI criteria[4,5]. The work showed that the materials utilization factor (MUF) was increased from 28 percent to 81 percent and cost savings of 25 to 40 percent could be realized for this part, depending on the quantity purchased. A production lot of this part has been ordered and is scheduled for flight in late 1980.

3.2.2 F-18 Arrestor Hook Support Fitting

Scale-up of the process to manufacture the F-18 arrestor hook support fitting (AHSF) will demonstrate the viability of HIP titanium P/M technology in the manufacture of large, complex, structurally critical airframe parts. This fitting provides the fuselage attachment of the arrestor hook used for cable braking in carrier landing.

Currently, in early production, the AHSF is machined from a hand-forged billet with a buy-to-fly weight ratio (raw weight/machined weight) of 15. However, a die forging with a buy-to-fly weight ratio of 9 will be used for producing this part. This compares with a ratio of 2 when produced from a near-net HIP shape. Machining of the P/M part will be limited to facing and drilling the lug and the finish machining of the interior of the AHSF along the upper flanges which mate with the keel structure. Three iterative changes in mold geometry are anticipated to establish the mold shape to produce an accurate near-net AHSF shape.

Due to the critical nature of the part, only the ELI (Extra Low Interstitial) grade Ti-6Al-4V powder will be used. The powder is being produced by the rotating electrode process. A heat treatment study using witness blocks was performed to select the heat treatment for the HIP AHSF[5]. In this investigation, three different heat treatments as well as the as-HIP condition were investigated to select the heat treatment which would provide satisfactory mechanical properties with the minimum amount of distortion during the heat treatment cycle. The three heat treat conditions were beta anneal, recrystallize anneal, and mill anneal. Tension, fracture toughness, precracked charpy, fatigue-crack growth, and low cycle fatigue tests supplemented by microstructural analysis were performed on all four conditions. Based on the heat treatment investigation results, the mill anneal condition has been selected for the HIP AHSF. The tensile properties of the annealed P/M material meet the specification requirements of wrought titanium material and the heat treatment response of the P/M material is similar to that for the wrought material.

The first iteration of the AHSF is complete and the second trial is nearing completion. The first iteration shape proved the process to be feasible in producing a large and complex shape such as the AHSF. Figure 7 shows the first shape trial AHSF with a comparison to the actual machined part. The HIP AHSF was heat treated to a mill-anneal condition

and chemically milled following heat treatment. In addition, microstructural examination, nondestructive inspection, and mechanical tests were performed, and the results are discussed below.

Radiography of the entire AHSF was performed at a minimum of two angles. Radiographic and penetrant examination revealed several surface voids as well as several areas which were not completely filled. Indications of dense particles (probably tungsten) were found throughout the part by radiography. These indications averaged approximately 0.035 inch in size with the largest particles being about 0.085 inch. Even though some of the areas on the HIP AHSF were not completely filled, overall dimensions and the shape of the part were considered very good. No significant dimensional changes were noted due to the heat treatment.

Test specimens from several locations were excised from the heat treated and chem-milled first iteration AHSF. Tensile properties of specimens (average of three) excised from the first iteration AHSF are as follows: YS = 121 ksi, UTS = 133 ksi, and elongation = 16 percent. These properties met the specification requirements for the wrought Ti-6Al-4V ELI material being used for the AHSF. Two precracked charpy specimens excised from the first iteration AHSF were tested in slow bend. The nominal crack strength obtained from these tests (255 ksi) were 15 percent below the values obtained for the

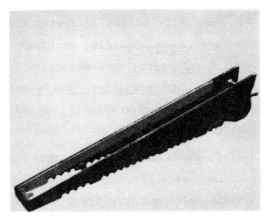

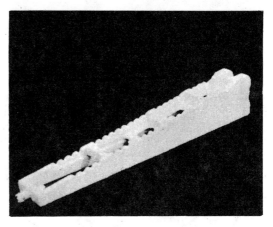

a. MACHINED WAX PATTERN OF THE ARRESTOR HOOK SUPPORT FITTING

b. CERAMIC SHELL MOLD MADE FROM THE WAX PATTERN IN (a)

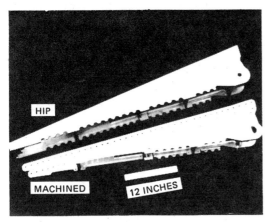

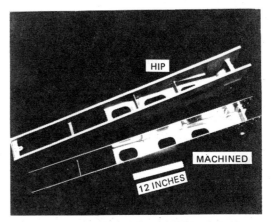

c. COMPARISON OF MACHINED AND FIRST ITERATION HIPED ARRESTOR HOOK SUPPORT FITTING

Figure 7. First Iteration Shape Trial Results for Arrestor Hook Support Fitting

mill annealed P/M material (295 ksi) tested during the heat treatment study[5]. However, the equivalent fracture toughness of 75 ksi-in.$^{1/2}$ exceeded the minimum specification requirements of 70 ksi-in.$^{1/2}$. A fatigue-crack growth specimen excised from the AHSF was also tested. The results were virtually identical to the values obtained during the heat treatment study. Besides some unfilled and uncompacted areas, the first iteration results also showed other problem areas. These were (1) powder contamination, (2) lack of temperature uniformity in the autoclave over the length of the AHSF, and (3) hydrogen pick-up. All these problem areas are being addressed in future shape trials.

Preliminary cost-analysis indicates a significant cost-savings by the use of the HIP AHSF which increases even more with time because of the increasing price of titanium. A final cost-analysis with the breakdown of each cost element will be performed once the shape trials are complete.

3.2.3 F-14A Nacelle Frame

The third naval aircraft part is a nacelle frame weighing 53 pounds which will be fabricated by electron-beam (EB) welding four HIPed powder subcomponents into an oval-shaped frame section (Figure 8). Scale-up

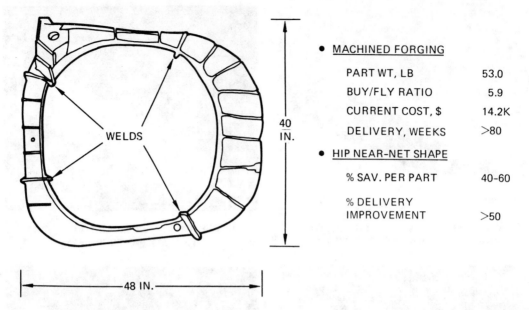

● MACHINED FORGING	
PART WT, LB	53.0
BUY/FLY RATIO	5.9
CURRENT COST, $	14.2K
DELIVERY, WEEKS	>80
● HIP NEAR-NET SHAPE	
% SAV. PER PART	40-60
% DELIVERY IMPROVEMENT	>50

Figure 8. Hybrid Welded F-14 Nacelle Frame (Artist's Conception)

to parts ranging from 20 to 50 pounds such as the AHSF promises significant MUF (around 40-80 percent), cost savings over 40 percent and potential for reducing delivery times by about 50 percent or more. In this case, an attempt is being made to increase the MUF from 14 percent for machined forgings to over 90 percent for the HIP component. Qualification of this type of part should increase autoclave throughput and result in reduced autoclave charges per part.

Preliminary EB welding parameter work performed on transverse butt welds on HIP Ti-6Al-6V-2Sn indicates that acceptable quality and mechanical properties are attainable. Tensile ultimate strength, yield strength and elongation for the EB butt welds were 158 ksi, 149 ksi and 16 percent, respectively.

The program, which is just starting, consists of three phases: powder certification, design and processing of EB-welded hybrid nacelle frames, and reproducibility study and structural component tests.

4. SUPERPLASTIC FORMING/DIFFUSION BONDING (SPF/DB)

Superplastic forming of titanium sheet for structural shape making is a technology that has been widely investigated and has now advanced to a production process. When superplastic forming is combined with diffusion bonding it becomes possible to fabricate a number of structural shapes based on sandwich construction. An example of a basic sandwich geometry would be a "truss-core" sandwich consisting of two cover sheets and a zig-zag shaped core. Sine-wave and four-sheet construction are other examples.

4.1 SPF/DB OF F-14 GLOVE VANE

To maximize the benefit of this new technology, redesign is necessary. Such redesign is planned in the program with Grumman for an F-14 glove vane. This glove vane is a 3-inch deep, 7-foot long, 3-foot wide triangular 2024 aluminum assembly, consisting of 64 parts and over 1700 fasteners of various types. Although labor costs have been reduced somewhat by semi-automatic fabrication techniques, manual operations remain high and are typical of many similar structures. In addition, because the material is aluminum, corrosion protection in the form of coatings and sealants is required. The redesigned glove vane will be made by an expanded sandwich truss core of four sheet construction. This approach is particularly attractive because it will replace aluminum built-up structure with its inherent vulnerability to environmental corrosion and associated high life-cycle maintenance costs while at the same time reducing structural weight by 10 percent and assembly costs by 25 percent. The program, in addition to making the glove vane, is expected to qualify the process for numbers of other F-14A details and, ultimately, similar structures for DoD procurement.

4.2 SPF/DB OF CORONA-5

Another innovation in the SPF/DB area is the SPF/DB of new titanium alloy, CORONA-5. This alpha-beta titanium alloy (Ti-4.5Al-5Mo-1.5Cr) has excellent toughness, greater than 100 ksi-in.$^{1/2}$ at 135 ksi UTS. The advantage for the SPF/DB sandwich application, however, is that some of the "excess" toughness can be traded off for strength. The ultimate strength level of 160 ksi was chosen because it can be attained by a sim-

ple air cool without distortion, whereas to achieve this strength level in Ti-6Al-4V would require a water quench with high distortion. Thus, CORONA-5 for this application would far surpass Ti-6Al-4V in strength, toughness, and ease of fabrication; and would even compete favorably with the beta alloys in strength and toughness when the differences in modulus and density are accounted for.

To develop the necessary parameters for definition of SPF/DB CORONA-5 a contract was let to Rockwell International Science Center by the Navy using Air Force funding. The objective of the program was to heat treat a typical three sheet sandwich geometry to 160 ksi UTS without water quenching.

First titanium slabs of two different oxygen contents were hot rolled to 0.170-inch thick sheet. These were cold rolled 50 percent to 0.073 inch with a 1400F/5 minute strand line anneal and another 1400F final anneal. The 0.085 percent oxygen material had an average longitudinal and transverse bend test result of 3.0t while the average for the 0.153 percent oxygen material was 4.0t; meeting the specification requirements from MIL-T-9046. These results indicate that CORONA-5 can easily be supplied as cold rolled strip at a savings of up to 30 percent over Ti-6Al-4V for the process. The relatively low ductility would be fully restored in the SPF/DB heat cycle. Hence the indications are that CORONA-5 will be applicable in low cost titanium sand-

Source: Materials 1980, October 1980, 539-553

wich construction having high strength and stiffness to weight ratios, intermediate in properties between the conventional metals and metal matrix composites yet lower in cost than composites and with higher confidence in reproducible minimum mechanical property values.

5. SUMMARY

The Naval Air Systems Command has several ongoing programs that are expected to bring about substantial economies in the manufacture of titanium parts. Three of these are near-net shapes processes for three typical shapes - 1) isothermal rolling for high aspect ratio parts, 2) powder metallurgy and powder metallurgy/electron beam welding for more equiaxed complex parts, and 3) an advanced forming/assembly procedure, SPF/DB and SPF/DB combined with a new, high-strength, high-toughness titanium alloy, CORONA-5. These technologies are being applied to parts on Navy aircraft: F-14 wing beam and F-18 fuel bay floor support by isothermal rolling, F-14 fuselage brace and F-18 arrestor hook support fitting by powder metallurgy and F-14 nacelle frame by powder metallurgy/electron beam welding, and the F-14 glove vane by SPF/DB. SPF/DB with CORONA-5 is being demonstrated on a four-sheet sandwich structure. These programs will lead to reduced acquisition cost by reducing material input and/or piece count.

6. REFERENCES

1. A.G. Metcalfe, W.J. Carpenter, and F.K. Rose, Presented at National Aerospace Engineering and Manufacturing Meeting, San Diego, CA, Oct. 1974.

2. V.C. Petersen and V.K. Chandhok, Proceedings of the Symposium "Powder Metallurgy of Titanium Alloys," 109th Annual AIME Meeting, Las Vegas, Nevada, Feb. 1980.

3. J.H. Schwertz, V.K. Chandhok, V.C. Petersen, V.R. Thompson, AFML-TR-78-41, 1978.

4. R.H. Witt, et. al., Final Reports on Contract No. N00019-70-C-0598, Dec 1971, Contract No. N00019-74-C-0301, April 1975; and Contract No. N00019-76-C-0143, June 1977.

5. G.R. Chanani, R.H. Witt, W.T. Highberger, and C.A. Kelto, Proceedings of the Fourth International Conference on Titanium, May 1980, Kyoto, Japan.

6. G.R. Chanani, W.T. Highberger, C. Kelto, and V.C. Petersen, Proceedings of the Symposium "Powder Metallurgy of Titanium Alloys," 109th Annual AIME Meeting, Las Vegas, Nevada, Feb. 1980.

SECTION V
Hot Die Forging

RECENT DEVELOPMENTS IN HOT-DIE FORGING OF TITANIUM ALLOYS

C. C. Chen and J. E. Coyne

Wyman-Gordon Company, Worcester, Massachusetts   01613

Introduction

In recent years, considerable effort has been made on the manufacturing
technology for hot-die forging of titanium alloys (1-4). The major contribution of the hot-die forging is to reduce or to eliminate the influence of die
chilling and material strain-hardening. As a consequence, more refined shape,
better utilization of costly input material, substantial reduction of machining cost, and reduced number of forging operations are achievable.

The main objective of this paper is to present the recent developments
and advancements in the manufacturing technology for hot-die forging of
titanium alloys. The general content of the paper includes the current technology for hot-die forging to near-net shapes, metallurgical fundamentals to
hot-die technology, and economic and technological considerations for both
$(\alpha+\beta)$ and β-titanium alloys. Some of the current and potential manufacturing
applications for the technology are also illustrated.

Hot-Die Forging to Near-Net Shapes

In conventional forging, most of the forging operations employ die temperatures below 427C (800F) and ram rates above 4 mm/sec. (10 in./min.). At
these die temperatures and ram rates, the influence of die chilling and strain
hardening is very significant; the forgeability of the alloys at a given forge
temperature is then appreciably reduced. The manufacture of hot-die forgings
at relatively low ram rates appears to improve this situation; the forging of
near-net shape to significantly decrease the traditional material allowance
between the as-forged shape and the finish-machined shape becomes achievable.

From a broad sense, the hot-die forging process is defined as a deformation process during which the forging dies are maintained at the same or a
temperature slightly below that of the alloy being deformed. For the case of
Ti-6Al-4V alloy (designated as Ti-6-4), the forging temperature for $(\alpha+\beta)$
forging is about 954C (1750F). As the die temperature approaches 954C (1750F),
the metal flow can be closely controlled by the processing variables, and as
the material allowance between the forging and the machined part approaches
zero, it may become possible to forge a shape to the machine outline with only
a small amount per surface allowed, which can be removed by chem-milling or
very limited machining. Note that the terminology "hot-die forging" in this
paper covers both isothermal and hot-die (near-isothermal) processes.

The forging design attainable in conventional forging is primarily a
function of the alloy and its forging temperature, along with the number of
forging sequences processed through separate tool impressions. Increased web
thicknesses, large fillet radii, increased rib and flange widths, along with
decreased depths are design features generally required for adequate filling
of the die cavity using unit forge pressures of reasonable magnitudes. However, with a die system operating at increased temperatures, the decreased
differential between the forging stock temperature and the tool temperature

allows a more refined shape of the forged component to be produced in a given operation. Further control of the major process parameters such as the forge temperature, ram rate, preform microstructure, strain, strain rate, forge pressure, and dwell time shows that additional refinement is possible still maintaining reasonable forging forces at the tool temperature. Therefore, in hot-die forging, extensive metal flows are possible within one die cavity providing the preform shapes have previously distributed the necessary volume of material in localized zones from which the new shape is generated.

In the earlier development programs [2-4], the hot-die forging of ($\alpha+\beta$) titanium alloys concentrated on the concept of isothermal forging. In order to maintain truly isothermal forging, the technical efforts in these programs were primarily made to resolve the problems of designing, heating, and operating a tooling system above 927C (1700F). As a result, the IN-100 cast dies and induction heating tooling were developed for producing small and moderate size forgings. However, both structural and property stabilities of the IN-100 cast die alloys become serious problems at these operating temperatures, and the cost-effectiveness of the process to manufacturing applications becomes questionable. Based on the results of the experiments on forge temperature/die temperature/strain rate interactions, a more recent work demonstrated that by reducing the die temperature and increasing the ram rate, the hot-die (near-isothermal) approach is beneficial for increased die stability and strength so that a more dilute, less expensive Ni-base alloy could be used as die material (Figure 1). As a result, a tooling system using Astroloy die inserts and modular design concepts to accommodate a range of part sizes up to a maximum plan view of 3871 cm^2 (600 in^2) was developed. To further substantiate the hot-die technology to the manufacturing applications, very extensive effort was subsequently made to determine the forgeability, structure, and properties of various titanium alloys [5-15].

Metallurgical Fundamentals to Hot-Die Technology

Major metallurgical factors to hot-die forging technology are deformation processes, structural characteristics, and mechanical properties. Since the manufacturing capability for hot-die processing to near-net shapes requires the knowledge of the material response to forging deformation and since the forging of titanium alloys is a structure-sensitive problem, the understanding of the nature of metallurgical response under hot-die conditions is of great importance to hot-die technology. It has been demonstrated that the near-net achievable of the forgings depends strongly on forge temperature, die temperature, strain rate, and preform microstructure [5-8]. Adequate control of these variables can maximize the forgeability of the alloy and optimize the resultant microstructure and properties of the forgings.

Figure 2 illustrates the basic flow stress versus strain rate and strain-rate-sensitivity (m) versus strain rate plots for a typical superplastic material such as titanium alloys. As can be seen, three types of forging deformation, namely creep deformation, superplastic deformation, and conventional hot-working can be closely exercised in hot-forming of titanium alloys. However, the selection of the deformation type in manufacturing applications often depend on the factors such as forge pressure required, production rate, and other metallurgical qualifications. Because of significant chilling-effect from the dies, the conventional forging cannot provide close control of the deformation processes.

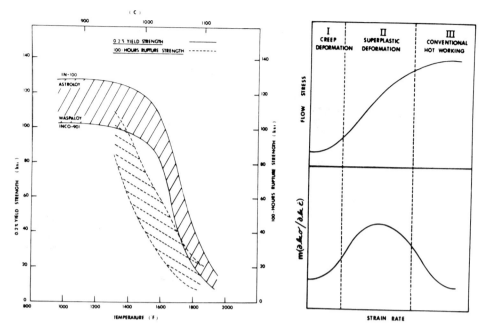

Figure 1   Effect of temperature on the yield
strength and 100-hour rupture life
of available nickel-base die alloys

Figure 2   A schematic illustration of the
deformation types for a superplastic material

The forging deformation properties under hot-die conditions were extensively investigated for various titanium alloys in recent years [5-8].   In
particular, the effect of preform microstructure, forge temperature, and ram
rate on the deformation behavior was emphasized (Figure 3).   It was found that
the deformation characteristics under forging could be quantitatively related
to the hot deformation properties of the alloy, and the rate-controlling
deformation process in $(\alpha+\beta)$ forgings was attributed rather to the dynamic
softening than to the more conventional rate-controlling mechanisms.   For
current commercial $(\alpha+\beta)$ titanium alloys, the deformation of titanium alloys
in the practical range of forge temperatures (871-982C or 1600~1800F) and ram
rates (0.4~4 cm/s or 0.1~1.0 in/min.) was found to associate with high strain-
rate-sensitivity and large strain softening; the deformation process is attributed to pseudo-superplastic behavior.

The structural characteristics of the forgings vary significantly with
die temperature, strain rate, stock temperature, strain and preform microstructure.   Figure 3 also presents examples of the as-forged microstructures
at various forge conditions illustrating the influence of preform microstructure and forge temperature on structural characteristics of the hot-die
forgings.   It is seen that the microstructural response due to forge processing is extremely sensitive to the forge temperature for both $(\alpha+\beta)$ and β
preforms.   At 816 to 899C (1500 to 1650F) temperature range, forging results
in a structure characterized by dynamic recrystallization, preferably occurring at and near the α/β interface boundaries.

The influence of die temperature and section thickness of the forgings on
the as-forged microstructures of Ti-6-4 alloy is given in Figure 4.   A significant difference in the nature of transformed-α regions can be clearly seen
for both $(\alpha+\beta)$ and β-forgings.   Also, a martensitic transformed-α' for conventional forging and a Widmanstatten transformed-α for isothermal forging are
seen for $(\alpha+\beta)$ forgings.   Note that the fracture toughness and ductility
properties are generally degraded by the presence of martensitic-α' microstructure.   For β-forgings, the hot-die approach to develop pseudo-β microstructure can significantly improve the ductility of the forgings.   It has

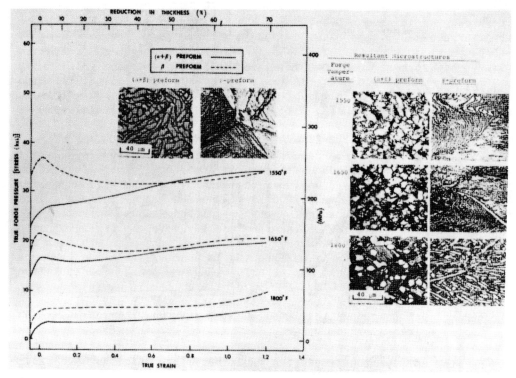

Figure 3   Examples of true stress-strain curves for isothermal forging of Ti-6Al-4V
            alloy pancakes using α+β and β-preforms at various forge-temperatures;
            ε̇=0.4 min.⁻¹.

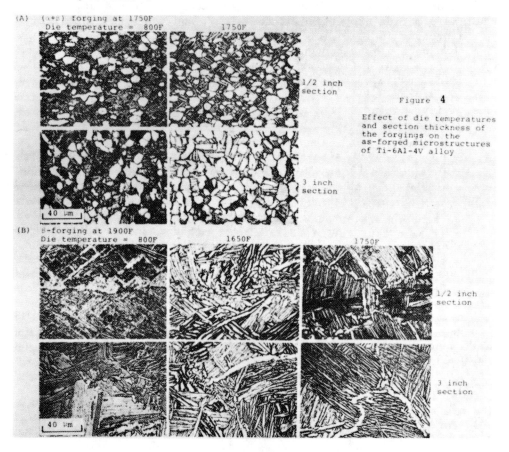

(A) (α+β) forging at 1750F
     Die temperature = 800F   1750F

1/2 inch
section

3 inch
section

Figure **4**

Effect of die temperatures
and section thickness of
the forgings on the
as-forged microstructures
of Ti-6Al-4V alloy

(B) β-forging at 1900F
     Die temperature = 800F   1650F   1750F

1/2 inch
section

3 inch
section

been previously shown that excellent uniformity of macrostructures is achievable for isothermal forgings. Very extended surface layers are observable for conventional forgings; the increased shear band and reduced metal flow due to die chill are very profound.

It can be generally stated that hot-die forgings result in a more uniform and improved mechanical properties of the forgings if proper forging variables are used [10-13]. The resultant properties depend critically on the forge temperature, die temperature, and preform microstructure. In general, fracture toughness, tensile ductility, and creep properties of the forgings can be significantly benefited by hot-die forging [11, 12]. The fracture toughness of the forgings generally increases, as the die temperatures increase (Figure 5). By controlling the die temperature slightly below the β-transus temperature, the relatively low ductility of the β-forgings can be significantly improved by hot-die approach (Figure 6). The creep resistance of Ti-6Al-2Sn-4Zr-2Mo∿0.1Si alloy (designated as Ti-6242Si) is seen to significantly increase by isothermal forgings (Figure 7). Metallurgically, such an improvement can be attributed to a more uniform, precise control of the nature and distribution of both globular-α and transformed products.

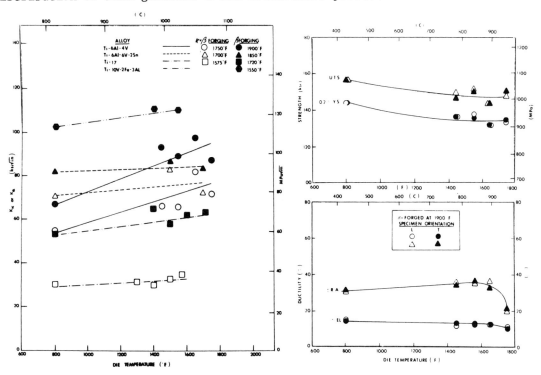

Figure 5 Variation of fracture toughness with die temperature for Ti-6Al-4V, Ti-6Al-6V-2Sn, Ti-17, and Ti-10V-2Fe-3Al alloys

Figure 6 Room temperature tensile properties vs. die temperature for β-forged Ti-6Al-4V structural shape forgings

Economic and Technological Considerations

The precise cost analysis for hot-die forging is difficult to provide because it varies with the particular forging selected; however, the major cost elements for hot-die forgings are associated with die material and operating die temperature, forging stock, forge temperature, and lubrication effectiveness.

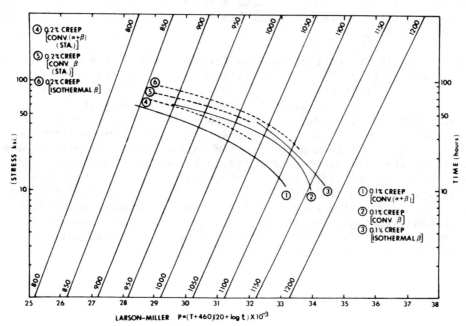

Figure 7   Comparison of creep strength of Ti-6Al-2Sn-4Zr-2Mo 0.1Si
            alloy forgings at 0.1% and 0.2% creep strain

Current cast IN-100 and wrought Astroloy for hot dies in air may permit
satisfactory properties for (α+β) forging at die temperatures at or below 899C
(1650F), but both structural and property stabilities of these alloys become
serious problems at higher operating temperatures. The increase in die temperatures is necessary if one attempts to maximize the achievable near-nets and
considers the potential die alloys for β-processing of (α+β) titanium alloys.
This requires the upgrading of strength capability and high temperature stability of the die alloys.

The TZM-alloy is very satisfactory for strength and stability as a die
material in service temperatures up to 1204C (2200F) and is being satisfactorily used in production applications as hot-die material. However, the
alloy requires a protective atmosphere or vacuum around the die system to
prevent oxidation of the dies. The costs for basic material and the required
oxidation protection for using TZM dies equipment are very high. Furthermore,
the press capacity is limited in the controlled atmosphere, and many elements
need extremely precise control and integration during forging. It appears to
be more practically and economically acceptable if other high oxidation resistant die alloy could be developed to satisfy technical requirements as hot-
dies in air.

Unlike conventional forging for titanium alloys, a recent work on hot-die
forging has demonstrated that the forge pressure required for hot-dies depends
strongly on the flow stress of the alloys [6-8]; it means that, at a given
temperature, the alloy having lower flow stress will require less energy for
forging deformation. It was further demonstrated that the use of the β-
titanium alloy lowers the flow stress required at 704 to 982C (1300 to 1800F)
temperature range, as compared to the (α+β) alloy. For example, the yield
stress of Ti-10V-2Fe-3Al alloy (designated as Ti-10-2-3) at 816C (1500F) is
only about one-quarter of that for Ti-6-4 alloy [8].

Figure 8 gives a direct comparison of the forge pressure between Ti-10-2-3 and Ti-6-4 alloys for experimental structural shape forgings. It is seen that the forgeability of Ti-10-2-3 alloy under hot-die conditions at part/die temperatures of 843/760C (1550/1400F) are comparable to those for the Ti-6-4 alloy at part/die temperature of 954/899C (1750/1650F). In addition, the current hot-die technology can be employed for both (α+β) and β-forgings of β-titanium alloys. A direct comparison of the hot-die forgeability between the two alloys for two structural shape forgings is further presented in Figure 9; the forge temperature/die temperature/ram rate, unit pressure applied, and forging dimension are included. Also, it has been recently demonstrated that the newly developed metastable-β titanium alloys may offer significant improvement in strength-toughness combination of the forgings from that of (α+β) Ti-alloys and may result in improvement in structural efficiency of aircraft structural titanium forgings. The suitability of Ti-10-2-3 alloy forgings for structural applications at 1242 MPa (180 ksi) tensile strength level has been achieved for large production forgings [13, 14].

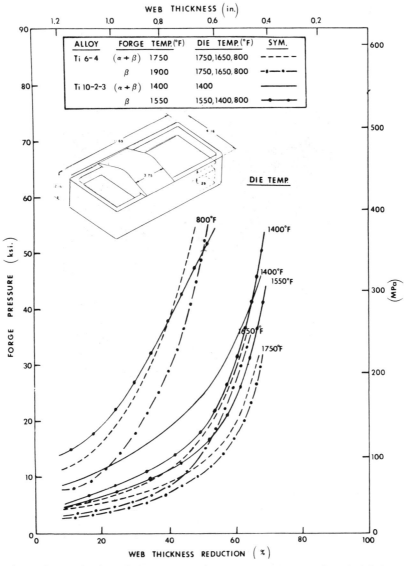

Figure 8   Variation of forge pressure with web thickness for Ti-6Al-4V and Ti-10V-2Fe-3Al experimental structural shape forgings

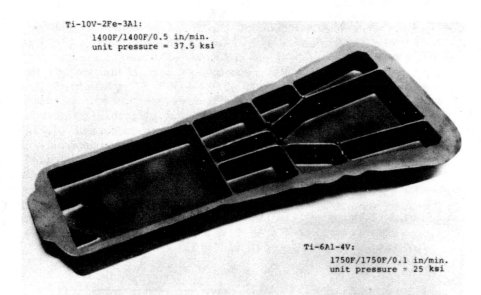

Ti-10V-2Fe-3Al:

1400F/1400F/0.5 in/min.
unit pressure = 37.5 ksi

Ti-6Al-4V:

1750F/1750F/0.1 in/min.
unit pressure = 25 ksi

Figure 9a   Isothermally forged F-15 bulkhead center body:

(∿16 in. long x 5.5 in. wide x 1.5 in. deep)
Plan view ≈80 in^2 web ≈0.18 in.
Ship weight ≈ 5.5 lbs.   rib ≈0.08 to 0.14 in.

Figure 9b Isothermally forged F-100 1st stage compressor:
Ti-6Al-2Sn-4Zr-6Mo

Diameter   ≈ 14 in.   Height (max.) ≈ 12 in.
Plan view ≈ 152 in^2

It has been very recently demonstrated [15] that none of the available lubricants for hot-die forging of (α+β) titanium alloys could provide a satisfactory balance of the lubricity and adhesion properties for manufacturing applications, and there is a need for new lubricant formulations with improved lubrication effectiveness. Most of the commercial lubricants provided excellent anti-friction characteristics, but displayed unfavorable adhesion properties. From a production basis, any failure in removing the forgings under the extreme thermal environments encountered by hot-die systems in the forge shop will cost many times the amount that can be achieved by improving the anti-friction features. However, at the present state-of-technology, commercial lubricants with excellent combinations of lubricity, adhesion characteristics, and environmental inertness are available for hot-die forging applications in the 704-816C (1300-1500F) temperatures. Figure 10 gives a comparison of structural component forgings produced by hot-die forging using three most promising lubricants for each of (α+β) and β-titanium alloys.

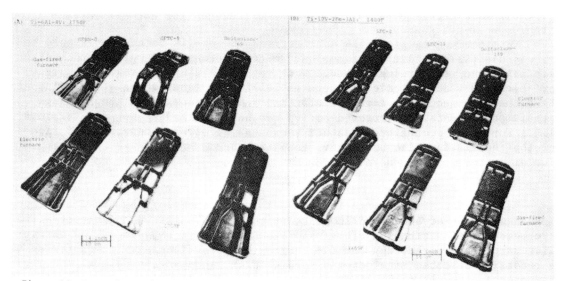

Figure 10 Comparison of structural component forgings produced by isothermal forging using **three** most promising lubricants for each of (α+β) and β-titanium alloys; blast-cleaned condition.

Another major obstacle to the current hot-die technology for (α+β) titanium alloys is the high energy consumption necessary for heating and tooling times at high die temperatures. Here the energy consumption for the processing depends not only on the heat-up to the temperature, but also on energy and time necessary to maintain the forge temperature after each forging operation. From a forger's standpoint, hot-die forging of Ti-10-2-3 alloy at lower die temperatures reduces the energy consumption, and consequently reduces heat-up times and temperature-maintaining costs. It also possesses practical optimization in forging operations at relatively low temperatures, reduces the handling difficulties and increases the operation safety in the forge shop.

Manufacturing Applications

The application of hot-die processes to produce near-net forgings can be realized from two major areas: (1) economic justification of the process, and (2) improved structural efficiency of the forgings. Although the ability of hot-die process to produce cost-effective near-net shapes depends upon the die materials and their fabrication costs, the forgeability of the alloy, the die and forge temperatures used, and the effectiveness of the lubrication systems, it can be generally stated that the hot-die technology has been established as a reliable manufacturing method for the production of airframe and engine components. The technology can be cost-effective, depending on the size, shape, and quantity of a given alloy and forging, and has been demonstrated in producing moderate and large size forgings, as well as complex shape components.

Examples of current applications for hot-die technology to manufacture near-net shapes of titanium alloys at Wyman-Gordon Company are F-15 Bulkhead Center Body (Ti-6-4), CFM-56 engine Fan Disks (Ti-17), JT-8D first and second stage Compressors (Ti-6-4), and F-100 first stage Compressor (Ti-6Al-2Sn-4Zr-6Mo). It is believed that with recent increases in raw material costs and material availability problems for titanium alloys, the popularity of hot-die process to manufacture near-net shape forgings will continue to grow. In particular, the use of β-titanium alloys could be keys to further improvements of both cost-effectiveness and structural efficiency for hot-die forging of titanium near-net shapes.

As discussed earlier, the applications of hot-die technology should be very beneficial to forging technology where a precise control of processing variables and resultant microstructures of the forgings is required. Thus, the technology should have great potential in the manufacture of dual property titanium components for improved efficiency and economics of the forgings [11]. Furthermore, the press capacity can be maximized by hot-die forgings, and the tensile ductility, fracture toughness and creep properties of titanium alloy forgings can be significantly improved.

Summary

Significant advancements in near-net shape technology by hot-die forgings have been made for titanium alloys in recent years. Although the cost-effectiveness of this technology for (α+β) titanium alloys depends on the particular forging selected, the hot-die forging has been generally demonstrated as a readily acceptable manufacturing process for producing structural and engine components. The improvement in the cost-effectiveness of the technology is continuously increasing with recent increases in titanium costs and titanium availability problems.

Current problems for hot-die forging of (α+β) alloys are: high costs of die material and their fabrication, poor die stability at high temperature, poor lubrication effectiveness of available lubricants, slow heating times and temperature-maintaining costs. However, the use of β-titanium alloys will significantly improve both economic and technological factors for hot-die forging. In addition to the improved and simplified forging process, excellent combinations of strength, toughness, and other properties for structural applications are achievable for β-titanium alloys.

References

1. C. C. Chen et al: Air Force Technical Report AFML-TR-77-136, 1977.
2. T. Watmough et al: Air Force Technical Report AFML-TR-70-61, 1970.
3. K. M. Kulkarni et al: Air Force Technical Report AFML-TR-74-138, 1974.
4. A. J. Vazquez and A. Hayes: Air Foce Technical Report AFML-TR-74-123, 1974.
5. C. C. Chen and J. E. Coyne: Metallurgical Transactions, 7A (1976), 1931.
6. C. C. Chen: Third Int. Conf. on "Titanium", May 1976, Moscow, USSR.
7. C. C. Chen: Wyman-Gordon Company Report RD-77-110, 1977.
8. C. C. Chen: Wyman-Gordon Company Report RD-75-118, 1975.
9. C. C. Chen: Wyman-Gordon Company Report RD-75-112, 1975.
10. C. C. Chen and C. P. Gure: Wyman-Gordon Company Report RD-74-120, 1974.
11. C. C. Chen: Wyman-Gordon Company Report RD-79-116, 1979.
12. C. C. Chen: Wyman-Gordon Company Report RD-79-119, 1979.
13. C. C. Chen: Wyman-Gordon Company Report RD-77-108, 1979.
14. C. C. Chen and R. R. Boyer: Journal of Metals, 31 (1979), 33.
15. C. C. Chen: Air Force Technical Report AFML-TR-77-181, 1977.

Computer Aided Design and Manufacturing (CAD/CAM) of Hot Forging Dies

TAYLAN ALTAN

This paper reviews the status of CAD/CAM applications in forging and the significant variables of the forging process, which must be considered in developing and applying computer technology in forging. In addition, the principles of CAD/CAM for design and manufacture of finisher and blocker dies are reviewed. Examples are given to illustrate the computer simulation of metal flow to design blocker shapes for forging applications. Finally, the paper discusses the present status of CAD/CAM in forging, the immediate economic and practical applications of this technology, and the expected future research and development.

INTRODUCTION

The practical design of a forging process involves:

a) The conversion of the available machined part geometry into a forging geometry by using guidelines associated with design of forgings and limitations of the forging process.

b) Design of finisher dies, including determination of flash dimensions, forging stresses and forging load. In some cases it may even be appropriate to calculate die stresses and modify the die geometry in critically stressed areas of the die to reduce the probability of premature die failure.

c) Design of blocker or pre-blocker dies; this includes the calculation of forging volume including flash allowance and the estimation of blocker and pre-blocker die geometry including web thicknesses, rib heights, fillet, and corner radii.

d) Design of the preform and estimation of stock size; this includes the prediction of desired metal distribution in the stock (by preforming or busting operations) prior to forging in the blocker die.

Traditionally, the above process and die design steps are carried out using empirical guidelines, experience, and intu-

TAYLAN ALTAN, Battelle Columbus Laboratories, Columbus, Ohio, 43201.

ition. Once the die design steps are concluded, then the forging dies are conventionally manufactured by a) directly machining from a die block, b) making a solid model and copy milling, or c) making a graphite electrode and electro-discharge machining (EDM) the dies. The graphite electrodes, in turn, can be manufactured by a) copy milling, b) abrading using a special abrading machine, or c) numerical control (NC) machining.

STATUS OF CAD/CAM APPLICATIONS IN FORGING

During the last decade, computers are being used increasingly for forging applications. The initial developments started at Martin Marietta of Torrance, California, where NC machining of forging dies and copying models was pioneered.[1,2] In the mid-seventies, computer aided drafting and NC machining were also introduced at Alcoa (Cleveland) for structural forgings and at Westinghouse (Winston-Salem) for forging steam turbine blades. During the last five years, several companies started to use stand-alone CAD/CAM systems; normally used for mechanical designs, drafting and NC machining; for design and manufacturing of forging dies. These companies include Graphics

Manufacturing Systems in Santa Ana, California, Red Oak Forge in Red Oak, Iowa, Klein Tools in Chicago, Illinois, Caterpillar Tractor Company in East Peoria, Illinois,[3] and Battelle Columbus Laboratories in Columbus, Ohio. At Battelle, forging process analysis and metal flow simulation software are also being developed under United States Air Force and United States Army sponsorship and utilized together with the drafting and NC capabilities of a stand-alone CAD/CAM system.[4-7] Application of CAD/CAM to forging has been also investigated and applied at the University of Birmingham in England.[8,9]

These recent applications and developments of new methods for simulating forging operations indicate that CAD/CAM can augment significantly the productivity and the skill of the die designer. This is accomplished primarily by computerizing area and volume calculations, predicting the stresses and forging loads for a given die geometry, and, in some relatively simple cases, by simulating metal flow during forging.

HOT FORGING AS A SYSTEM

The application of CAD/CAM in hot forging should consider each forging operation (such as finish or blocker) forging) as a "system", as illustrated in Figure 1.[10] For the purpose of die design and manufacture, the major interactions between the variables of the system must be known. These interactions are schematically illustrated in Figure 2. Ideally, the use of CAD/CAM should be capable of simulating the forging process in terms of metal flow (kinematics), stresses, and temperatures. The details of metal flow and the temperature influence the microstructures or properties of the forging and the force/energy requirements. In addition, the simulation of metal flow allows for ascertaining whether the initial billet or blocker geometry is adequate for filling the die cavity, i.e., for producing the desired forging, without any forging defects. As seen in Figure 2, to simulate metal flow and to predict forging load and energy it is necessary to know:

a) the flow stress and the forgeability of the forged material under processing conditions,

b) the conditions of friction and heat transfer at the tool-material interface and the friction shear factor (or the coefficient of friction),

c) the geometry of the product or of the die cavity under consideration.

If the finisher die is being investigated, the die geometry is known from the forging drawing. If a blocker or a pre-blocker die is being designed, then the geometry is initially estimated using empirical rules and is later modified based on the results of metal flow simulation. The value of friction factor is obtained from the ring test or other appropriate tests. The flow stress data are determined most commonly by conducting a uniform compression test or a torsion test.[11]

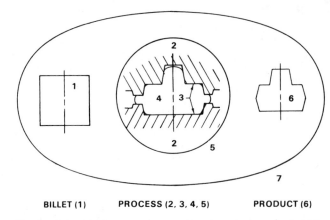

BILLET (1)   PROCESS (2, 3, 4, 5)   PRODUCT (6)

Fig. 1—A hot forging operation considered as a system.[1] 1. Billet 2. Tooling 3. Tool-Material Interface 4. Deformation Zone 5. Forming Equipment 6. Product 7. Plant and Environment.

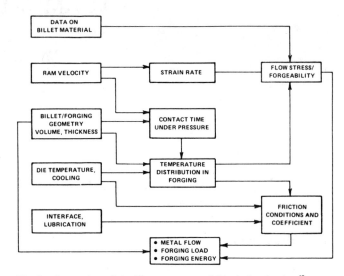

Fig. 2—Interaction of significant system variables in hot forging.[10]

PRINCIPLES OF CAD/CAM APPLICATION IN FORGING

A brief outline of an integrated CAD/CAM approach to hot forging is shown in Figure 3. This approach is general in nature and can be applied to most forgings. The most critical information necessary for forging die design is the geometry of the forging to be produced. The forging geometry, in turn, is obtained from the machined part drawing by modifying this part geometry to facilitate forging. In the process of conversion, the necessary forging envelope, corner and fillet radii, and appropriate draft angles are added to the machined part geometry. Further, difficult-to-forge deep recesses and holes are eliminated and thin and tall ribs are thickened.

This geometric manipulation is best done on a stand-alone CAD/CAM system. These systems are commercially available and have the necessary software for computer aided drafting and NC machining. A typical CAD/CAM system

146

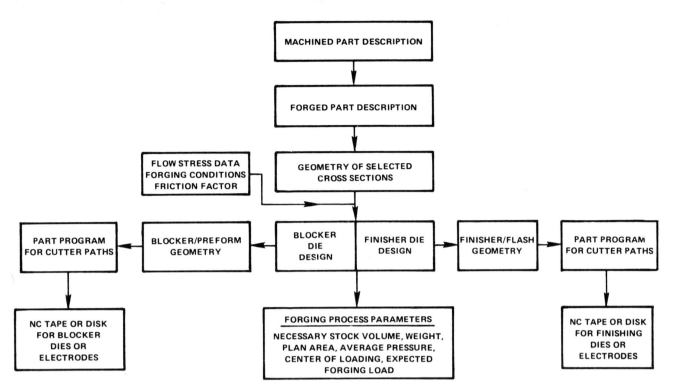

```
                    ┌─────────────────────────┐
                    │ MACHINED PART DESCRIPTION│
                    └─────────────────────────┘
                                 │
                                 ▼
                    ┌─────────────────────────┐
                    │  FORGED PART DESCRIPTION │
                    └─────────────────────────┘
                                 │
                                 ▼
┌──────────────────┐   ┌─────────────────────────┐
│ FLOW STRESS DATA │   │  GEOMETRY OF SELECTED   │
│ FORGING CONDITIONS│──│     CROSS SECTIONS      │
│ FRICTION FACTOR  │   └─────────────────────────┘
└──────────────────┘
```

Fig. 3—Outline of an integrated CAD/CAM approach for hot forging.[4]

consists of a minicomputer, a graphics display terminal, a keyboard, and a digitizer with menu for data entry, an automatic drafting machine, and hardware for information storage and NC tape punching or floppy disk preparation. Such stand-alone CAD/CAM systems are expensive, in the order of (U. S.) $200,000 to $400,000. However, they can increase the productivity in mechanical drafting four to 10 times, depending on specific applications. Such CAD/CAM systems also allow, at various levels of automation, a three dimensional representation of the forging, the possibility of zooming and rotating the forging geometry display on the graphics terminal screen for the purpose of visual inspection. These systems, ideally, should also allow the sectioning of a given forging, *i.e.*, the description, drawing, and display of desired forging cross sections for the purpose of die stress and metal flow analyses. An example of a three dimensional representation of a connecting rod forging die is seen in Figure 4.[12] In this figure, hidden lines (behind the displayed viewing surfaces) are not removed. There are CAD/CAM systems and color graphics terminals which allow hidden line removal or display of lines on various surfaces in different colors.

In designing the finisher and blocker dies for hot forging it is best to consider critical cross sections of a forging, where metal flow is plane strain or axisymmetric. Such a sectioning is illustrated, schematically, in Figure 5[7] and it is exactly the same procedure as the methods used in today's design practice. The cross sectioning approximates the complex 3-dimensional geometry and metal flow, present in a practical forging, with a 2-dimensional metal flow. Thus,

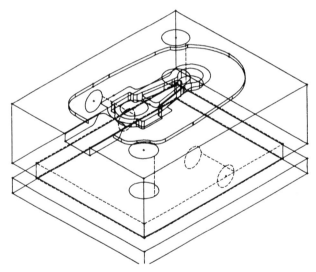

Fig. 4—Three dimensional display of a connecting rod die, prepared on a Computervision CAD/CAM system.[12]

the stresses and pressures can be calculated and the blocker and preform sections can be designed for each section. Using an interactive graphics terminal, the results can be displayed for easy interaction between the designer and the computer system. Thus, modifications on die design can be made easily and alternatives can be explored.

DESIGN OF FINISHER DIES

In a typical multiple die forging set-up, the stresses and the load are higher in the finisher die than in blocker and pre-blocker dies. Therefore, it is necessary to predict these

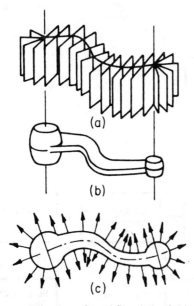

Fig. 5—Planes and directions of metal flow during forging of a relatively simple shape. (*a*) Planes of flow, (*b*) Finish Forged Shapes, (*c*) Direction of Flow.

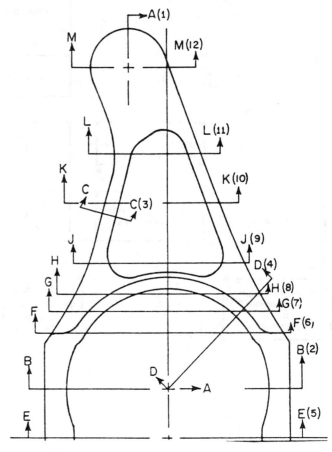

Fig. 6—Plan View of a titanium alloy connecting link and the sections analyzed in a previous study.[7]

stresses and the forging load so that the appropriate forging machine can be selected, and the dies are designed to avoid breakage. In addition, the dimensions of the flash should be optimized. The flash geometry is selected such that the flash is encouraged to restrict metal flow into flash gutter. This results in increasing the forging stresses. Therefore, the designer must make a compromise. On one hand he increases the die stresses by restricting the flash dimensions (thinner and wider flash on the dies); on the other hand, he does not want the forging pressure to reach a certain value which may cause die breakage due to mechanical fatigue.

To analyze stresses, the computerized "slab method of analysis" has been found to be most practical. This method, described in earlier publications,[5–7] can be used to estimate forging stresses for plane strain as well as axisymmetric metal flow sections. In applying this technique, selected sections of a forging are considered, as seen in Figures 5 and 6.[7] A section is considered as an assembly of several connected components, or deformation units, each having a characteristic metal flow behavior. The boundaries for these deformation units, within a particular region, may consist of the die cavity surface or by surfaces within the material along which internal shearing occurs, as illustrated in Figure 7.[7]

Toward the end of the forging stroke, material does not flow along the die surface at all points, as the dies come together. In areas such as under tall ribs, material in the rib stays stationary once the rib is filled. Below the rib, material continues to flow but by shearing within itself rather than by sliding across the die surface (Figure 6). To predict the forging stresses at the end of the forging operation, the computerized slab method uses the geometry of the flow

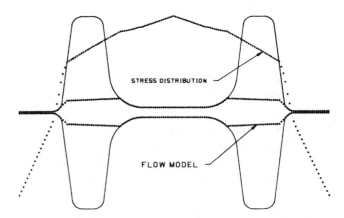

Fig. 7—Stress distribution on cross section J-J of the connecting link for an assumed flow stress of 1000 PSI (calculated by new CV software)[7] (Section width including flash = 7.6 inch).

surfaces (sliding or shearing) rather than the actual die surfaces. The contour of these flow surfaces is determined by minimizing the calculated maximum forging stress with respect to the height and taper angle of the internal shear surfaces. Once the flow surface is known, the stress distribution is calculated for the entire section using modular equations, derived for each deformation unit, describing the flow surface. The flow model and the stress distribution, calculated for an example section, are shown in Figure 7. These calculations take into account:

a) the friction factor for the material forged, die used and lubricant applied,

b) the geometry of the upper and lower die surface at the cross section considered, including flash width and thickness,

c) the flow stress of the material in the die cavity and in the flash.

By "modifying" the flash dimensions, die and material temperature, the press speed, and the friction factor, the die designer is able to evaluate the influence of these factors upon forging stress and load. Thus, he can select conditions which appear most favorable to him. In addition, the calculated forging stress distribution can be utilized for estimating the local die stresses in the dies by means of elastic finite element method (FEM) analysis. After these forging stresses and loads are estimated for each selected section, the loads are added and the center of loading for the forging is determined.

At Battelle, the computerized "slab method" for predicting stresses and load were originally implemented in a large frame CDC 6500 computer,[5] as well as in a PDP 11/40 minicomputer.[6] Recently, this software has also been developed for a Computervision (CV) stand-alone CAD/CAM system.[7] Thus, the CV system can now be used for a) preparing forging and die drawings, b) generating forging cross sections, and c) calculating forging loads and stresses. As an example, Fig. 7 shows the flow model and the stress distribution obtained with this new CV software.[7]

DESIGN OF BLOCKER DIES

Design of blocker and preform geometries is the most critical part of forging die design. The blocker operation has the purpose of distributing the metal adequately within the blocker (or preform) to achieve the following objectives:

a) fill the finisher cavity without any forging defects,

b) reduce amount of material lost as forging flash,

c) reduce die wear by minimizing metal movement in the finisher die,

d) provide required amount of deformation and grain flow so that desired forging properties are obtained.

Traditionally, blocker dies and preforms are designed by experienced die designers and they are modified and refined by die tryouts. The initial blocker design is based on several empirical guidelines. These guidelines depend on the material used and the forging machine utilized. They can be summarized in a general form as:

a) In plan view, the blocker is slightly narrower than the finisher, about 0.5 to 1 mm on each side, so that it can fit into the finisher die.

b) The blocker has usually larger fillet and corner radii to enhance metal distribution.

c) The areas of blocker cross sections are slightly larger, one to three pct, than those of the finisher.

d) For forging high ribs in the finisher, it is at times necessary to have lower ribs in the blocker. At the same time the web thickness in the blocker is larger than in the finisher.

e) In forging, in order to enhance metal flow toward the ribs, it is useful to provide an opening taper from the center of the web toward the ribs.

f) In steel forgings, whenever possible, the ribs in the blocker sections should be narrower but slightly higher than those in the finisher sections. This reduces die wear.

At the present, the computer aided design (CAD) of the blocker cross sections can be carried out using interactive graphics. However, this method still uses the same empirical relationships listed above, but stored in quantitative manner in the computer memory. The main advantages of CAD of blocker dies at the present are:

a) The cross sectional areas and volumes are rapidly and accurately calculated.

b) The designer can easily modify geometric parameters such as fillet and corner radii, web thickness, rib height and width, etc. He can immediately review the alternative design on the screen of the computer graphics terminal as seen in Figure 8.[13]

c) The designer can zoom in to investigate a given portion of the forging, as seen in Figure 9, and he can perform

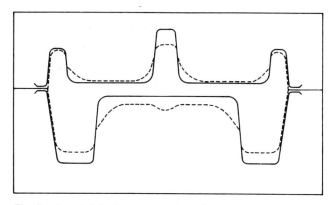

Fig. 8 — A typical forging cross section and a possible blocker design displayed on a computer graphics terminal.[13]

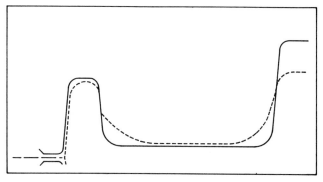

Fig. 9—"Zooming" to examine a small portion of the blocker/finisher cross sections in computer-aided design.

sectional area calculations for a given portion of the forging, where the metal flow is expected to be localized; *i.e.,* the metal would not flow into neighboring regions.

d) If necessary, the designer may review the blocker position in the finisher dies at various opening positions to study the initial die blocker contact point during finish forging, as seen in Figure 10.[13] All these geometric manipulations and drawings are basically not different from what is being done manually today, except they are done with the computer much more accurately and faster than it can be done manually.

COMPUTER AIDED SIMULATION OF THE FORGING PROCESS

The ultimate advantage of computer aided design in forging is achieved when reasonably accurate and inexpensive computer software is available for simulating metal flow throughout a forging operation. Thus, forging "experiments" can be run on the computer by simulating the finish forging of an "assumed" or "selected" blocker design. The results can be displayed on a graphics terminal. If the simulation indicates that the selected blocker design does not fill the finisher die or if too much material is wasted, then another blocker design can be selected and the computer simulation or the "experiment" can be repeated. This computer aided simulation will reduce the amount of necessary and expensive die tryouts.

The plastic deformation phenomenon in hot forging is very complex and involves: a) nonsteady state flow, b) nonuniform distribution of strains, strain-rates and temperatures in the deforming metal, c) difficulties in estimating the flow stress in various parts of the forging during deformation, and d) difficulty in estimating the friction factor. Consequently, only limited amount of progress has been made so far in simulating an entire forging operation. At this time, computer codes are available for two well-defined part families: Turbine and compressor blades, where metal flow is predominantly plane strain, and round forgings, where the metal flow is axisymmetric. In more complex forgings, such as crankshafts, connecting rods, structural aircraft parts and other hardware type forgings, similar to that seen in Figure 6, metal flow simulation can be done for selected sections (plane strain or axisymmetric) of the given forging. However, the mathematical simulation techniques necessary for this purpose are still being developed. At this time, these simulation techniques can be easily and economically used only for sections with relatively simple geometries.

Simulation in Blade Forging Using the Slab Method

Blades represent a well-defined geometric family. Consequently, the application of CAD/CAM in blade forging is economically feasible and is being practiced to various levels of sophistication.[14,15] Recently, a system of computer programs capable of analysis and design of forging dies for turbine and compressor blades, was developed at Battelle.[16] This system has a modular structure based on the various functions it is required to perform and thus is extendable into an integrated CAD/CAM system.

The major functions performed by this system may be summarized as:

1) Read in and preprocess the blade geometry. Check for errors in the input data and convert into internal standard form (canonical form). Calculate the cross sectional area.

2) Determine the flow stress under the forging conditions lubrication, temperature, equipment characteristics, based on available data. An average flow stress is calculated on the basis of an average temperature in the cavity and in the flash.

3) Determine the forging plane based on minimum side loading of the dies during forging. The turbine blade is rotated around its stacking axis in small increments until the position with minimum resultant horizontal force is determined. Load calculations are based on the "Slab Method" and deformation elements as seen in Figure 11. The stress distribution, the expected vertical and horizontal loads, and the center of loading are calculated. The summations of the horizontal and vertical loads on all cross sections give the resultant loads acting on the two die halves. The "optimum" forge plane position is where the horizontal forces on the dies are minimized. This, in turn, minimizes the die shift during forging, resulting in improved tolerances on the product.

4) Modify leading edge flash geometry so as to bring all neutral surfaces on the same line. Thus, a more uniform metal flow with uniform flash losses will result. (The neutral surface is the plane where the stress distribution has its peak, as seen in Figure 11.)

5) Simulate the forging process, as illustrated in Figure 12, in order to determine the best position of the preform in the

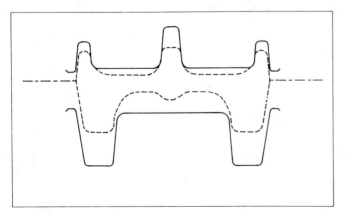

Fig. 10—Blocker, designed by computer, shown with finisher dies in separated position.[7]

STRESS DISTRIBUTION ON A BLADE CROSS SECTION DURING FORGING

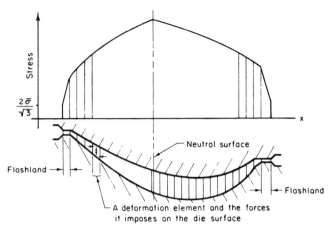

Fig. 11 — An airfoil die cross section, its deformation elements and the expected stress distribution at die closure.[16]

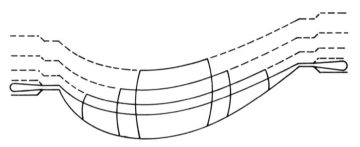

Fig. 12 — Metal flow and die fill in blade forging as simulated by a computer program.[10]

die cavity. This position not only insures die cavity fill during forging, but also flash losses may be minimized by using the minimum preform dimensions indicated.

6) Summarize the results. Provide a printout of all the results obtained in a concise format in engineering terms.

7) Provide a 3-D display of the die surfaces together with the flashland as designed by the computer programs. The user can rotate the blade in space and view in perspective or orthographic projection and modify the calculated forge-plane position, if deemed necessary.

After the user is satisfied with the design, the die cavity geometry together with boundary stress state can be made available to specialized finite element elastic deflection analysis routines. These consist of, in principle, two major modules. The first one prepares the grid system and the second one performs the actual finite element deflection analysis. Where necessary, the local elastic deflections may be superimposed on the gross deflection of the particular press bed. Using these routines, the die surface coordinates are corrected for the expected deflection during forging and are output in a special format for use by part programming routines for NC machining of the upper and lower die surfaces.

This CAD/CAM procedure, although it still needs some further development, has been found to be very valuable in the establishment of an integrated production system for forged blades.

Simulation of Round Forgings Using the Rigid Plastic Finite Element Method

Recently, the rigid plastic finite element formulation, developed by Kobayashi and his co-workers,[17] has been further developed at Battelle by Oh[18,19] for practical application in forging die design. This advanced FEM based program, called ALPID (Analysis of Large Plastic Incremental Deformation) uses a) high order elements, b) a general description of the die and automation of the boundary conditions, and c) an automatic initial guess generation. Initial but highly promising results, so far obtained by ALPID, indicate that this software is able to simulate a large number of two-dimensional forging operations with reasonable accuracy and at acceptable cost. Figure 13 shows the display of results obtained by ALPID in simulating an axisymmetric forging operation.[18] In this case, the calculations were made for assumed isothermal conditions for a titanium alloy (Ti-6242) forged at 950 °C. Although not given here, ALPID is capable of predicting the strains, strain-rates, and stresses at grid points of the deformed material. In fact, using special plotting software, these results are displayed on a graphics terminal and plotted on paper in form of contour plots.

Another example of simulation with ALPID is seen in Figure 14 for a backward extrusion operation.[19] In this case, the initial billet has a 149 mm diameter and a 345 mm height. The material was 1045 steel forged at 1100 °C.

SUMMARY AND FUTURE OUTLOOK

CAD/CAM is being applied increasingly in forging technology. Using the three-dimensional description of a machined part, which may have been computer designed, it is possible to generate the geometry of the associated forging. For this purpose, it is best to use a stand-alone CAD/CAM system with software for geometry handling, drafting, dimensioning, and NC machining. Thus, the forging sections can be obtained from a common data base. Using well-proven analyses based on slab method or FEM techniques, forging load and stresses can be obtained and flash dimensions can be selected for each section, where metal flow is approximated to be two dimensional (plane strain or axisymmetric). In some relatively simple section geometries, a computer simulation can be conducted to evaluate initial guesses on blocker or preform sections. Once the blocker and finisher sections are obtained to the designer's satisfaction, then this geometric data base can be utilized to write NC part programs to obtain NC tapes or disks.

Source: Journal of Applied Metalworking, January 1982, 77-85

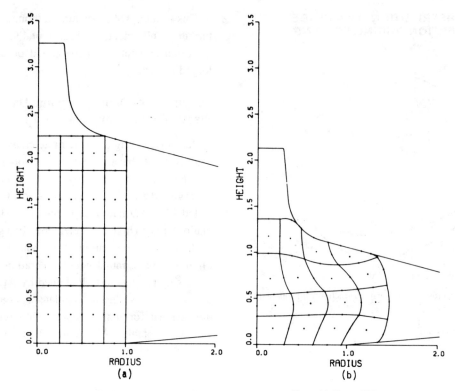

Fig. 13—Simulation of axisymmetric spike forging.[18] (m = 0.3, Temp = 1750F)
(a) Undeformed FEM Grid, (b) Deformation at Die Stroke = $0.5H_0$ where
H_0 is the Initial Billet Height.

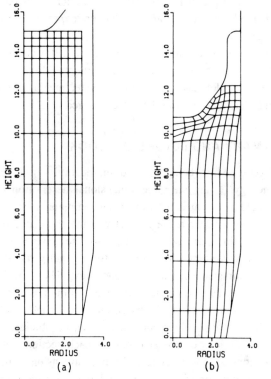

Fig. 14—Simulation of a backward extrusion operation.[18] (a) Undeformed
FEM Grid, (b) Deformation at Die Stroke = $0.302H_0$.

This CAD/CAM procedure is still in a stage of development. In the future, this technology can be expected to evolve in two main directions:

a) handling of the geometry of complex forgings, *e.g.*, 3-dimensional description, automatic drafting and sectioning, NC machining,

b) utilization of design analysis, *e.g.*, calculation of forging stresses, stresses and stress concentrations in the dies, prediction of elastic deflections in the dies, metal flow analysis, and blocker/preform design.

Geometry of Complex Forgings

The software, available for NC machining, needs further improvements for handling complex sculptured surfaces, encountered in many forgings. However, most CAD/CAM systems, available in the market today, have reached such capabilities that they can describe most, if not all, forgings encountered in today's practice. These systems are capable of describing complex 3-dimensional surfaces and blending of two or more adjacent surface patches, as well as NC machining of an object, described by such surfaces. Several of these systems have good man-machine interface, using interactive color graphics with built-in capabilities for zooming, rotating, translating, and mirror imaging the display on the graphics screen. In addition, entry of data and program commands into the computer are facilitated by the use of a digitizer tablet with menu area.

Based on the availability of such geometry handling software, we can expect that the present application of CAD/CAM in forging, mainly for drafting and NC machining of forging dies (electrodes or models), will continue to increase at a rapid rate. The principal barriers to widespread acceptance of this application appears to be a) apparent high cost of introducing CAD/CAM, b) management inertia, and c) lack of trained personnel. However, the forging industry, worldwide, is under considerable pressure to modernize and to increase the productivity of skilled die makers, who are becoming increasingly scarce. In addition, computer aided systems for drafting and NC machining are becoming relatively inexpensive. Consequently, we can expect to see a very significant increase in the number of forge and die shops that will utilize CAD/CAM in the next decade.

Design Analysis

At this time, it is possible to estimate the forging stresses, average pressure, the total forging load, and the center of loading, as described in this paper. The technology which requires additional research and development is the simulation of metal flow for optimum blocker/preform design. Some significant advances have been made in this direction through the use of plastic FEM analysis. However, there are still a considerable number of questions to be answered before this analysis can be a practical and cost-effective design tool.

For 2-dimensional metal flow simulation, it is necessary to develop procedures/techniques for:
- automatic generation of forging cross section (plane strain or axisymmetric)
- selecting initial FEM mesh and the distribution of the mesh size
- remeshing and for deciding when, after a certain amount of deformation is simulated, remeshing is necessary
- assembling together the blocker sections, obtained from analysis by metal flow simulation.

At this time, there are no methods available for describing and/or simulating 3-dimensional metal flow, encountered in practical forging operations. Considerable effort is necessary to develop these techniques.

In summary, it can be expected that the application of CAD/CAM in forging will continue to increase. Initially CAD/CAM will be used for drafting and NC machining. Next the geometric capabilities of available CAD/CAM systems will be augmented by analysis software to a) calculate forging stresses and load and b) design blocker shapes for relatively simple forgings, using metal flow simulation. At a later date, we can expect that 3-dimensional metal flow simulation and design of practical complex forging shapes will also be possible. As a result, necessary and expensive die try-out trials in the forge shop floor will be reduced. In addition, material utilization will be improved by optimizing the geometry of blockers and preforms by computer aided simulation and by optimizing the flash design.

ACKNOWLEDGMENTS

The CAD/CAM and analysis techniques summarized in this paper were developed to a large extent by the author's colleagues, all members of Battelle's CAM group. The author gratefully acknowledges these contributions as follows: Dr. Nuri Akgerman (forging of blades and rib/web aircraft parts), Dr. Aly Badawy and Mr. Carl Billhardt (load and stress calculations and analysis software developed for the Computervision system), and Dr. Soo-Ik Oh (development of FEM based simulation techniques).

REFERENCES

1. N. F. Wood: "Three-Pronged Attack on Die Sinking Problems", *Machine and Tool Blue Book*, November 1973, p. 87.
2. C. T. Post: "The Strong Link That Ties CAD to CAM", *Iron Age*, August 22, 1977, p. 29.
3. K. G. Dawson: "Meeting the Diesinker's Challenge of the Future", *Precision Metal*, February 1980, p. 17.
4. N. Akgerman: "Design and Manufacture of Forging Dies: Computer Aided Methods", SME Technical Paper, MF_72-531, 1972.
5. T. L. Subramanian and T. Altan: "Application of Computer Aided Techniques to Precision Closed-Die Forging", *Annals of CIRP*, November 1, 1978, vol. 27, p. 123.
6. T. Altan, C. F. Billhardt, and N. Akgerman: "CAD/CAM for Closed-Die Forging of Track Shoes and Links", SME Paper, MS 76-739, 1976.
7. A. Badawy, C. F. Billhardt, and T. Altan: "Implementation of Forging Load and Stress Analysis on a Computervision CADDS-3 System", *Proceedings of the Third Annual Computervision Users Conference*, Dallas, TX, September 1981.
8. Y. K. Chan and W. A. Knight: "Computer Aided Design and Manufacture of Dies for Long Hot Forgings", *Proceedings of the 6th NAMRC*, Gainesville, FL, 1978, p. 455.
9. Y. K. Chan and W. A. Knight: "Computer Aided Manufacture of Forging Dies by Volume Building", *Journal of Mechanical Working Technology*, 1979, vol. 3, p. 167.
10. T. Altan and G. D. Lahoti: "Limitations, Applicability and Usefulness of Different Methods in Analyzing Forming Problems", Keynote Paper, *Annals of the CIRP*, February 1979, vol. 28, p. 473.
11. J. R. Douglas and T. Altan: "Flow Stress Determination for Metals at Forging Rates and Temperatures", *ASME Transactions, J. Engr. for Industry*, February 1975, p. 66.
12. S. L. Semiatin and G. D. Lahoti: "Forging of Metals", *Scientific American*, August 1981, p. 82.
13. T. L. Subramanian and T. Altan: "Application of Computer Aided Techniques to Precision Closed Die Forging", *Annals of CIRP*, January 1978, vol. 27, p. 123.
14. O. Voigtlander: "The Manufacturing of Blades for Turbines and Compressors: Precision Forging of the Airfoil" (in German), *Industrie-Anzeiger*, May 13, 1969, vol. 91, no. 40, p. 908.
15. J. T. Winship: "Third Screw Press for Blades", *American Machinist*, October 1976, p. 133.
16. N. Akgerman and T. Altan: "Application of CAD/CAM in Forging Turbine and Compressor Blades", ASME paper 75-GT-42, published in *ASME Trans., Journal of Engineering for Power*, April 1976, vol. 98, series A, no. 2, p. 290.
17. C. H. Lee and S. Kobayashi: "New Solutions to Rigid-Plastic Deformation Problems Using a Matrix Method", *Trans. ASME, J. Engr. for Industry*, 1973, vol. 95, p. 865.
18. S. I. Oh: "Finite Element Analysis of Metal Forming Problems with Arbitrarily Shaped Dies", to be published.
19. S. I. Oh, G. D. Lahoti, and T. Altan: "ALPID-A General Purpose FEM Program for Metal Forming", *Proceedings of NAMRC IX*, State College, PA, May 1981, p. 83.

SECTION VI
Isothermal Forging

ISOTHERMAL FORGING--FROM RESEARCH TO A PROMISING NEW MANUFACTURING TECHNOLOGY

by

K. M. KULKARNI, Manager, Metalworking Technology, IIT Research Institute

ABSTRACT. In the past ten years, many investigations have centered on hot die or isothermal forging of titanium alloys. This paper describes isothermal forging of three large components at IIT Research Institute. This includes a nose wheel for F-111 aircraft, a centrifugal impeller, and a bulkhead for F-15 aircraft. The 0.277 m^2 (430 sq. in.) plan area bulkhead is the largest and the most complex isothermal forging made thus far. The paper also refers to work at other companies. Finally, criteria for selection of components suitable for isothermal forging are discussed with reference to commercial exploitation of the technology.

INTRODUCTION. This paper is about isothermal or heated die forging of titanium alloys. From the pioneering work in the mid 60's at IIT Research Institute (IITRI), this net or near-net forging process has seen a significant amount of technical development and is now reaching commercial applications. In this paper, first the historical background is reviewed briefly. Then examples are given of three main components forged by IITRI--a nose wheel for F-111 aircraft, a centrifugal impeller for T53 engine, and a large and complex bulkhead for F-15 aircraft. The examples illustrate the advancement in the technology and how the process is now ripe for commercial exploitation. Brief reference is made to work in other organizations and criteria are discussed for selection of components for production by isothermal forging.

HISTORICAL BACKGROUND. It was in the late 50's that IITRI investigated [1] usage of superalloy dies heated to 870°C (1600°F) for precision forging of steel. The heated die forging process for titanium alloys was studied in a program [2] at IITRI in mid-1960. The process is particularly applicable to such materials because the flow stress of titanium alloys increases rapidly with speed of deformation. [3,4] Figure 1 schematically compares the conventional and isothermal forging processes. In conventional forging the large difference in the workpiece and the die temperatures necessitate a higher speed to minimize heat loss from the workpiece to the dies. Then, the forging pressure increases substantially because of the sensitivity of the flow stress to higher deformation speeds.

In isothermal forging the problem of heat loss is eliminated by using heat resistant dies which could be operated at about the same temperature as the workpiece. Then, relatively slow deformation speed can be employed to make a forging with much greater complexity even when using only a fraction of the load that is required for conventional forging. The technique preferred by IITRI involves setting a

maximum limit on the load. With this approach the press moves rapidly initially and slows down only in the final stages when the resistance to deformation builds up. With this method, the duration of forging load is only about 3 to 5 minutes long, whereas usage of a constant slow press speed may make the forging cycle unnecessarily much longer.

The main advantages of the isothermal forging process include its ability to produce complex profiles beyond the capability of conventional techniques, the need for a lower forging pressure, and the resultant ability to produce larger forgings using existing forging equipment. The greater complexity possible with the process also reduces the cost of the input material and of finish machining. However, the additional cost of the dies and the heating system must be taken into account in evaluating the cost effectiveness for specific applications. On the other hand, the isothermal forging eliminates several intermediate forging steps and the cleaning and relubricating operations necessary in conventional processes. The criteria for selection of components suitable for isothermal forgings are discussed in more detail in a subsequent section.

The initial program [2] sponsored by the Air For Materials Laboratory at IITRI saw considerable development of the technology, and led to IITRI's patent [5] for the isothermal forging process for titanium alloys. The tooling and the die heating system as well as the variety of forgings made in this program are discussed in References 6 and 7. The precision casting technique utilized for making large superalloy castings to be used as forging dies is described in Reference 8. The considerations necessary in selecting the die material were reviewed in Reference 9. In this initial effort, dies weighing up to 640 kg (1400 lb) per half were precision cast in IN-100 and MAR-M 200 nickel-base superalloys. Five different forging configurations were forged, the last being a 0.071 m^2 (110 sq. in.) plan area nose wheel described in greater detail in the next section.

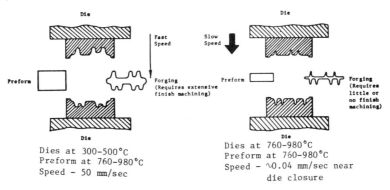

Dies at 300-500°C
Preform at 760-980°C
Speed - 50 mm/sec

Dies at 760-980°C
Preform at 760-980°C
Speed - ∿0.04 mm/sec near die closure

Fig. 1. Comparison of Conventional (left) and Isothermal (right) Forging Processes.

Table 1. Summary of Initial Work on Ti-6Al-6V-2Sn Alloy Isothermal Forging (2)

Shape (vertical cross-section)	Forging Representative Dimensions, mm		Plan Area, mm²	Weight, kg	Die-Set Material	Weight per Die Half, kg	Isothermal Forging Die and Workpiece Temp., °C	Press Speed mm/s†	Forging* Pressure, MPa
	x	y							
	95	7·9	6 800	0·9	IN-100	90	980 / 870	1 / 1	262 / 655
	95	7·9	6 800	0·9	MAR-M 200	90	980 / 760	0·04 / 1	83 / 655
	140	9·1	10 000	1·6	IN-100	180	980 / 870	1 / 1	262 / 538
	380	10·3	36 000	7·0	IN-100	640	980 / 900	0·04 / 0·04	76 / 124
	190	1·6	24 000	1·5	MAR-M 200	180	980 / 870	0·04 / 0·04	83 / 131
	320	3·2	71 000	10·0	IN-100	640	900	0·04	124

† The press speed was 0.04 mm/s only in the final stages of forging.

* For conversion to in-lb system, use 1 MPa = 145 psi.

The summary of the entire program is provided in Table 1.

NOSE WHEEL FOR F-111 AIRCRAFT

This nose wheel is made from Ti-6Al-6V-2Sn alloy and has a diameter of 320 mm (12 1/2 in.) and a plan area of 0.071 m² (110 sq. in.). It is characterized by a thin web of only 3.2 mm (1/8 in.) and the circular symmetry of the wheel is hampered by 8 bosses for bolts, both on the interior and exterior of the web. Such thin web for this particular geometry is impossible to obtain by conventional forging techniques. Consequently, it would have been necessary to make the forging with an initial thick web. Then, because of the non-symmetry casued by the bolt bosses, it would have required extensive end milling operations to produce the contoured thin web. The objective of the isothermal forging effort was to forge the web net so that only simple machining operations could be employed.

The isothermal forging dies for the nose wheel were made (2) by precision casting from IN-100 nickel-base superalloy. The top die and the bottom die each weighed nearly 640 kg (1400 lb) excluding riser, and on both dies the main die cavity profiles were as-cast. A central ejection pin--also made from IN-100--defined the inside diameter of the wheel hub on the forging, helped alignment of the two dies, and served to eject the forging from the lower die. In contrast with the usual conventional forging practice, the isothermal forging had a central through-hole and was forged in an orientation such that the outside of the wheel was made in the lower die. The dies were heated in situ by means of two cylindrical induction coils. Details of process development are given elsewhere.(2,6,10)

The nose wheel was successfully isothermally forged (Figure 2) in a single operation from a simple washer-shaped preform. It is important to mention that starting from the washer-

Fig. 2 A.

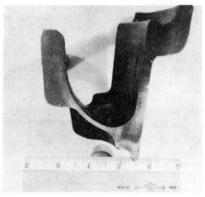

Fig. 2 B.

shaped preform, the forging of the thin web required nearly 90% reduction in thickness in a single operation--something well beyond the capability of any conventional forging methods. Also, as desired, the web area was forged net so that end milling of the web was eliminated and finish machining primarily comprised simple turning operations. The weight of this forging was approximately 10 kg (22 lb). Typically it was forged with die-workpiece temperature in the range of 900°-925°C (1650° to 1700°F) using a forging load of about 8.9 MN (1000 tons) corresponding to a forging pressure of only 124 MPa (8,000 psi). The forging operation--from placing the preform in the dies to removal of the forging--required nearly 3 to 5 minutes.

Fig. 2. Full-scale nose wheel blank and isothermal forging of Ti-6Al-6V-2Sn (a), and a cross section of the forging (b).

The nose wheel successfully underwent extensive qualification testing. Since October, 1973, the nose wheels are being field tested for corrosion resistance on four F-111 aircraft at Mellis Air Force Base.[11] The improved corrosion resistance of the titanium alloy wheels offers the potential for reduced life cycle cost in comparison with conventional aluminum alloy wheels. The titanium alloy nose wheel is thus the first component that successfully demonstrated the applicability of the isothermal forging process for production of aerospace hardware.

CENTRIFUGAL IMPELLER FOR T-53 ENGINE

A finish-machined impeller is shown in Figure 3. This Ti-6Al-4V alloy component is approximately 330 mm (13 in.) in diameter and is characterized by 36 thin blades on one side. The conventional forging for the component weighs nearly 5.4 kg (12 lb); thus the conventional forging requires extensive machining in production of this forging. The plan area of this forging is approximately 0.078 m² (120 sq. in.). In order to isothermally forge this impeller, a two-phase program was sponsored at IITRI by the Army.[12] In the first phase, a half-scale impeller was produced with four groups of straight blades ranging in thickness from 2.5 mm (0.1 in.) to 5.1 mm (0.2 in.) with 19 mm (3/4 in.) in length. The experience gained in forging the half-scale impeller was valuable in designing the tooling for and selecting the processing parameters for the full-scale impeller forging in the second phase.

Fig. 3. View of finish-machined Ti-6Al-4V impeller.

The single piece upper die for the full-scale impeller was made from MAR-M 200 nickel-base superalloy. It weighed 213 kg (470 lb) excluding the riser. The lower die had a three-piece construction. An existing 640 kg (1400 lb) IN-100 casting was machined to form a cylindrical cavity and employed as the die holder. A 118 kg (260 lb) IN-100 precision casting was made as the lower die insert. The blade detail was electrical discharge machined in the insert and it was assembled into the die holder. A central ejection pin was the third component in the lower die, and the pin functioned in much the same way as that for the nose wheel described above. The impeller dies were heated in situ using induction coils. Note that this is the first application of a large die holder with a replaceable die insert which would allow usage of the same basic die system for a variety of forgings with different configurations within a certain size range.

The initial geometry of the preform was selected on the basis of the Phase I work for

the half-scale impeller and simulated metal flow studies with a clay-type of material and Plexiglas dies. The geometry could then be quickly finalized on the basis of a few actual forging trials--first with an aluminum alloy and then with the titanium alloy. The dimensions finally selected for the preform were approximately 286 mm (11 1/4 in.) in outside diameter, 165 mm (6 1/2 in.) in inside diameter, and 70 mm (2 3/4 in.) in height with a 38 mm (1 1/2 in.) x 45° taper on one corner. The preform was typically preheated to 955°C (1750°F), and the forging load was 8.9 MN (1000 tons) corresponding to a forging pressure of only 117 MPa (17,000 psi). The die temperature was generally 55 to 85 C degrees (100 to 150 F degrees) lower than the preform temperature.

Excellent quality impellers were forged (Figure 4) in either one or two isothermal operations. It is anticipated that, with some minor modifications to the die and preform geometries, good die filling can be consistently obtained in one isothermal operation. As can be seen in Figure 4, much of the blade detail was achieved in the forging so that this 11 kg (24 lb) forging will need much less finish machining than the 17 kg (38 lb) conventional forging. The thinnest portion of the blade on the isothermal forging was only 4.1 mm (0.16 in.) in thickness with the blade height ranging to 19 mm (3/4 in.). The web thickness near the outside diameter was typically 5.3 mm (0.21 in.).

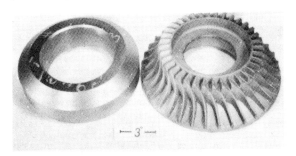

Fig. 4 A

Fig. 4 B

Fig. 4. Preform and isothermally forged impeller (A) and close-up view of the forging (B).

In a separate program,[13] the impellers isothermally forged by IITRI were evaluated by Avco. This extensive evaluation included ultrasonic inspection, metallography, dimensional inspection, tensile testing, and fatigue testing. The isothermal forgings were judged quite satisfactory in all these regards. The Avco program also involved a preliminary cost analysis.

Source: 6th North American Metalworking Research Conference Proceedings, 1978, 24-32

The three elements of cost considered were machining of vane area, machining of other portions, and procurement of forging. Unfortunately, as discussed below on the basis of Reference 13, the analysis did not include similar types of cost for the conventional and isothermal forgings.

The T53 impeller has vanes which are radial with respect to the axis of rotation. Avco machines the vane area using specially designed and built three-axis cam controlled milling machines. The vane machining cost for a conventional forging is nearly $130 per impeller. The isothermal forging has 3.1 kg (6.9 lb) or 67% less excess stock in the vane area than a conventional forging. Unfortunately, the cam controlled machine cannot speed up to take advantage of this. Avco estimated that N/C machining of isothermal forgings could save $69 per impeller in comparison with cam controlled milling of conventional forgings. But then, a $500,000 capital investment was assumed for N/C machining, and it was stated that this leads to a breakeven point of 7,630, which is beyond the anticipated total impeller production of about 2800 in eight years. Avco estimated that one N/C machine could replace up to six cam controlled mills, but no attempt was made to offset the investment for the N/C machine by the cost for engineering and fabrication of the special mills.

In portions apart from vanes, an isothermal forging has 3.0 kg (6.6 lb) less stock than a conventional forging. This was projected to lead to a saving of $71 per impeller. As for the cost of the isothermal forging, in consultation with Wyman-Gordon, Avco stated it to be $550 per forging plus a tooling charge of $38,000. The former is $61 more than that for a conventional forging. Thus, these two factors show a net saving of $10 per impeller which gives a very high breakeven point when the tooling charge is amortized. Again Avco did not take into account the cost of the forging dies for conventional forging. Moreover, the cost assumed for isothermal forging tooling is far in excess of that actually incurred by IITRI, even though Wyman-Gordon planned on a somewhat oversize isothermal forging.

Clearly the main point here is that, in comparing two processes and especially a new technology of the type of isothermal forging, it is important to consider a part which is yet to reach production. This will allow a more realistic comparison since all the costs must be taken into account for both the alternatives. When such even basis is employed, many components will prove to be cost-effective and suitable candidates for fabrication by isothermal forging.

BULKHEAD FOR F-15 AIRCRAFT

The particular component investigated was the fuselage section 749 bulkhead for F-15 aircraft. It is made from Ti-6Al-4V alloy. The finish-machined component has a plan area of approximately 0.277 m^2 (430 sq. in.) and a thin central web of only 1.5 mm (0.06 in.). The part is characterized by long thin ribs on both sides of this central web. The rib thickness ranges from 1.3 to 2.0 mm (0.05 to 0.08 in.) and height ranges to 64 mm (2 1/2 in.). There are 23 pockets on one side of the web and 19 on the other. The component weighs only 10 kg (22 lb), and is machined from a conventional blocker forging which weighs 154 kg (340 lb) because of the complexity of the part and allowance for any design changes. Because of

the high cost of this component, the Air Force sponsored a program[14] at IITRI with the objective of making the bulkhead as a precision isothermal forging so that it will need only a single finish machining cut. A machining envelope of 2.3 mm (0.09 in.) per surface was recommended by McDonnell Douglas.

The die system designed for the bulkhead comprised a one-piece top die and a lower die with a die insert. The die material was IN-100 nickel-base superalloy. The top and the lower dies are probably the largest ever precision castings in a nickel-base superalloy. They were made by Cast Masters Division of Latrobe Steel Company and are shown in Figures 5 and 6.

Fig. 5. As-cast IN-100 top die for F-15 bulkhead forging.

Fig. 6. As-cast IN-100 lower die for F-15 bulkhead forging.

The trimmed casting for the top die weighed 1720 kg (3800 lb) and that for the lower die weighed 2680 kg (5900 lb). The lower die insert weighed nearly 195 kg (430 lb). The usage of precision casting minimized the extent of required finish machining. The extensive rib detail was obtained by Alcoa by electrical discharge machining. The detail was primarily in the top die and in the lower die insert. Initially, a completely trapped die design was selected. Eventually a small 6 mm (1/4 in.) thick flash was allowed on one side of the web along the two long sides.

The schematic of the die heating system is shown in Figure 7. The burners in this gas heating system were located in the main system. The dummy furnace was employed only during preheating, and the two halves of this system were retracted just prior to the forging operation. During die closure, the main heating system covered part of the bottom die and minimized heat loss from the latter. The heating

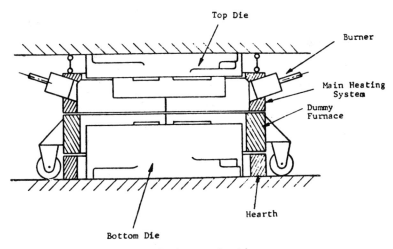

Fig. 7. Schematic of die heating system during preheating.

system was constructed by Selas Corporation of America. An independent supplementary system by Alcoa provided heat to the bases of the dies for better temperature uniformity in the vertical direction.

The preforms (Figure 8) for the forging were made by upsetting and subsequent conventional blocker die forging. Nearly half the capacity of a 310 MN (35,000 ton) hydraulic press was employed for blocker forging. The preforms were coated with a commercially available glass-based lubricant. They were then heated in an electric furnace and transferred to the heated isothermal forging dies for the finishing operation. Practically all the detail was obtained in one isothermal forging operation, though in a few cases a second operation was used for some added improvement in die filling.

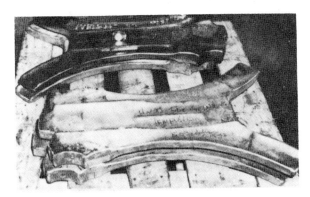

Fig. 8. Close-up views of blocker die forged preforms for F-15 bulkhead forging.

The optimum conditions for forging were preheat furnace temperature of 940°-970°C (1725°-1775°F), die temperature of 900°-940°C (1650°-1725°F), and a forging load of 39 MN (4400 tons)* maintained for about 5 minutes. This corresponds to a very low forging pressure of approximately 138 MPa (20,000 psi). It is interesting to note that the extensive detail in isothermal forging (Figure 9) was achieved with only a quarter of the load needed for the simple preforms (Figure 8) made by conventional

---

*When two isothermal forging operations were used in the finishing die, a lower load of only 31 MN (3500 tons) was adequate.

al blocker forging. In the course of eight forging series, nearly 70 isothermal forgings were produced. Also, progressive improvements were made in the preform design and, to a minor extent, in the die geometry. The isothermal forgings showed excellent and very nearly complete die filling. They had a thin web of only about 6.4 mm (1/4 in.) thickness over practically the entire plan view. The ribs were also only 6.4 mm (1/4 in.) in thickness and had heights ranging to 64 mm (2 1/2 in.) on each side of the thin web. This is the largest and most complex bulkhead isothermally forged thus far.

Fig. 9 A.

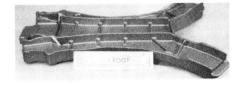

Fig. 9 B.

Fig. 9. Two views of isothermally forged F-15 bulkhead. (Notice the excellent filling of rib detail in spite of the large variation in designed rib heights).

All in all, the tooling system worked well. However, the complexity of the bulkhead made removal of the forging from the die difficult, expecially in the earlier forging series. This necessitated heating of the lower die in the closed position which was not initially planned. In the closed position of the die, the burners in the main system were very close to the die surface and, consequently, large cracks occurred on the bottom die outer surface. These were ground off and did not show any tendency of propagation. Because of a combination of high thermal and mechanical stresses, cracking

Table 2. Tensile properties of bulkhead forgings.[14]

Specimen Type		Tensile Strength, psi	Yield Strength, psi	Elong., %	Reduction in Area, %
Data by MCAIR on Isothermal Forging					
22 round specimens from thicker portions	Max	148	140	19	49
	Min	140	126	13	38
	Avg	144	132	16	45
9 flat specimens from thin web	Max	146	134	16	
	Min	138	125	8	
	Avg	142	129	12	
Data by Alcoa on Isothermal Forging					
22 specimens from thicker portions	Max	154	138	19	48
	Min	140	128	15	39
	Avg	145	132	17	43
9 specimens from thin web	Max	146	134	19	49
	Min	143	130	17	38
	Avg	144	132	17	44
Data by MCAIR on Conventional Forging					
(Unspecified)	Max	146	138	17	40
	Min	137	128	13	25
	Avg	141	132	15	34
MIL-F-83142, Minimum Properties					
--		130	120	10	25

also occurred at the bottom of the deepest sections in the dies corresponding to the high ribs. After 70 forging operations, twenty dimensions ranging to 483 mm (19 in.) were inspected on the top die and the lower die insert. They showed an average increase of 0.3% because of die usage and occasional benching.

The isothermally forged bulkheads were subjected to extensive qualification testing. A summary of the tensile test data is provided in Table 2 and, as can be seen, the data meet the forging specifications. Crack growth testing was also carried out by McDonnell Douglas and the data for the isothermal forging followed the same trends as that for the conventional forgings.

McDonnell Douglas compared the isothermal forging with a conventional blocker forging and a hypothetical 5° conventional forging for cost analysis. It was projected that the raw isothermal forging will cost about 40% more than the present blocker forging and about 10% more than the 5° conventional forging. On the other hand, the isothermal forging machining cost will be 52% lower than that for the blocker forging and 38% lower than the 5° forging. Thus, excluding the die costs, the isothermal forging would show a small overall cost saving in relation to the conventional forgings. The breakeven point with existing tooling would be only a few hundred pieces. But if new tooling was allowed for because of the condition of the dies, the breakeven point was far above the anticipated production run. In any case, an important point is that the isothermal forging will reduce the machining cycle time which is significant if the production rate for the aircraft increases.

This program demonstrated the potential of the isothermal forging process for production of large and complex aerospace components. How-

ever, it also brought forth the difficulties in the operation of tooling system for such parts. More work is needed for developing heating systems capable of obtaining uniform temperature in large dies for long durations. Better techniques will be required if large parts with thin webs are to be made without much distortion of the forgings during ejection. Finally, for cost effectiveness, it is very important to develop materials and manufacturing processes that can reduce the cost of isothermal forging dies.

Fig. 10. Comparison of isothermal (right) and conventional (left) forgings for bulkhead.

The significant accomplishment of this program was the successful forging of the largest and most complex bulkhead to date. The 0.277 m2 (430 sq. in.) plan area with only a 6.4 mm (1/4 in.) web thickness over much of this area and ribs of up to 64 mm (2 1/2 in.) on both sides of the web represents substantially higher complexity than any part isothermally forged before. The work also showed that nickel-base superalloy precision castings weighing up to 2700 kg (6000 lb) can be made by a commercial vendor. Heating systems capable of heating such large dies to 960°C (1760° F) were also fabricated. The greater complexity achieved by isothermal forging is obvious in Figure 10 which compares the isothermal forging weighing only 29 kg (65 lb) with the current conventional forging which weighs 154 kg (340 lb). Thus, the isothermal forging will reduce machining by some 125 kg (275 lb) per forging.

GROWTH AND COMMERCIALIZATION OF THE TECHNOLOGY

Currently there is an immense interest in the isothermal forging process and additional work continues in several organizations. Under an Air Force program, Ladish Company investigated isothermal forging of a torque rib for F-15 aircraft.[11,15] This has become the first part to reach commercial production. The isothermal forging weighs 11.5 kg (25.4 lb) as compared to the conventional forging weight of 20.5 kg (45.3 lb) and the finished part weight of 2.6 kg (5.7 lb). The general shape and straight cross ribs make the part an excellent candidate for fabrication by isothermal forging.

The Air Force has also funded programs[11] on the process at Wyman-Gordon Company for development of better lubrication techniques and for isothermal forging of a connecting link and a spindle bearing support. It is planned to forge nearly half the surface area of the connecting link net. The bearing support isothermal forging is expected to weigh 21.3 kg (47 lb) as opposed to a 54 kg (119 lb) conventional forging. The company has also suggested interest in making parts ranging up to 0.387 m2 (600 sq. in.) in plan area.[16]

Some of the newer beta titanium alloys can be forged at lower temperatures than the alpha-beta alloys referred to in this paper. This could permit usage of lower cost die materials. Unlike the alpha-beta alloys, the beta alloys require specific heat treatment and microstructure for them to be advantageously isothermally forged. Such work is continuing at the Air Force Materials Laboratory.[17]

In recent years, many small to intermediate size forging companies are also taking an increasing interest in the technology. IITRI considers this a very promising development and is actively pursuing such technology transfer. Undoubtedly, the recent selection by McDonnell Douglas of some 100 parts of small plan area ranging to 0.065 m2 (100 sq. in.) for isothermal forging will help this trend.[18]

Commercialization of isothermal forging begins with the selection of suitable components. (See next section.) It also requires careful attention to such key factors as forging design, design and fabrication of forging dies, heating system and support system, optimization of process parameters, preform design, lubrication system, and development of proper procedures for forging. Increasing usage of titanium alloy isothermal forging will go hand-in-hand with an increase in the number of forging vendors who build up expertise in the technology with likely specialization in certain types of products.

CRITERIA FOR SELECTION OF COMPONENTS

Successful application of the new technology is greatly dependent on selection of proper components. This demands careful consideration of many different factors with reference to specific components. It is hoped that a discussion of some key factors here will facilitate this process. They include component size, shape, and complexity, precision required on the finish-machined component, available vendors, and total expected production run.

Components that are expensive to produce, either because of their geometry or because of the extensive finish machining, are good candidates. In the aerospace industry, conventional forgings may often weigh five to ten times or more than the finish-machined component weight. However, it must be remembered that, on a per-pound basis, the finish machining operation is far more expensive than rough machining. Consequently, cost-effectiveness of the isothermal forging process increases greatly if some of the surfaces can be forged net.

Component size and complexity dictate whether some of the surfaces can be forged net. Smaller size and lower complexity make it easier to forge the surfaces net. In large forgings of over about 0.097 m2 (150 sq. in.) plan area, net surface forging becomes increasingly difficult because of the manufacturing tolerances on die dimensions, variability in material properties, and variation and nonuniformity of die temperatures. Long parts with thin webs are also susceptible to distortion during ejection. If all of the rib detail is on one side of the web, the distortion can be minimized by forging the web with excess stock on the other side without comparable penalty in increased machining cost.

Parts with pronounced circular symmetry such as aircraft wheels are particularly well suited for isothermal forging. As was demonstrated in the case of the nose wheel, this process can allow production of thin webs to net dimensions, thus eliminating expensive end milling operations. Moreover, such wheels are generally required in large quantities which helps to amortize the high tooling costs generally associated with isothermal forging. Further, the symmetry of the component aids in getting longer die life and may allow resinking the die cavity several times.

Selection of die material and fabrication techniques are of major importance because of the high cost of isothermal forging tooling. Detailed discussion of such factors is beyond the scope of this paper but, naturally, availability of reusable die holders, modular dies, or common die heating and support systems can greatly reduce the tooling cost per forging. Such possibilities are easier to find for small to intermediate size components. Also, short runs may justify usage of cheaper, lower strength dies whereas long production runs or close tolerances may warrant usage of stronger but more expensive TZM dies. This is true even though the TZM dies, because of their poor oxidation resistance, necessitate that isothermal forging be carried out in vacuum or in inert atmosphere and thus increase processing cost.

Isothermal forgings of very large plan area of about 0.194 m2 (300 sq. in.) and over are likely to be primarily of near-net type--meaning, requiring finish machining. Conventional forging for such components--mainly large airframe structural parts--often weigh more than ten times the finish-machined part weight as was the case for the F-15 bulkhead. The ability of the isothermal forging process to substantially reduce the machining allowance as well as many intermediate forging dies and processing needed in conventional forging tend to make isothermal forging attractive. But the reduction may be largely in rough machining. Also, the die costs for such large parts are high, the processing is difficult, required production quantity is normally low, and only a few major forging companies can produce the parts. Such factors adversely affect the cost-effectiveness of the process for such applications.

An important consideration is for the part designer and forging vendor to get together at an early stage. Dimensional tolerances are often selected arbitrarily because they can be achieved easily by machining. In many cases, such as for example the location of stiffening ribs, the positional tolerances can be loosened without compromising the end usage. Wider tolerances increase the possibility of net forging. Another related factor is to consider isothermal forging before the part is in production by some other technique. Otherwise, unfavorable cost comparison may result, as in the case of the Avco analysis referred to earlier.

By and large, isothermal forging is likely to compete most often with conventional forging as a process alternative. But other processes which may compete with isothermal forging include diffusion bonding, electron beam welding, precision casting, and direct hot isostatic pressing from powder. Much will depend on the advancement in the state of each technology and the size, geometry, and required quality of the specific application under consideration. Generalized guidelines are difficult, and the forging vendor would be well advised to follow the progress in competing processes.

CONCLUDING REMARKS

The isothermal forging technology is ripe for commercial exploitation. The increasing costs of raw materials, labor, and energy should further help the trend towards net and near-net shape isothermal forging of titanium alloys. The parallel development of the isothermal forging process for superalloys will also help the application for titanium alloys since many facility-related items are common for forging of the two types of alloys.

ACKNOWLEDGEMENTS

The isothermal forging work described in this paper was sponsored at IITRI by the Air Force Materials Laboratory and the Army Mechanics and Materials Research Center. Throughout these programs, significant contributions were made by Mr. T. Watmough, Mr. D. Stawarz, and Dr. N. M. Parikh--all formerly my co-workers at IITRI. Alcoa was a major subcontractor for forging of the nose wheel and the bulkhead, and Mr. A. Favre and Mr. C. Yohn were the key Alcoa personnel. McDonnell Douglas participated in decisions about bulkhead forging design and evaluated the bulkhead isothermal forgings. Mr. W. Richards was the main contact there. Mrs. M. Dineen typed the paper.

REFERENCES

1. H. R. Nichols, W. H. Graft, V. Pulsifer, and P. R. Gouwens, "Development of High Temperature Die Materials," Armour Research Foundation Report AMC-TR-59-7-579 for AMC Aeronautical Systems Center, Wright-Patterson Air Force Base, Ohio, 1959.

2. T. Watmough, K. M. Kulkarni, and N. M. Parikh, "Isothermal Forging of Titanium Alloys Using Large Precision-Cast Dies," Air Force Materials Laboratory Technical Report AFML-TR-70-161, prepared by IIT Research Institute, Chicago, July 1970.

3. J. E. Coyne, G. H. Heitman, J. McClain, and R. B. Sparks, "The Effect of Beta Forging on Several Titanium Alloys," Metals Eng. Quart., Vol. 8, August 1968, pp. 10-15.

4. A. J. Griest, A. M. Sabroff, and P. P. Frost, "Effect of Strain Rate and Temperature on the Compressive Flow Stresses of Three Titanium Alloys," Trans, ASM, Vol. 51, 1959, pp. 935-945.

5. T. Watmough and J. A. Schey, "Hot Forming of Titanium and Titanium Alloys," U. S. Patent No. 3,635,068, Jan. 18, 1972.

6. K. M. Kulkarni, T. Watmough, and N. M. Parikh, "Isothermal Forging of Titanium Alloys Using Precision-Cast Dies," Technical Paper MF70-122, Society of Manufacturing Engineers, 1970.

7. K. M. Kulkarni, N. M. Parikh, and T. Watmough, "Isothermal Hot-Die Forging of Complex Parts in a Titanium Alloy," J. Inst. Metals, Vol. 100, 1972, pp. 146-151.

8. J. S. Prasad and T. Watmough, "Precision-Cast Superalloy Dies for Isothermal Forging of Titanium Alloys," paper presented at 73rd AFS Casting Congress, Cincinnati, May 5-9, 1969.

9. K. M. Kulkarni, "Die Materials for Isothermal Forging," paper presented at 1977 Symposium on New Developments in Tool Materials and Applications, Chicago, March 21-22, 1977.

10. K. M. Kulkarni, T. Watmough, N. M. Parikh, D. Stawarz, and M. Malatesta, "Isothermal Forging of Titanium Alloy Main Landing Gear Wheels," Contract No. F33615-70-C-1533, Air Force Materials Laboratory, Technical Report AFML-TR-72-270, December 1972.

11. "Putting the Squeeze on Titanium Forging Costs," Iron Age, Dec. 1, 1975; also "Air Force Material Lab Develops New Isothermal Forging Process for Titanium," Light Metal Age, February, 1976.

12. T. Watmough, "Development of Isothermal Forging of Titanium Centrifugal Compressor Impeller," IIT Research Institute Report for Contract No. DAAG46-72-C-0067, Army Materials and Mechanics Research Center, Technical Report AMMRC CTR 73-19, May, 1973.

13. E. K. Knauf and L. J. Fiedler, "Evaluation of Isothermal Forgings for T53 Impellers," Avco Lycoming Report for Contract No. DAAG46-75-C-0078, Army Materials and Mechanics Research Center and U. S. Army Aviation Systems Command, Technical Report AMMRC CTR 76-14, January 1976.

14. K. M. Kulkarni, T. Watmough, D. Stawarz, and N. M. Parikh, "Isothermal Forging of Titanium Alloy Bulkheads," IIT Research Institute Report for Contract No. F33615-71-C-1167, Air Force Materials Laboratory, Wright Patterson Air Force Base, Technical Report AFML-TR-74-138, August, 1974.

15. A. Hayes, "History on the Production of Isothermal Forged Titanium Parts," paper presented at the 1977 International Engineering Conference and Tool & Manufacturing Exposition, Detroit, May 9-12, 1977.

16. R. Larsen, "Wyman-Gordon Estimates 1976 Capital Budget at $30-Million," Amer. Metal Market, Metalworking News Edition, Vol. 83, April 26, 1976.

17. A. Adair, "The Impact of Beta Alloys on Isothermal Forgings," paper presented at the 1977 International Engineering Conference and Tool & Manufacturing Exposition, Detroit, May 9-12, 1977.

18. J. Thornton, "McDonnell Readying an Advance in Titanium Forgings for Its F-15," American Metal Market, Metalworking News Edition, Vol. 83, Oct. 11, 1976; also "McDonnell Seeking Bids on F-15 Titanium Forgings," American Metal Market, Metalworking News Edition, Vol. 83, No. 135, July 12, 1976.

Isothermal forming

a low-cost method of precision forging

Hot forging can have serious limitations when it is used on some materials. Dr G H Gessinger, of Brown Boveri & Co Ltd's Research Centre in Dättwil, Baden, Switzerland, describes a technique which could prove to be the answer

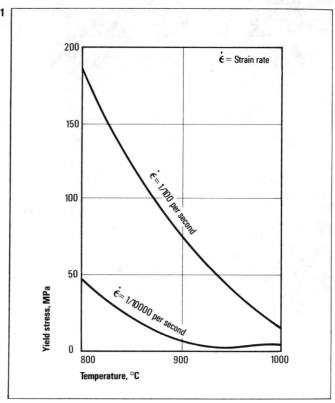

1 Yield stress of Ti6Al4V as a function of temperature and strain rate

Conventional hot forging of iron, nickel, titanium and aluminium alloys exploits the characteristic common to all metallic materials that their resistance to deformation decreases rapidly with increasing temperature. Hence, for a given forging installation, larger diameter blanks can be formed, provided the workpiece temperature is increased. This extrapolation, however, is limited by a number of natural factors:
● The temperature at which an alloy begins to melt.
● Possible phase transformations which can lead to a deterioration in the properties of the finished product (e.g. forging of Ti alloys at 1000°C, at which the high-temperature phase occurs).
● The poor workability of many high-temperature materials such as high-alloy nickel castings.
● Changes in temperature during the deformation process as a result of heat transfer from the hot workpiece to the relatively cold forging die set.
● The stresses to which the tooling may be safely subjected.

Added to these restraints is the fact that plastic deformation is a function of pressure, temperature and time specific to a material. Under the practical conditions of hot forging, it is therefore necessary to reduce the material properties, boundary conditions and cost considerations to a common formula. In the past, the search for such an optimum combination has led in different directions.
a) In the case of high-speed forging with a forging hammer, pressure is increased and, at the same time, the duration of the deformation process is reduced. The disadvantage here is the high die stresses encountered when manufacturing precision parts.
b) Higher pressures with short and medium forging times can be achieved by building increasingly large mechanical and hydraulic presses. Here, the disadvantages are the high capital investment and the problem of keeping such installations working at capacity when small production batches are involved. Because of the high deformation pressures, die temperatures have to be kept low and, as a consequence, a specific, short deformation time may not be exceeded. To date, this is the most widespread process.
c) A reduction in pressure combined with a decrease in the deformation rate can be attained by bringing the forging dies to a temperature close to that of the workpiece (e.g. hot-die forging, isothermal forming, superplastic forming).

The processes listed under c) are based on a specific characteristic of certain alloys, i.e. the marked dependence of yield stress on the strain rate. The origin of this characteristic is to be found in phenomena such as creep, i.e. slow plastic deformation of materials at high temperatures, and superplasticity. Taking the most common titanium alloy as an example – titanium alloyed with 6% aluminium and 4% vanadium (Ti6Al4V) – Fig. 1 illustrates that the stress at which plastic deformation commences, decreases with both the stress at which plastic deformation commences, decreases with both increasing temperature and diminishing strain rates. The fact that this point, known as the yield stress, begins to rise again with low strain rates at temperatures above 950°C, can be attributed to grain growth, phase transformations and the related loss of superplasticity. The isothermal forming processes are based on the phenomenon of superplasticity which is discussed in the following paragraphs.

Superplasticity

A characteristic of certain materials is that the stress σ required to deform them becomes very low when deformation takes place at comparatively high temperatures in the region of 0.5 T_m (T_m = melting temperature in kelvin) and at low strain rates. Under these conditions, such materials have a very high ductility. This is

Reprinted with permission from Engineering, July 1979, 926–929, © 1979 Design Council

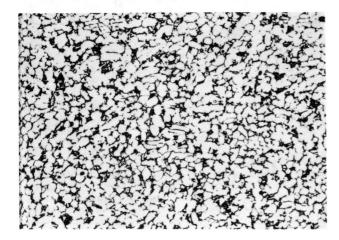

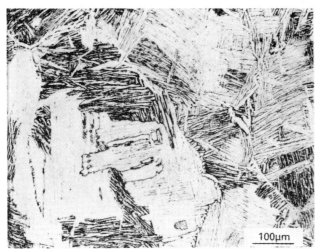

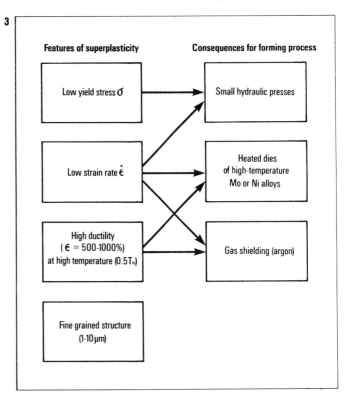

2 Micrographs of Ti6Al4V; left, fine-grained material, right, coarse-grained material

3 Features of superplasticity and consequences for the forming process

demonstrated by an elongation after fracture of several hundred per-cent during creep testing. These properties, which constitute superplasticity, are subject to a fine-grained structure having a grain size of less than 10 μ, and which remains stable at the deformation temperature. Superplasticity itself is based on grain boundary sliding mechanisms.

As grain growth will occur at deformation temperatures, measures have to be taken in order to maintain a fine-grained structure. This can be achieved by having either a high volume fraction of particles up to 10 μm of a second phase, or a small volume fraction of a finely dispersed second phase.

Many titanium and high-temperature nickel alloys (superalloys) inherently fulfil the essential requirements of superplasticity, or can be converted to this fine-grained state by means of a special heat treatment process. The micrograph in Fig. 2 compares a fine-grained second phase structure and a coarse-grained structure in Ti6Al4V.

Fig. 3 is a tabulation of the salient features of superplasticity and the derived requirements of a manufacturing process. The low yield stress permits simple, inexpensive, low-power presses to be used. Low deformation rates mean long contact times in the die which should, therefore, have the same temperature as the workpiece. The choice of tool materials is determined by the forming temperature and pressure. At temperatures of around 1000°C, only cast nickel-base superalloys, with a coarse-grained structure to improve their high-temperature strength, and molybdenum alloys are acceptable. Because of the high deformation temperatures involved, either the workpiece or the dies would oxidize, depending on the material employed. Consequently, the forming process is often done in a vacuum or under argon gas. The high ductility of superplastic alloys enables considerably finer details to be forged than would be possible by conventional hot-forming processes. Nickel superalloys and titanium alloys have proved suitable for the forming process.

Isothermal forming

It is difficult to compare isothermal forming with the traditional forging techniques (Fig. 2). A special 300 t press having little in common with a conventional forming press has been developed at the BBC Research Centre (Fig. 4). The press ram and the die set, which is induction heated, are located in a vacuum chamber. This can be filled with argon and is accessible through a vacuum lock only.

4 300 t experimental press

Source: Engineering, July 1979, 926-929

The answers to three questions, still open at the time, were obtained with the aid of this press:
1. How does the material flow during isothermal forming?
2. What are the mechanical properties of the forging produced and, in particular, do those properties deteriorate through workpiece contamination which is unavoidable during the forming process?
3. Under which conditions is the process economically viable?

Superplastic material flow

The first question was of significance owing to the fact that many of the characteristic quantities obtained for superplastic alloys had been determined by uniaxial tensile testing only. It was therefore first necessary to prove that material flow in a complicated workpiece occurs at the same slow strain rates; otherwise the deformation energy would have to be considerably greater. The answer was sought by taking the eutectic of PbSn which is superplastic

at room temperature and by continuously observing the deformation of a grid photo-etched on it. Fig. 5 shows that when a cylindrical blank of suitable diameter is chosen, the grid deforms uniformly enough.

Point-by-point measurement of the local deformation rate showed that all the volume elements were within the range of superplasticity when the speed of the male die remains constant (equivalent to increasing strain rate). Fig. 6 illustrates the

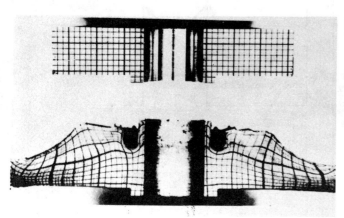

5 Deformation grid before and after isothermal forming of a PbSn compressor wheel

7 Initial and final geometry of a compressor wheel in Ti6Al4V, 160 mm in diameter

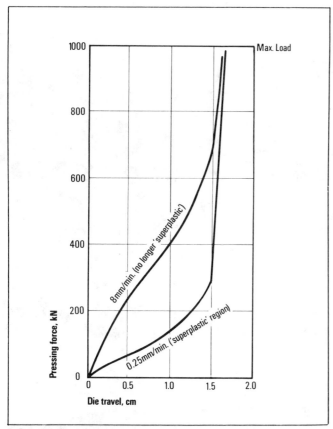

6 Relationship between pressing force and die travel when pressing a PbSn wheel at two different deformation rates

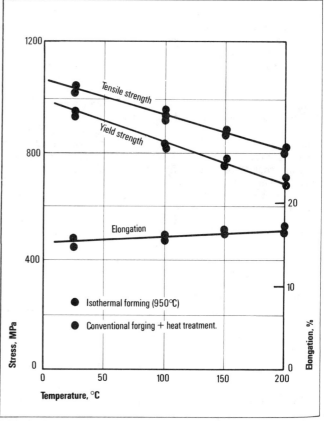

8 Mechanical properties of Ti6Al4V formed conventionally and isothermally

dependence of the pressing force on the distance travelled by the male die when pressing the PbSn model shown in Fig. 5. This dependence was determined at two different strain rates; the lower rate describes the deformation in the superplastic range. The blank used was a hollow cylinder with a cross-sectional area smaller than that of the compressor wheel pressed. The blank is shaped by the descent of the male die and the cross-sectional area increases. Therefore the pressing force rises as the die advances, the die speed remaining constant until the maximum pressing force specified has been reached.

The geometric correlations determined on the PbSn eutectic are also valid for the isothermal forming of Ti6Al4V titanium alloy, because the deformation and yield stress of all superplastic alloys are bound by the same mathematical interrelationship. It also follows that the high forming temperature of 950°C creates a number of inherent technical problems as regards diemaking and the choice of lubricants. Fig. 7 shows the blank and the final geometry of a 160 mm diameter compressor wheel formed in Ti6Al4V. This example of a potential application in the turbine engineering field was formed in a single operation using a heated molybdenum die set. The forming operation took 20 minutes and was done at 950°C under an argon shield.

Maintaining the mechanical properties

The second question was concerned with the mechanical properties. Due to the pronounced affinity of oxygen for titanium at 950°C, it is impossible to avoid the formation of a thin film of oxide, even when working in a vacuum or under argon. It is known that this type of 'forging skin' can reduce the fatigue strength of a Ti6Al4V workpiece to a level considerably below that of an equivalent component having a machined surface. This factor could therefore seriously reduce the cost effectiveness of the process. Systematic trials produced a surface treatment process which completely eliminates this forging-skin effect, thus overcoming the major obstacle to the precision forging of titanium.

The mechanical properties of a finished forging are determined primarily by its grain size. An essential condition of isothermal forming is superplasticity, i.e. fine grain. An isothermally formed workpiece therefore remains superplastic, i.e. relatively soft, at the forming temperature. If, however, the intended operating temperature of a component lies within its own deformation range and the material is liable to be too soft for the operating conditions, then its grain structure has to be coarsened by heat treatment. This

applies in the case of nickel superalloy forgings. Titanium alloy components on the other hand are (normally) employed at temperatures well below that of deformation and are unlikely to require subsequent heat treatment. Conventionally forged components, however, always need subsequent heat treatment and in this respect isothermal forming is more economical than existing methods, at least in the case of titanium alloys.

Cost effectiveness

In order to assess fairly the cost effectiveness of the process as a whole, it is necessary to compare all aspects of four different processes. In the case of conventional upset-forging there is no saving of material possible; on the other hand, there is generally no need for dies. Although conventional closed-die forging definitely economises on material, allowances are required over the entire surface and for fine detail. Whilst the cost of machining is lower than when upset-forging, it is nevertheless still very high. As conventional steel dies are employed, the tooling costs are about average.

Isothermal forming with allowances provides a further saving in material but the entire surface has to be finish machined. Tooling costs are high because high-temperature alloys are needed to make the dies. Savings in material are greatest when isothermal forming to net shape is employed. An added advantage is that critical surfaces do not require finishing. However, the demands placed on tooling and the tooling costs are higher than any of the other processes considered. When the total costs of material, forming, tooling and machining are analysed, the cost-saving potential is only then evident when the high machining and material cost factors can be held so low that the tooling costs do not predominate. In the case of many titanium and nickel-alloy forgings, cost analysis of this kind will show that isothermal forming is the least expensive of the processes compared.

Powder metallurgy: a rival and complement

In this branch of technology, isothermal forming as a process is not without rivals. Particularly in the case of nickel superalloys, power-metallurgy manufacturing processes such as hot isostatic pressing (h.i.p.) have made their presence felt. Superalloy powders are considerably more expensive than the cast materials normally used in the conventional forming processes. The h.i.p. process however, can reduce macrosegregation, i.e. possible non-uniform distribution in highly-alloyed nickel alloys, to the scale of the powder particle diameter ($\sim$50-100 μm). In order to obtain

superior mechanical properties, h.i.p. preforms are often finish-formed by isothermal forming. The object of future development work is to eliminate the forming process altogether.

This does not apply to titanium alloys. First, the price of titanium alloy powders is very high and secondly, the formability of conventional titanium alloys under isothermal conditions is so good that only in exceptional cases could powder metallurgy be expected to bring about a reduction in costs.

Unquestionably, the hot pressing of powders in an isothermal press is an interesting complement to the isothermal-forming process. Development work in this direction is being continued at the BBC Research Centre, initially by way of basic trials. In many cases, this process could eliminate the expensive hot isostatic pressing operation.

Summary

Extensive trials conducted at the BBC Research Centre over a period of years have shown that in combination with certain alloys, isothermal forming is an economical forging process. The process centres around the use of forging dies heated to the same temperature as the workpiece. As such, it exploits the low resistance to deformation of these alloys at low strain rates. The major advantage of the process lies in its being more economical than competitive processes. Isothermal forming is suitable for materials which are expensive to machine and also for highly complex forgings. This manufacturing method, as devised by BBC, produces forgings which require virtually no finishing. The mechanical properties of the titanium components produced are at least equal to those of conventionally forged and heat-treated alloys●

Advantages
● Precision forging operation using few dies (in the optimum case, one)
● Low forming pressure, hence simple, compact presses, or production of larger parts on conventional machines
● Uniformly distributed deformation results in products with homogeneous mechanical properties
● Fewer forging defects (rejects) cut the cost of testing
● No surface cracking
● Heat treatment after forging is simplified, or even eliminated
● Closer maintenance of tolerances

Drawbacks
● High die costs: to offset these a minimum of 70 to 100 forgings must be guaranteed

A STUDY OF THE ISOTHERMAL FORMING OR CREEP FORMING OF A TITANIUM ALLOY

by

J.W.H. Price and J.M. Alexander, Department of Mechanical Engineering, Imperial College, London

The basic mechanisms of isothermal forming with particular reference to the titanium alloy 6 Al 4V (6% Aluminium 4% Vanadium remainder Titanium) are examined by means of compression and die filling experiments. Lubrication at 800°C-980°C is studied for the forming of this alloy and the load-displacement relation for the tests is predicted from upper-bound calculations. A steel is tested under isothermal forming conditions as well. A finite element solution suitable for studies of the high deformations encountered in the process is presented.

INTRODUCTION

Isothermal forming (or creep forming) is a technique of forging in which use is made of the fact that metals generally are softer the higher the temperature and the lower the strain-rate imposed upon them. In isothermal forming the dies are heated to the same temperature as the specimen and loading is at a slow speed. As a result the material has time for extensive deformation, and can be made to conform very accurately with the shape of the dies. The dimensional accuracy of the product is thus very good and only a minimum of further working or machining is required to obtain a finished product. The process has been used in particular for the production of difficult shapes in titanium alloys and is discussed and examples given in numerous recent references (1).

The study undertaken here was to examine isothermal forming from a more fundamental viewpoint, since researches so far have been concerned with complicated, commercially important, practical forgings. The present research has thus been directed towards compression tests and die filling experiments using dies of simple axisymmetric shape, and also with obtaining some theoretical predictions. This paper is mainly devoted to a description of the experimental section of this work. There is also an outline of a finite element solution that is suitable for application in a situation having the high strains involved in the process. The method enables accurate solutions to be obtained without an unreasonable use of computer resources.

EXPERIMENTS

i   Equipment

The experimental equipment was designed to model industrial isothermal forging. An induction furnace of 13kWatt and 3kH$_z$ frequency was used which enabled fairly even heating of axisymmetric shapes up to 75mm diameter. The time for heating from room temperature up to 950°C (1740°F) was about 15 minutes. A 500kN (50 ton) standard Denison testing machine with controllable speeds was used and a continuous autographic recording of load, temperature of dies and displacement of crosshead was made. The dies were made from a nickel alloy designated IN-100 which has the chemical composition 10Cr, 15Co, 3Mo, 7Ti, 5.5Al, 1.0V, 0.06Zr, 0.015B and balance Ni. These dies are extremely hard up to and beyond the temperature of 950°C and as a result their main machining was carried out by EDM (spark erosion) with the final finish being obtained from conventional methods (i.e. turning with tungsten carbide tipped tools then 'fine ground'). The experiments were carried out using specimens of titanium 6 Al 4V unless otherwise noted.

ii   Cylindrical Compression Tests

Figures 1 and 2 are typical results from the compression of a cylinder of titanium alloy to a very high strain, the graphs being the output of a computer program which processes the data. The material was held at a constant temperature of about 800°C, a graphite lubricant (in water base) was applied lightly to it and the crosshead speed was set at 3mm/sec. The original cylinder was 19.05mm high and 12.8mm diameter but after compression it

was only 1.6mm high and about 45mm in diameter. This represents a uniform logarithmic strain of -2.4 along the axis or a reduction of 91.5% in height.

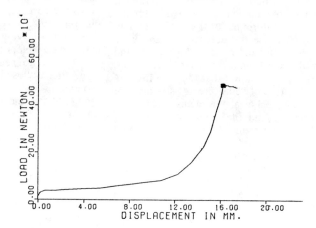

Figure 1. Compression Test; Load versus displacement

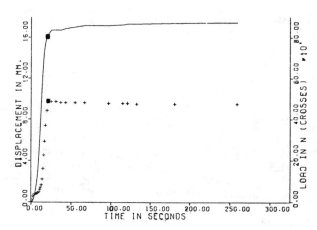

Figure 2. Compression Test;
Load (Crosses) and displacement versus Time

On figures 1 and 2 two stages of deformation might be discerned separated by the squares on the figures. The first stage is a smooth curve during which the crosshead moves with an approximately constant speed of 3mm/sec. Most of the deformation was achieved in the lower 10% of the load range. In the region of the square the machine approached its maximum load of 500kN (50 tons) and the crosshead speed slowed down so that it was barely moving. In this region it can be considered that the deformation continues by creeping. On figure 2 it can be seen that the creeping regime occupies most of the time of the experiment but in fact there is very little further decrease in height of the specimen during this period.

On inspection of the specimen after the experiment it was found that there had been almost no relative movement between specimen and dies. Marks on the ends of the original specimen did not change their radius during the experiment. The very large increase in area of the ends that occurs during the experiment is provided by the material from the sides stretching and rolling onto the dies. The graphite thus serves very little purpose in promoting sliding along the die-metal interface.

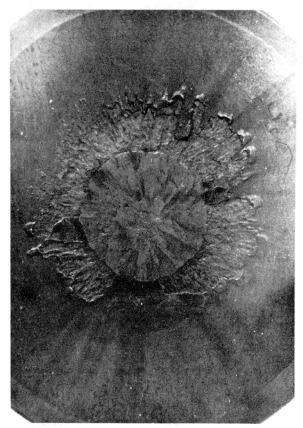

Plate 1. A specimen over-lubricated with graphite

Application of too much graphite can, however, cause considerable damage to a specimen as illustrated on Plate 1. This plate illustrates a specimen originally 12mm high and 12.8mm in diameter now deformed to 2.1mm thick and 4)mm in diameter. There are several features of this features of this picture that require explanation.

The top of the specimen is marked by a 'sunburst' appearance with lines radiating from the centre of the specimen. This feature is apparently caused by the layer of graphite lubricant being extruded radially out between specimen and die. The lubricant forms a layer of comparatively low viscosity between specimen and die and the amount of it extruding increases with pressure. This phenomenon is probably a measure of the 'friction hill' on these specimens, as discussed by Avitzur (2). In the centre of the specimen there is a circular area which comprises its original top end. This area has basically not changed in size, though movement of lubricant across it is obvious. Outside this is a region where the surfaces that were originally on the outside of the cylinder have rolled up onto the die. Towards the centre of this region the outflow of lubricant has been so strong that in most areas it has torn up the oxide layer on the specimen and forced it outwards. Near the outside circumference of the specimen there is a lighter bend. This is probably a surface, most of which has been newly created. Figure 3 illustrates an experiment with plasticine which could explain this phenomenon, 3(a) illustrates the initial, two colour cylinder. Figure 3(b) illustrates a cross-section of the cylinder after having imposed a heavy deformation without lubricant whilst figure 3(c)

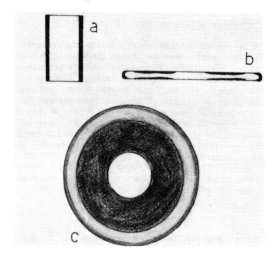

Figure 3. Compression of a plasticine cylinder

shows a plan view. These diagrams illustrate that the light band could be caused by material from the interior of the specimen approaching the surface very closely thus forming a band of mainly fresh surface. The original outside surface is mainly trapped onto the dies at this high strain.

In situations where damage occurs, such as that illustrated here, the die surface is quite likely to be damaged as well as the specimen and die may require regrinding. The amount of damage that occurs depends on the relative strengths of the die and the specimen. The strengths of the nickel alloy dies and the titanium alloy specimen both vary with temperature, but in the range 800-980°C it appears that the nickel alloy changes at a much slower rate than the titanium alloy. In experiments it was found that the die was much more likely to be marked by the tests at 800°C than at 980°C.

There was no need to proceed to a load as high as 500kN to achieve a large strain in the titanium cylinders. In some experiments the load was held at a much lower level and almost as much strain was achieved although in a longer time. A strain of -2.0 was obtained after holding a specimen, similar to the one reported in figures 1 and 2, for 3 minutes at 50kN (5 tons). Stresses can also be reduced by raising the temperature of the tests but in titanium 6 Al 4V the usual maximum temperature for hotworking is 980°C, for above this temperature the alloy becomes single phase and this can cause deleterious effects on properties when the specimen is cooled.

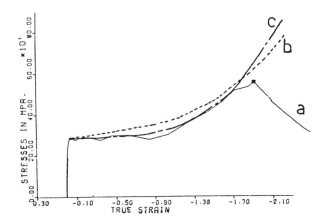

Figure 4. Stress-strain curves for a compression Test
Curve a. Experimental results
Curves b and c. Upper bound solutions

Source: 4th North American Metalworking Research Conference Proceedings, 1976, 46-53

iii   The Stress Strain Curve

Figure 4 is a plot of the average axial stress (pressure at interface) versus average logarithmic axial strain (curve A).   The period of creeping which starts in the region indicated by the square is seen to involve a considerable strain, although the specimen does not decrease in height greatly.   This is because the specimen is now a thin disc with a height to diameter ratio of less than 1:20.   This curve is better understood (at least up to the region of the square) when it is plotted in conjunction with an upper bound solution for axisymmetric compression .   There is an upper bound solution which assumes no barrelling and friction governed by a shear coefficient at the boundary (see for example reference (2)).   The velocity field for this situation is

$$u = \frac{r\,V_o}{2\,a} \qquad \& \qquad v = \frac{z\,V_o}{a} \qquad (1)$$

where the symbols are described on figure 5.

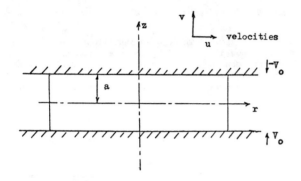

Figure 5. Notation for upper bound solutions

A shear coefficient, m, is used such that

$$T = m\,\frac{\bar{\sigma}_o}{\sqrt{3}} \qquad (2)$$

where T is shear stress at the boundary and $\bar{\sigma}_o$ is the flow stress of the material (which is assumed to be constant). For the curve illustrated m was chosen as its maximum possible value of 1.

However the curve for this solution (curve B) does not conform very closely to the data.   In view of the fact that there is no sliding at the boundary and that there is extreme barrelling during the test this is not a very satisfactory velocity field.   A special case of a velocity field proposed by Kobayashi and Thomsen (3) can be used.   In this special case there is no relative movement between the platens and the specimen and the barrelling can be taken to be parabolic in form.   This is written

$$u = \frac{3\,V_o}{4\,a}\,r\left(1 - \frac{z^2}{a^2}\right),$$
$$v = -\frac{3\,V_o}{2\,a}\left(z - \frac{z^3}{3a^2}\right). \qquad (3)$$

The solution is much better as shown by curve C on figure 4.   In the region of the square, however, the solution fails.   This means that in this region the basic assumptions of the upper bound technique must no longer be applicable.   Thus either the velocity field is no longer accurate or the flow stress of the material is changing. Of these the latter seems most relevant for, in the region of the square, the speed of the crosshead is decreasing considerably and it has been shown by Lee and Backofen(4) that the material has a flow stress that is strongly strain-rate dependent.   This region can thus be termed the region of creep forming.

iv   Lubrication

The graphite lubricant used in the compression tests described above serves no apparent function in the process. In fact, isothermal forming of titanium can be performed without any lubricant without any change in loads or behaviour of the specimen.   However, in these experiments a thin film of graphite was normally applied to the specimens and dies because it is possible that it might aid the removal of the specimen from the dies or reduce die wear (though no direct evidence was obtained to these effects).   Thus, with graphite lubricant the contract between the specimen and the dies is not characterised by the normal processes of metal to metal lubrication seen at lower temperatures but is probably dominated by the hard oxide coatings on both the dies and specimens. Glass is another lubricant which is readily available and suitable for metal working at these high temperatures. In contrast to graphite, glass has a profound effect on the type of deformation that is seen.   If a thick layer is applied to the ends of a cylinder in a compression test it can even produce a reverse barrel or hour-glass shaped specimen as shown by Douglas and Altan (5).   In that case the layer of molten glass has such a low viscosity that its radial extrusion during the compression test causes the specimen to deform outwards more at the die-specimen interface than at the centre of the specimen.

Comparison of ring tests of the lubricants used in these studies is made on Plate 2.   At the top is the starting ring which is 6.45mm high, with an outer diameter of 25.5mm and an inner diameter of 12.9mm.   The two specimens on the left are not lubricated and in this case the friction is so high that the hole in the middle eventually closes up at high reductions.   Using the calibrations of Hawkyard and Johnson (6) the shear coefficient, m in equation 2 above, appears equal to one (actually the points fall above the line for m=1 but this value for m is the maximum according to the theory).   The centre specimen was lubricated by a thick layer of graphite.   Apart from the damage on the surface of the specimen there is no apparent difference in performance from the specimens with no lubricant.   The two specimens are not so circular as those without lubricant, an effect which is caused by the uneven flow of the glass lubricant.   This flow is often limited preferentially to small areas and the material in these areas can be distorted more than in others.   This effect is also seen to a lesser extent on thickly lubricated graphite specimens. For the glass lubricated rings Hawkyard and Johnson's graphs give a shear coefficient of 0.06.

In general the glass lubricant is probably not satisfactory for isothermal forming when compared to a thin film of graphite.   The surface finish on the specimen can be rougher with glass lubricants and the specimens are less likely to be circular after forming.   The glass solidifies as the specimen cools and thus has to be removed physically from the specimen.   The specimens are also more difficult to remove from the dies for at high temperatures the molten glass forms an adhesive mass which can cause an air lock and retard the removal of the specimen, while at low temperatures the solid glass must be fractured.

v   Profile Dies

The main profile die experiments have been carried out on two different dies.   Both dies have similar dimensions but one die has a more generous fillet radius and taper angle than the other.   The basic dimensions of the two dies are illustrated on figure 6.   The final product thus has a truncated conical region, hereafter termed the 'cap' around which there is formed a flange as the dies close.   There was no noticeable difference in performance between these two dies (P die and R die).   In view of the fact that there is no slipping of the titanium along the horizontal surfaces of the dies the fillet radius makes virtually no difference.   There was no problem extracting the finished product from the dies when using a thin film of graphite lubricant, though glass provided considerable

Plate 2. *Rings for lubricant tests.*
        Top - Original specimen, ratio 1:½:¼
        Left - No lubricant
        Centre - Graphite
        Right - Glass
        Speed - 3mm/sec; Material - Titanium 6Al 4V
        Temperature - 950°C.
        Grid on plate 2 is inches.

Plates 3-5

Profile Die (P die).
Speed - 3mm/sec
Metal - Titanium 6Al 4V
Lubricant - Graphite
Temperature - 950°C.
Specimen volume - 16,500mm³
*   (1 cubic inch)*
Original specimen diameters -

    Plate 3 19.2mm (¾ in.)
    Plate 4 25.4mm (1 in.)
    Plate 5 35.0mm (1.38 in.)

Plate 3

Plate 4

Plate 5

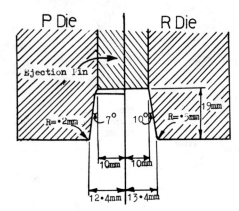

Figure 6. Dimensions of the two profile dies

TABLE 1

Alloy	Steel	Titanium
Final load	500kN	400kN
Final thickness	4mm	3mm ± .2mm
Final diameter	63mm	72mm
Reduction	84%	88%
Axial strain	-1.9	-2.1
Stress at -.18 strain	51.5MPa	18.1MPa
	(7.46ksi)	(2.62ksi)
At maximum load	No apparent creeping	Appreciable creeping

problems. A die with a taper angle of 2° was also used but removal of the specimens from this die was more difficult requiring somewhat more force and even heat expansion of the die. With dies P and R a complete fill of the cap was always possible and a thin flash (0.05-0.2mm) was extruded up where the ejection pin did not fit exactly inside the die.

Plates 3-5 demonstrate the differences in forming stages when starting with three different billets. Die P was used with graphite lubricant at 950°C, and the three starting shapes all have the same volume (16,500mm³). The first billet (on plate 3) has a diameter that is a little smaller than the ejection pin (19.2mm) and thus it fits into the bottom of the cap of the die. In the first stage the cap is completely filled and the bottom of the specimen is compressed and is thickening. At this stage the load begins to rise and rest of the test is occupied by what is virtually the compression of a cylinder to a thin disk, since the material in the cap barely deforms. Plate 4 illustrates a specimen of starting diameter 25.4mm which is only slightly smaller than the diameter of the bottom of the cap. As the load is applied this specimen slips at the edges where it contacts the inside of the cap in the die. Such a situation causes a quite severe situation for die wear, and the press shudders a little at this time which possibly indicates that a stick-slip friction situation is occurring. As deformation continues a lip is formed on the specimen which indicates a region of quite severe shearing. The lip disappears in the final specimen which is also the only specimen to show complete filling of the cap. Plate 5 illustrates a specimen of 35mm in diameter. In this specimen there is no rising into the cap during the early stages of the experiment, in fact, the total height of the specimen slightly decreases. Finally, however, the flange becomes sufficiently thin for the material to be forced up into the cap, and the height of the specimen increases. At constant crosshead velocity this billet is the quickest of the three to be finished for it requires the least travel of the crosshead.

vi   The Isothermal Forming of Steel

A steel of composition 0.23C, 2V, 4Cr, 5.1Mo, 5.2W was used to test the possibilities of the isothermal forming of steel. The basic purpose of the additions is to provide a durable oxide that could withstand the conditions of isothermal forming. It is probable that a steel with 4% Chromium as its only addition would probably form satisfactorily in an air environment as discussed by Alexander and Turner (7) and, of course, steels with different additions may also prove useful.

Compression tests at 980°C with thin film graphite lubrication using billets 25.4mm height and 25.4mm diameter were performed with the steel alloy and the titanium alloy. The two experiments are compared in Table 1.

The main conclusion from these tests is that for steel the stress levels were about 2.8 times higher and that very little useful deformation occurs by 'creep forming'.

Profile die P was also used successfully for forming the steel. The billet diameter of 25.4mm is smaller than the opening of the cap so that filling of the cap was easily achieved. However the minimum flange thickness of 3.25mm that was obtained at 500kN was not sufficiently thin to ensure that good rising into cavities could be obtained as illustrated for titanium on plate 5. By comparison the titanium easily achieves a flange thickness of 2.75-2.50mm and filling of the cap at 400kN. The steel was unable to form a vertical flash between the ejector pin and die as thin as that of the titanium, the minimum flash achieved having a thickness of 0.5mm. The steel conformed well to little marks on the dies and also to the sharp fillet radii. However the oxide coat was not quite as good as that of the titanium, being less tough and slightly friable, so the steel does not fit quite so sharply into the detailed features. The steel specimen was easy to extract from the die, but presumably because the oxide coat is not so strong as titanium oxide there was more evidence of the specimen slipping over the die during forming.

FINITE ELEMENT STUDIES OF ISOTHERMAL FORMING

An outline of the approach used for obtaining finite element solutions for large deformations is given here. A fuller description of the method used in this work will be given elsewhere (8). Several features of the isothermal forming of graphite lubricated titanium alloy must be considered in the approach.

(i) The material is strongly strain-rate sensitive but at the temperature of forming there is little strain hardening. The constitutive relationship must therefore be such that flow stress is a function of both strain rate and temperature but not necessarily of strain.

(ii) There are high deformations obtained in the process and in most cases all the material is substantially deformed. The plastic strains that occur will be many times larger than any elastic strains. Hence a method in which the elastic strains are neglected will be satisfactory.

(iii) The die-specimen interface normally exhibits no slipping but some boundaries that originally are free eventually come into contact with the die.

(iv) The dies can be considered to be rigid.

From consideration of these properties a finite element scheme was chosen that would give a satisfactory modelling of the process without too great a use of computer time.

(a) An increment of plastic work per unit volume, dW_p, can be written, following Hill (9), as

$$dW_p = \sigma'_{ij} d\Sigma^p_{ij} \qquad (4)$$

where σ'_{ij} are the deviatoric stresses, $d\Sigma^p_{ij}$ are the increments of plastic strain, and the summation convention applies. This can be rewritten in terms of a plastic strain rate per unit volume

$$\dot{W}_p = \sigma'_{ij}\dot{\Sigma}^p_{ij} \ . \qquad (5)$$

The normal elastic finite element formulation can be converted to using strain rates and deviatoric stresses instead of strains and stresses respectively (11). The nodal variables are now considered to be velocities and the increments of deformation are really time steps. It is a work rate that is minimized throughout the volume.*

(b) For the constitutive relationship the Levy-Mises flow rule can be used (9)

$$d\Sigma^p_{ij} = \sigma'_{ij}d\lambda \qquad (6)$$

where $d\lambda$ is an instaneous constant. In order to calculate $d\lambda$ the equation must be written to involve time:

$$\dot{\Sigma}^p_{ij} = \sigma'_{ij}/2\eta \ . \qquad (7)$$

It is written in this form in order to emphasize the connection between this formulation and that for the flow of viscous fluids at very low Reynolds numbers. Thus η can be interpreted as an instantaneous viscosity. In the present program provision was made for the 'viscosity' to be estimated at each iteration for each gauss point in each element. The value was found from interpolation directly from experimentally determined stress-strain rate-temperature data (as is provided in ref 4) or from an equation giving viscosity in terms of temperature and strain rate. The strain rate and temperature from the last iteration is used to estimate the viscosity for the current iteration.

(c) Incompressibility must be ensured. Of the techniques which may be used to achieve incompressibility the most efficient and simplest is the penalty function approach (10). In this approach a large positive number,α, is multiplied by the volume integral of the square of the volumetric strain rate:

$$\alpha\int_v\dot{\Sigma}^2_v dv \qquad (8)$$

and this is added to the work rate functional,χ, for the finite element virtual work rate formulation

$$\chi = \int\dot{W}_p dv \ + \alpha\int_v\dot{\Sigma}^2_v dv \ + B. \qquad (9)$$

B represents the contributions of body forces (normally considered to be unimportant in slow speed plasticity problems) and surface tractions.

(d) The boundaries of the specimen are either free or geometrically determined by the dies if sticking friction, as described in the experiments above, is assumed. Frictional conditions have been allowed for in the present program but are not normally useful for isothermal forming. Provision is made for boundary nodes that were originally free to roll onto the dies and become fixed.

(e) During the deformation the mesh must be continuously updated to reduce the problem as far as is possible to one of infinitesimal strain. This is achieved by using small time or displacement steps and each node is moved after each increment. The movement of each node is calculated by multiplying the velocity calculated at that node in the previous increment by the time elapsing between increments. The distorted elements that are produced after a number of increments represent the only memory that the model material has of its previous shape.

As a result the mesh must be arbitrarily reformed only when it is absolutely necessary. This occurs at times when elements become excessively distorted or negative areas are created. An example of this is shown on figure 7.

---

* The original program was kindly provided by Professor O.C. Zienkiewicz. It is an elastic program with isoparametric quadrilateral elements and a high speed elimination solution for banded matrices. Technical details of this program are described by Zienkiewicz (10).

(f) A subroutine was developed to calculate the mean stress, σ_m, at the nodes using the deviatoric stresses and shear stresses calculated in the main program. This is achieved by use of the equilibrium equations and this can be demonstrated by reference to the x direction as follows:

First $\qquad \sigma_m = \sigma_x - \sigma'_x \qquad (10)$

where σ_x is stress in the x direction and the dash indicates a deviatoric stress. The equilibrium equation, for the case where the only shear stress (T) is in the x,y plane, is given by

$$\frac{\partial\sigma_x}{\partial x} = -\frac{\partial T_{xy}}{\partial y} \ . \qquad (11)$$

Thus using equation (10)

$$\frac{\partial\sigma_m}{\partial x} = -(\frac{\partial\sigma_x}{\partial x} + \frac{\partial T_{xy}}{\partial y}) \ . \qquad (12)$$

Similarly it can be shown

$$\frac{\partial\sigma_m}{\partial y} = -(\frac{\partial\sigma_y}{\partial y} + \frac{\partial T_{xy}}{\partial x}) \ . \qquad (13)$$

Hence an integration in the x,y plane can be carried out to find σ_m throughout that plane. The initial values of σ_m are calculated at a free boundary, for there the deviatoric stress normal to the boundary must be equal to the negative of the mean stress.

This method of mean stress calculation has the chief advantage over other methods (12) that the stiffness matrix still requires only two variables at each node, and that the mean stress calculation need only be performed at times during the calculation when it is specifically required. However, the method relies on the values of deviatoric stresses calculated at the nodes in order that the integration can be carried from one element to another. In the formulation used the stresses calculated at the nodes are not so accurate as the stresses calculated at the four (2 x 2) gauss points in each element. As a result a small but cumulative error occurs during integration across the specimen and the method is not suited to regions of extreme strains where nodal stresses are often unsatisfactory.

Some example results from this program are presented in figures 7-9. The results compare well with experiments. The computer time taken for a single increment is about 20 seconds on a CDC 6400 computer and this is sufficiently economic to consider for routine use. The largest practical increment considering the conflicting demands of economy and accuracy of assumptions and difficulties of mesh and boundary distortions is probably about 10-15% reduction in height.

CONCLUSIONS

The experimental evidence presented leads to several main conclusions about isothermal forming.

i   The strains that can be obtained are very high and the loads are low. High strains can be achieved by holding the material for extended periods at low loads. A successful upper-bound solution is possible.

ii   Satisfactory results are best obtained with a thin layer of graphite lubricant or no lubricant at all. However there is no slipping between dies and specimens except in extreme circumstances.

iii   Sharp fillet radii and low taper angles can be used without difficulty on profile dies. Die filling is easy, even into sharp corners if a suitably shaped starting billet is used.

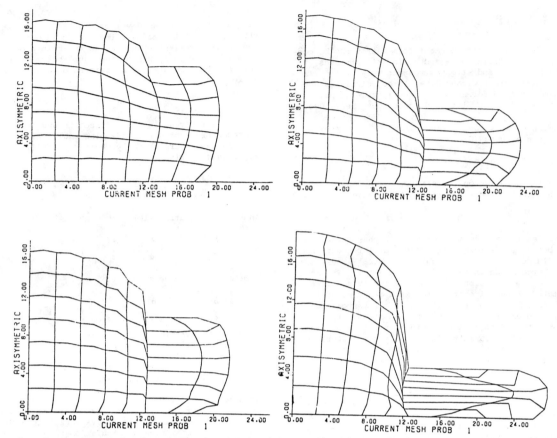

Figure 7.   *The isothermal forming of Titanium 6Al 4V with a profile die as calculated by the finite element method.   Compare with Plate 5.*

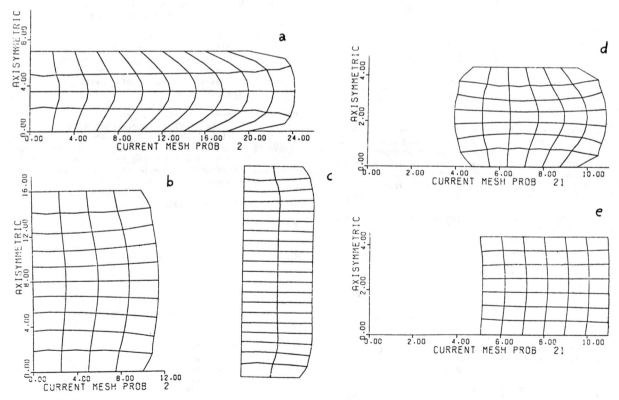

Figure 8.   *Five examples of forging between flat dies as calculated by the finite element method. a, b and c are cylindrical compression specimens of height to diameter ratios of ¼:1:4 respectively.   Note double bulging in specimen c.   d and e are ring test, d with sticking friction, e with a friction coefficient of 0.16.   Original sizes were 6.37mm high, 4.77mm inner and 9.54mm outer radii.*

iv Forming to quite high strains is possible even if
    the material does not exhibit any creeping or super-
    plastic properties, as illustrated by the isothermal
    forming of steel.

With the finite element method described it is possible
to obtain a theoretical representation of the forming
process with less time taken than with many finite
element solutions to non-linear problems.

ACKNOWLEDGEMENTS

One of the authors (JWHP) acknowledges the support of a
scholarship from the Commonwealth Scientific and
Industrial Research Organization, Australia.   Provision
of specimen materials by Rolls Royce (1971) Ltd. and
British Steel Corporation is gratefully acknowledged.

REFERENCES

(1) "Isothermal Hot Die Forging of Complex Parts in a
     Titanium Alloy", K.M. Kulkarni et.al., Journal of
     Institute of Metals, 100, p 146-51, (May 1972).

     "Isothermal Forming of Titanium Alloys", S.Z. Figlin
     et.al., Kurznechno-Stanpovnochoe, (9), p 6-9,
     (Sept. 1972).

     "Isothermal Forging and other Emerging Processes",
     F.W. Bolger, Forging Equipment Materials and
     Practices, US National Technical Information Service,
     MCIC-HB-O3, p 213-45, (1973).

(2) "Metal Forming: Processes and Analysis", B. Avitzur,
     McGraw Hill, New York, (1968).

(3) "Solutions for Axisymmetric Compression and Extrusion
     Problems", S. Kobayashi and E.G. Thomsen, Int.
     Journal of Mechanical Science, 7, p 131, (1965).

(4) "Superplasticity in some Titanium and Zirconium
     Alloys", D. Lee and W.A. Backoffen, Trans. AIME,
     239, p 1084, (1967).

(5) "Flowstress Determination for Metals at Forging Rates
     and Temperatures", J.R. Douglas and T. Altan, Trans.
     ASME (B), Journal Eng. for Ind., 97(1), p 66 (fig 2),
     (Feb. 1975).

(6) "An Analysis of the Changes in Geometry of a Short
     Hollow Cylinder during Axial Compression",
     J.B. Hawkyard and W. Johnson, Int. Journal of
     Mechanical Science, 9, p 163, (1967).

(7) "A Preliminary Investigation of the Dieless Drawing
     of Titanium and some Steels", J.M. Alexander and
     T.W. Turner, 15th Machine Tool and Design Conf.,
     ed. S.A. Tobias and S. Koenigsberger, Macmillan,
     London, p 525, (1975).

(8) "The Finite Element Analysis of Two High Temperature
     Metal Deformation Processes", J.W.H. Price and
     J.M. Alexander, Second Symposium on Finite Elements
     in Flow Problems, Rappalo, Italy, (June 1976).

(9) "The Mathematical Theory of Plasticity", R. Hill,
     Clarendon Press, Oxford, (1952).

(10) "The Finite Element Method in Engineering Science",
     O.C. Zienkiewiez, McGraw Hill, London (1971).

(11) "Visco-Plasticity, Plasticity, Creep and Visco-
     Plastic Flow", O.C. Zienkiewicz, Int. Conference on
     Computational Methods on non-linear Mechanics,
     University of Texas, Austin, Texas, (1974).

(12) "Viscous Incompressible Flow with Special Reference
     to Non-Newtonian (Plastic) Fluids", O.C. Zienkiewicz,
     and A.N. Godbole, Int. Symposium on F.E.M. in Flow
     Problems, Swansea, (1974), to be published by
     J. Wiley.

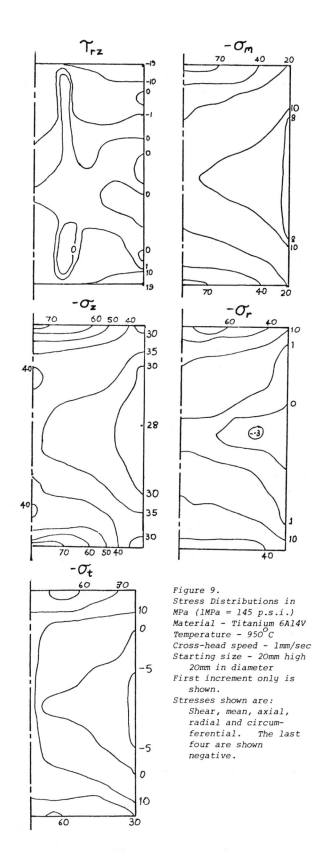

Figure 9.
Stress Distributions in
MPa (1MPa = 145 p.s.i.)
Material - Titanium 6A14V
Temperature - 950°C
Cross-head speed - 1mm/sec
Starting size - 20mm high
    20mm in diameter
First increment only is
    shown.
Stresses shown are:
    Shear, mean, axial,
    radial and circum-
    ferential.   The last
    four are shown
    negative.

Isothermal Hot-Die Forging of Complex Parts in a Titanium Alloy

K. M. Kulkarni, N. M. Parikh, and T. Watmough

The paper describes experimental work performed to develop the technique of isothermal forging of titanium alloys. In this process heat loss from the blank is prevented and low press speeds can then be employed to take advantage of the strain-rate-sensitivity of titanium alloys. In all, five different types of forging—three structural shapes ranging up to 36 000 mm² in plan area and two wheel shapes of up to 71 000 mm² plan area—were forged. The dies were made from nickel-base high-temperature alloys by precision casting and were heated in the press by induction to the forging temperature of up to 1000° C. Starting from simple preforms, it was possible to make the various complex shapes in one stroke of the press. With a press speed of the order of 0·04 mm/s in the final stages of forging, the forging pressures were one-fifth to one-tenth of those in conventional forging. Compared with conventional titanium alloy forging methods, the principal advantages of the isothermal process include an ability to make very complex forgings from simple preforms, the possibility of producing larger forgings using available equipment, more efficient utilization of material, improved control over metallurgical structure in finished forgings, and reduced surface contamination because of shorter exposure of the forging blank at high temperature. In addition, in isothermal forging, the pressure requirements appear to be insensitive to the shape and size of a forging.

In conventional titanium alloy forging practice, the forging temperature varies from the β transus to ∼ 150 degC below this temperature. Since the die temperature normally ranges between 300 and 500° C, a considerable temperature gradient exists between the die and the billet, and to reduce the amount of workpiece cooling, fairly high forging speeds are utilized. Thus, the flow stress of the material increases owing to the rapid heat loss from the billet, and also to its high strain-rate-sensitivity. Because of these considerations, complex titanium alloy forgings require a number of forging steps in several sets of dies. Even then, these forgings usually have to be machined to a final shape and size, and so the net cost of the forging tends to be quite high.

In the isothermal process the die and the forging billet are heated to the same temperature; there is no danger of cooling of the workpiece, and any suitable press speed can be employed during forging.

Before undertaking forging, experiments were conducted to select the die materials, temperature, and the speed of forging. A variety of shapes, including a 310 mm-dia. full-scale aircraft nose wheel, were then forged to final dimensions.

Strain-Rate-Sensitivity of Titanium Alloys

In any forging operation, most of the deformation takes place under compressive stress conditions. Simple upsetting tests were conducted under isothermal conditions to study the effect of strain rate on the flow stress of a Ti–6Al–6V–2Sn alloy (see Table I for composition). Fig. 1 shows the flow-stress/strain-rate results at three temperatures. It can be seen that at 982° C by changing the strain rate from 0·4 to 0·005 s⁻¹ the flow stress was decreased by as much as a factor of five.

TABLE I

Composition (wt.-%) of Die and Workpiece Materials

Element	IN-100	MAR-M 200	Ti-6Al-6V-2Sn
Carbon	0·17	0·15	0·05*
Silicon	0·40	0·15	—
Manganese	0·40	0·15	—
Sulphur	0·015*	0·015*	—
Phosphorus	0·015*	0·015*	—
Tungsten	—	12·50	—
Chromium	10·00	9·00	—
Vanadium	1·00	—	5·0–6·0
Nickel	Bal.	Bal.	—
Molybdenum	3·00	—	—
Cobalt	15·00	10·00	—
Boron	0·010	0·015	—
Titanium	4·30	2·00	Bal.
Aluminium	5·50	5·00	5·0–6·0
Niobium	—	1·00	—
Zirconium	0·05	0·05	—
Iron	1·00*	1·50*	0·35–1·00
Copper	†	0·10*	0·35–1·00
Tin	—	—	1·5–2·5
Nitrogen	—	—	0·04*
Oxygen	—	—	0·20*
Hydrogen	—	—	0·15*

*Maximum
†Lowest possible amount

Therefore, in isothermal forging a reduction in the forging pressure could be expected to result from the use of low strain rates.

Manuscript received 20 July 1971; in revised form 19 November 1971. K. M. Kulkarni, BS, MS, N. M. Parikh, BS, MS, ScD, and T. Watmough, AMCT, AIM, are in the Metals Division of the IIT Research Institute, Chicago, Illinois, USA.

Production of Forging Dies

To permit isothermal forging at a low strain rate it was essential to produce dies that could operate satisfactorily at temperatures up to 1000° C. Nickel-base high-temperature alloys IN-100 and MAR-M 200 (Table I) were selected as the die materials on the basis of their compressive yield stress, hot hardness, and oxidation characteristics. Since these materials are difficult to machine, the forging dies were precision-cast by a ceramic moulding process. Complete details of die-material selection, moulding, and casting are given elsewhere.[1] Five sets of such dies ranging in weight from 90 to 640 kg per die half, were made for forging various shapes; as an example, Fig. 2 shows 180 kg MAR-M 200 dies for a scaled-down nose wheel forging ready for installation in the press.

Forging Experiments

The work material was Ti–6Al–6V–2Sn alloy. The forging dies were induction-heated *in situ* (in the press) at the supply frequency of 60 Hz. The smallest (90 kg) dies were heated to 980° C in ~ 4 h, whereas the largest (640 kg) dies required ~ 10 h. Two water-cooled stainless-steel base plates satisfactorily insulated the press bed from the heated dies. (Additional details of tooling are presented in Ref. 2.) The smaller forgings were made in a 9 MN hydraulic press at IITRI, and a 27 MN hydraulic press at Alcoa, Cleveland, was utilized for the large forgings in the 640 kg dies.

Titanium alloys must be protected from oxidation during heating and forging, and the billet must also be lubricated for forging. Markal CRT-22, a commercial glass-based titanium coating, was found satisfactory for both these purposes. No lubricants were applied to the dies.

A summary of the experimental conditions and results is provided in Table II. In all, six different die sets were utilized to produce forgings of five different shapes ranging in weight from 0·9 to 10 kg and in plan area from 6800 to 71 000 mm². The two 90 kg die sets used in the initial part of the work were identical except for the die material, and the forging tests showed that both IN-100 and MAR-M 200 were satisfactory as die materials for the isothermal forging process.

The effect of die temperature in improving die filling under otherwise identical conditions and with a press speed similar to that in conventional forging can be observed by the difference between the isothermal and non-isothermal forgings shown in Fig. 3. The influence of other important variables in isothermal forging tests is discussed in the following sections.

Effect of Press Speed

Most of the forging tests with the first three die sets were conducted at a press speed of 1 mm/s, and the pressure requirements for isothermal forging are shown by the upper curve in Fig. 4. In the forging test with the last three die sets the press was advanced gradually by manual control. The press speed was essentially determined by the resistance to deformation offered by the work material. Thus, the press speed was much higher initially when the resistance was low, and then the speed gradually decreased as the resistance offered by the workpiece material increased. In the final stages of forging, the press speed was < 0·04 mm/s. The entire forging operation was completed in ~ 3–5 min. This technique of forging will henceforth be referred to as 'creep forging'.

Thus, in creep forging, the load was gradually increased to, and then maintained at, a preselected value. The forging

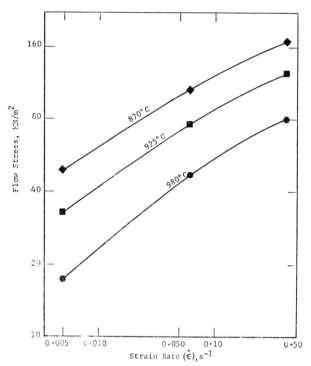

Fig. 1 *Effect of strain rate and temperature on flow stress of Ti–6Al–6V–2Sn alloy under isothermal forging conditions.*

Fig. 2 *180 kg MAR-M 200 dies for scaled-down nose wheel ready for installation in the press.*

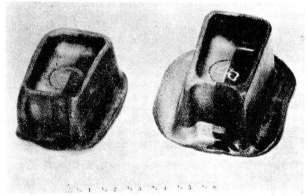

Fig. 3 *Comparison of forgings produced by non-isothermal and isothermal techniques at nominal 4·36 MN press load and 21·2 mm/s press speed, with a workpiece temperature of 320° C (left) and 980° C (right).*

TABLE II

Summary of Ti–6Al–6V–2Sn Alloy Isothermal Forging Data

Shape (vertical cross-section)	Forging Representative Dimensions, mm x	y	Plan Area, mm²	Weight, kg	Die-Set Material	Weight per Die Half, kg	Isothermal Forging Die and Workpiece Temp., °C*	Press Speed mm/s†	Forging Pressure, MN/m²
	95	7·9	6 800	0·9	IN-100	90	980 870	1 1	260 600
	95	7·9	6 800	0·9	MAR-M 200	90	980 760	0·04 1	83 860
	140	9·1	10 000	1·6	IN-100	180	980 870	1 1	260 580
	380	10·3	36 000	7·0	IN-100	640	980 900	0·04 0·04	76 124
	190	1·6	24 000	1·5	MAR-M 200	180	980 870	0·04 0·04	83 130
	320	3·2	71 000	10·0	IN-100	640	900	0·04	130

*Only the data for maximum and minimum temperatures are shown; see Fig. 4 for additional data.
†The press speed was 0·04 mm/s only in the final stages of 'creep forging'. (See text.)

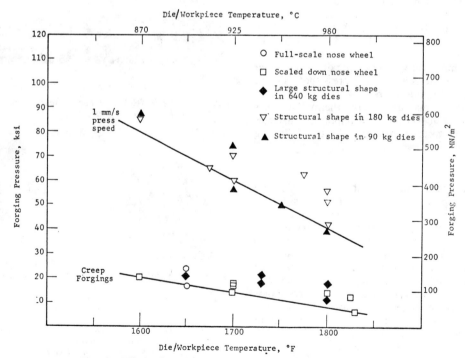

Fig. 4 *Effect of temperature on forging pressure in isothermal forging of Ti–6Al–6V–2Sn alloy. (The lines indicate the minimum pressure required for die-filling.)*

180

Fig. 7 *Full-scale nose-wheel blank and isothermal creep forging of Ti–6Al–6V–2Sn. Forging temp. 900° C; load 8·0 MN; plan area ~ 0·07m².*

Fig. 5 *Sections of Ti–6Al–6V–2Sn isothermal forgings of scaled-down nose wheel produced at different press speeds. Die/work-piece temperature 980° C; press load 2·2 MN.*

Blank   1 min   3 min   5 min

Fig. 6 *Two views of blank and incremental large structural shape forgings produced by maintaining load for different periods. Die and blank temp. 980° C; load 2·67 MN; press speed 0·04 mm/s.*

operation was completed when the dies reached the closed position (usually in 3–5 min). In forgings made at higher press speeds, ~ 1 mm/s, the top die was retracted when the load reached a preselected value (chosen to give die closure at the press speed used). In the latter case the load was not maintained at die closure for any appreciable period of time.

The data presented in Table II and Fig. 4 show that the forging pressure required to achieve die filling was reduced substantially by creep forging, where in the final stages the press speed was only 0·04 mm/s instead of 1 mm/s. Another way of looking at the effect of press speed is that, for a given forging pressure, more complicated forging geometry was achieved by creep forging. This is illustrated in Fig. 5 for scaled-down nose-wheel forgings produced isothermally at 980° C. For a forging load of 2·2 MN, the forging made at a press speed of 1 mm/s had a web thickness of 6·3 mm, whereas, with the same maximum load, a very thin web of 1·52–1·87 mm was achieved on the creep forging.

Effect of Forging Shape and Size

Three different structural shapes were forged in this work, the first two from simple, rectangular blanks and the largest from a flat blank, as shown in Fig. 6. Flat washer-like blanks were used for making the two wheel-shaped forgings, the larger being of 310 mm dia. (see Fig. 7). The wheels had a complicated cross-section, as shown in Fig. 8 for the 310 mm-dia. full-scale nose wheel, and with the latter forging the reduction in the thin web area was nearly 90%.

In spite of the great differences in plan area, shape, and size, the various forgings were made from relatively simple blanks in a single stroke of the press. In addition, Table II shows that, for any given temperature and press-speed combination, the forging pressure was not very sensitive to the shape and size of the forging.

Mechanical and Metallurgical Evaluation

Property evaluation was carried out on the full-scale nose-wheel forgings, since these represented the most complex geometry in the programme. For this particular forging, the specifications for mechanical properties were as follows: 0·2% offset YS 1170 MN/m² min.; UTS 1240 MN/m² min.; elongation 3% min. Since some variation can be expected in any forging from place to place, tensile-test specimens were machined from the different locations indicated in Fig. 9. The various treatments given to the tensile-test specimens and the resulting mechanical properties are shown in Tables III and IV, respectively. In the isothermal process, the hot forging was removed from the dies and was water-quenched to facilitate handling and observation. The main interest was in heat-treatments D and E in Table III, which are identical except for the ageing time and are similar to those commonly used for conventional titanium alloy forgings. It can be seen from Table IV that both these treatments met the minimum property requirements on the forging. As expected, there was also some tendency towards lower elongation values at higher strength levels. One interesting observation during the heat-treatments was that the isothermal forgings did not undergo any measurable distortion, whereas in many conventionally forged items distortion can be a severe problem.

The microstructures of the isothermal nose-wheel forgings were studied in essentially the same locations as shown in Fig. 9. Microstructures were observed in forgings produced at 900° C, water-quenched directly from the forging dies, solution-treated at 900° C for 30 min, water-quenched, and

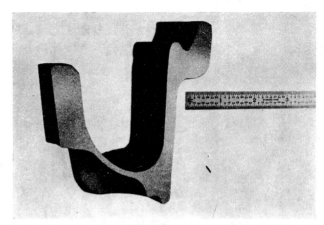

Fig. 8 Cross-section of full-scale nose-wheel isothermal forging.

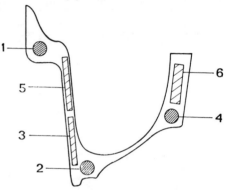

Fig. 9 Diagram showing locations of specimens for mechanical and metallurgical evaluation of full-scale nose-wheel forgings.

	TABLE III
	Heat-Treatment or Processing Schedule for Full-Size Titanium Alloy Nose-Wheel Forgings Isothermally Forged at 900° C

Treatment Type	Details
A	Water-quenched directly from forging press
B	Water-quenched directly from forging press, aged 565° C/24 h
C	Water-quenched directly from forging press, solution-treated 930° C/1/2 h, water-quenched, aged 565° C/24 h, air-cool
D	Water-quenched directly from forging press, solution-treated 900° C/1/2 h, water-quenched, aged 565° C/24 h, air-cool
E	Water-quenched directly from forging press, solution-treated 900° C/1/2 h, water-quenched, aged 565° C/8 h, air-cool

aged at 565° C for 24 h (treatment D in Table III). The microstructure was generally fine-grained with primary α and acicular α and some orientation due to metal flow. The isothermal forgings had a finer grain size than similar forgings conventionally produced, because of the much heavier reductions imposed in the single-step isothermal forging.

In the entire experimental work on isothermal forging the

TABLE IV

Tensile Properties of Specimens Removed from Isothermally Forged Ti–6Al–6V–2Sn Nose-Wheel Forgings at Sites Indicated and after Various Processing Treatments

Site*	Processing Schedule†	0·2% Offset Yield Strength, MN/m²	UTS, MN/m²	Elong. over 1/2 in Gauge-Length, %	Reduction in Area, %
1	E	1330	1380	4·5	8
2	E	1240	1290	5·5	15
3	E	1280	1310	7·0	21
4	E	1240	1310	5·0	20
1	D	1240	1280	5·0	21
2	D	1310	1350	8·0	24
3	D	1380	1420	3·5	7
4	D	1280	1350	5·5	11
5	B	1170	1280	11·5	33
5	A	1300	1430	5·5	7
5	B	1210	1290	10·0	23
5	B	1170	1260	9·5	25
6	C	1230	1330	5·5	16
6	D	1170	1270	12·0	43
blank (radial direction)	similar to D	1170	1260	11·0	24
blank (tangential direction)	similar to D	1240	1280	3·0	6
(Specification properties)		1170	1240	3·0	—

*No. refers to location in Fig. 9.
†Letter refers to type of treatment in Table III.

total heating period for the forging blank was between 20 min and 2 h, depending on blank size. Since the forgings were made in a single step from forging blanks that were originally machined to the required dimensions, this was the only high-temperature exposure to which they were subjected. Therefore, as anticipated, little α-phase formation was observed on the surface of the forging.

Conclusions

The important conclusions of this work can be summarized as follows:

(1) It is feasible to produce dies and a die-heating system for isothermally forging titanium alloys at temperatures up to 1000° C.

(2) Isothermal creep (or slow-speed) forging can eliminate or reduce the number of intermediate forging stages and can produce complex forgings at very low forging pressures even when starting with simple preforms. Up to 90% reduction can be achieved in a single die set in one operation.

(3) In the isothermal slow-speed forging process the forging pressure appears to be insensitive to the size and complexity of the forging.

(4) Because of the lower forging-pressure requirements in isothermal slow speed forging, much larger forgings than are currently possible could be made with available forging equipment.

(5) The ability of the isothermal-forging technique to produce complex forgings from simple preforms can greatly increase the net utilization and reduce scrap losses of expensive titanium alloy stock.

(6) Isothermally forged parts have satisfactory mechanical properties and microstructures. The shortened exposure to high temperature in isothermal forging reduces surface contamination considerably.

(7) Applications for isothermal creep forging should be viewed in the light of the above advantages as against the need for expensive dies and a die-heating system.

Acknowledgements

This research was conducted at the IIT Research Institute, Chicago, for the US Air Force, under Contract No F33(615)–67–C–1722. Thanks are due to Mr. H. A. Johnson and Mr. L. C. Polley of the Air Force Materials Laboratory, Wright-Patterson Air Force Base, for their support and for permission to publish the results. The authors are grateful to Mr. D. A. Stawarz, Experimentalist, for his contribution to this work.

References
1. J. S. Prasad and T. Watmough, *Trans. Amer. Found. Soc.*, 1969, 77, 289.
2. K. M. Kulkarni, T. Watmough, and N. M. Parikh, 'Isothermal Forging of Titanium Alloys Using Precision-Cast Dies,' paper **MF70-122**, presented at a meeting of the Society of Manufacturing Engineers, Dearborn, Mich., USA, May 1970.

SPECIMEN GEOMETRIES PREDICTED BY COMPUTER MODEL OF HIGH DEFORMATION FORGING

J. W. H. PRICE† and J. M. ALEXANDER‡

Imperial College, London

(*Received* 28 *October* 1978; *in Revised form* 26 *January* 1979)

Summary—This paper presents the specimen geometries predicted by a finite element model of isothermal forging. In isothermal forging the specimens are subjected to very high deformations and considerable flow of material results. The geometries are shown by a number of computer produced plots of finite element meshes which represent the flow of the material at various stages during the forging process. The predicted geometries are found to conform well with experiments for a variety of specimen and die configurations.

NOTATION

B a constant
m shear coefficient
V volume
$\dot{W}_p$ plastic work rate
α a large constant
ϵ_{ij}^p an element of the plastic strain sensor
$\dot{\epsilon}_{ij}^p$ an element of the plastic strain rate sensor
$\dot{\epsilon}_v$ volumetric strain rate
η instantaneous viscosity
$d\lambda$ an instantaneous constant
σ_{ij}' an element of the deviatoric stress sensor
$\bar{\sigma}_0$ equivalent flow stress
τ shear stress
χ a functional.

INTRODUCTION

The work presented here was developed in conjunction with the experimental study of a process known as isothermal forming. In this process use is made of the fact that at high temperatures and low strain rates metals generally deform at low stress levels. In the process of isothermal forming the dies are heated to the same temperature as the specimen and the forging is carried out at low speed. This contrasts to normal hot forging where the dies are much colder than the specimen, the forging occurs at higher speeds, and the load is often applied several times using graduated dies. In isothermal forming the specimens can deform very extensively during the single loading and conform closely with the shape of the dies. The dimensional accuracy of the product is very good and thus a minimum of further machining is required to obtain the finished product. The process has been used particularly for materials which are difficult to work and where scrap is undesired such as titanium alloys. Examples of the use of the process are given in a number of references[1] and the experimental arrangements for the work described in this paper are discussed in another paper[2]. The experiments concentrated on titanium or steel alloys (of types which resist oxidation at high temperatures) and the dies were made of IN 100.

This paper concentrates on the diagrams produced by a computer program which was developed to model the experiments. These diagrams show the material undergoing very large deformations such as are obtained in isothermal forming, and it is found that the flow of the metal predicted by the program represents those seen in the process very closely. The basic modelling technique used the finite element method[3] to determine deformations and material properties were numerically interpolated from the data of Lee and Backoffen[4].

†Now at Nuclear Power Company (Risley) Ltd., Warrington.
‡Now at Mechanical Engineering Department, University College, Swansea.

THE COMPUTER MODEL OF ISOTHERMAL FORMING

Only an outline of the approach used for obtaining finite element solutions for large deformations is given here; a fuller description of the method used in this work is given in another paper[5]. Several special features of the isothermal forming of graphite lubricated titanium alloy specimens are included in this approach. (i) The material is strongly strain-rate sensitive but at the temperature of forming there is little strain hardening. The constitutive relationship must therefore be such that flow stress is a function of both strain rate and temperature but not necessarily of strain. (ii) There are high deformations obtained in the process and in most cases all the material is substantially deformed. The plastic strains that occur will be many times larger than any elastic strains. Hence a method in which the elastic strains are neglected will be satisfactory. (iii) The die-specimen interface normally exhibits no slipping with graphite, but some boundaries that originally are free eventually come into contact with the die by rolling onto the die surface. If lubrication such as molten glass is used then relative movement is observed. (iv) The dies can be considered to be rigid.

From the consideration of these properties a finite element scheme was chosen that would give a satisfactory modelling of the process without too great use of computer time.

(a) An increment of plastic work per unit volume, dW_p, can be written, following Hill[6] as:

$$dW_p = \sigma'_{ij}\, d\epsilon^p_{ij} \tag{1}$$

where σ_{ij} are the deviatoric stresses, $d\epsilon^p_{ij}$ are the increments of plastic strain, and the summation convention applies. This can be re-written in terms of a plastic strain rate per unit volume.

$$\dot{W}_p = \sigma'_{ij}\epsilon^p_{ij}. \tag{2}$$

The normal elastic finite element formulation can be converted to use strain rates and deviatoric stresses instead of strains and stresses respectively. The nodal variables are now considered to be velocities and the increments of deformation are really time steps. It is a work rate that is minimised throughout the volume.

(b) For the constitutive relationship the Levy–Mises flow rule can be used[6].

$$d\epsilon^p_{ij} = \sigma'_{ij}\, d\lambda \tag{3}$$

where $d\lambda$ is an instantaneous constant. To facilitate calculations the equation must be written to involve time:

$$\dot{\epsilon}^p_{ij} = \sigma'_{ij}/2\eta. \tag{4}$$

It is written in this form in order to emphasise the connection between this formulation and that for the flow of viscous fluids at very low Reynolds numbers. Thus η can be interpreted as an instantaneous viscosity. In the present program provision was made for the "viscosity" to be estimated at each iteration for each Gauss point in each element. The value was found from interpolation directly from experimentally determined stress–strain rate-temperature data (as is provided in Ref.[4]). The strain rate and temperature from the last iteration is used to estimate the viscosity for the current iteration.

(c) Incompressibility must be ensured. Of the techniques which may be used to achieve incompressibility the most efficient and simplest is the penalty function approach[7]. In this approach a large positive number, α, is multiplied by the volume integral of the square of the volumetric strain rate:

$$\alpha \int \dot{\epsilon}_v^2\, dV \tag{5}$$

and this is added to the work rate functional, χ, to obtain the finite element virtual work rate formulation

$$\chi = \int \dot{W}_p\, dV + \alpha \int \dot{\epsilon}_v^2\, dV + B \tag{6}$$

B represents the contributions of body forces (normally considered to be unimportant in slow speed plasticity problems) and surface traction rates.

(d) The boundaries of the specimen are either free or geometrically determined by the dies if sticking friction, as described in the experiments above, is assumed. Frictional conditions have been allowed for in the present program. Provision is made for boundary nodes that were originally free to roll onto the dies and become fixed.

(e) During the deformation the mesh must be continuously updated to reduce the problem as far as it possible to one of infinitesimal strain. This achieved by using small time or displacement steps and each node is moved after each increment. The movement of each node is calculated by multiplying the velocity calculated at that node in the previous increment by the time elapsing between increments. The distorted elements and a high speed elimination solution for banded matrices. Technical details of this program are material has of its previous shape.

As a result the mesh must be arbitrarily reformed only when it is absolutely necessary. This occurs at times when elements become excessively distorted or negative areas are created. (The original program was kindly provided by Prof. O. C. Zienkiewicz. It is an elastic program with isoparametric quadilateral elements and a high speed elimination solution for handed matrices. Technical details of this program are described by Zienkiewicz[3].)

APPLICATION OF THE PROGRAM

1. *Cylindrical compression*

A test situation of a cylindrical compression test with no friction at the interface is a useful check on whether constant volume is obtained in the program. On carrying out such tests it was found that there was no detectable departure from constant volume, the mesh deforming uniformly and remaining always completely cylindrical. Such a test is shown on Fig. 1.

A more difficult situation occurs with sticking friction. Examples are shown on Fig. 2 of the deformation for cylinders of Ti6A14V at 950°C, using material data from Lee and Backoffen[4]. The height to diameter ratios were originally $\frac{1}{4}$, 1 and 4 respectively and as yet in none of these examples has any material rolled onto the dies. The most interesting example shown is the tall thin cylinder which exhibits a double bulge effect, a phenomenon seen in experiments on specimens of this ratio. The deformation in the central region of the tall specimen is quite uniform because the effects as the friction at the ends becomes reduced.

Fig. 3. illustrates a specimen originally of height to diameter ratio $\frac{2}{3}$ which has been taken to a higher strain of 83% reduction. The free surface of this specimen has proceeded to roll onto the dies and also the mesh has been straightened out or reformed on one iteration because the movement of the nodes can cause the elements to cross one another in highly deformed regions. The original volume of this specimen was 14137 mm^3 but now it is approx. 13504 mm^3 measured from the diagram which is a loss of about 4.3%.

The loss of volume is due to the fact that constant volume or zero volumetric strain rate is only met

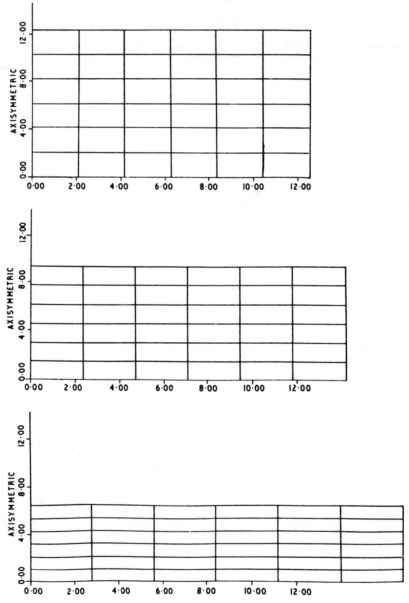

FIG. 1. Compression of a cylinder without friction, one quadrant shown. Initial dimensions 25 mm high, 25 mm dia.

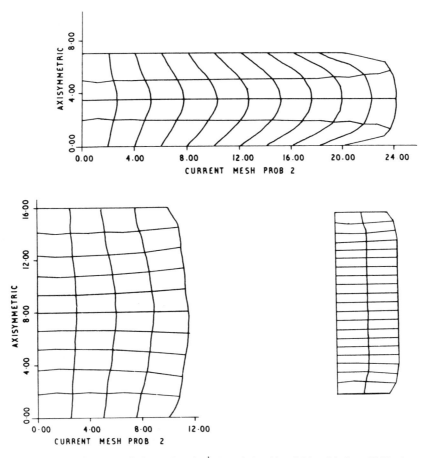

FIG. 2. Compression of cylinders of ratio $\frac{1}{4}$, 1 and 4 with sticking friction. Half of each specimen is shown.

exactly at certain points within each of the eight noded quadilateral elements. This is because these elements can only satisfy C^0 continuity across their boundaries, that is, only the velocities have been calculated with a continuous value throughout the body. The derivatives of the velocities are not continuous through the element boundaries and thus the strain rates, which are a function of the derivatives of the velocities must be discontinuous at the element edges. To achieve continuous derivatives, elements which enforce a C^0 continuity are required, but such elements involve considerably more complication and many extra nodal variables. The trade-off in computer effort to achieve C^1 continuity is not justified in most problems (see Strang and Fix[8]). However, it is found that the derivatives of the nodal variables are accurately calculated at certain points within the elements and that the 2×2 Gauss points are normally very close to such points (see Naylor[9]). In the cases being investigated it is found that the volumetric strain rate is indeed very close to zero when calculated at the 2×2 Gauss points but finite values normally occur at the nodes. The values at the nodes can be either negative or positive and tend to cancel each other out.

Figs. 3–5 illustrate high deformations of specimens of ratios 2/3, 1 and 3/2. Here the operation of the subroutines which enables nodes that are on the free surface at the beginning of the deformation to stick on to the dies is fully illustrated. The computing times for these high deformations is quite economical with the penalty function approach and these results (without stress calculations) were all obtained in less than two minutes of central processor time on the CDC 6400 at Imperial College.

2. Ring tests

Fig. 6 illustrates a ring test; the compression of an annulus between flat dies. The starting geometry has the ratio 4:2:1 (outer dia.:inner dia.:thickness) and the thickness of the specimen was 6.45 mm. The frictional conditions are those of sticking friction and could be represented by a value of shear coefficient of unity, i.e. m equals one in the equation

$$\tau = m \frac{\bar{\sigma}_0}{\sqrt{3}} \qquad (7)$$

where τ is the shear stress at the interface and $\bar{\sigma}_0$ is the equivalent flow stress of the deforming material. The factor of $\sqrt{3}$ is a consequence of the Von Mises flow criterion. The value of $\bar{\sigma}_0$ is not a constant and is dependant on strain rate and temperature.

Fig. 7 illustrates a similarly shaped ring, but in this case the friction subroutine was introduced into the program and a value of 0.06 was used for the shear coefficient. Fig. 8 illustrates a plot of reduction in inner

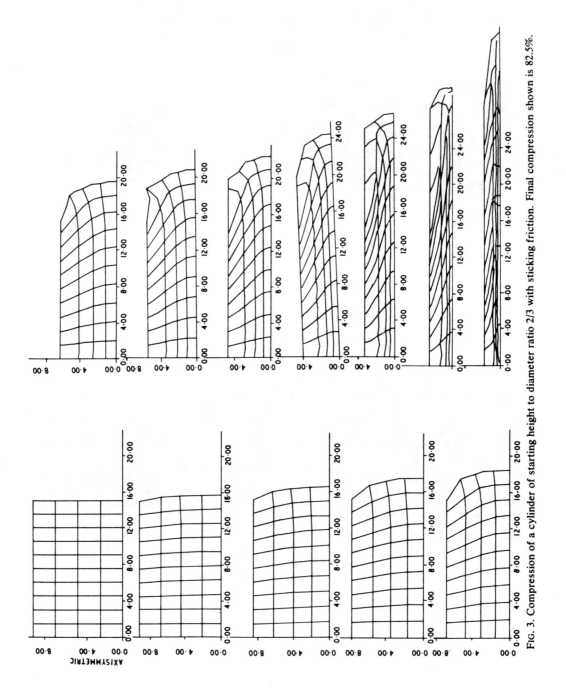

FIG. 3. Compression of a cylinder of starting height to diameter ratio 2/3 with sticking friction. Final compression shown is 82.5%.

AXISYMMETRIC

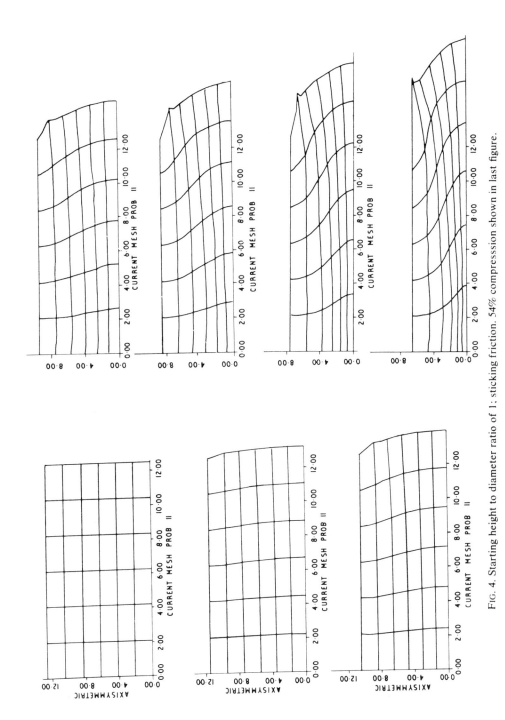

FIG. 4. Starting height to diameter ratio of 1; sticking friction. 54% compresssion shown in last figure.

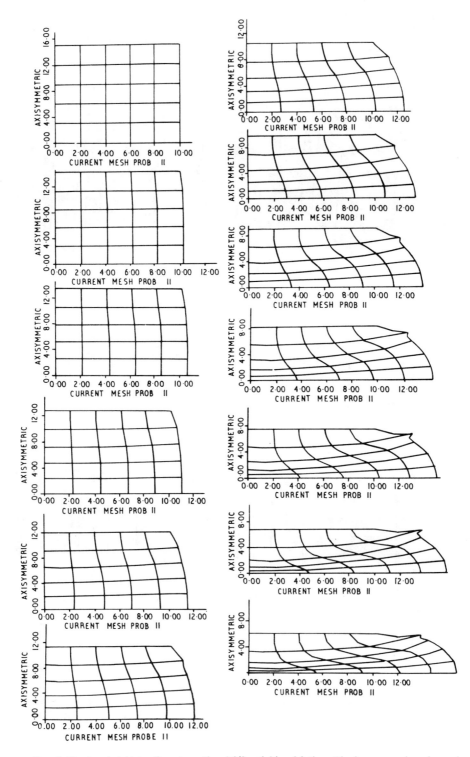

FIG. 5. Starting height to diameter ratio of 3/2; sticking friction. Final compression shown is 60%. One quadrant only is shown.

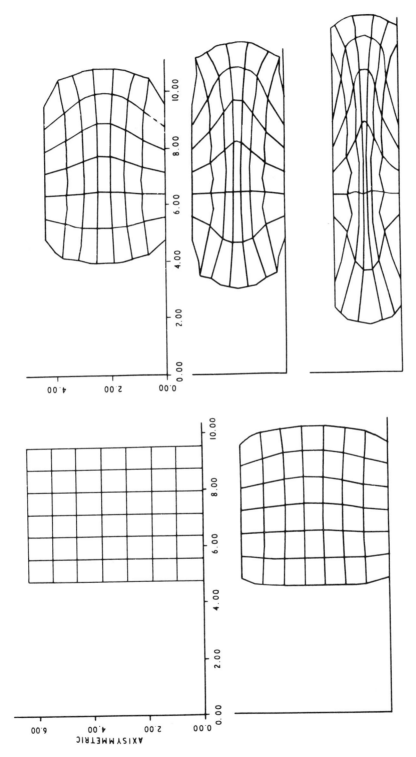

AXISYMMETRIC

FIG. 6. Ring test with sticking friction, half a specimen is shown. Final compression is 38.8%

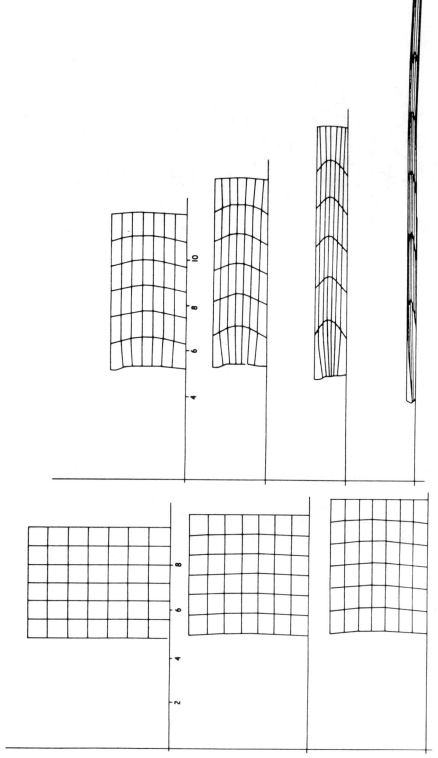

FIG. 7. Ring test with a shear coefficient is 0.06; half a specimen is shown. Final compression is 92%.

192

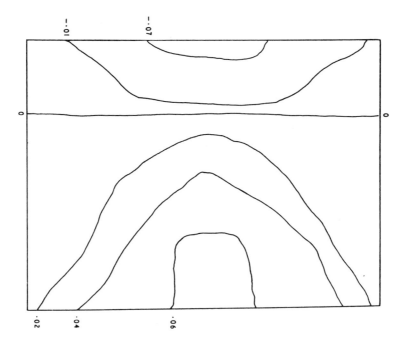

Fig. 9. Tangential strain rate distribution in the sticking friction ring test of Fig. 7.7 (first increment). Units are sec^{-1}.

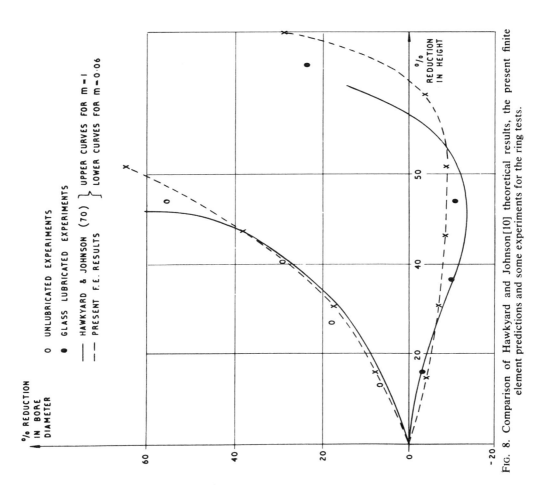

Fig. 8. Comparison of Hawkyard and Johnson[10] theoretical results, the present finite element predictions and some experiments for the ring tests.

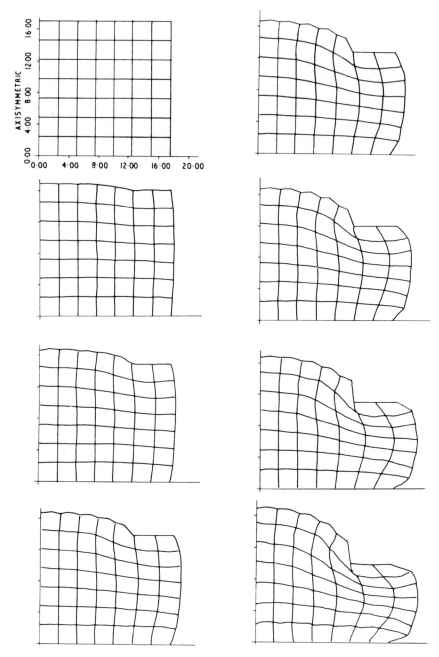

FIG. 10(a)

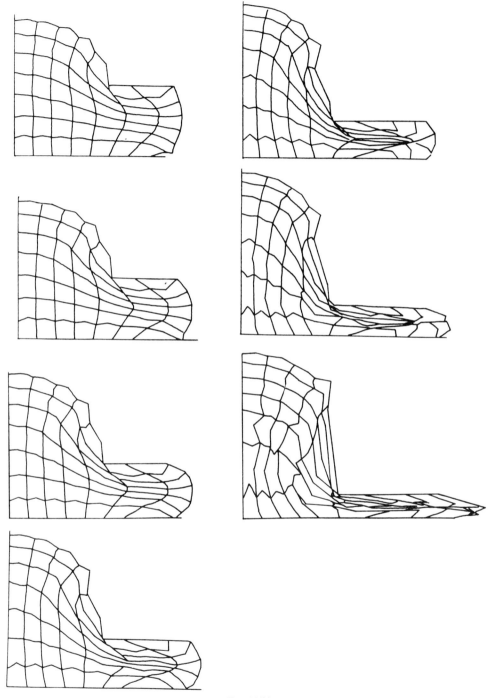

F<small>IG</small>. 10(b)

F<small>IG</small>. 10. Compression of 17.5 mm high, 17.5 mm dia. specimen with a profile die. The mesh is not reformed at any stage.

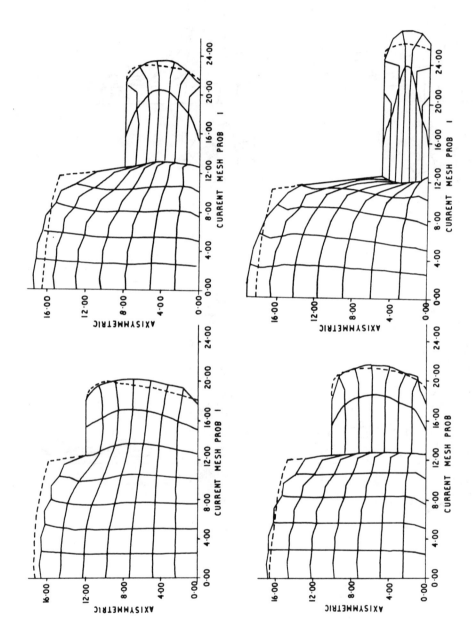

Fig. 11. Compression with a profile die (see Fig. 10). The mesh in this case is reformed at increments 7, 12. The dashed line gives some close experimental results for the outside profile of comparable specimens.

bore diameter versus reduction in height and compares the results of some experiments, the current finite element results and the predictions of Hawkyard and Johnson[10]. The lubricant used for the low friction case in the experiments was molten glass, the high friction case used no lubricant. The mid-section radii have been used for all the comparisons (Hawkyard and Johnson assume plane sections remain plane and thus the mid-section radius for their analysis is the same as the end radii). Complete agreement with experiments cannot be expected because the real friction effects are not likely to be precisely the same as indicated in equation (7) throuthout the test, nor are the specimens accurately round after the compression. The net result suggests there is little to choose between the two theoretical methods of predicting ring behaviour.

The neutral surface (the surface that does not move) can be determined for the sticking friction case using the hoop strain rate distribution from the finite element analysis. The neutral surface does appear to be very nearly cylindrical as is shown on Fig. 9.

3. *Profile dies*

Fig. 10 illustrates the results for an extensive compression of a finite element model of the profile die experiments reported in Ref.[2]. Subroutines have been included in the program which allow not only the nodes from the outside surface to stick on to the dies, but also allow the nodes on the top surface of the specimen to stick to the conical surface of the top die. As will be noted in the latter stages of this compression, some of the element boundaries tend to cross each other and the die surface in the later stages of compression.

Fig. 11 illustrates a modification of the program which allows the mesh to be reformed at stages during the compression to remove this problem. Whenever the elements appear to be about to cross, in this modification the whole mesh is reformed in an arbitrary manner, but the exterior surfaces are not altered. The results of the finite element method are compared with experimental profiles that were measured from approximately similar specimens using a projection microscope. As can be seen the comparisons are quite satisfactory.

The most important important indicator which occurs during this forming is the reversal of flow when material, instead of flowing into the flanges from the bulk of the specimen, flows from the flanges into the conical region in later stages of compression. On the experimental specimens this is detected by the specimen increasing in height, a feature which can be seen to occur in the last stage of the forming. The finite element results on Fig. 11 are presented for a specimen that is as close as possible to experimental specimens. As can be seen, if the element boundaries joining the corner of the top die to the surface of the lower die are inspected the flow indeed does reverse in the finite element model and the height does increase in the last stage shown.

CONCLUSIONS

A finite element method has been described which can model the very high deformations seen in isothermal forging. The geometries shown in the various stages of the forging, and for a variety of specimens, are in good agreement with experimental results.

Acknowledgement—One of the authors (J.W.H.P.) acknowledges the support of a scholarship from the Commonwealth Scientific and Industrial Research Organisation, Australia.

REFERENCES

1. K. M. KULKARNI, *J. Inst. Metals* **100**, 146–151 (1972); S. Z. FIGLIN, *Kurznechno-Stanpovnochoe* **9**, 6–9 (1972); F. W. BOLGER, *Forging Equipment Materials and Practices*. U.S. National Technical Information Service, MCIC-HB-03, pp. 213–245 (1973).
2. J. W. H. PRICE and J. M. ALEXANDER, *4th North American Metalworking Res. Conf.*, Columbus, OH, p. 46 (1976).
3. O. C. ZIENKIEWIEZ, *The Finite Element Method in Engineering Science*. McGraw-Hill, London (1971).
4. D. LEE and W. A. BACKOFFEN, *Trans. AIME* **239**, 1084 (1967).
5. J. M. ALEXANDER and J. W. H. PRICE, *17th Machine Tool Design and Research Conf.* Macmillan, London (1977).
6. R. HILL, *The Mathematical Theory of Plasticity*, Clarendon Press, Oxford (1952).
7. O. C. ZIENKIEWICZ and A. N. GODBOLE, *Int. Symp. on F.E.M. in Flow Problems*, Swansea (1974). Wiley,
8. G. STRANG and G. J. FIX, *An Analysis of the Finite Element Method*. Prentice Hall, N.J. (1973).
9. D. J. NAYLOR, *Int. J. Num. Methods Engng* **8**, 443 (1974).
10. J. B. HAWKYARD and W. JOHNSON, *Int. J. Mech. Sci.* **9**, 163 (1967).

The Occurrence of Shear Bands in Isothermal, Hot Forging

S. L. SEMIATIN and G. D. LAHOTI

The occurrence of shear bands in isothermal, hot forging of metals has been investigated using a variety of experimental and analytical tools. Isothermal sidepressing tests on Ti-6Al-2Sn-4Zr-2Mo-0.1Si were conducted at a variety of hot working temperatures and deformation rates to develop a data base on the occurrence of shear bands as a function of material properties and processing conditions. An instability criterion was developed, and the tendency to form shear bands was shown to depend strongly on the ratio of the nondimensional work-hardening (or flow-softening) rate to the strain-rate sensitivity of the flow stress. This analysis and supporting experimental data formed the basis for the development of workability maps, which delineate safe regimes for hot working in terms of strain rate and temperature. Employing a sophisticated computer code, an extensive series of simulations was also run for the sidepressing operation to obtain an idea of the detailed process by which shear bands initiate. These simulations were conducted using actual Ti-6242 material properties as well as hypothetical properties which are representative of those of other metals deformed in the hot working regime. From these simulations, predictions of the influence of flow softening and rate sensitivity on flow localization were made and found to be similar to experimental results and predictions obtained from the instability analysis. However, process simulation gave a more complete insight into the initiation and growth stages in the development of shear bands that was not possible using only the instability analysis.

I. INTRODUCTION

INCREASING emphasis is being placed on materials conservation. This emphasis translates into tighter controls in manufacturing practice in areas such as metalworking, powder metallurgy, and casting. For example, there is a strong interest in forging of parts to near-net shape, a goal which, if realized, would minimize subsequent machining time and material waste.[1] Success in near-net-shape-metalworking operations such as forging will result only when accurate control of metal flow *via* preform design, lubrication, and so forth can be achieved without introducing defects in the worked parts. In cold forging, these defects include free-surface fractures, die-contact fractures, and central bursts.[2] In warm and hot forging, defects such as brittle intergranular fractures, hot shortness, cavitation, and shear bands must be avoided.[3,4,5]

Besides promoting better metal flow and die filling in hot forging, isothermal or hot-die forging (dies and workpiece at or near the same temperature)[6] eliminates the source of many shear band defects, namely, die chilling. During forging, chilling leads to the development of dead metal zones (regions of limited metal flow in the workpiece) whose extent depends on die temperature, working speed, lubrication, and the material's flow stress dependence on temperature. However, material properties and geometry may cause shear bands to occur even in isothermal or near-isothermal-forging situations.[7] In these instances, techniques must be developed to predict shear band occurrence.

The prediction of shear bands in isothermal, hot forging can be studied from a gross viewpoint[5,8,9] (in which only the *event* of failure onset is usually considered) or from a viewpoint employing process simulation[10] (in which failure is studied as a total *process*). In the former case, it is typical that some sort of instability criterion is examined in a manner very analogous to the prediction of tensile uniform elongation or sheet metal forming limits using maximum load conditions.[11,12] On the other hand, process simulation to predict failure is much more complex and has received limited attention in cold metalworking operations,[10,13–16] let alone hot-working operations.

The present paper is a continuation of previously reported work on the deformation and unstable flow of Ti-6Al-4Zr-2Mo-0.1Si (Ti-6242) in hot forging[7] and hot torsion.[17] In the previous work, an instability criterion was used to gage flow localization tendencies and shear band occurrence during isothermal deformation. The predictions compared favorably to observations. In the present work, the instability criterion is again employed to interpret data and shear band observations from a more complete test matrix of isothermal, hot forging variables. In addition, results from computer simulation, run to obtain detailed strain, strain rate, and stress histories, are reported. Shear band formation predictions from these simulations demonstrate the power of analytical techniques.

II. MATERIALS AND PROCEDURES

A. *Materials*

The material used in the investigation was Ti-6242 (chemistry in wt pct: 6.0 aluminum, 2.2 tin, 4.1 zirconium, 2.0 molybdenum, 0.09 silicon, balance titanium). Specimens were cut from the same forging bar stock used in previous investigations,[7,17] and details of fabrication can be found elsewhere.[7] Briefly, part of the material was used in the as-received condition and had a structure of globular α (α grain size = 11 microns) in a transformed β matrix, hereafter referred to as the $\alpha + \beta$ microstructure.[6] Another part of the bar stock was beta annealed at 1010 °C (1850 °F) for 30 minutes and air cooled to produce an acicular, transformed β microstructure (prior β grain size = 350 microns), hereafter referred to as the β microstructure.[6]

S. L. SEMIATIN, Principal Research Scientist, and G. D. LAHOTI, Senior Research Scientist, are both with the Metalworking Section, Battelle's Columbus Laboratories, Columbus, OH 43201.
Manuscript submitted March 25, 1981.

Reprinted with permission from Metallurgical Transactions A, 13A, February 1982, 275-288, © 1982 American Society for Metals and The Metallurgical Society of AIME

The hot compression behavior of the $\alpha + \beta$ and β microstructures was documented for temperatures of 816 to 1010 °C (1500 to 1850 °F) and strain rates of 0.001 to 10.0 second^{-1} and has been reported in detail before.[7] Although the stress-strain curves for a given microstructure show little difference when measured on specimens cut longitudinally or transversely from the bar stock, comparison of flow curves for the two microstructures at the same temperatures and strain rates exhibit striking differences. The β microstructure curves have generally higher yield stresses and larger amounts of flow softening[18] when compared to $\alpha + \beta$ microstructure curves. On the other hand, the general dependence of flow stress on temperature is similar for the two microstructures as are the strain rate sensitivities of the flow stress. A complete summary and interpretation of these results is contained in Reference 7.

B. Experimental Procedures

Hot, isothermal sidepressing (lateral compression between flat dies) of cylindrical rods of the program material was selected to simulate a typical hot-forging operation for titanium alloys — the forging of jet engine compressor blades — and to establish the material variables that control shear band development in isothermal forging situations. Sidepressing is a useful operation to investigate from a heuristic standpoint also, because it is a plane-strain operation, and results from it may be extrapolated to predictions of workability in other operations such as rolling and extrusion of certain shapes. Moreover, plane-strain operations offer a maximum amount of constraint to metal flow (limiting flow to two directions) as compared to axisymmetric operations (in which flow occurs in all three directions). With this two-dimensional constraint, the conditions for "block-type" shearing are found, and it appears that occurrence and persistence of shear bands is more likely than in axisymmetric deformations.[19] Lastly, experience has shown that die misalignment appears to play a relatively minor role in shear band formation in the sidepressing operation when compared to axisymmetric operations such as uniaxial compression.[20]

Measuring 0.71 cm. (0.280 inch) diameter × 2.54 cm. (1.00 inch) long, cylindrical sidepressing specimens were cut longitudinally as well as transversely from bars of both the $\alpha + \beta$ and β microstructures. Tests were run on lubricated specimens at temperatures of 843, 913, and 982 °C (1550, 1675, and 1800 °F). Deltaglaze 93 was used for lubrication at the two lower temperatures and Deltaglaze 69 at the highest temperature. Both lubricants were manufactured by Acheson Colloids Company of Port Huron, Michigan. Prior to testing, specimens were preheated for 15 minutes (five minutes to reach temperature, 10 minutes for soaking) in a hot, isothermal compression fixture, described previously,[7] which was mounted in an MTS machine. After preheating, the specimens were deformed between the flat dies of the fixture which moved in response to a programmed constant ram speed of the MTS machine. The ram speeds were nominally either 0.081, 9.1, or 63.5 mm per second (0.0032, 0.36, or 2.5 inches per second). Load-stroke, load-time, and stroke-time data were obtained using standard X-Y recorders and high-speed recorders (Gould, Inc., Model No. 2007-4290-00) for low and intermediate

ram speed tests and a Data Memory System (Zonic Technical Laboratories, Inc., Cincinnati, Ohio, Model No. MP509) for the highest ram-speed tests.

To study the development of shear bands during deformation, sidepressing tests were run to various reductions in height. After deformation, specimens were air cooled. Metallography was then used to establish the presence of shear bands in the deformed sidepressing specimens and validate analytical models to predict their occurrence and persistence. To this end, transverse metallographic sections of these samples were made using standard mounting, grinding, polishing, and etching (etchant: 95 ml. H_2O, 3.5 ml. HNO_3, 1.5 ml. HF) techniques.

III. RESULTS

A. Load-Stroke Curves

All load-stroke curves from the constant-crosshead-speed-sidepressing tests had similar shapes. As was noted from the curves for the longitudinally-cut specimens (Figure 1) and the nearly identical curves for transverse specimens, these loads exhibit a slow increase at the beginning of deformation, followed by a much more rapid increase at height reductions (relative to the initial specimen diameter) greater than approximately 30 to 40 pct. Because the test materials show little or no work hardening and even flow softening in some cases (particularly β microstructure specimens tested at 843 and 913 °C [1550 and 1675 °F])[7,21] it can be deduced that the major source of the load increase is the rapid increase in cross-sectional area. A secondary source is the increasing effect of friction with deformation (i.e., the "friction hill" effect). However, for the good lubrication conditions that were obtained (friction factor was 0.2 to 0.3) and the relatively small transverse-section aspect ratios that were imposed, the overall friction effect was deduced[22] to be small. In fact, load estimates obtained by multiplying the measured flow stresses and the contact area calculated assuming that the deformation geometry was similar to that in plane-strain compression of rectangular bars were usually within 20 pct of the load data that were spot checked. These estimates were always lower than measurements partially because of neglect of friction and partially, it is believed, because of geometrical effects that were overlooked.

Added credibility for the load-stroke data is obtained by comparing curves at the same temperature and crosshead speed for the two different microstructures, $\alpha + \beta$ and β. At the beginning of the deformation, the curves for the β microstructure are characteristically higher than the corresponding $\alpha + \beta$ microstructure curves (Figures 1a and 1c). At strokes of 3 to 4 mm. (0.12 to 0.16 inch), which correspond to approximately 50 pct reduction, the curves approach each other and, in some instances, cross over each other. These observations are similar to those for the stress-strain curves.[7] The β microstructure has yield stresses higher than those for $\alpha + \beta$ microstructure, but the remaining portions of the flow curves approach each other and at times even cross over at large strains on the order of 0.7 (approximately 50 pct reduction in height in uniaxial compression).

B. Metallography

Shear bands were found to be a common feature of the

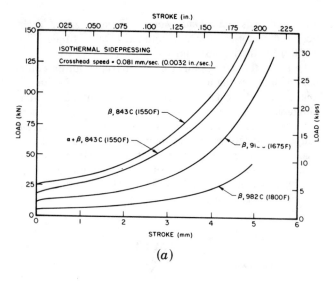

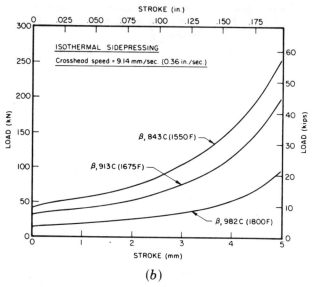

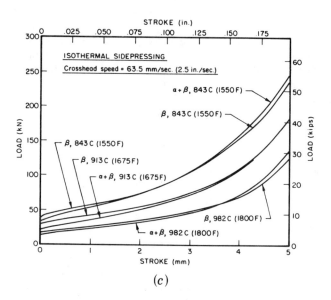

Fig. 1—Load-stroke curves from isothermal sidepressing tests run on longitudinal specimens at constant crosshead speeds of (a) 0.081 mm per second (0.0032 inch per second), (b) 9.14 mm per second (0.36 inch per second), and (c) 63.5 mm per second (2.5 inch per second).

isothermal sidepressing deformation of both microstructures at a variety of deformation rates and nominal test temperatures. Several criteria were used to establish that the observed deformation features were indeed shear bands. These were as follows:

1. The deformation within the shear bands must be considerably more intense than outside the shear bands. In the $\alpha + \beta$ microstructure material, intense deformation was marked by extreme shearing of the globular-alpha grains.[7] For the β microstructure material shear bands led to gross destruction of the Widmanstätten alpha platelet structure or localized regions in which a recrystallized microstructure of globular-alpha grains was found.[7]

2. The directions of the two complementary shear bands in sidepressing sections must not be perpendicular where they intersect. Because of symmetry, two intersecting shear bands are always formed in sidepressing. These shear bands originate along the slip lines of classical plasticity theory,[23,24,25] the orthogonal net of the two directions of maximum shear stress (and maximum shear strain). However, since shear bands must be associated with material elements and not spatial, or geometry-related directions, their directions must rotate away from the compression axis once they are initiated. Thus, although the two incipient shear bands in sidepressing may be initially perpendicular at their intersection because they are along sliplines, further deformation must lead to rotation of one or both of these bands of intense deformation.

For $\alpha + \beta$ microstructure sidepressings, shear bands were observed only for the highest deformation rate at the 843 °C (1550 °F) test temperature. In contrast, for the β microstructure, shear bands were observed for all three deformation rates at temperatures of 843 and 913 °C (1550 and 1675 °F) and for the two higher deformation rates at the test temperature of 982 °C (1800 °F) (Figures 2, 3, and 4). In all cases, the shear bands that were observed began as regions of intense deformation in the form of X's, the legs of which rotated away from the primary compression axis with increasing deformation. With further deformation, the intersecting shear straining led to the formation of "flat" regions of intense deformation at the center of the specimens, which eventually bowed toward one of the die surfaces with yet further deformation. This process is shown schematically in Figure 5.

The intensity and initiation strain for shear bands in the two microstructures varied with temperature and deformation rate. A qualitative rating of the degree of localization was obtained from the metallographic sections (Table I). For the β microstructure, the general trend was that the degree of localization increased with increasing deformation rate and decreasing temperature (Figures 2, 3, and 4). The strain levels at shear band initiation were also estimated from metallographic sections (Table I). These strain levels were were based on the thicknesses (h) of the specimens with the least reductions that exhibited shear bands. Average effective strain $\bar{\varepsilon}$ was calculated for these specimens from $\bar{\varepsilon} = 1.15 \ln(h/d_0)$, where d_0 denotes the initial specimen diameter. This equation is based on modeling of the deformation as being analogous to the forging of a rectangular bar, the factor of 1.15 arising from the plane-strain condition. It is recognized that this approximation is crude, but it allows the observed trends to be compared to a first order.

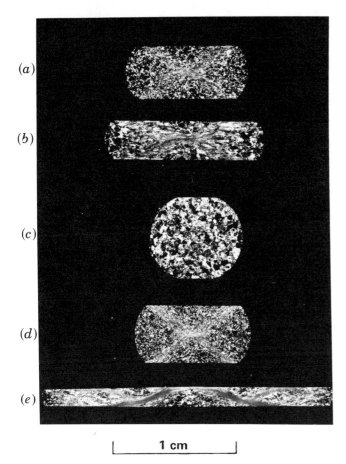

Fig. 2 — Transverse metallographic sections of β microstructure specimens isothermally sidepressed at 843 °C (1550 °F) using a crosshead speed of (a), (b) 0.08 mm per second (0.0032 inch per second) and (c), (d), (e) 63.5 mm per second (2.5 inch per second). Reductions (relative to the initial diameter) are (a) 40 pct, (b) 60 pct, (c) 20 pct, (d) 40 pct, and (e) 75 pct.

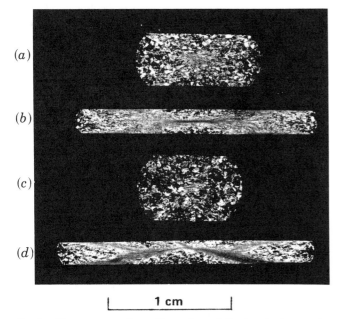

Fig. 3 — Transverse metallographic sections of β microstructure specimens isothermally sidepressed at 913 °C (1675 °F) using crosshead speeds of (a), (b) 0.08 mm per second (0.0032 inch per second) and (c), (d) 63.5 mm per second (2.5 inch per second). Reductions (relative to the initial diameter) are (a) 40 pct, (b) 75 pct, (c) 30 pct, and (d) 75 pct.

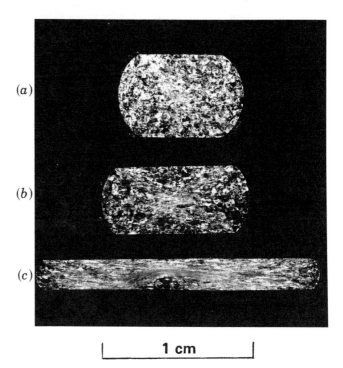

Fig. 4 — Transverse metallographic sections of β microstructure specimens isothermally sidepressed at 982 °C (1800 °F) using a crosshead speed of 9.1 mm per second (0.36 inch per second). Reductions (relative to the initial diameter) are (a) 25 pct, (b) 40 pct, and (c) 75 pct.

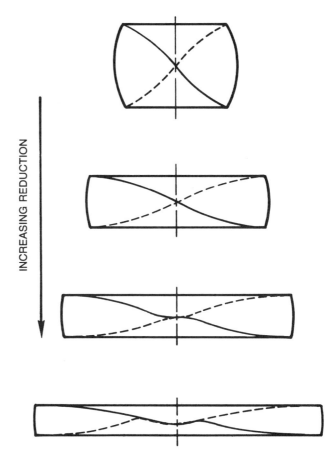

Fig. 5 — Schematic representation of the mechanism of shear band formation in isothermal sidepressing.

Table I. Shear Band Occurrence in Isothermal Sidepressing Tests

Microstructure	Test Temperature, C (F)	Crosshead Speed, mm/sec. (in./sec.)		Shear Band Localization Rating*	Maximum Thickness, h, at Which Shear Bands Were Observed, mm (in.)		Average Effective Strain, $\bar{\varepsilon}$, at h
$\alpha + \beta$	843 (1550)	0.081	(0.0032)	0	—		—
$\alpha + \beta$	843 (1550)	63.5	(2.5)	3	5.1	(0.201)	0.38
$\alpha + \beta$	913 (1675)	0.081	(0.0032)	0	—		—
$\alpha + \beta$	913 (1675)	63.5	(2.5)	0	—		—
$\alpha + \beta$	982 (1800)	0.081	(0.0032)	0	—		—
$\alpha + \beta$	982 (1800)	63.5	(2.5)	0	—		—
β	843 (1550)	0.081	(0.0032)	1	4.3	(0.170)	0.57
β	843 (1550)	9.1	(0.36)	2	4.1	(0.162)	0.63
β	843 (1550)	63.5	(2.5)	3	5.0	(0.196)	0.41
β	913 (1675)	0.081	(0.0032)	1	4.3	(0.168)	0.59
β	913 (1675)	9.1	(0.36)	2	4.6	(0.183)	0.49
β	913 (1675)	63.5	(2.5)	2	5.0	(0.197)	0.40
β	982 (1800)	0.081	(0.0032)	0	—		—
β	982 (1800)	9.1	(0.36)	1	4.2	(0.164)	0.62
β	982 (1800)	63.5	(2.5)	1	4.2	(0.165)	0.61

*Qualitative Rating System:
    0: No shear bands observed.
    1: Diffuse shear bands observed.
    2: Moderately localized shear bands observed.
    3: Strongly localized shear bands observed.

IV. DISCUSSION

A. *Occurrence of Shear Bands*

Workability Maps for Shear Bands. The occurrence of shear bands in isothermal forging is strongly dependent on factors such as material properties, geometry, and friction. In all tests discussed in the Results Section, friction was minimal and test geometry was identical. Hence, it is not surprising that the shear bands that were observed followed similar directions in all cases. In fact, these directions are identical to those predicted by slip line field theory.[26] The major factor determining the occurrence of shear bands in the present investigation, therefore, was material properties.

The effect of material properties on the tendency to form flow localizations during isothermal forging has been discussed before.[7] In this previous work, a criterion developed by Jonas, *et al*[27] and Dadras and Thomas[28] to predict unstable bulging in uniaxial compression was extended to predict shear band formation.

Unstable flow in forging is assumed to require a maximum in flow stress σ at some point (or points) in the flow field:

$$d\sigma = 0 = \left(\frac{\partial\sigma}{\partial\varepsilon}\bigg|_{\dot{\varepsilon},T}\right)d\varepsilon + \left(\frac{\partial\sigma}{\partial\dot{\varepsilon}}\bigg|_{\varepsilon,T}\right)d\dot{\varepsilon} + \left(\frac{\partial\sigma}{\partial T}\bigg|_{\varepsilon,\dot{\varepsilon}}\right)dT. \quad [1]$$

In this expression, ε, $\dot{\varepsilon}$, and T denote strain, strain rate, and temperature, respectively. (Although this analysis is done in terms of uniaxial loading, similar arguments in terms of effective stress, effective strain, and effective strain rate would apply to multiaxial situations.) Eq. [1] can be manipulated to yield an expression for the expected rate of flow localization, $1/\dot{\varepsilon}\ (d\dot{\varepsilon}/d\varepsilon)$, once instability sets in:[7]

$$\frac{1}{\dot{\varepsilon}}\frac{d\dot{\varepsilon}}{d\varepsilon} = -\frac{\gamma'}{m} = \alpha, \quad [2]$$

in which,

$$\gamma' = \frac{1}{\sigma}\frac{d\sigma}{d\varepsilon}\bigg|_{\dot{\varepsilon}}, \quad [3]$$

and,

$$m = \frac{\partial\ln\sigma}{\partial\ln\dot{\varepsilon}}\bigg|_{\varepsilon,T}. \quad [4]$$

It is seen that the flow localization tendency (in terms of the fractional change of strain rate with strain) is dependent on the nondimensional work-hardening (or softening) rate of the material and its strain-rate-sensitivity parameter. The former is determined from flow curves measured in constant strain-rate compression or tension tests (data from which include implicitly the effect of deformation heating on flow properties), and the latter from step-strain-rate-change tests run over various strain-rate ranges. From Eq. [2], it is obvious that large flow softening rates (*i.e.*, negative work-hardening rates), which are often found in hot working, and small strain-rate sensitivities favor the localization of strain.

Jonas, *et al*,[27] suggest that materials with α parameters of five or greater are particularly susceptible to *persistent* flow localizations. Earlier observations for Ti-6242[7] supported this hypothesis, and the present observations add further confirmation for it. The α parameters (Eq. [2]) for this material were calculated from flow curves (not corrected for deformation heating) and strain-rate sensitivities measured at various temperatures and strain rates (Tables II and III). From these calculated data, temperature and strain rate regimes in which α was equal to or greater than five for at least one strain level were determined. This procedure forms the basis for the development of workability diagrams for the two Ti-6242 microstructures (Figure 6). Analogous to workability diagrams for other kinds of defects found in hot working,[29] these diagrams delineate temperature and strain-rate domains in which shear bands would not be expected (the "SAFE" domains) and domains in which they should be

Table II. Flow Localization Parameter, α, for Ti-6242 of $\alpha + \beta$ Microstructure

Nominal Test Temperature (C (F))	Strain Rate (Sec.$^{-1}$)	$\alpha(\varepsilon = 0.10)$	$\alpha(\varepsilon = 0.225)$	$\alpha(\varepsilon = 0.40)$	$\alpha(\varepsilon = 0.575)$
816 (1500)	0.001	3.4	2.6	2.1	1.8
816 (1500)	0.1	0.4	3.0	2.9	2.0
816 (1500)	1.0	<0	1.9	5.9	6.2
816 (1500)	10.0	<0	1.3	5.1	9.7
871 (1600)	0.001	2.4	2.2	1.6	1.1
871 (1600)	0.1	2.1	2.6	2.2	1.8
871 (1600)	1.0	1.6	4.0	4.8	4.4
871 (1600)	10.0	<0	2.7	4.2	4.4
913 (1675)	0.001	3.1	2.7	1.9	0.9
913 (1675)	0.1	2.1	1.7	1.9	1.2
913 (1675)	1.0	2.0	3.1	2.9	2.8
913 (1675)	10.0	0.3	2.4	2.5	4.0
954 (1750)	0.001	0.9	1.3	1.0	1.4
954 (1750)	0.1	1.6	0.9	1.1	1.0
954 (1750)	1.0	1.0	1.2	1.4	1.1
954 (1750)	10.0	<0	0.7	2.1	3.6
1010 (1850)	0.001	2.1	1.4	0.9	0.0
1010 (1850)	0.1	0.6	0.0	0.3	0.4
1010 (1850)	1.0	<0	<0	0.2	0.6
1010 (1850)	10.0	<0	<0	0.9	1.4

Table III. Flow Localization Parameter, α, for Ti-6242 of β Microstructure

Nominal Test Temperature (C (F))	Strain Rate (Sec.$^{-1}$)	$\alpha(\varepsilon = 0.10)$	$\alpha(\varepsilon = 0.225)$	$\alpha(\varepsilon = 0.40)$	$\alpha(\varepsilon = 0.575)$
816 (1500)	0.001	6.8	4.6	4.2	3.8
816 (1500)	0.1	9.9	7.1	4.6	4.0
816 (1500)	1.0	7.8	12.0	10.6	9.0
816 (1500)	10.0	<0	14.2	20.2	16.2
871 (1600)	0.001	5.4	4.7	3.8	2.6
871 (1600)	0.1	10.2	5.8	3.9	2.6
871 (1600)	1.0	11.9	8.5	6.5	5.1
871 (1600)	10.0	4.9	8.0	1.5	6.1
913 (1675)	0.001	6.2	4.1	3.3	2.3
913 (1675)	0.1	10.1	5.1	3.2	1.6
913 (1675)	1.0	10.1	6.3	4.4	3.7
913 (1675)	10.0	4.5	4.9	4.7	4.3
954 (1750)	0.001	7.2	4.5	3.6	2.0
954 (1750)	0.1	7.7	4.2	2.7	2.0
954 (1750)	1.0	7.4	4.6	2.9	2.3
954 (1750)	10.0	4.7	4.6	3.1	2.5

expected ("FAIL" domain). For the most part, the shear band observations reported in the Results Section support the theoretical predictions of the workability diagrams (Figure 6). The major discrepancies occur in observations taken near the boundaries between the SAFE and FAIL regions and may be due to the use of α equal to five, rather than some other value. On the other hand, however, the *degree* of localization (which was noted in the Results Section) does indeed vary with temperature and strain rate for the β microstructure in the manner suggested by the magnitude of the tabulated flow localization parameters (Table III). This agreement adds further credence to the general formulation of the flow localization parameter concept.

Shear Band Initiation. The derivation of workability diagrams for shear band occurrence, just described, uses the criterion that the flow localization parameter be equal to or

greater than five for some strain level. When the material properties are such that α is equal to or greater than five over a range of strains (as is the case for much of the data in Tables II and III), one may logically expect that the strain at which α first reaches this value (for a particular temperature-strain rate combination) is the initiation strain for the shear band. Inspection of the data in Tables II and III establishes these strain levels to be generally in the range of 0.1 to 0.2. Comparison of these strains with observations (Table I) shows large discrepancies, however. Part of the discrepancy between observations and predictions results from observation of shear bands after the initiation event. Since shear bands are internal defects which can be observed only after deformation is completed and since no other phenomenon (*e.g.*, some characteristic of the load-stroke behavior) can be positively identified with shear band

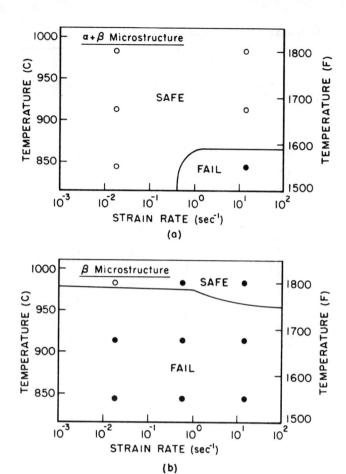

Fig. 6—Workability maps for occurrence of shear bands in hot forging of Ti-6242 with (a) $\alpha + \beta$ microstructure and (b) β microstructure. Workability predictions (—) and forging conditions in which shear bands were (●) and were not (○) observed are noted.

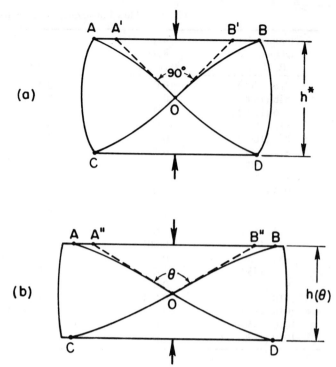

Fig. 7—Construction used to determine reduction at which shear bands initiate in sidepressing: (a) geometry corresponding to reduction at which shear bands initiate and (b) geometry corresponding to reduction at which shear bands first observed in metallographic sections.

initiation, it is not surprising that shear bands noted in metallographic sections have initiated at lower strains.

A better estimate of the strain at initiation is obtainable from the metallographic sections, though, by using a simple geometric construction (Figure 7). This construction makes use of the assumption that once shear bands initiate, deformation proceeds more or less by block shear. For the purposes of argument, assume that shear bands initiate as shown in Figure 7a. Deformation proceeds by shearing along the shear bands AOD and COB and deformation of the "blocks" AOB, BOD, DOC, and COA which lead to rotation and "bending" of the shear bands. At higher levels of deformation, the geometry assumes the form shown in Figure 7b. The "block" AOB has deformed into the shape shown. For the geometry studied, the deformation of "block" AOB may be approximated by the deformation of the area A′ OB′ (Figure 7a) into the area A″ OB″ (Figure 7b). Thus, by measuring the angle θ and height $h(\theta)$ (Figure 7b) at which shear bands are first observed in metallographic sections, the height at initiation $h*$ (which corresponds to $\theta = 90$ deg) may be estimated:

$$h* = h(\theta) \sqrt{\tan (\theta/2)}. \qquad [5]$$

Calculations based on Eq. [5] and measurements from metallographic sections are given in Table IV. Once again effective strain $\bar{\varepsilon}$ has been calculated from $\bar{\varepsilon} = 1.15 \times$

$\ln (h*/d_0)$. The use of the geometrical construction lowers significantly the strain levels at which shear band initiation may be postulated to have occurred. However, these strains are still substantially higher than predicted. This could be due to the fact that shear bands need some strain to develop (once initiation occurs). This requirement is analogous to that for the development of diffuse necking in sheet metal tensile testing.[30] Furthermore, it must be realized that the strain distribution in sidepressing is very inhomogeneous, and the definition of overall effective strain is very approximate. Therefore, the strains near the initiating shear bands may be somewhat different from the bulk strain. For these reasons, process simulation of isothermal sidepressing was performed to clarify the discrepancy and offer a basic insight into how shear bands develop during this and other deformation modes.

B. Process Simulation to Predict Shear Band Occurrence

Modeling Procedures. Using an advanced computer program, ALPID, developed by Oh,[31] deformation in isothermal, lateral sidepressing was modeled to predict the occurrence of shear bands in forging of Ti-6242. Combining features of the upper-bound and finite-element methods[32,33] of plasticity analysis, the code requires significantly less computation time for metalworking simulations than conventional methods. This is done partly by employing advanced types of elements with linear, quadratic, or cubic displacement distributions. With these elements, far fewer elements are required for simulation. Furthermore, the program can be applied to a wide range of workpiece and tool geometries and treat materials with rigid-plastic ($\bar{\sigma} = \bar{\sigma}(\bar{\varepsilon})$) or rigid-viscoplastic ($\bar{\sigma} = \bar{\sigma}(\bar{\varepsilon}, \dot{\bar{\varepsilon}})$) properties.

(Bars over field quantities are used to denote effective quantities, such as effective stress, effective strain, or effective strain rate.)

In the present application, the plane-strain capabilities of ALPID were utilized for the viscoplastic deformation of Ti-6242. Since ALPID does not have capabilities for calculation of temperature changes during forging, flow stress data from constant-strain-rate-compression tests, which were not corrected for deformation heating,[7] were used in the form $\sigma = \sigma(\varepsilon, \dot{\varepsilon})$. By using these data, the effect of temperature rises on flow stress is implicitly included in the independent variable of strain, since $\overline{\sigma} = \overline{\sigma}(\overline{\varepsilon}, \dot{\overline{\varepsilon}}, T) = \overline{\sigma}(\overline{\varepsilon}, \dot{\overline{\varepsilon}}, T(\overline{\varepsilon}))$, to a first approximation.*

---

* For material elements which undergo large *changes* in strain rate during sidepressing, this assumption is valid only approximately when flow stress data from constant strain-rate tests are used.

Because of symmetry, the deformation in only one quadrant of the cylindrical sidepressing sample cross section needed to be examined. The discretization of this problem required 11 elements, each with nine nodal points (Figure 8). Runs with this quadrant divided into fewer or more elements showed this mesh to be optimal. For boundary conditions, the die was assumed to move at a constant crosshead speed, as was done in the experimental part of the program. The die-workpiece interface friction was characterized by a friction factor of 0.2 (τ, the interface shear stress $= 0.2\overline{\sigma}/\sqrt{3}$), a value based on lubricant evaluations using the ring test.[34]

The applicability of the computer analysis to model the sidepressing operation was judged from two sets of simulations. One set of simulations was run for actual cases for which experimental data were measured. The other set was obtained for hypothetical materials whose flow properties bound those of the Ti-6242 at hot working temperatures. These hypothetical situations allow an insight into the effect of specific material parameters, such as α, on flow stability. These results will now be discussed.

Sidepressing Simulations at 913 °C (1675 °F). Several sidepressing simulations were run at 913 °C (1675 °F) using the measured flow stress data.[7] For these cases, observed load-stroke characteristics and microstructural changes were employed to judge the capabilities of the program. Typical predicted and measured load-stroke curves are shown in Figure 9 for the β microstructure deformed at 913 °C (1675 °F) for a crosshead speed of 9.14 mm per second (0.36 inch per second). The "herky-jerky" nature of the predicted load-stroke curve results from the large distances between adjacent nodal points. Abrupt load increases occur as individual nodes come into contact with the die, the forces over which are summed to give the net load. In spite of this, experiment and theory show good agreement for this as well as the other simulations, suggesting that the modeling procedure is accurate.

Comparison of predicted effective strain-rate fields to observed deformation patterns further substantiates the ability of ALPID to model the sidepressing problem. Strain rate was chosen as the primary field quantity to inspect, as opposed to strain, because it gives the best idea of the *instantaneous* tendency toward shear band formation. It does this by showing deformation rate gradients or discontinuities which characterize shear bands. On the other hand, strain,

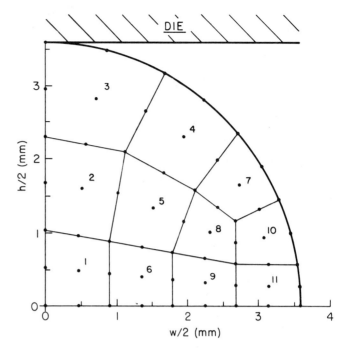

Fig. 8 — Finite-element grid used in simulation of sidepressing.

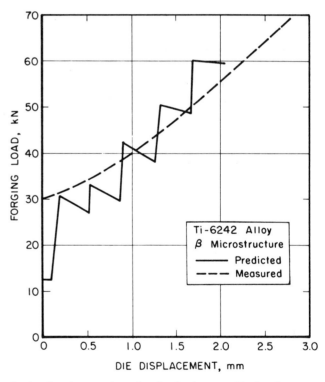

Fig. 9 — Sample comparison of predicted and measured load-stroke curves in isothermal sidepressing of cylinders, 913 °C (1675 °F), crosshead speed ≃ 9.14 mm per second (0.36 inch per second).

which is a cumulative variable, can mask behavior if localization occurs at large strains at which differences in strain from one region to another may be difficult to detect.

Effective-strain-rate contours for the two microstructures deformed at a crosshead speed of 9.14 mm per second (0.36 inch per second) at 913 °C (1675 °F) show striking differences (Figure 10). It will be recalled from the Results

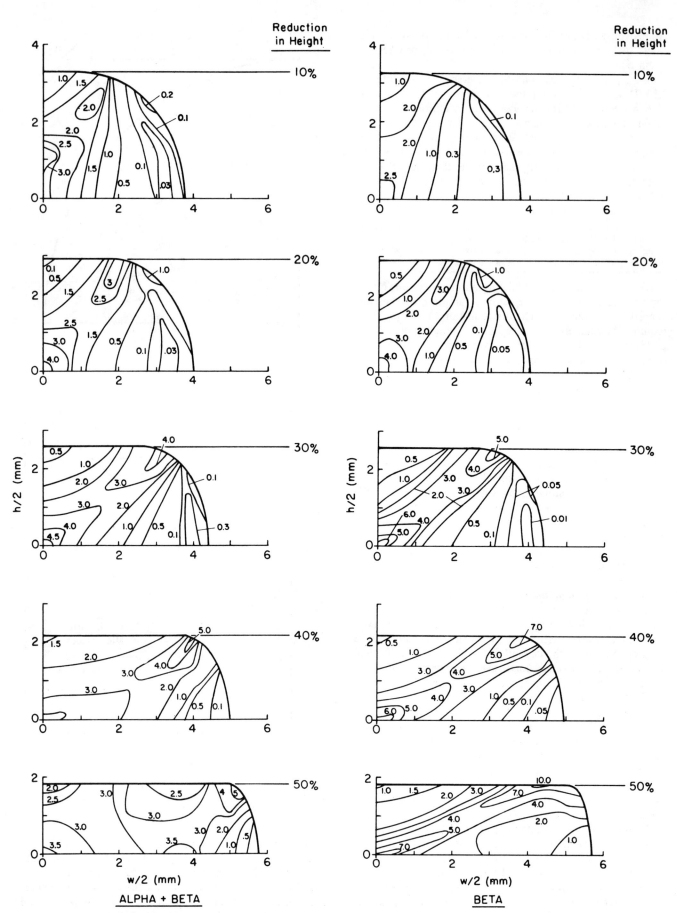

Fig. 10—Predicted strain-rate contour plots for isothermally sidepressed Ti-6242 with $\alpha + \beta$ and β microstructures. Specimen temperature–913 °C (1675 °F), crosshead speed–9.14 mm per second (0.36 inch per second).

Section that the $\alpha + \beta$ microstructure deformed without forming shear bands, and the β microstructure formed shear bands under these processing conditions. Simulation results are consistent with these observations. For the $\alpha + \beta$ microstructure, strain-rate contours do not show large gradients across the cross section (Figure 10). Somewhat larger values of strain rate are apparent along the lines of velocity discontinuity in the classical slip-line field[26] at height reductions of 20 and 30 pct, but these disappear with increasing reduction.

In contrast, the β microstructure simulation shows large strain-rate gradients across the cross section (Figure 10). The regions of highest strain rate originate along the velocity discontinuities, but in this case, they persist to higher reductions and are observed to be related to individual material elements, not particular spatial directions determined by geometry. The high strain-rate region begins to rotate away from the ordinate at a reduction between 20 and 30 pct (in good agreement with the reduction at which shear bands were observed to initiate experimentally (Table IV)). This is believed to be the indication of shear band initiation in the process simulation. It should be emphasized that this flow localization in the process simulation occurs as a natural consequence of the material properties and geometry. No instability criterion was inserted or needed to be inserted in the program to generate this deformation pattern. After 50 pct reduction in height, the deformation pattern for the β microstructure simulation became so nonuniform that further computer simulation was not possible without remeshing. On the other hand, the simulation for the $\alpha + \beta$ case was run to 70 pct reduction without encountering any problems whatsoever of this kind.

Parametric Study of Sidepressing. In order to obtain an in-depth idea of the capabilities of the computer code, isothermal simulations were run from a parametric viewpoint using a wide range of hypothetical material properties typical of hot-forged metals. The hypothetical materials were assumed to have flow stresses σ which depend on strain ε and strain rate $\dot\varepsilon$ as $\sigma = f(\varepsilon)g(\dot\varepsilon)$. At a constant strain rate, the flow stress dependence on strain, $f(\varepsilon)$, of most metals at hot-working temperatures lies between two extremes (Figure 11). (Note in Figure 11 that the normalization factor σ^* includes a constant term and the strain rate dependence $g(\dot\varepsilon)$ of the flow stress.) One of the extremes of $f(\varepsilon)$ is a flow curve which shows an initial work-hardening interval followed by a flow-stress plateau (Curve A, Figure 11). This

curve is typical of much of the $\alpha + \beta$ microstructure data. The other extreme is a curve showing little or no initial work-hardening followed by a large amount of flow softening (Curve B, Figure 11), which is typical of much of the β microstructure flow stress data. The large strain behavior of these two flow curves is characterized by $|\gamma'|$'s (Eq. [3]) of 0.0 for Curve A and 1.25 for Curve B. The rate sensitivity of the flow stress of the hypothetical materials, $g(\dot\varepsilon)$, was assumed to vary in a power-law fashion, namely, $g(\dot\varepsilon) \sim \dot\varepsilon^m$, and the rate sensitivity exponent m was assumed to have values of either 0.0, 0.125, or 0.30. These values span those typically measured in compression tests.[7] With these assumptions, the flow stress is completely described by $\sigma = (\text{const.})\ \dot\varepsilon^m f(\varepsilon)$. As before, the temperature dependence of the flow stress is implicitly included in $f(\varepsilon)$. With these properties, α's (Eq. [2]) between 0 and ∞ were obtained, suggesting a wide range of tendencies to form shear bands.

Sidepressing simulations were run using ALPID and the element grid shown in Figure 8. In all cases, an arbitrary crosshead speed of 9.14 mm per second (0.36 inch per second) was selected. The crosshead speed was not important because (1) the flow curves had preselected values

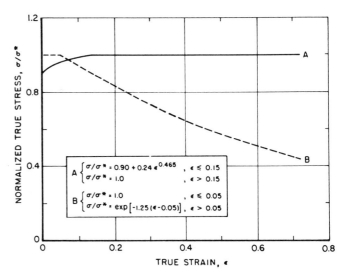

Fig. 11—Nondimensionalized flow curves for metals deformed in the hot-working regime (not corrected for deformation heating). Curves show extremes of no flow softening (Curve A) and large amount of flow softening (Curve B).

Table IV. Strains at Shear Band Initiation Estimated from Block Shear Hypothesis

Microstructure	Test Temperature, C (F)	Crosshead Speed, mm/sec. (in./sec.)		Estimated Thickness at Shear Band Initiation, mm (in.)		Effective Strain, $\bar\varepsilon$, at Shear Band Initiation
$\alpha + \beta$	843 (1550)	63.5	(2.5)	5.6	(0.219)	0.28
β	843 (1550)	0.081	(0.0032)	4.9	(0.191)	0.44
β	843 (1550)	9.1	(0.36)	5.0	(0.195)	0.42
β	843 (1550)	63.5	(2.5)	4.9	(0.192)	0.43
β	913 (1675)	0.081	(0.0032)	5.3	(0.210)	0.33
β	913 (1675)	9.1	(0.36)	4.7	(0.187)	0.46
β	913 (1675)	63.5	(2.5)	5.3	(0.208)	0.34
β	982 (1800)	9.1	(0.36)	5.2	(0.204)	0.36
β	982 (1800)	63.5	(2.5)	4.9	(0.192)	0.43

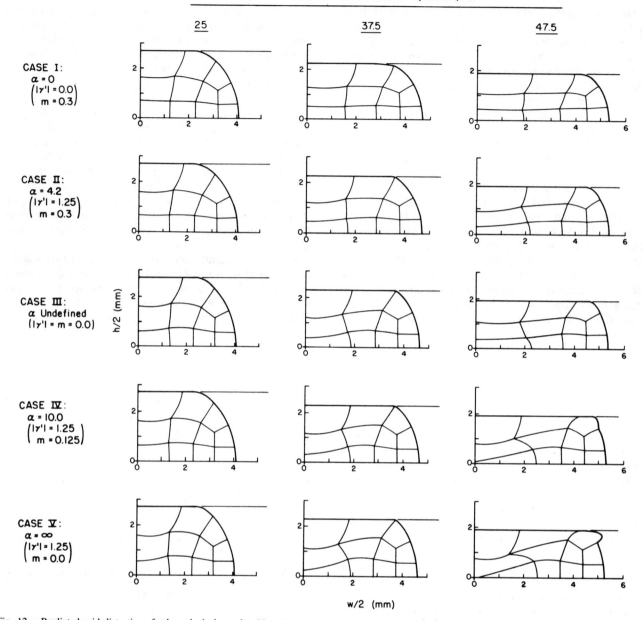

Fig. 12—Predicted grid distortions for hypothetical metals with varying degrees of flow softening ($|\gamma'|$) and strain-rate sensitivity (m).

of rate sensitivity m independent of strain rate, and (2) the effects of speed and heat transfer on flow properties (*e.g.*, flow softening due to deformation heating) are implicitly included in the flow curves measured (or, in this case, postulated) at the constant strain rates in nominally-isothermal-compression tests. The latter point follows from the previous argument that to a first approximation $\overline{\sigma}(\overline{\varepsilon},\dot{\overline{\varepsilon}},T) = \overline{\sigma}(\overline{\varepsilon},\dot{\overline{\varepsilon}},T(\overline{\varepsilon})) = \overline{\sigma}(\overline{\varepsilon},\dot{\overline{\varepsilon}})$.

Shown in Figure 12, predicted grid distortions for five different cases exhibit a strong influence of the material properties. For instance, the two cases for which α is less than five (Case I, $\alpha = 0$ and Case II, $\alpha = 4.2$) exhibit relatively uniform deformation at reductions of 25, 37.5, and 47.5 pct. The simulation for these two cases was taken to 75 pct reduction, with only minor nonuniformities appearing in the second case at reductions greater than 40 pct. For Case III (α undefined), Case IV ($\alpha = 10$), and Case V

($\alpha = \infty$), grid distortions are increasingly nonuniform with increasing α at a given reduction, on the one hand, and with increasing reduction for a given α on the other. For Cases IV and V, deformation became so nonuniform after approximately 50 pct reduction that continuation of the simulation was impossible. The grid distortions in Figure 12 also pinpoint the regions of higher-than-average deformation. These may be noted by looking at changes in the angles between intersecting grid lines from which it is observed that deformation is indeed predicted to localize along the directions observed in metallographic section (Figures 2, 3, and 4).

Predicted effective-strain-rate contours (Figure 13) exhibit a trend similar to that for the grid distortions. Strain-rate concentrations increase with increasing α. Modest ones are seen in Cases II and III and strong ones in Cases IV and V. In addition, localizations observed in Cases IV and V occur along lines which rotate toward the abscissa with

REDUCTION IN HEIGHT (Percent)

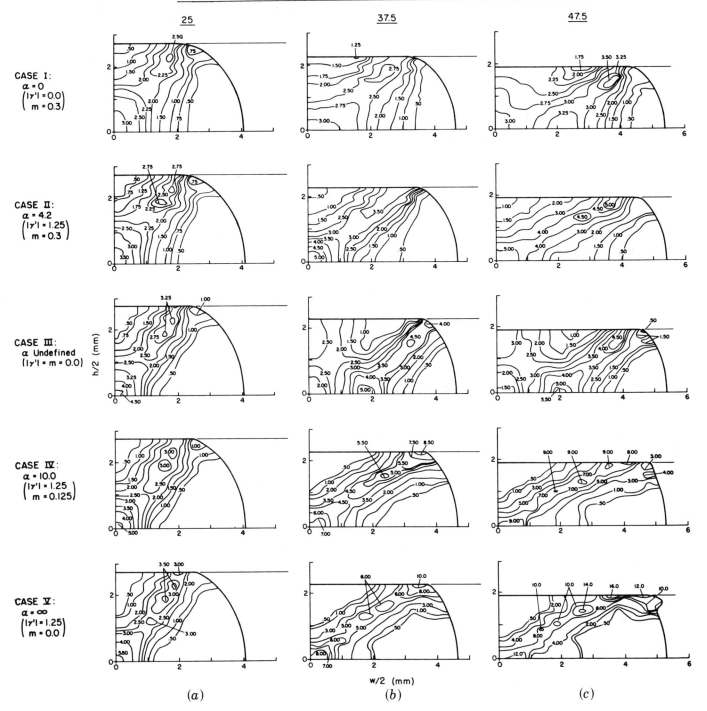

Fig. 13—Predicted strain-rate contours for hypothetical metals with varying degrees of flow softening ($|\gamma'|$) and strain-rate sensitivity (m). Contours for height reductions (relative to the initial diameter) of column (a): 25 pct, column (b): 37.5 pct, and column (c): 47.5 pct.

increasing reduction. For these two cases (and for the others as well), regions of highest strain rate lie at 45 deg to the ordinate at a reduction of 25 pct. Rotation of these regions occurs at higher reductions. Hence, the initiation event (marked by the reduction at which the angle between intersecting flow localizations deviates from 90 deg) has been predicted to occur at reductions greater than 25 pct. Another interesting observation is the tendency of the flow localizations to form "flats" near the center of the specimen,

much like those actually observed. This can be seen by comparing the 47.5 pct reduction patterns for Cases IV and V to observations in Figures 2, 3, and 4. A related phenomenon is seen in Case III, in which the strain-rate localization even moves away from the center. These and other phenomena are highly suggestive of the block shearing required for shear band occurrence. Velocity maps for these cases further confirm and illustrate the physical situation involved in shear band formation and persistence.

Source: Metallurgical Transactions A, 13A, February 1982, 275-286

Comparison of α Parameter and Simulation Predictions. The simulation results for the wide range of material properties just discussed allow a more complete evaluation of the accuracy of the α-parameter method of predicting shear bands. As shown by simulation, intense shear bands should definitely be expected when α is 10 or greater. Furthermore, modest localizations may be expected for α's between four and 10. Thus, the use of α = 5 as a critical value for shear band occurrence is a useful rule of thumb to obtain a first-order idea of when intense shear bands may be expected. However, it must be kept in mind that it is only approximately valid and that at best it indicates the *degree* of tendency toward localization, and not the occurrence *per se*.

Another feature illustrated by the process simulation results is the fact that flow localization is a process and not an event. Strain and strain rate concentrations do not occur instantaneously. For this reason flow localization cannot be expected to occur when α first reaches some critical value (such as five) at some point in the flow field. Application of this premise for Cases IV and V, for example, would require flow localization to be noticed at the reduction at which some material element has undergone an effective strain of $\bar{\varepsilon} = 0.05$, which, because of inhomogeneous deformation, should occur at a reduction less than that corresponding to an effective thickness strain of 0.05. Simulation shows, though, that localizations are first developed at reductions between 25 and 37.5 pct, which correspond approximately to effective thickness strains between 0.33 and 0.54. For the actual cases studied experimentally, a remarkably similar divergence of initiation strain based on the strain at which the α parameter first attains a value of five (Tables II and III) and that observed (Table IV) is noted.

V. SUMMARY AND CONCLUSIONS

Isothermal sidepressing of Ti-6242 cylinders at a variety of hot-working temperatures and deformation rates were investigated experimentally and theoretically to establish criteria for the occurrence of shear bands in hot forging of this and other alloys. The occurrence of shear bands in isothermal sidepressing was observed to depend strongly on starting microstructure, test temperature, and deformation rate. With a starting microstructure of globular alpha in a transformed beta matrix (α + β microstructure), shear bands were observed only at low hot-working temperatures and high deformation rates. In contrast, shear bands were observed at all hot-working temperatures and most deformation rates when the starting microstructure consisted of Widmanstätten alpha (transformed β, or simply, β microstructure). Using these observations, the following conclusions were drawn:

1. The overall occurrence of shear bands in hot, isothermal sidepressing can be predicted from a phenomenological perspective using flow stress data and strain-rate sensitivities measured in uniaxial compression tests. An instability criterion based on the necessity of a maximum in stress at some point in the flow field was developed, and the tendency for subsequent flow localization was shown to depend on α, the ratio of the nondimensional work-softening (or hardening) rate to the strain-rate sensitivity parameter.

2. Workability maps delineated in strain rate – temperature space establish the applicability of the α parameter for predicting shear band occurrence. Shear band observations for both microstructures were in good agreement with predictions based on α ≥ 5 for shear band occurrence.

3. Because of the inherent inhomogeneity of deformation involved in sidepressing, it is not surprising that attempts at predicting shear band *initiation* using the α-parameter formulation are unsuccessful. Process simulation using an advanced computer code is very successful in this regard, yielding predicted reductions for initiation which are in good agreement with observations. These predictions are made without the use of an instability criterion inserted into the computer code, and demonstrate the program's capability to model properly the deformation mechanics.

4. Process simulation predictions of shear band occurrence in hypothetical materials whose properties bound those of the Ti-6242 are in quantitative agreement with α-parameter predictions for these same materials. The simulations show an increasing degree of flow localization with increasing α, a trend which is observed experimentally for Ti-6242. In addition, many of the shear band features observed experimentally (such as the formation of "flats" where the shear bands intersect at the center of the sidepressing specimen) are seen in the process simulation. Moreover, process simulation underlines the formation of shear bands as a process and not as a single event.

5. The successful application of both an instability criterion and process simulation to predict the occurrence of shear bands in hot, isothermal sidepressing demonstrates the complementary nature of the two methods. From an engineering viewpoint, the occurrence of shear bands in other simple hot-forging operations can probably be predicted using solely the α parameter. On the other hand, for complicated forging geometries, the α parameter may give some insight, but a full process simulation may be required by the designer in order to predict the occurrence of the shear band defect.

ACKNOWLEDGMENTS

This research was graciously supported by the Air Force Office of Scientific Research, Air Force Systems Command, USAF, under grant no. AFOSR-79-0048 (Dr. A. Rosenstein, Program Manager). Many technical discussions with Drs. A. Hoffmanner and T. Altan helped in the formulation of much of the work reported here. In addition, the assistance of Messrs. T. Merriman, W.W. Sunderland, and C.R. Thompson in the performance of the experimental work, and the assistance of Dr. S.I. Oh and Messrs. C.H. Tuan and J. Starkey in computer simulation was highly appreciated. Dr. C.C. Chen, formerly of Wyman-Gordon Company, kindly supplied the materials used in the investigation.

REFERENCES

1. R. H. Witt and W. T. Highberger: *Proc. Tenth National SAMPE Technical Conference*, Kimesha Lake, NY, October 17-19, 1978, Society for the Advancement of Material and Process Engineering, 1978, p. 282.

2. S. K. Suh and H. A. Kuhn: *Modern Developments in Powder Metallurgy*, Met. Powder Ind. Fed., Princeton, NJ, 1977, vol. 9, p. 407.
3. F. N. Rhines and P. J. Wray: *Trans. ASM*, 1961, vol. 54, p. 117.
4. C. M. Sellars and W. J. McG. Tegart: *Inter. Met. Reviews*, 1972, vol. 17, p. 1.
5. H. Rogers: *Ann. Rev. Mat. Sci.*, 1979, vol. 9, p. 283.
6. C. C. Chen: *Metallurgical Fundamentals to Ti-6Al-2Sn-4Zr-2Mo-0.1Si Alloy Forgings. I. Influence of Processing Variables on the Deformation Characteristics and the Structural Features of Ti-6242 Si Alloy Forgings*, Report RD-77-110, Wyman-Gordon Co., North Grafton, MA, October 1977.
7. S. L. Semiatin and G. D. Lahoti: *Metall. Trans. A*, 1981, vol. 12A, p. 1705.
8. R. S. Culver: *Metallurgical Effects at High Strain Rates*, R. W. Rohde, *et al.*, eds., Plenum Press, New York, NY, 1973, p. 519.
9. R. F. Recht: *J. Appl. Mech., Trans. ASME*, 1964, vol. 31E, p. 189.
10. Z. Marciniak and K. Kuczynski: *Inter. J. Mech. Sci.*, 1979, vol. 21, p. 609.
11. W. A. Backofen: *Deformation Processing*, Chapter 10, Addison-Wesley Publishing Company, Reading, MA, 1972.
12. R. D. Venter and M. C. de Malherbe: *Sheet Met. Ind.*, 1971, vol. 48, p. 656.
13. Z. Marciniak and K. Kuczynski: *Inter. J. Mech. Sci.*, 1967, vol. 9, p. 609.
14. P. W. Lee and H. A. Kuhn: *Metall. Trans.*, 1973, vol. 4, p. 969.
15. A. K. Ghosh and S. S. Hecker: *Metall. Trans. A*, 1975, vol. 6A, p. 1065.
16. N. M. Wang and B. Budiansky: *J. Appl. Mech., Trans. ASME*, 1978, vol. 45, p. 73.
17. S. L. Semiatin and G. D. Lahoti: *Metall. Trans. A.*, 1981, vol. 12A, p. 1719.
18. J. J. Jonas and M. J. Luton: *Advances in Deformation Processing*, Plenum Press, New York, NY, 1978, p. 215.
19. D. Hauser: Ph.D. Thesis, Ohio State University, Department of Metallurgical Engineering, Columbus, OH, 1973.
20. A. L. Hoffmanner: Battelle's Columbus Laboratories, Columbus, OH, private communication, 1980.
21. S. L. Semiatin, G. D. Lahoti, and T. Altan: *Process Modeling: Fundamentals and Applications to Metals*, T. Altan, H. Burte, H. Gegel, and A. Male, eds., ASM, Metals Park, OH, 1980, p. 387.
22. E. G. Thomsen, C. T. Yang, and S. Kobayashi: *Plastic Deformation in Metal Processing*, Macmillan, New York, NY, 1965, p. 234.
23. A. Mendelson: *Plasticity: Theory and Application*, Macmillan, New York, NY, 1968.
24. A. K. Chakrabarti and J. W. Spretnak: *Metall. Trans. A*, 1975, vol. 6A, p. 733.
25. W. Johnson, G. L. Baraya, and R. A. C. Slater: *Inter. J. Mech. Sci.*, 1964, vol. 6, p. 409.
26. S. Kobayashi, C. H. Lee, Y. Saida, and S. C. Jain: *Analytical Prediction of Defects Occurrence in Simple and Complex Forgings*, Technical Report AFML-TR-70-90, University of California, Berkeley, CA, July 1970.
27. J. J. Jonas, R. A. Holt, and C. E. Coleman: *Acta Met.*, 1976, vol. 24, p. 911.
28. P. Dadras and J. F. Thomas, Jr.: *Res. Mechanica Letters*, 1981, vol. 1, p. 97.
29. R. C. Koeller and R. Raj: *Acta Met.*, 1978, vol. 26, p. 1551.
30. A. K. Ghosh: *Metall. Trans.*, 1974, vol. 5, p. 1607.
31. S. I. Oh: *Int. J. Mech. Sci.*, 1982, vol. 24, in press.
32. B. Avitzur: *Metal Forming: Processes and Analysis*, McGraw-Hill, New York, NY, 1968.
33. S. Kobayashi: *Metal Forming: Interrelation between Theory and Practice*, A. L. Hoffmanner, ed., Plenum Press, New York, NY, 1971, p. 325.
34. C. C. Chen: *Evaluation of Lubrication Systems for Isothermal Forging of Alpha-Beta and Beta Titanium Alloys*, Technical Report AFML-TR-77-181, Wyman-Gordon Co., North Grafton, MA, November 1977.

COMPUTER-AIDED PREFORM DESIGN FOR PRECISION ISOTHERMAL FORGING

by

T. L. Subramanian, Staff Scientist, Battelle's Columbus Laboratories, Columbus, Ohio 43201
N. Akgerman, Principal Scientist, Battelle's Columbus Laboratories, Columbus, Ohio 43201
T. Altan, Research Leader, Battelle's Columbus Laboratories, Columbus, Ohio 43201

ABSTRACT

Developments in computer-aided design and manufacturing of forging dies have shown considerable potential for assisting the experienced die designer. In a recent development program, these CAD/CAM techniques were significantly improved to apply them to precision isothermal forging of titanium parts. The new CAD/CAM forging system, based on the original modular approach, is now capable of handling forging cross sections consisting of normal L's, step L's and tapered L's. In addition, the system can (a) design forging cross sections from given machined part geometries, (b) edit and modify interactively the final preform geometry, and (c) design preblockers using the blocker geometry as the input.

INTRODUCTION

The demand for high performance airframe structures at moderate costs has stimulated the development of sophisticated forging techniques, such as Isothermal Forging. In isothermal forging, the dies are heated to almost the same temperature as the forged material in order to reduce die chilling effects and the consequent increase in the forging pressure. Because of its high strength-to-weight ratio, titanium is one of the most preferred materials for aircraft structures, and isothermal forging has been found to be very effective to forge the strain rate and temperature sensitive titanium alloys.

In isothermal forging, the metal flow is very much enhanced by the reduced die chilling effects. Therefore, it was originally thought that any complicated shape can be forged using simple preform shapes. Recent manufacturing development studies[1-6], however, indicated that even in isothermal forging accurate design of the preform shapes, having a very carefully engineered and controlled distribution of stock, is very important to avoid cold shuts, die lock due to premature die fill, and other similar defects. Furthermore, because of the increase in production cost due to additional equipment necessary for die heating, the isothermal forging process, in order to be cost effective, must essentially be a precision forging process. Consequently, all the difficulties associated with precision forging, especially close control of volume in preforms, are also present in isothermal forging.

Traditionally, preforms are designed by experienced designers spending hours of valuable engineering time. Recent advances in computer-aided design and manufacturing (CAD/CAM) have shown how the experience-based intuition and skill of designers can be augmented by the computerized analyses and design[7-9]. Despite the considerable ease in the preform design and the manufacture of forging dies by the computerized methods, the CAD/CAM techniques require improvements so that they can be used under practical industrial conditions.

In a recent Air Force sponsored project, Battelle, in cooperation with Wyman-Gordon Company, and with McDonnell Aircraft Company, improved the existing computer-aided preform design procedure for use in precision isothermal forging. The result of this recent effort is the second generation of CAD/CAM system for forging dies. This paper describes the principles and advances in the developments of computerized preform design for precision isothermal forging.

MODULAR APPROACH IN PREFORM DESIGN

A detailed review of several aircraft structural forgings reveals that nearly all the rib-web type structural parts have cross sections which can be divided into a number of basic L shapes, as shown in Figure 1. Thus, once a generalized preform design is formulated for the basic L shape, preforms for most rib-web type forgings can be obtained without the necessity to write a separate program for every new cross section. Hence, the computerized preform design begins with a pattern recognition procedure. The coordinate data for each cross section, obtained manually or by APT part programs, are systematically analyzed to identify the presence of basic L modules. The cross section is divided into modular L shapes. The preform for each modular L shape is designed, using the same subprogram, and the modular preform geometries are then assembled in a building-block manner to obtain the preform for the whole cross section. This approach is basically the same as in the earlier work[7,8]. However, the present system has several improvements in the input/output procedure, in the preform design, in the stress analysis and load calculations, and in the display of the results. In addition, the new system has capabilities to recognize and then to design preforms for modules other than the basic L module.

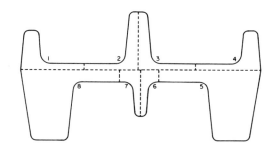

FIGURE 1. SEPARATION OF A CROSS SECTION INTO MODULAR L COMPONENTS

In a typical L shape, there is only one fillet radius. During pattern recognition, this basic criterion is used to determine the presence of modular L shapes. Almost all cross sections of the rib-web type parts are composed of the basic L modules only, but there are a few exceptions. For example, a "Step L", seen in Figure 2a, and a "Tapered L", shown in Figure 2b, have two fillets in each module. In the original design procedure, such deviations were not recognized and, consequently, the end results became unpredictable. In the present design, the pattern recognition algorithm identifies these two new modules and another feature (to be described later), and labels the modules by a coding procedure. These labels are then decoded during the preform design and during stress and load calculations to obtain the desired results for the individual modules.

Preform Design for the Modular L Shape

In the original computer-aided preform design, a logarithmic curve, obtained from the analysis of unrestricted metal flow, was used as a fillet in the preform for the modular L shape. Although theoretically sound, the logarithmic curve is not used in practice due to difficulties in die machining. Furthermore, the logarithmic curve, as a fillet, imposes severe restrictions on the preform forging geometry, especially for those with thin webs and tall ribs. Hence, to be compatible with the industrial design practices and flexibility, the present version of the preform design uses only circular arcs and straight-line configurations to obtain the preform geometry.

The new preform design procedure for the modular L shape is illustrated in Figure 3. The design parameters used to define the preform geometry are listed at the bottom of the same figure. Starting from Point I, which is the origin of the L module, the height of Point II is obtained as a multiple of the finish shape web thickness, Height 1-2 in Figure 3. In order to ensure optimum metallurgical properties, a certain amount of minimum reduction must be taken at the final blow of the forging. Thus, the multiplication factor to obtain the preform web thickness, called the "web ratio", is determined from this forging requirement. From Point II, Line II-III is drawn either parallel to the finish shape web Surface 2-3, or at a predefined inclination θ_{1P}. In simple forgings, this inclination will have practically no effect on die filling or defect formation. However, if there are two unequal height ribs adjacent to each other, such an inclination can be made to retard the metal flow into the easy-to-fill short rib and thereby tend to accomplish simultaneous filling of both ribs.

Line III-IV is drawn through what is called a "pickup point". The location of the pickup point, particularly in the X-direction, is very critical. Its location limits the preform rib thickness. More importantly, the location of the pickup point, in combination with the size of the fillet radius, determines the initial contact point between the die and the preform surfaces. In precision isothermal forging, precise distribution of the metal in the preform is more important than in conventional forging. The location of the pickup point, in combination with the two radii multiplication factors listed at the bottom of Figure 3, provides practically an unlimited choice for metal distribution. In addition to passing through the pickup point, Line III-IV

is drawn parallel to the finish shape rib surface 3-4. However, if necessary, this line may also be drawn at a predefined inclination θ_{2P}.

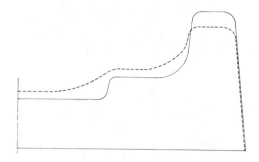

FIGURE 2a. STEPPED L AND A TYPICAL PREFORM SHAPE

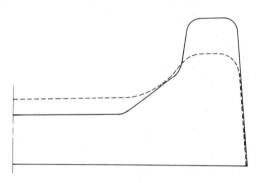

FIGURE 2b. "TAPERED SURFACE L" AND A TYPICAL PREFORM SHAPE

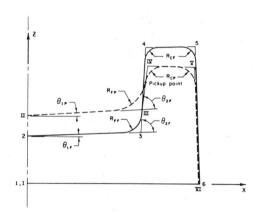

$$\text{Web Ratio} = \frac{\text{Height I-II}}{\text{Height 1-2}}$$

$$\text{Rib Ratio} = \frac{\text{Height of Point IV}}{\text{Height of Point 4}}$$

$$\text{Corner Radius Multiplication Factor} = \frac{R_{CP}}{R_{CF}}$$

$$\text{Fillet Radius Multiplication Factor} = \frac{R_{FP}}{R_{FF}}$$

FIGURE 3. PREFORM DESIGN PARAMETERS (Subscript F refers to finish shape, and P refers to preform)

Source: 5th North American Metalworking Research Conference Proceedings, 1977, 198-203

To facilitate easy positioning of the preform in the die cavities, the preform width is normally made shorter than the finish shape width. Hence, Points V and VI in Figure 3 are defined such that the Plane V-VI is parallel to the Plane 5-6, but inwards from Plane 5-6 by one-hundredth of the total width of the cross section. Finally, Line IV-V is drawn at a suitable height such that the preform volume is equal to the finish shape volume plus flash losses.

From the metallurgical viewpoint, the web ratio is the preferred design parameter. However, some designers prefer to limit the preform rib height based on the forgeability of the material and forging conditions. In such cases, instead of the web ratio, "rib ratio", which is the ratio of the preform rib height to the finish shape rib height, can be assigned. When the rib ratio is assigned, preform design begins with the height of the Surface IV-V and ends in adjusting the height of the web surface II-III to match the volume of the preform to the finish shape volume, plus flash losses. Corner and fillet radii of the preform shapes are usually larger than the finish shape radii. Especially for titanium forgings, the ratio varies between two and three. By assigning suitable multiplication factors, the designer may obtain the desired values for the preform radii.

Preform Design for Additional Modules

In addition to the basic L module, "Step L" and "Tapered L", shown in Figure 2, are the two modules that are most frequently encountered in practice. In the new system of computerized preform design, the pattern recognition algorithm identifies these two additional modules and labels them by a coding procedure. During the preform design, these labels are decoded to identify the type of the L shape, and preforms are designed with geometries comparable with finish shape geometries.

Preform Design for the Step L

A typical step L and a possible preform shape for the step L are shown in Figure 4. As seen in this figure, a step L is basically a normal L shape (polygon 5'-5-6-7-8-VI') plus an attachment to the web as indicated with broken lines (polygon 1-2-3-4-5'-VI'). Hence, the preform design for the step L is performed in two steps. The volume of metal that will go into the attachment portion of the preform (polygon I-II-III-IV-V'-VI') is first estimated from the assigned design parameters. The preform for the regular L shape is designed by the conventional procedure, but with due consideration for the volume of metal in the attachment (polygon I-II-III-IV-V'-VI'). When the volume matching is completed, the newly defined preform geometry is augmented with the geometry of the attachment by adding the appropriate coordinates to the preform coordinate data.

Preform Design for the Tapered L

Because of the close resemblance in the finish geometries of the tapered L and the normal L, preform geometries for both these modules are very likely to be the same. Hence, in the modified design procedure, the preform for the tapered L is designed using a ficticious equivalent L module.

In Figure 5, the true geometry of a tapered L is shown in full lines, while the equivalent L geometry is indicated with broken lines. In the equivalent L shape, the two fillet arcs of the tapered L are merged into one fillet with the radius equal to the average of the two fillets in the original geometry. Second, the volume of metal in the equivalent L is kept equal to that in the tapered L by redistributing the excess metal over the web top surface, as shown in Figure 5. After analyzing several possible variations, the present choice to redistribute the excess metal in a tapered wedge like shape was selected because this modification is expected to produce minimum amount of material displacement during forging.

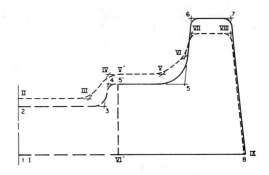

FIGURE 4. DIAGRAM ILLUSTRATING THE PREFORM DESIGN PROCEDURE FOR A STEP-L

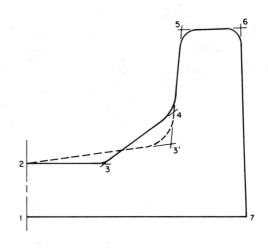

FIGURE 5. MODIFICATION OF THE TAPERED L INTO AN EQUIVALENT NORMAL L

Bulges in Preform to Avoid Suck-In Defects

While forging complicated shapes, bulges, as shown in Figure 6, are added in the preform geometry to avoid suck-in defects. Through a very simple interactive procedure, the designer indicates whether bulges are to be added to all ribs, to a selected few ribs, or to no ribs. During the separation of the cross section into modular L shapes, the bulge feature is also incorporated in the label of the L module by the same coding procedure.

Similar to the preform design for the step L, preforms with bulges are also designed in two steps. The volume of metal necessary to form the bulge is estimated first. The bulge geometry is defined using the design parameters supplied by the designer. If values are not assigned to bulge design parameters, default values are calculated from the rib geometry. The preform for the regular L is designed by the conventional procedure, but with due compensation for the volume of metal in the bulge. After the completion of the preform design and the rearrangement of the modular preform geometries (described in the next section), coordinate data defining the bulge geometries are inserted in appropriate locations in the arrays defining the preform geometry of the cross section.

Assembling the Modular Preforms

After the completion of the preform design for the modular L's, the modular preforms are rearranged to the original orientation of the L shapes, and assembled in the proper sequence to obtain the preform for the whole cross section. For this purpose of later assembly, during the separation of the cross section into L modules, (a) the origin of the L shape, (b) its orientation in the horizontal X-direction, and (c) its orientation in the vertical Z-direction for each L shape are stored in a separate file with suitable labels.

Assembling the modular preforms, although appears to be very simple, causes a problem. During the preform design for the modular L shape, either the web height or the rib height of the preform is prescribed from the forging requirements. Given one of the two heights, the other height is estimated from the volume constancy requirements of the preform design. Because the volume of the metal in any L shape depends on its geometry, the estimated height of the rib (or the web, if rib ratio is assigned) of any two adjacent L's need not be equal. Consequently, there may be a step on the top surface of the rib (or of the web). Because such an abrupt change in heights in the preform geometry is not acceptable, after assembling the modular preforms, relative heights of the ribs (or the webs) of adjacent L preforms are tested systematically. Any variation in the heights of the ribs (or the webs) is eliminated by selecting a common height which will blend the uneven surfaces while maintaining the total volume of both the preforms to be equal to their original volume. After blending the steps, the volume of the preform is precisely calculated and compared with the total volume of metal in the finish shape and flash losses. Because the distribution of metal in the preform is very critical in precision forging, any discrepancy in the preform volume is adjusted by uniformly increasing or decreasing the preform thickness.

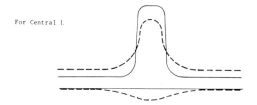

For Central L

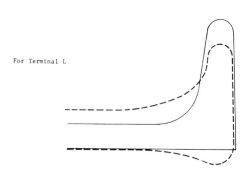

For Terminal L

FIGURE 6. ALTERNATIVE PREFORM SHAPES FOR THIN, TALL RIBS ON ONE SIDE OF THE WEB

COMPUTER PROGRAMS DIEFRG

To combine the benefits from the experience of the human designer with the high speed of the modern computers, the entire die design procedure has been computerized for interactive operation. Thus, the designer can conveniently control the course of the design process while the computer is still performing the computations. This system of computer programs is called DIEFRG. In its present structure of segmented loading, DIEFRG requires $50,000_8$ or 20,480 core memory. This system, although structured for Battelle's CDC-6400 computer, can be used even in minicomputers with facilities for additional core storage. In its present form, DIEFRG uses Tektronix's Terminal Control System software package to display the preform and finish shape geometries on the Tektronix's 4012 or 4014 CRT display terminals.

Starting with a set of coordinate data describing the various cross sections of the forging, DIEFRG processes, analyzes, and designs preform geometries to the specifications of the designer. The coordinate data are obtained either manually or from APT part programs and read from a storage file. If requested, the original cross section as read in, is displayed on the CRT terminal. The designer is then requested to input data on the design parameters and forging conditions.

Forging Design from the Machined Part Data

If the read in coordinate data are for a machined part cross section instead of a forging cross section, a forging section can be readily designed by prescribing the machining allowance and the draft angle. After adding the machining allowance and the draft angle to the machined part section, the computerized forging design procedure also modifies hard-to-forge deep holes and tall ribs. At the completion of the forging design, the forging section is displayed along with the original machined part section. If the results are not satisfactory, a different design may be obtained by reassigning the design parameters.

Preprocessing the Coordinate Data

The coordinate data are preprocessed next. The preprocessing includes:

(a) Arrangement of the data in a clockwise pattern

(b) Rotation of the data to start with a definite starting point

(c) Recognition of the corners and fillets and assignment of respective radii

(d) Recognition of the L modules

(e) Determination of the origin and the geometric orientations of the L shapes

(f) Separation of the cross section into modular L shapes

(g) Computation of volume, plan area, center of gravity, and perimeter of the cross section as well as those of the individual L modules.

After the preprocessing, the data is displayed again and any corrections, if necessary, can be implemented through the "editor" of the DIEFRG. In addition, several options, such as (a) modify the parting line location, (b) selectively modify the corner and fillet radii, (c) redefine the origins for the L modules, and (d) temporarily modify the finish shape geometry for the exclusive purpose of preform design, are also available.

Preform Design

Preform design starts with the display of the cross section divided into L modules as shown in Figure 1. The designer is requested to input design parameters. Although input data are necessary to obtain a specific design, stored default values can also provide a design which in many cases can serve as a start. The designed preform shape is displayed along with the original cross section, as seen in Figure 7. If necessary, the designer may review the displays with the dies at different opened positions and thereby evaluate the locations of the preform in the finish dies, Figure 8. He may even zoom part of the display to view the details of the critical zones. The preform geometry, if necessary, may be modified by editing the coordinate data defining the preform. During the course of design, all assigned values to various design parameters and all modifications are listed. By assigning different input values, the preform geometry of each modular L can be independently varied. If the displayed preform shape is not acceptable, alternate shapes can be obtained by modifying input parameters. When the results are satisfactory, coordinate data defining the preform geometry are written on a separate file for the NC machining of the preforming dies.

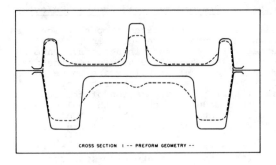

CROSS SECTION I -- PREFORM GEOMETRY --

FIGURE 7. A TYPICAL FORGING CROSS SECTION AND A POSSIBLE PREFORM DESIGNED BY THE IMPROVED CAD PROCEDURE

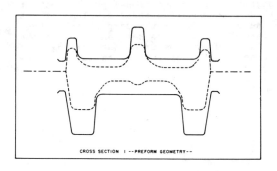

CROSS SECTION I -- PREFORM GEOMETRY --

FIGURE 8. PREFORM AND FINISH DIE GEOMETRIES WITH THE FINISH DIES IN AN OPENED POSITION

Preblocker Design and Other Capabilities

When the finish forging geometry is very complicated, a preblocker may be necessary to forge the preform shape. If so, the preblocker geometry can be designed by the same system DIEFRG, using the coordinate data defining the preform geometry as the input data base.

In addition to preblocker and blocker design, DIEFRG has capabilities to estimate the forging load, the peak pressure, and the stress distribution for a given set of material properties, forging conditions, and the cross-sectional geometry. If the estimated load or stress is excessively high or low, alternate designs can be readily obtained by changing one or more input parameters.

SUMMARY AND FUTURE OUTLOOK

A computerized system for the design and analysis of forging rib-web type structural components, especially for the design and manufacture of blocker and preblocker dies, has been developed. Because of the very flexible design scheme and very stringent volume estimation, this design system, called DIEFRG, is well suited for precision isothermal forging. In its present form, DIEFRG can handle almost all rib-web type structural cross sections with very few exceptions. Because of the extreme complexity of the analysis in the forging process, most default design parameters used in DIEFRG are based on prior empirical knowledge. However, DIEFRG is relatively simple to improve and update whenever more reliable and advanced data on preform design are available. Furthermore, in DIEFRG, the basic preform geometry (for normal L, step L or tapered L) is designed by one single subprogram for both blocker and preblocker designs. Hence, the preform design procedure can be readily modified by any forging company without divulging any proprietary information, by changing just one subprogram in the DIEFRG system. This system is available to companies and universities in the U.S.

ACKNOWLEDGMENTS

The work reported in this paper is based on a project sponsored by the Air Force Materials Laboratory under Contract No. F33615-75-C-5049 with Mr. Norman E. Klarquist as Program Manager. The authors gratefully acknowledge this support. The authors also thank the staff of Wyman-Gordon Company and McDonnell Aircraft Company for their technical assistance in formulating the principles of a practical preform design approach. Special thanks are due to Mr. Charles P. Gure of Wyman-Gordon for his valuable suggestions.

REFERENCES

(1) Kulkarni, K. M., Watmough, T., Parikh, N. M., Stawarz, D., and Malatesta, M., "Isothermal Forging of Titanium Alloy Main Landing Gear Wheels and Nose Wheels", IITRI, Chicago, Final Report AFML-TR-72-270, Air Force Materials Laboratory, Contract No. F33615-70-C-1533, December, 1972.

(2) Watmough, T., Kulkarni, K. M., and Parikh, N. M., "Isothermal Forging of Titanium Alloys Using Large Precision-Cast Dies", IITRI, Chicago, Final Report AFML-TR-70-161, Air Force Materials Laboratory, Contract No. F33615-67-C-1722, July, 1970.

(3) Kulkarni, K. M., Watmough, T., Stawarz, D., and Parikh, N. M., "Isothermal Forging of Titanium Alloy Bulkheads", IITRI, Chicago, Final Report AFML-TR-74-138, Air Force Materials Laboratory, Contract No. F33615-71-C-1167, August, 1974.

(4) Vazquez, A. J., and Hayes, A. F., "Isothermal Forging of Reliable Structural Forgings", Ladish Company, Cudahy, Wisconsin, Final Report AFML-TR-74-123, AFML Contract No. F33615-71-C-1264, June, 1974.

(5) Watmough, T., "Development of Isothermal Forging of Titanium Centrifugal Compressor Impeller", IITRI, Chicago, Final Report No. AD764-266, Army Materials and Mechanics Research Center, Contract No. DAAG46-72-C-0067, AMMRC CTR 73-19, May, 1973.

(6) Spiegelberg, W. D., and Lake, F. N., "Isothermal Forging of Ti-6Al-4V Alloy as an Improved Process for Fabricating Weapon Components", Quarterly Engineering Reports ER-7665 - 1 to 3, U.S. Army Weapons Command, Contract No. DAAF03-73-C-0093, September, 1973, through February, 1974.

(7) Akgerman, N., Subramanian, T. L., and Altan, T., "Manufacturing Methods for a Computerized Forging Process for High-Strength Materials", Battelle's Columbus Laboratories, Final Report No. AFML-TR-73-284, Air Force Materials Laboratory, Contract No. F33615-71-C-1689, January, 1974.

(8) Akgerman, N., Altan, T., "Computer-Aided Design and Manufacturing of Forging Dies for Structural Parts", presented at the First North American Metalworking Research Conference, May 14-15, 1973.

(9) Biswas, S. K., and Knight, W. A., "Preform Design for Closed-Die Forgings: Experimental Basis for Computer-Aided Design", Int. J. Mach. Tool Des. Res., Vol. 15, pp 179-193, 1975.

DETERMINATION AND PRESENTATION OF FLOW STRESS DATA FOR USE IN ISOTHERMAL FORMING ANALYSES

by

R.L. HEWITT, Research Officer
J-P.A. IMMARIGEON, Research Officer
P.H. FLOYD, Technical Officer

Structures and Materials Laboratory, National Aeronautical
Establishment, National Research Council Canada,
Ottawa, Ontario

ABSTRACT. A scheme of presenting high temperature flow stress data is described which clearly separates the effects of structure and strain rate, and allows the designer to incorporate both in the analysis of high temperature, rate-sensitive forging.

A brief description of a suitable facility for obtaining the data is presented together with a discussion of typical structure development in powder fabricated nickel base superalloys. A scheme for data determination is outlined to minimize the amount of data collection and a limited amount of experimental data presented to illustrate the concepts.

Finally, the prediction of structure and properties within a forging is discussed, which leads to the possibility of optimizing properties in different parts of the component by suitable choice of preform and processing parameters.

INTRODUCTION. Isothermal forming is becoming increasingly important in the aerospace industry, both for forming high strength nickel and titanium alloys for use in gas turbine engines (1,2) and for making airframe structural components (3,4). In this type of forging, the die temperature is maintained at the working temperature and because of the absence of die chilling of the workpiece, slower forging rates can be used to take advantage of the high rate sensitivity exhibited by some of these materials.

It is evident that different parts of a complex forging undergo varying amounts and rates of deformation and that there can therefore be a variation in properties throughout the part. It is important to know these variations and it is usual to determine them by 'cut-up' procedures (5), i.e. a trial forging is cut into sections and each piece is tested individually. This is an expensive undertaking and an analysis which could accurately predict the deformation pattern would perhaps reduce the number of sections that need to be tested.

Several attempts have been made to predict forming stresses and flow patterns in isothermal forging operations. Gessinger (6) used the visioplasticity technique with a eutectic Pb-Sn alloy (which is highly rate sensitive at room temperature) to determine the deformation behaviour in a simulated compressor wheel forging. He showed that there can be an order of magnitude difference in strain rate at a given location during the forging cycle and a similar difference in strain rate between parts of a

forging at a given time. Clearly this is very significant and may give rise to different structures and mechanical properties within the forging besides affecting the forming stresses and die filling capabilities of the material.

Price and Alexander (7) attempted to predict the deformation profiles of isothermally forged titanium billets (Ti-6Al-4V) using a simple finite element technique. In this method, suggested by Zienkiewicz and Godbole (8), the material is treated as a viscous non-Newtonian incompressible fluid, with incompressibility assured by use of a penalty function. The method appears to be useful for getting approximate solutions, but may not be suitable if it is necessary to have accurate values of strain rates throughout the part.

More complex procedures such as the matrix method (9,10) have been developed and these can give a very accurate description of the flow within a forging if suitable material property data is available. This is one of the major difficulties in any of these analyses. Very little material flow data is available for the materials of interest, and unfortunately much that is published is incomplete. This is because it is usually given in terms of flow stress versus strain rate at various temperatures. It has been shown (11,12) that this is quite inadequate since the flow stresses of these materials generally show a considerable sensitivity to structure. Furthermore, the structure is continually changing during forging because of grain refinement or grain growth. It is therefore important that flow stress data for these materials be determined and presented in terms of both strain rate and structure at a given temperature.

Immarigeon and Floyd (12) have performed extensive testing on several hot isostatically compacted nickel base superalloys and have developed a good understanding of the deformation process. Based upon that understanding it is now possible to suggest a convenient form of data representation and indicate how it may be determined. This is the purpose of the present paper.

DATA PRESENTATION AND USE. The flow stress, σ, of nickel base superalloys at isothermal forging rates and temperatures is a function of the temperature, T, strain rate, $\dot{\varepsilon}$, and structure, S. If the temperature variation within the forging is not significant, then at a given forging temperature, the flow stress can be written

$$\sigma = f(\dot{\varepsilon}, S)$$

Thus at any time during the forging cycle it is necessary to know the strain rate and structure distributions, which both vary with time, to calculate the flow stresses throughout the forging and hence the deformation.

In an incremental analytical approach it is usual to assume a strain rate within each element from which an approximate flow stress is obtained. The velocities of the nodal points are then calculated after a time interval Δt and hence the strain rate within the elements determined. This gives an improved value of flow stress and the calculations are repeated until a suitable convergence is obtained. Thus, if the material flow stress is considered to be independent of structure, it is only necessary to know the flow stress as a function of strain rate.

If the structure is important, then a strain rate is assumed for the first time increment as before. Since the starting structure is known, an approximate flow stress for each element can be found if the flow stress is known as a function of strain rate and structure. The strain rates can then be calculated as before until convergence is obtained. Now if the rate of change of structure is known as a function of the current structure and strain rate, the new structure after the time interval Δt can be calculated and the computation can proceed to the next time interval and

so on.

Thus the analyst requires two plots from the materials engineer, one of flow stress as a function of structure and strain rate at various temperatures, and one showing the rate of change of structure as a function of structure and strain rate at various temperatures.

Materials of interest for isothermal forging generally possess a fine grain size and exhibit some degree of superplasticity. It has been observed that, as a first approximation, the structure, S, in these materials can be characterized by the grain size, d (11,12,13,14), in the appropriate high temperature, low strain rate region. Thus the former plot for these materials is relatively straightforward, requiring only flow stress determinations at known strain rates and grain sizes. The latter plot requires a little more thought.

For powder processed nickel base superalloys, Immarigeon and Floyd (12) observed that at any given strain rate and temperature in the region of interest, there was an 'equilibrium' grain size to which the structure evolved during working. In addition, the instantaneous rate of change of grain size appeared to depend only upon the difference between the instantaneous and 'equilibrium'* grain sizes. Thus for computational purposes, the instantaneous rate of change of structure can be considered to be independent of prior history. It is therefore possible to construct the rate of change of structure plot for fine-grained powder processed nickel base superalloys at a given strain rate and temperature from a curve of grain size versus strain (or time) at that temperature. It is possible that similar methods will be applicable for other materials of interest.

DATA DETERMINATION. It is obviously not a simple matter to collect the data required for the plots described above. It is necessary to have a system capable of providing a constant true strain rate with precise temperature and strain rate control. In addition it must have a facility for rapid quenching in order to retain the microstructure at the termination of the test for grain size determination. Such equipment has been under continual development in this laboratory for several years and is

* The 'equilibrium' state we refer to need not necessarily be attained, but could be approached asymptotically.

Source: 8th North American Metalworking Research Conference Proceedings, May 1980, 129-134

fully described elsewhere (15). It consists of a hydraulic servo controlled test machine modified to produce constant true strain rates from 10^{-5} to 10 s^{-1}. High temperature compression tooling is provided with flat faced TZM or hot pressed silicon nitride anvils for testing at loads up to 100 kN (22,000 lbf) and temperatures up to 1200°C. This permits testing of 6 mm (.25 inch) diameter specimens of even the strongest superalloys. Testing is performed in a controlled atmosphere under conditions of very low friction and the specimens may be quenched in any suitable medium within one second of the completion of the test.

In order to get the most information from the least number of specimens and with the minimum effort, an understanding of the structure development is useful. Figure 1 shows three curves of flow stress versus strain at different true strain rates for the same initial structure and test temperature. For the

increases very rapidly, but at a relatively low strain, the material recrystallizes and the flow stress decreases as the average grain size of the material is refined. Again after a large strain the flow stress reaches a steady state value, which is higher than that of C, where the structure has again reached 'equilibrium'. At some intermediate strain rate a curve such as B is obtained where neither flow softening nor hardening occurs. This is the strain rate for which the 'equilibrium' structure is similar to the starting structure.

It is clear from these curves that the steady state flow stress increases with increasing strain rate while the 'equilibrium' grain size decreases with increasing strain rate. For practical forging rates and temperatures, the majority of hot isostatically pressed nickel base superalloys produce flow stress curves between those of A and B. Typical examples are shown in Figure 2.

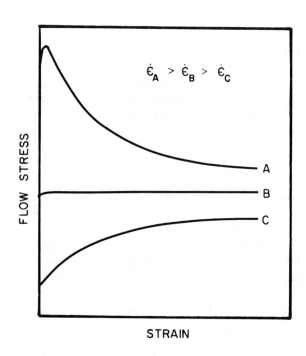

Fig. 1 Schematic flow stress curves at different strain rates for the same starting structure.

slowest strain rate, curve C, the flow stress of the material increases with strain as the grains coarsen. After a fairly large strain, the flow stress reaches a steady state where the structure is no longer changing. This structure (and grain size) is what we have referred to as the 'equilibrium' structure. At the highest strain rate a curve such as A is obtained. Here the flow stress initially

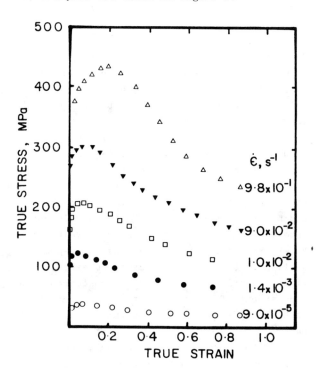

Fig. 2 Flow stress versus strain for nickel base superalloy IN713LC at 1050°C and various strain rates.

To obtain the rate of change of grain size versus grain size data, it is first necessary to produce a plot of grain size versus time at each strain rate and temperature. This is achieved by testing several specimens at the desired strain rate and stopping the test after different amounts of deformation, as indicated in Figure 3a, and then measuring the grain size.

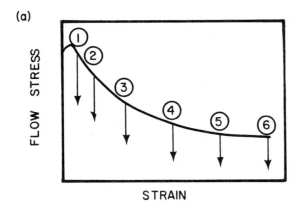

(a)

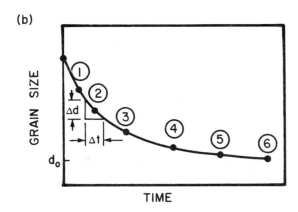

(b)

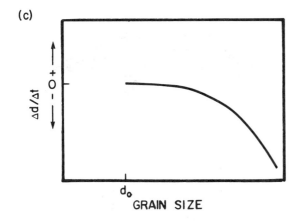

(c)

Fig. 3 Construction of the rate of change of
        grain size plot for one temperature &
        strain rate.

Since the strain is proportional to time in
these tests, the grain size versus time plot
of Figure 3b can be constructed and hence the
rate of change of grain size plot produced as
in Figure 3c. If these tests are repeated
at different strain rates with the same starting grain size, and the peak flow stress noted,
these tests will yield the curve of flow stress

versus strain rate at the starting grain size
in addition to completing the growth rate data.
This assumes that the grain size at the peak
flow stress is very similar to the starting
grain size.

     It is then necessary to obtain flow stress
versus strain rate data for other grain sizes.
This is most easily achieved by starting a
test at a high strain rate and interrupting it
after a certain level of deformation corresponding to one of the previous tests at that
strain rate, e.g. point 2 in Figure 3a. The
grain size of the specimen will then be known
at this point and the test is continued in
situ at a different strain rate and the new
peak flow stress measured. No additional
grain size measurement is required. By using
different secondary strain rates the curve
of flow stress versus strain rate can be
obtained for this grain size. Data for other
grain sizes are obtained similarly by taking
the initial deformation to points 3 and 4 etc.
in Figure 3a.

     Figure 4 shows a rate of change of grain
size plot for a powder processed nickel base
superalloy IN713LC at one temperature and
strain rate, while Figure 5 shows the flow
stress versus strain rate for various grain
sizes at one temperature.

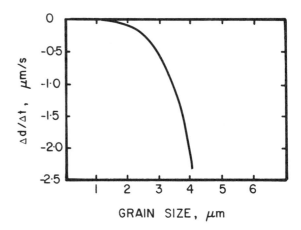

Fig. 4 Rate of change of grain size plot for
        IN713LC at 1050°C and a strain rate
        of 0.09 s^{-1}.

DISCUSSION. Although the ideas expressed in
this paper have been demonstrated only for
the fine-grained powder processed nickel-base
superalloys, it is probable that similar techniques can be used for other materials of
interest. The first requirement is that the

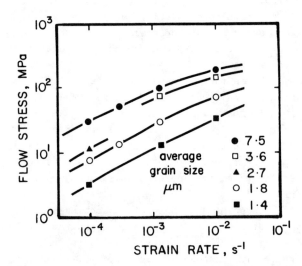

Fig. 5 Flow stress versus strain rate for
IN713LC as a function of grain size
at 1050°C

structure can be adequately described by the
average grain size, and the second is that the
instantaneous rate of change of structure can
be considered to depend only on the difference
between the instantaneous and an 'equilibrium'
grain size. The first is generally met by
materials which exhibit superplastic properties, which includes all the materials currently being isothermally forged commercially.
The second requirement will need confirmation
for materials other than those specifically
considered in this paper. However, even
materials not meeting these requirements can
be dealt with in a similar manner if some
quantity, S, can be found which will adequately
characterize the structure.

With the availability of this kind of
data it should then be possible with a suitable
analytical method (e.g. finite elements) to
follow the development of the structure within
a component during forging, besides being
better able to predict the forming stresses
etc., and thus predict the structure in the
finished forging. If this structure can be
related to mechanical properties, it then
becomes possible to predict the properties
of various parts of a complex forging.

The relationship between mechanical properties and structure is not well developed
for these materials. It is being pursued in
this laboratory by isothermally forging small
uniform billets that are large enough to permit
mechanical property evaluation of the forged
material. By systematically varying the

forging parameters it should be possible to
determine if the mechanical properties can be
related to the simple description of structure
used in the analytical method.

If this is so, it then becomes possible,
within limits, to obtain different preferred
properties in various parts of the forging
by a suitable choice of forging preform shape
and forging schedule. Using powder metallurgy
techniques allows great flexibility in the
choice of preform shape, and the preform need
not be limited to conventional forging stock.
For example, the preform shapes shown in
Figures 6a and 6c might be used rather than
the conventional form to produce the shape
shown in Figure 6b. These preforms will
clearly experience very different deformation
histories with a resultant difference in
final structure distribution.

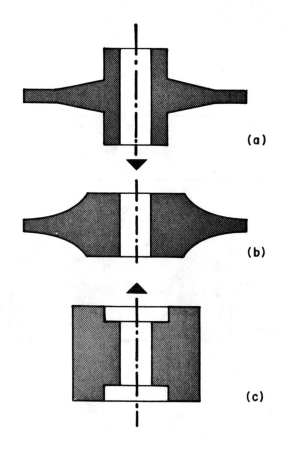

Fig. 6 Hypothetical preform shapes

REFERENCES

(1) "Progress Report on the Gatorizing[TM]
Forging Process", R.L. Athey and J.B.
Moore, SAE Paper No. 751047, Society for

Automotive Engineers, (1975).

(2) "Iso-Forging of Powder Metallurgy Super-Alloys for Advanced Turbine Engine Application", N.M. Allen, Proceedings of Specialist Meeting on Advanced Fabrication Processes, AGARD Conference Proceedings CP-200, p.5, (1976).

(3) "Isothermal Forging Makes a First Good Impression", anon, Iron Age, p.43, October (1977).

(4) "Isothermal Hot Die Forging of Complex Parts in a Titanium Alloy", K.M. Kulkarni, N.M. Parikh and T.M. Watmough, J. Inst. Metals, Vol. 100, p. 146, (1972).

(5) "Design Philosophy For Engine Forgings", J.D. Alexander, Forging and Properties of Aerospace Materials, The Metals Society, Book 188, p.1, (1978).

(6) "Die Anwendung der Superplastizität in der Umformtechnik", G.H. Gessinger, Radex-Rundschau, p. 124, (1977).

(7) "A Study of the Isothermal Forming or Creep Forming of a Titanium Alloy", J.W.H. Price and J.M. Alexander, Proceedings of Fourth North American Metalworking Research Conference, p. 46, (1976).

(8) "Flow of Plastic and Visco-Plastic Solids with Special Reference to Extrusion and Forming Processes", O.C. Zienkiewicz and

Alloys", H.W. Hayden and J.H. Brophy, Trans. ASM, Vol. 61, p. 542, (1968)

(15) "Isothermal Forging at the National Aeronautical Establishment", R.L. Hewitt, J-P.A. Immarigeon, W. Wallace, A. Kandeil and M.C. de Malherbe, DME/NAE Quarterly Bulletin, National Research Council Canada, No. 4, p.1, (1978) P.N. Godbole, Int. J. Num. Methods in Eng., Vol. 8, No. 1, p.3, (1974).

(9) "New Solutions to Rigid-Plastic Deformation Problems Using a Matrix Method", C.H. Lee and S. Kobayashi, J. Eng. for Industry, Vol. 95, p.865, (1973).

(10) "Rigid-Plastic Analysis of Cold Heading by the Matrix Method", S.N. Shah and S. Kobayashi, Proc. 15th Int. MTDR Conference, p. 603, (1974).

(11) "Mechanical Behaviour and Hardening Characteristics of a Superplastic Ti-6Al-4V Alloy", A.K. Ghosh and C.H. Hamilton, Met. Trans, Vol. 10A, p.699, (1979).

(12) "Microstructural Instabilities During Superplastic Forging of a Nickel-Base Superalloy Compact", J-P. A. Immarigeon and P.H. Floyd, Submitted for publication to Met. Trans.

(13) "Superplastic Deformation of Ti-6Al-4V Alloy", A. Arieli and A. Rosen, Met. Trans, Vol. 8A, p.1591, (1977).

(14) "The Interrelation of Grain Size and Superplastic Deformation in Ni-Cr-Fe

DIE MATERIALS FOR ISOTHERMAL FORGING

K. M. Kulkarni
Manager, Metalworking Technology

IIT Research Institute
10 West 35 Street
Chicago, Illinois 60616

ABSTRACT

Isothermal forging is a new process for producing
net or near-net shape titanium alloy and superalloy
components. It utilizes heat-resistant dies which can
be heated to or near the same temperature as the temperature of the preform to be forged. Then slow
deformation speed is employed to produce complex forgings at low loads. Since the process puts unusual
demands on the die materials, the search continues for
a more suitable material. This paper describes the
requirements for the die materials and the techniques
of their evaluation with specific reference to isothermal forging of titanium alloys. Some of IIT
Research Institute's work on selection of the die material is then described, along with a comparison of the
representative materials that have been considered.
The future prospects are then briefly mentioned.

INTRODUCTION

Titanium alloys and nickel-base superalloys are
used in many aerospace applications. These materials
are very difficult to forge conventionally and, consequently, oversized forgings must be made which are
then finish-machined extensively. With the recent
emphasis on cost reduction, new techniques are being
developed to produce near-net shape components that
would require less finish machining. The isothermal
or heated die forging process is one such technique.
In this paper the principle and the advantages of the
process are pointed out with specific reference to
titanium alloys. The usage of the dies at high temperatures subjects them to unusually difficult working
conditions. The resulting requirements for die materials are then described along with some of the techniques of evaluation. A number of candidate die
materials representative of such classes as wrought
and cast nickel-base superalloys, refractory materials,
and ceramics are compared.

ISOTHERMAL FORGING AND ITS ADVANTAGES

Conventionally, titanium alloys are exceedingly
difficult to forge because of two counteracting effects. Firstly, the titanium alloy work material is
in the range of 1600° to 1900°F, whereas the die steel
dies are at the lower temperature of 600° to 900°F;
thus during forging, the work material cools rapidly
and increases in strength. Secondly, the strength of
the titanium alloys is extremely sensitive to speed
of deformation (Fig. 1). Consequently, high deformation speeds cannot be employed to counteract fully the
cooling effect mentioned earlier. The strength and
hence the forging load increase because of cooling
between the dies, or, if attempt is made to use high
speeds, the load requirements go up because of the
strain rate effect. As a result, large structural
forgings rarely have section thicknesses of less than
1/4 in. over major portions.

The isothermal forging technology pioneered by
IITRI (1-3) involves usage of dies that can be operated in the temperature range of 1600° to 1800°F,
thus minimizing or eliminating the temperature difference (and hence the heat transfer) between the titanium alloy workpiece and the dies. This permits a slow
speed of deformation to produce complex forgings. The
speed can be controlled independently or, preferably,
the load can be preset to the estimated maximum value.
With the latter approach, the hydraulic press ram
moves quickly initially and then slows down as the resistance to deformation offered by the forging increases. The entire operation takes place in about 3 to 5
min. Although this is much longer than the time required for conventional forging, one isothermal
forging operation gives more detail than what could be
achieved with three or more conventional forging
operations with several die sets and much intermediate
cleaning and trimming. The process* is schematically
compared to conventional forging in Fig. 2.

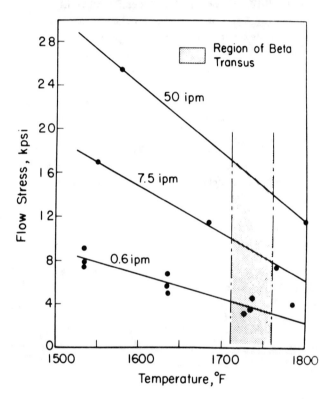

Fig. 1. Effect of Strain Rate and Temperature on Flow
Stress of Ti-6Al-6V-2Sn Alloy Under Isothermal
Forging Conditions.

---

*The discussion in this paper is mainly for isothermal
forging of the high strength alpha-beta titanium alloys
such as Ti-6Al-4V and Ti-6Al-6V-2Sn. They constitute
much of the forging weight in use in the mid-seventies.
The majority of the considerations in this paper are
equally applicable to forging of other titanium alloys
and nickel-base superalloys. The main difference is
that for titanium alloys forged at lower temperatures
(say) 1300°-1500°F, the requirements on dies are less
severe. On the other hand, nickel-base superalloys
forged at 1900°F or higher usually necessitate usage
of high-strength refractory alloys as die materials.

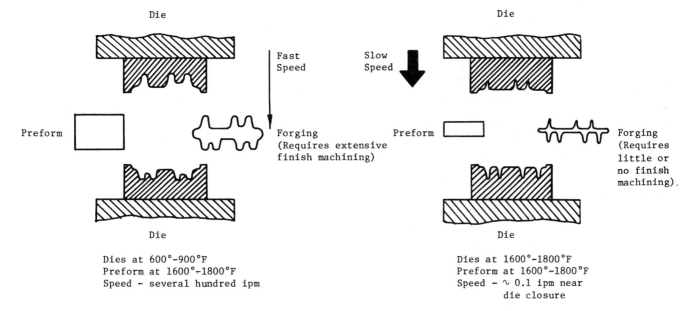

Fig. 2. Comparison of Conventional (left) and Isothermal (right) Forging Processes

Illustrations of some of the isothermal forgings made (1-3) are shown in Figs. 3 and 4. The forgings in Fig. 3 could have been made conventionally, but it would have taken several different stages, whereas each of the isothermal forgings was made in one single operation from a single block. The 12 1/2 in. diameter nose wheel (Fig. 4) for F-111 aircraft represents a reduction of over 90% to produce the 1/8 in. thick web. Such thin web and high deformation are completely beyond the scope of conventional forging techniques. The nose wheel is the first isothermal forging to be put on a plane and tested extensively (4). Even larger forgings, such as an F-15 bulkhead with 430 sq in. plan area, have been made by IITRI. The forging has been cited in Ref. 4, and the details of its production will be described elsewhere (5). In this case the conventional forging weighed 330 lb, whereas the isothermal forging weighed only 66 lb.

The isothermal forging process offers several important advantages. First, thin webs and complex sections can be produced which are completely beyond the capability of conventional techniques. Secondly, the degree of finish machining can be decreased substantially, thus reducing finish machining costs. Also, because of the complex detail, the input weight can be decreased substantially, reducing the material cost. Furthermore, the isothermal forging, in spite of its capability to produce thinner sections, requires only 1/4 or less of the forging load required in conventional practice. Consequently, existing hydraulic presses can be used for making far larger components.

REQUIREMENTS FOR DIE MATERIALS

The operation of the dies for forging titanium alloys at high temperature, up to 1800°F, subjects the dies to very severe conditions, and only a few select materials are capable of giving any reasonable performance. The requirements can be put in two main

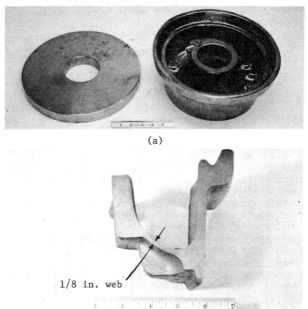

(a)

1/8 in. web

(b)

Fig. 3. Cross Sections of Two Ti-6Al-6V-2Sn Alloy Isothermal Forgings (left - plan area 10 in.^{2}; right - plan area 15 in.^{2})

Fig. 4. Full-Scale Nose Wheel Blank and Isothermal Forging of Ti-6Al-6V-2Sn (a), and a Cross Section of the Forging (b).

categories: those pertaining to dimensional stability and those pertaining to fabricability. The cost is a composite of these two main factors.

The isothermal forging process is used for making high precision net or near-net components. In making such components, the dimensional precision of the die affects the precision of the forging and, hence, the former is very important. Whereas in conventional forging a slight creep or deformation of the die would pose little problem, a comparable deformation in isothermal forging dies would be totally unacceptable. For example (Fig. 5), consider a forging having two ribs 20 in. apart. Assume that the finish-machined tolerance on the relative rib locations is ±0.030 in. In conventional forging the machining envelope may be 1/4 in. or more per surface so that, if the dies creep or deform by even 0.2 in., a forging can still be machined to give an acceptable part. Assuming, however, that the machining envelope in a near-net shape isothermal forging is only 0.060 in., a die dimension change of 0.2 in. will make the forging a reject. This is true even if the individual ribs show excellent detail.

The severe demand placed on the die material, especially when attempting to make a net part, can be illustrated by another example. Assume that a 5 in. long titanium alloy part is to be isothermally forged net to a tolerance of ±0.010 in. The titanium alloy parts are primarily used in aerospace industry where the total quantity required is generally fairly small. Suppose that, in all, 500 airplanes are to be made with one such component on each plane. Based on a forging temperature of 1750°F, the forging pressure may be approximately 15,000 psi and the corresponding induced stress in the die might be estimated as nearly three times the forging stress. If economic considerations dictate that all the forgings be made in a single die set without any repairs, the die material property requirements can be estimated as follows:

$$\text{Allowable creep} = \frac{\text{total dimensional variation}}{\text{dimension}} \times 100$$

$$= \frac{.010 \times 2}{5} \times 100 = 0.4\%$$

Total time under load = Time per operation x number of forgings

$$= 5 \text{ min} \times 500 = 2500 \text{ min}$$

Assumed factor of safety = 2

Then, allowable die material creep = 0.4% in 5000 min or nearly 0.1% in 20 hr.

Thus the satisfactory die material may show a maximum creep of only 0.1% in 20 hr at 45,000 psi stress at 1750°F. Note that this stiff condition was arrived at without considering other factors mentioned below.

In addition to creep, the die dimensions may also change because of local plastic deformation, die wear, and any cracking in deep sections of the dies. Furthermore, the initial precision with which the dies are made also influence the precision that can be expected on the forging. The problem is further compounded by the dimensional changes and thermal stresses generated by any nonuniformity in the die and work material temperatures or their thermal expansion characteristics. The requirement for the die material is then that it must be strong enough to withstand the stresses generated during forging, to avoid or minimize cracking, creep, and wear. Since the die life is defined in terms of the unacceptability of the forging produced in the dies, any deficiency in the material in these factors would cause the die life to be poor and increase the element of die cost per forging.

There are other desirable features in the die material which are less important than the dimensional stability provided some minimum properties are there. For example, the die material should have good oxidation resistance and should not react with the workpiece or the lubricant that it may come in contact with. Table 1 summarizes the various requirements.

Table 1

DESIRABLE ATTRIBUTES FOR DIE MATERIALS

Resistance to creep at temperatures of up to 1800°F
High tensile and compressive strength and resistance to indentation
Good oxidation resistance
Ease of fabricability by forging, casting, or P/M and availability of a reasonable number of die fabricators
Good machinability
Feasibility for weld repair
Absence of excessive reaction with lubricants or work material
Reasonable cost

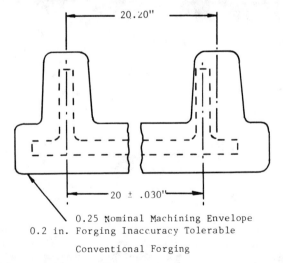

0.25 Nominal Machining Envelope
0.2 in. Forging Inaccuracy Tolerable

Conventional Forging

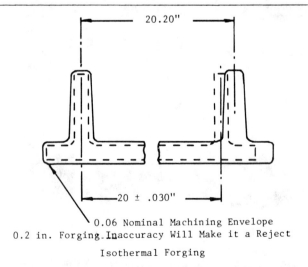

0.06 Nominal Machining Envelope
0.2 in. Forging Inaccuracy Will Make it a Reject

Isothermal Forging

Fig. 5. More Severe Effect of Dimensional Inaccuracy in Isothermal Forging as Compared to Conventional Forging (dotted lines show desired finish-machined part geometry).

The performance requirements on the die material naturally dictate that it should have a high strength. By and large, materials in this category are also extremely difficult to fabricate. Moreover, different materials require different methods of fabrication and, in fact, the properties of the materials may also depend on the technique of die fabrication. In addition, the number of suppliers available also depends on the die material and fabrication method. Consequently, in the isothermal forging process the die fabrication is an important element of the total die cost. Ideally, the fabrication process should be such that there are several available vendors, the cost is reasonable, and the technique leads to good properties of the product.

The dies for isothermal forging can be made by several different methods. Precision casting (6)--as the name implies--often eliminates the need for finish machining the die cavity. In other cases, the dies could be rough or precision cast and then further machined by conventional or electrical discharge machining. If wrought materials are to be used for dies, suitable forgings or rolled plates must be made which are then finish-machined. Other alternatives include powder metallurgy processing. The optimum technique is dependent on die material, component geometry, and die size. Naturally, vendor capabilities and time and cost constraints are related major considerations. Detailed discussion of these factors is, however, beyond the scope of this paper.

TECHNIQUES OF EVALUATION AND EXAMPLES OF DATA

A number of techniques are available for evaluating the suitability of the candidate die materials for isothermal forging. Creep is a major reason for change in die dimensions because of prolonged usage. Therefore, it is essential to obtain some data about the creep of the die materials at the temperature and pressure conditions that the dies would be subjected to. Long-term data are available for common materials (7,8). The data are usually expressed as so many percent of creep in a certain duration at a given temperature for a given stress or, alternatively, the creep life to rupture is expressed in terms of number of hours at a particular temperature and stress level.

Tensile test data are useful for estimating the resistance of the die materials to cracking because of locally high tensile stresses generated in the deep cavities of the die. Such data about the high temperature tensile strength are more easily available (7,8) than data about many other high-temperature properties. Hot compression tests are basically simple upsetting tests conducted at high temperatures. This type of information is useful to estimate the resistance of the die material to upsetting or plastic deformation

because of locally high compressive stresses. The hardness of the material at high temperature gives some indication of its capability to withstand indentation. The mutual indentation hot hardness test used in IITRI's work (6) employs small cylindrical specimens (Fig. 6) mounted in a holder. The assembly of specimens and holders can be heated in situ in a hardness testing maching with a built-in furnace. This test has many advantages, such as simple specimen preparation, ability to test large areas of specimens, and no need for a separate indenter suitable for high temperatures. The hot hardness values calculated from this test and hot compression yield strengths for a number of candidate die materials are shown in Figs. 7 and 8.

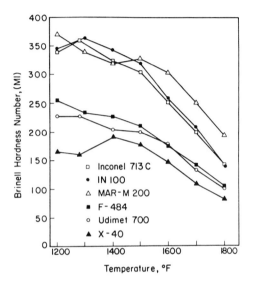

Fig. 7. Brinell Hardness Calculated from Mutual Identation Hardness of Air-Melted and Cast Nickel-Base and Cobalt-Base Superalloys.

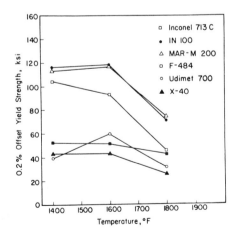

Fig. 8. Hot Compression Yield Strength of Air-Melted and Cast Superalloys.

Fig. 6. Specimen with Holder for Mutual Indentation Hot Hardness Testing of the Candidate Die Materials.

The oxidation resistance of a die material can be estimated by preparing small samples with freshly machined surface and then measuring the change in weight after subjecting them to the higher temperature and whatever environment is expected for various amounts of time. As mentioned earlier, provided the oxidation resistance is adequate, it is not a major factor in isothermal forging. The reactivity of the die material with the work material or the lubricant can also be readily checked by keeping small pieces of the die materials in contact with lubricant or the workpiece materials for various durations at high temperatures, and then inspecting the microstructure at the die surface and any surface pitting. The reactivity and oxidation resistance become important only when they are so inferior that they could influence the lubrication characteristics, and hence die filling, or alternatively, when the die surface may deteriorate and give a poor surface appearance on the forging.

The fabrication techniques that may be suitable for a die material are extremely difficult to judge without actual testing. For example, in IITRI's initial work, precision casting was considered as a method for die fabrication, and an extensive amount of work (6) was conducted on development of the technique before it could be demonstrated to be satisfactory. An example of a precision-cast die half weighing nearly 300 lb net is shown in Fig. 9. A similar but much larger die was used for forging the nose wheel shown in Fig. 4.

COMPARISON OF CANDIDATE DIE MATERIALS

In this section, representative members of four different categories of materials that have been considered as die materials for isothermal forging are compared. More attention is given to IN-100 since this is the die material used in much of IITRI's work and perhaps in most other work conducted on this process thus far. The properties of this material that are satisfactory and others which are not are pointed out since they usually form the basis for comparison with other new materials that become available.

In talking about properties, it must be pointed out that much of the published information is based on testing of thin samples such as sheets. The dies for forging are generally thicker and their properties are often inferior to the handbook data. This must be taken into account in judging the suitability of a material for a particular application. Sometimes actual testing may be desirable, with samples taken

Fig. 9. Precision-Cast 300 lb, MAR-M 200 Upper Die Block Casting for the Half-Scale Nose Wheel Forging.

from section sizes comparable to those contemplated for the particular die. Table 2 provides a summary of the important data about the materials considered in the subsequent discussion.

Cast Nickel-Base Superalloys

The nickel-base superalloys have compositions carefully developed to obtain superior high-temperature properties. They are characterized by very complex chemical compositions and microstructures. The high strength is usually obtained by gamma prime precipitation and solid solution strengthening. Processing of these materials is also quite difficult. It is not the purpose of this paper to go into the details of metallurgical structure or processing. References 9 and 10 provide excellent details of this type, and also cite numerous references for additional information.

In IITRI's initial work (1,2,6), MAR-M 200 and IN-100 were shown to be the most promising materials as indicated also in Figs. 7 and 8. Of these, IN-100 was selected for much of the recent work because of its somewhat lower cost than, but equivalent performance to, MAR-M 200. Dies weighing up to 1400 lb per die half net were precision cast in IN-100 for dies, such as those used for the full-scale nose wheel shown in Fig. 4. The dies performed satisfactorily in IITRI's work where some machining was allowed on the forging. However, the deficiency in this material becomes apparent when the requirements for net shape forging estimated in the illustration in section on Requirements for Die Materials are compared to the properties of the IN-100 material at high temperature. For example, at 1700°F, 0.1% creep occurs in 20 hr at a stress level of only about 32,000 psi. At 1750°F, the corresponding stress level would be only about 25,000 psi (8).

IN-100 material has good castability (6) and, as demonstrated above, fairly large precision castings can be made in this material. The technology has advanced even more in the last few years in terms of capability for making larger dies. As for machining, the material is very difficult to machine and electrical discharge machining must be used in making complicated profiles. Thus, the die costs go up rapidly if precision casting alone is not adequate for obtaining the required die geometry. The material has quite satisfactory oxidation resistance. An indication of its oxidation resistance is shown by data of static tests where in 1000 min the weight gain observed was only 0.2 mg/cc^2 at 1600°F and about 1 mg/cc^2 at 2000°F (11). Dies made from this material can be employed for isothermal forging in air without any protective atmosphere. Since IN-100 is used extensively in the gas turbine industry, a reasonable number of foundries are available for making small castings to any required degree of precision. However, only a few foundries can handle large castings weighing in excess of a ton or so.

Wrought Nickel-Base Superalloys

Udimet 700 and Astroloy exemplify high-strength wrought nickel-base superalloys. These are, once again, used in the gas turbine industry and are fairly well established materials. The properties of Udimet 700 are considerably lower than those of IN-100 (Table 2). The advantage of wrought material is that it tends to show somewhat better reliability because of more uniformity of the structure and fineness of grain size. The high-strength wrought nickel-base superalloys are again difficult to machine by conventional techniques, but perhaps somewhat easier than

Table 2

COMPARISON OF CANDIDATE DIE MATERIALS

Material	Composition, w/o	Category Represented	0.2% YS, ksi		100 hr Rupture Strength, ksi		Protective Atmosphere Needed?
			1600°F	1800°F	1600°F	1800°F	
IN-100	15Co, 9.5Cr, 5.5Al, 5Ti, 3Mo, 0.95V, 0.015B, 0.18C. Max: 1Fe, 0.1Mn, 0.015S, 0.15Si; bal. Ni	Cast nickel-base superalloy	101	54	55	25	No
Udimet 700	18Co, 15Cr, 5Mo, 4.5Al, 3.5Ti, 0.03B, Max: 0.15C, 0.1Cu, 4Fe, 0.15Mn, 0.2Si, 0.015S, 0.06Zr; bal. Ni	Wrought nickel-base superalloy (used sometimes as casting also)	92	44	42	16	No
TZM[a]	0.5Ti, 0.08Zr, 0.03C. Max: 0.0005H, 0.01Fe, 0.002Ni, 0.002N, 0.00250, 0.008Si; bal. Mo	Refractory alloy	95	85	(85)	70	Yes
Si_3N[b]	As implied by formula	Ceramics	(120)	(110)	--	--	No

Note: The data are compiled from Refs. 8, 12, and 16.

[a]Properties in stress-relieved condition. 1600°F rupture strength estimated by extrapolation.

[b]Data in parentheses are for flexural strength which may be two to three times the tensile yield stress. Creep resistance is excellent. At 2300°F, 10,000 psi shows less than 0.5% creep in 100 hr. All data are for hot pressed silicon nitride.

cast nickel-base superalloys. They are available in small sizes fairly easily, whereas large forgings can be made only by few of the large forging vendors. Their oxidation resistance is adequate for usage as isothermal forging dies. The data in Ref. 8 suggest that at 1750°F only 20,000 psi stress would cause 1% creep in 20 hr. Naturally, 0.1% creep would occur at considerably less stress level.

Higher strength wrought nickel-base superalloys have become recently available, such as AF2-1DA (12) which shows creep and yield stress properties similar to IN-100. However, for a new material of this type the availability tends to be quite poor and the cost very much higher than any well established alloys.

There is very little published information about performance of wrought nickel-base superalloys as dies for isothermal forging. However, when comparing cast and wrought materials with similar properties, the performance is also likely to be similar. If powder metallurgy (P/M) processing is employed to produce wrought nickel-base superalloy dies, special heat treatment may be necessary to coarsen the grain size for improved creep resistance. The main difference in cast and wrought superalloys is that in the former the die cavity can be at least partially obtained during casting, whereas in wrought materials the die cavity is generally obtained by machining. P/M processing and hot isostatic pressing (HIP) may, however, allow direct production of the die cavity. The decision about using cast or wrought materials will depend primarily on considerations of cost and delivery time. In general, when all or much of the die cavity can be produced by casting and the die size is in excess of several hundred pounds, cast IN-100 dies are likely to be much cheaper than dies made of high strength wrought nickel-base superalloys. In large sizes the cast dies will also often require less time for fabrication.

Refractory Materials

Alloys of high melting metals such as tungsten, molybdenum, and columbium form this class of materials. They can be used at very high temperatures well in excess of the melting ranges of most nickel-base superalloys. TZM is an example of such an alloy.

TZM is a molybdenum-base alloy and has strength and creep resistance superior to IN-100 as shown in Table 2. Creep data for 1750°F were not available for comparison with data mentioned above for IN-100 and Udimet 700, but TZM is known to have much better creep resistance. At a high temperature of 2000°F, 0.1% creep occurred (8) in 5 hr at a stress level of 30,000 psi. Its oxidation resistance, however, is poor at temperatures above approximately 1000°F. TZM has been employed as die material in heated die forging of nickel-base superalloys (13). It is mentioned that, to protect the die material, the entire die stack, heater assembly and material handling system were enclosed to allow processing in vacuum at temperatures of up to 2075°F.

TZM is somewhat more expensive than nickel-base superalloys. The forging operation in TZM dies is likely to be costly because of the vacuum processing mentioned above. On the other hand, its higher creep resistance should lead to considerably better die life. Consequently, it is likely to prove a better material for isothermal forging dies when undertaking long production runs or making net shape forgings. It is relatively easy to machine. The alloy is fairly well established and easy to obtain in small sizes. Billets of large sizes such as a ton or larger can be obtained only on special order. The material is normally used in wrought condition only.

Ceramic Materials

Ceramic materials have excellent resistance to high-temperature corrosion and can be utilized at temperatures well above those that the nickel-base superalloys can withstand. Ceramics are characterized by relatively low ductility and show larger variability in properties. Consequently, statistical techniques must be used in defining useful strength levels. The properties of ceramic materials are influenced very substantially by the fabrication techniques (14,15) which affect their microstructure. Silicon nitride is an example of a strong ceramic material.

Hot-pressed silicon nitride has a high compressive strength of 500,000 psi at room temperature, and it can be used at temperatures of up to 3000°F (16). However, like most ceramics, it has low tensile strength. Generally, for ceramics flexural strength is specified instead of tensile strength. As has been pointed out (15), the flexural strength may be nearly three times the tensile strength. For silicon nitride, the flexural strength is about 120,000 psi at 1600°F and reduces only slightly to about 110,000 psi at 1800°F.

High strength ceramics are currently not available in large sizes, and billets of only about 12 in. linear dimension are likely to be available. Further, in view of their low tensile strength, any inserts made from ceramics should be carefully supported by metallic materials to prevent catastrophic breakage. This is difficult because of the problem of manufacturing tolerances and the large difference in thermal expansion characteristics of the ceramic and metallic materials. At present, it seems unlikely that ceramics will play any significant role in the near future as materials for isothermal forging dies.

FUTURE PROSPECTS

It is anticipated that titanium alloy isothermal forgings will find increasing applications in the years to come. Already an F-15 torque rib is being produced by Ladish Company (4). Also, McDonnell Douglas has identified (4,17,18) some 100 parts as potential candidates for production by isothermal forging. Most of these are smaller in size than 100 sq in. plan area. Large forging companies have suggested (4,19) capability for producing parts of up to 600 sq in. plan area. Parts with some degree of symmetry, such as aircraft wheels, and components with potential for eliminating nearly all finish machining seem particularly good candidates for fabrication as isothermal forgings.

The new applications will place increasing importance on finding better and cheaper die materials. Work continues on improving the high-temperature characteristics of many different alloy categories (20). Newer developments such as mechanical alloying (20,21) and dispersion-strengthened materials being developed for the gas turbine industry may have some applicability as die materials for isothermal forging of titanium alloys. On the other hand, some of the newer beta titanium alloys with processing temperatures of 1500°F may allow usage of existing die materials as satisfactory die materials.

REFERENCES

1. T. Watmough, K. M. Kulkarni, and N. M. Parikh, "Isothermal Forging of Titanium Alloys Using Large Precision-Cast Dies," Air Force Materials Laboratory Technical Report AFML-TR-70-161, July 1970, prepared by IIT Research Institute, Chicago.

2. K. M. Kulkarni, T. Watmough, and N. M. Parikh, "Isothermal Forging of Titanium Alloys Using Precision-Cast Dies," Technical Paper MF70-122, Society of Manufacturing Engineers, 1970.

3. K. M. Kulkarni, N. M. Parikh, and T. Watmough, "Isothermal Hot-Die Forging of Complex Parts in a Titanium Alloy," J. Inst. Metals, Vol. 100, 1972, pp. 146-151.

4. "Putting the Squeeze on Titanium Forging Costs," Iron Age, Dec. 1, 1975; also "Air Force Material Lab Develops New Isothermal Forging Process for Titanium," Light Metal Age, February, 1976.

5. K. M. Kulkarni, "Isothermal Forging--From Research to a Promising New Manufacturing Technology," to be presented at the SME 1977 Intern'l Engrs. Conf. & Tool & Mfg. Exposition, May 9-12, 1977, Detroit.

6. J. S. Prasad and T. Watmough, "Precision-Cast Superalloy Dies for Isothermal Forging of Titanium Alloys," paper presented at 73rd AFS Casting Congress, Cincinnati, May 5-9, 1969.

7. "High Temperature, High Strength Nickel-Base Alloys," (2nd ed.), Internat'l Nickel Company, Inc., 1968.

8. Aerospace Structural Metals Handbook, Vol. 4, Mechanical Properties Data Center, Belfour Stulen, Inc., Traverse City, Michigan, 1976.

9. C. T. Sims and W. C. Hagel (eds.), The Superalloys, John Wiley & Sons, New York, 1972.

10. Superalloys--Processing, MCIC-72-10, Proc. of the Second Internat'l Conf., Battelle Columbus Laboratories, Columbus, Ohio, September 1972.

11. "Engineering Properties of IN-100 Alloy," Internat'l Nickel Company, Inc., New York, 1968.

12. "Description and Engineering Characteristics of Eleven High-Temperature Alloys," DMIC Memo. 255, Battelle Memorial Inst., Columbus, Ohio, June 1971.

13. D. J. Evans, "Powder Metallurgy Processing of High Strength Turbine Disk Alloys," ASME publication 76-GT-96, March 1976.

14. E. A. Fisher, "Technical Ceramic Fabrication Processes," ASME publication 75-GT-110, March 1975.

15. F. F. Lange, "Structural Ceramic Materials Under Development," ASME publication 75-GT-107, March 1975.

16. D. B. Herbert, "Beyond Superalloys: Ceramics for High Temperatures," Machine Design, Vol. 46, No. 24, October 3, 1974.

17. J. Thornton, "McDonnell Seeking Bids on F-15 Titanium Forgings," American Metal Market, Metalworking News Edition, Vol. 83, No. 135, July 12, 1976.

18. J. Thornton, "McDonnell Readying an Advance in Titanium Forgings for Its F-15," American Metal Market, Metalworking News Edition, Vol. 83, Oct. 11, 1976.

19. R. Larsen, "Wyman-Gordon Estimates 1976 Capital Budget at $30-Million," Amer. Metal Market, Metalworking News Edition, Vol. 83, April 26, 1976.

20. "Hotter Engines Prompt Metals Advances," Aviation Week & Space Technology, January 26, 1976.

21. J. S. Benjamin, "Mechanical Alloying," Scientific American, Vol. 234, No. 5, May 1976, pp. 40-48.

Superior powder-metallurgy molybdenum die alloys for isothermal forging

L Paul Clare, Russell H Rhodes
GTE Sylvania Inc., Chemical and Metallurgical Division, Towanda, Pennsylvania, USA
Presented at the 9th Plansee Seminar, 23–26 May 1977, Reutte, Austria

Abstract. Mechanical properties of TZM, TZC, and Mo–Hf–C alloys prepared by powder metallurgical method are considered, with particular reference to 'net shape' isothermal forging.

Maturing of the isothermal forging industry has resulted in a desire to forge to 'net shape'. Net shape is interpreted as a finished forging with no machining required. To obtain these forged dimensions and surface finish, particularly with superalloys, either higher pressures or higher temperatures will be required. The temperatures proposed will be up to 1200 °C. It is reported that for an isothermal forging temperature of 955 °C, the forging pressure may be approximately 103 MPa, and the corresponding induced stress in the die might be estimated as nearly three times the forging stress (Kulkarni 1977). It was felt that the creep strength of powder-metallurgy TZM would not stand up to these requirements. Therefore, we started to explore the possibility of making higher strength powder-metallurgy molybdenum alloys. Two alloys which had been previously cast were investigated via the powder metallurgy route: TZC (Schmidt and Ogden 1963; Climax Molybdenum Company 1964) and Mo–Hf–C (Raffo 1970). Physical properties of these two powder-metallurgy alloys are considered in this paper. The compositional limits are shown in table 1.

4·45 cm diameter × 6·35 cm billets were fabricated of the three alloys. To see if there was any effect on physical properties and forgeability, we upset forged the 4·45 cm diameter billets to ~75% work (reduction in height) at three forging temperatures, 1260, 1200, and 1150 °C. All billets forged well, although there did seem to be a little edge cracking with TZC. There was no discernible difference in the tensile properties as a result of the different forging temperatures. Figure 1 is a plot of typical tensile properties taken at room temperature, 1204, and 1316 °C. Figure 2 is a plot of the DPH hardness values taken with a 10 kg load. Values represent samples forged and stress relieved for $\frac{1}{2}$ h at 1150 °C and then annealed 1 h at either 1400, 1450, or 1500 °C.

Samples of the forged and stress-relieved material were taken for creep testing. Test temperature used was 1204 °C, and stress used ranged from 170 to 330 MPa. Table 2 shows the results for the three alloys at the 330 MPa stress.

The data indicate a definite advantage for the TZC alloy over the Mo–Hf–C. We subsequently have made 6·4 cm diameter × 13 cm and 15 cm diameter × 25 cm billets of TZC, and they have been forged successfully, but we do not have any physical property data. However, we are concerned with upscaling the TZC alloy to

Table 1. Composition limits of TZM, TZC, and Mo–Hf–C alloys.

Alloy	Content (wt%)			
	Ti	Zr	Hf	C
TZM	0·40–0·55	0·06–0·12	–	0·01–0·04
TZC	1·0–1·4	0·25–0·35	–	0·07–0·13
Mo–Hf–C	–	–	1·0–1·12	0·06–0·0675

larger diameters because of its high carbon level and the potential effect of the latter on the forgeability. Therefore we are considering the fabrication advantages of the Mo–Hf–C alloy, even though its physical properties are not nearly as advanced as those of the TZC. The 27% decrease in the creep rate of Mo–Hf–C over TZM may be just enough to allow it to withstand the more stringent conditions inflicted by the 'net-shape' isothermal forging. We are continuing to look at properties and alloys, for it is anticipated that the use of isothermal forging for aerospace components and other hot-die pressing applications requiring the strength of refractory metal alloys will increase significantly in the coming years.

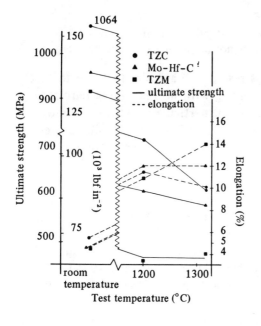

Figure 1. Tensile properties of the three alloys measured at room temperature, 1204 and 1316 °C.

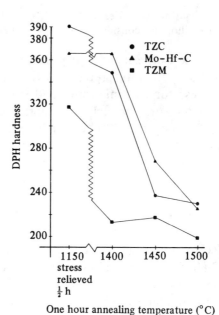

Figure 2. DPH hardness (10 kg load) of the three alloys.

Table 2. Creep-test results at 1200 °C (stress 330 MPa) and 1316 °C (stress 207 MPa).

Material	Time to failure (h)		Elongation (%)		Reduction in area (%)		Creep rate (%)	
	1200 °C	1316 °C	1200 °C	1316 °C	1200 °C	1316 °C	1200 °C	1316 °C
TZM	39	9·4	13	16·5	68	80	0·052	0·18
Mo–Hf–C	44	16·7	3	3·9	1	4·9	0·038	0·07
TZC	89	16·0	7	19·2	19	74·4	0·021	0·14

References

Climax Molybdenum Company, 1964 Climelt News No. 2, July
Kulkarni K M, 1977 "Die materials for isothermal forging" *Proceedings of 1977 Symposium on New Developments in Tool Materials and Applications, Chicago, Ill. 21–22 March*
Raffo P L, 1970 NASA Technical Note, D-5645, February
Schmidt F F, Ogden H R, 1963 DMIC Report 190, 20 September

LUBRICATION IN HOT ISOTHERMAL METALWORKING

Introduction

Hot isothermal metalworking processes are those in which both workpiece and tooling are heated to temperatures in the range 1400-2000°F. This technique is now being used primarily in the forging of titanium alloys and, to a small degree, in the forging of super-alloys and tool steels. Some development work is underway in the isothermal extrusion of these hard-to-deform alloys.

Under isothermal or approximately isothermal metalworking conditions, the heat transfer between die and workpiece is minimized so that slower strain rates can be used. At hot metalworking temperatures, most metals and alloys exhibit strain rate sensitivity of yield stress and, sometimes, of ductility; i.e., as strain rate increases, the yield stress increases and the ductility decreases. Also, some alloys become superplastic under conditions of low strain rate and high temperature. The combined effect of low strain rate and reduced heat transfer in isothermal metalworking improves metal flow and decreases forming forces.

Isothermal forging of titanium alloys has had much press recently, with many short articles appearing in metals and materials magazines. At least three articles[1-3] have described the application of isothermal forging of titanium alloys in the manufacture of U.S. Air Force F-15 torque ribs, F-111 nose wheels, and F-14 airframe stiffeners. Wyman Gordon, Ladish, Alcoa, Pratt and Whitney and other forge shops have all been involved in applying this new process, developed by the Air Force Materials Laboratory.

Furthermore, three articles have described some of the Russian work done on the hot isothermal forging and extrusion of titanium alloys. The foreign and U.S. developments have a high degree of commonality, in that nickel-base superalloy tooling and glass lubrication are used in the process. It appears that these features are necessary to withstand the high temperatures used and the long contact times prevalent at the tool/workpiece interface.

In preparing this article, information from several consultants involved with this process was pooled. Also, the international literature on the subject was reviewed. It is hoped that the information presented here will give our readers a first hand understanding of the role that lubrication plays in hot isothermal forming of titanium alloys, superalloys, and steels.

****

Glass Lubrication in Hot Isothermal Forging of Titanium Alloys

Glasses or glass-like materials appear to be the most satisfactory lubricants for hot isothermal forging of not only titanium alloys but also of superalloys and steels. It should be noted that isothermal forging of titanium alloys in nickel-base superalloy tools (IN-100) is the most prevalent process of the genre.

The role of glass lubrication in hot isothermal forging of titanium alloys is similar to that of any metalworking lubricant in conventional processing: it reduces friction and decreases or eliminates tool/workpiece contact. Typically, the IN100 tooling and titanium alloy workpiece in this process are maintained at approximately $1700^{\circ}F$ during forging as a compromise between adequate tool strength and workpiece formability. Strain rates used range from 0.005 to 0.1 sec^{-1}. At these conditions, a medium to light viscosity glass does an excellent job of lubrication. As shown in Fig. 1, glasses 7052, 0010, 8392, 8871, 9774 and 9773 are all reasonable lubricant candidates for this process, depending on pressure and temperature used. Approximate compositions of some of these glasses are presented in Table 1.

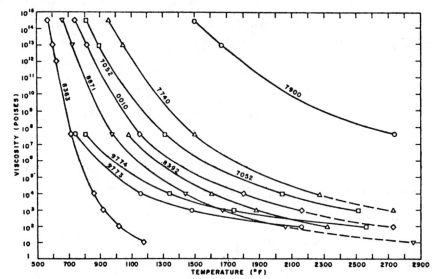

Fig. 1. Viscosity-Temperature Curves of Glasses.
(Corning Glass Works Data)

Reprinted with permission from Metalworking Interfaces, March-April 1976, 21-26, © 1976 Joseph Benedyk

Table 1

Compositions of Some Corning Glasses for
Isothermal Metalworking

Corning No.	Glass Type	Composition (wt%)
7052	Borosilicate	70 SiO_2, 0.5 K_2O, 1.2 PbO 28 B_2O_3, 1.1 Al_2O_3
0010	Potash-Soda-Lead	63 SiO_2, 7.6 Na_2O, 6 K_2O, 0.3 CaO, 3.6 MgO, 21 PbO, 1 Al_2O_3
8392	--	--
8871	Potash-Lead	35 SiO_2, 7.2 K_2O, 58 PbO
9774	Borate	--
9773	Borate	--

Probably the most important factor affecting good lubrication
in isothermal forging of titanium alloys is the viscosity of the
glass at processing temperatures ($1700^{\circ}F$) and pressures (10,000-
40,000 psi). The optimum viscosity of glass lubricants for titanium
alloy forging appears to be within the range 10,000-25,000 poise.
For glasses, typically, the viscosity-temperature curves at atmospheric
pressure (Fig. 1) are adequate to describe conditions during isothermal
forging, as it appears their viscosity is not sensitive to the low
level of pressures acting in the process. It should be noted that
a range of 100-1000 poise viscosity at billet temperature is suggested
for steel extrusion under conventional, i.e., nonisothermal conditions. The interface viscosity of the glass lubricant in steel extrusion rapidly rises due to heat loss.

Ordinary window glass (72 SiO_2, 1 Al_2O_3, 9 CaO, 4 MgO, 13 Na_2O,
1 K_2O) is also suitable as a lubricant for this application and is
the least expensive. About the only glasses that should be avoided

are sulfide containing glasses, because of the potentially harmful effects of sulfur on the tooling and workpiece.

Commercially available glass lubricants for isothermal forging of titanium alloys include the Deltaglaze series of glass lubricants (water or alcohol based suspensions) developed by Acheson Colloids Company, Port Huron, Michigan. Deltaglaze 69 is a water base suspension; Deltaglaze 347M and 349M are alcohol base suspensions. These are applied to the titanium workpieces by either spraying or brushing before the workpieces are heated. During heating, a uniform coating of glass forms on the workpieces. These suspensions contain modifiers to promote wetting of the workpiece.

The coefficient of friction for glass lubricated isothermal forging of titanium alloys, according to the conical piercing method[4], is 0.04-0.06. By comparison, the coefficient of friction for unlubricated forge upsetting of titanium alloys is 0.5.

During forging, some glass does remain in die cavities; however, a steady state condition develops after repeated forgings in which the glass adhesion to the workpiece promotes removal of enough lubricant from die recesses to alleviate any substantial buildup. In fact, the glass sometimes glues the forging to the die, especially in very deep impressions, so that appropriate tongs are necessary for extraction.

Glass Lubrication for Isothermal Forging of Superalloys and Heavily Alloyed Steels

Glass lubrication of superalloys poses few problems, in that these materials do not oxidize heavily at the forging conditions stated above. Somewhat higher pressures are involved, so that higher viscosity glasses (15,000-25,000 poise) might need to be used. The method of application, either spraying or brushing on of a glass powder suspension, is essentially the same.

Glass lubrication of steels is more demanding than for titanium alloys or superalloys. The reason for this is that the steels oxidize and decarburize during heating, and the glass lubricant must provide some means of protection. Inadequately protected steels exhibit pockmarks of accelerated oxidation/decarburization. No current glass lubricant will afford complete protection for steels. Furthermore, grease or oil deposits on initial workpieces aggrevate the problem of glass wetting. Korotkikh, et al.[7] describe a method of simul-

taneously heating and applying glass lubricant to a steel workpiece
by induction heating of the steel within a fluidized bed of 600-
800 micron glass particles (Fig. 2).

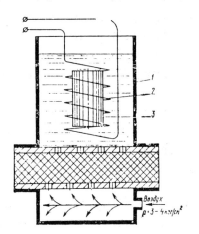

Fig. 2. Schematic Drawing of
Equipment for Applying Glass
Coatings on a Workpiece.

This equipment was successfully used for glass coating a variety of
steels (stainless steels, tool steels, etc.). On the average, the
coating thickness increases from 0.2 to 1.8 mm as heating time increases
from 16 to 36 sec.

Glass Lubrication for Isothermal Extrusion

Isothermal extrusion of titanium alloys has been tried with some
success, but the process is not as widely used as is isothermal forging. Essentially, the same types of glass lubricants used for forging
are also used for extrusion. Because of the simpler billet shape,
some other methods of applying lubricant can be used other than spraying or brushing of a glass powder suspension. Titanium alloy billets
can be rolled in powdered glass (window glass being suitable for many
applications) as they are removed from furnaces. A Sejournet-type
glass pad need not be used if a thick enough glass coating is applied
to a titanium billet. Generally thicker films are required for
isothermal extrusion, since the surface area increases substantially
relative to, say, isothermal forging.

Boyutsov, et al.[5] report that a coefficient of friction of 0.04-
0.06 was achieved in the isothermal extrusion of titanium alloys
at temperatures of 1500-1650°F using a borosilicate glass.

Because of the slower strain rates and the consequentially greater
time of exposure of steel surfaces to oxygen, isothermal extrusion of
steels, even with the best glass lubricants, is not considered viable
by many experts in the field.

****

References

(1) "Air Force Materials Lab Develops New Isothermal Forging Process for Titanium," Light Metals Age, Feb., 1976, p. 15.

(2) D. Yager, "McDonnell Makes Greater Use of Isothermal Titanium Forging," American Metal Market/Metalworking News, Dec. 15, 1975, pp. 5, 36.

(3) "Heated Dies Improve Isothermal Forging," Light Metal Age, Oct., 1973, p. 12.

(4) S. Z. Figlin, et al., "Advantages of Hot Isothermal Forging," Kuzn. Shtampov. Proiz., No. 9, 1972, pp. 6-8 (in Russian).

(5) V. V. Boyutsov, et al., "Character of Metal Flow during Isothermal Extrusion of Titanium Alloys," Kuzn. Shtampov. Proiz., No. 11, 1971, pp. 3-4 (in Russian).

(6) V. V. Boyutsov, et al., "Thermomechanical Condition of Isothermal Deformation of Titanium Alloys," Kuzn. Shtampov. Proiz., No. 12, 1975, pp. 1-3 (in Russian).

(7) E. L. Korotkikh, et al., "Application of Glass Lubricant on Workpieces during Hot Metalworking," Kuzn. Shtampov. Proiz., No. 4, 1975, pp. 16-17 (in Russian).

HIPing the high-performance alloys

Now tool steel, superalloy, and titanium-alloy parts can benefit from HIP

By John H. Moll
Technical Director,
PM and Welding Section
Colt/Crucible Research Center
Pittsburgh, Pa.

While powder metallurgy (PM) technology has existed for many years, it is only in recent times that PM has been applied to produce shapes of tool steels, superalloys, and titanium in an effort to conserve materials that are in short supply. In addition, hot-isostatic-pressing (HIP) of powder-filled shaped containers offers benefits, such as greater uniformity of product, property advantages of finer microstructures, and fewer finishing operations.

For example, superalloy disks of Rene 95 can be made by HIP with room-temperature and elevated-temperature properties comparable to wrought products.

Large PM turbine disk made using metal container.

241

Conventional processing of metals often results in considerable waste of material and energy. As a result, there is a need throughout industry for the direct production of net or near-net parts. This is particularly important in the production of high-speed tool steels, superalloys, and titanium alloys that contain high concentrations of critical elements, such as cobalt, chromium, tantalum and titanium. As an example, the radial compressor rotor for the F-107 gas turbine requires only 5.5 lb (2.5 kg) of material by the HIP process compared to 32 lb (14.5 kg) for a forged billet to yield a part that weighs 3.6 lb (1.6 kg).

How to shape up

Tool steel and superalloy powders are gas atomized. In this process, high-pressure gas is used to atomize a molten stream of air or vacuum-induction-melted metal. The resulting powder falls into a liquid/gas quench at the base of a chamber. The spherical nature of this powder is ideal for producing shaped PM parts. Subsequent processing of the powder involves screening and blending to obtain a controlled size distribution. The processed powder is loaded into shaped containers. Two basic types of containers are currently used, steel and ceramic. Ceramic molds are used for very complex shapes. Production containers are made from metal and may be produced from welded pipe to produce simple cylindrical shapes. Somewhat more complicated shapes may be produced from low-carbon or stainless steel sheet. Once metal containers are fabricated and cleaned, they are loaded with powder, evacuated and sealed.

The final step in the PM shape process is HIP. Sealed containers are placed in an autoclave, heated to temperatures ranging from 1600 to 2200 °F (870 to 1203 °C), depending upon the alloy, and pressed at pressures up to 15,000 psi (103 MPa) gas pressure. The product is fully dense.

Room temperature tensile and toughness properties for HIP Ti-6A1-4V (standard) powder

Condition	Orien-tation	Yield strength (10^3 psi)	Ultimate strength (10^3 psi)	Elonga-tion (%)	Reduc-tion of area (%)	W/A[a] (in-lb/in.^{2})	Predict-ed K_Q[b] (10^3 psi $\sqrt{}$ in.)
As-HIP	LT	135	142	12	40	410	63
HIP + vacuum anneal (1300°F/8hr/ slow cool)	LT	132	139	13	38	565	73
	TL	130	138	12	35	578	74
	ST	132	139	12	36	541	72
AMS spec 4928 H		120	130	10	15	—	—

[a]Tested in compression mode.
[b]Predicted from $K_Q^2/E = 0.57$ (W/A)

Source: V. C. Petersen and V. K. Chandhok, Interim Report AFML - IR - 184 - 7 T(I), June 1979.

Weight comparisons of parts selected for PM titanium

Sub-contractor	Part	Alloy	Part weight (lb)			
			Forging			Final part
			Billet	As-forged	HIP	
Boeing	Walking beam support (747)	Ti-6-4(std)	55	50	30	21
GD/Fort Worth	Horizontal stabilizer pivot shaft (F-16)	Ti-6-4(ELI)	148	120	65	32
GE/AEG	Compressor spool 3-8 (TF-34 Engine)	Ti-6-4(std)	147	104	70	15
MCAIR	Drop-out link (F-15) keel splice former (F-15), near-net	Ti-6-4(std)	115 6	94 4.7	54 0.9	13.8 0.4
Northrop	Arrestor hook support fitting (F-18)	Ti-6-4(ELI)	—	175	60	28.4
P&WA	3rd stage pan disk (F-100 engine)	Ti-6-2-4-6	120	94	70	27
Williams Research	Radial compressor rotor (F-107 gas turbine)	Ti-6-4(std)	32	30	5.5	3.6

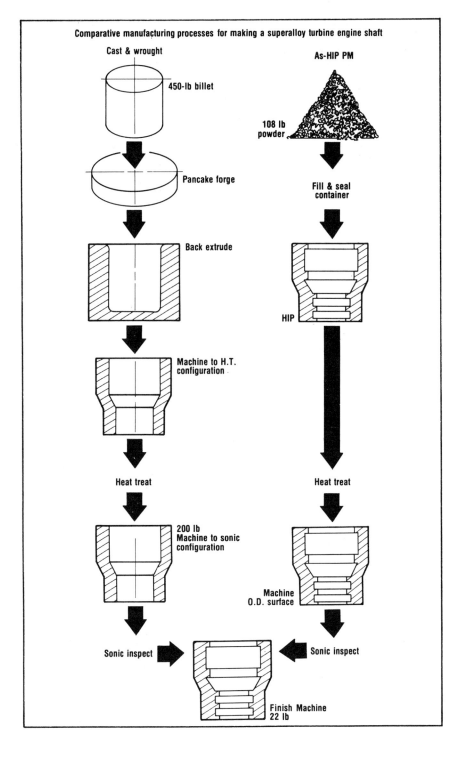

Comparative manufacturing processes for making a superalloy turbine engine shaft

Cast & wrought

450-lb billet

Pancake forge

Back extrude

Machine to H.T. configuration

Heat treat

200 lb
Machine to sonic configuration

Sonic inspect

As-HIP PM

108 lb powder

Fill & seal container

HIP

Heat treat

Machine O.D. surface

Sonic inspect

Finish Machine
22 lb

A drop-out link for the McDonnell-Douglas F-15 fighter is a large, deep-pocketed part typical of many titanium airframe parts. The weight of the HIP part is 54 lb (24.5 kg) compared to a forging weight of 94 lb (42.6 kg).

The process for producing PM titanium alloy shapes is similar to that described for tool steels and superalloys, except that powder is produced by the rotating electrode process. A tungsten-arc or plasma-arc provides localized melting at the end of a titanium alloy bar while the bar is rapidly rotated. The rotating motion of the bar spins off molten particles that subsequently solidify in an enclosed chamber. The resulting powder is spherical and similar to gas-atomized powder.

Tough tool steels

Tool-steel products made from gas-atomized powder offer a number of advantages over their conventionally produced counterparts. These include 1. Superior grindability. 2. Improved toughness. 3. Cross-sectional uniformity of hardness. 4. More uniform size change. Tool steels are highly alloyed materials, often containing substantial quantities of critical elements, such as chromium and cobalt.

In addition to improved performance, the PM process offers the ability to conserve critical materials used in tool steels. Typical applications include cutting tools, punches, shafts, broaches, and tube drawing preforms. Using conventional ingot technology, hollow shapes would be made by metalworking and/or machining.

Tool material can be saved with composite, near-net shapes made by the PM process. For instance, with this technique, highly alloyed material can be used only where required (i.e., at the outside cutting surface).

Shaping up superalloys

Superalloys are widely used in turbine engines. These alloys, normally nickel or cobalt-based, contain substantial quantities of critical materials, such as chromium, cobalt, titanium, and in some instances, tantalum. Less material is required with PM shape technology. Amount of starting material, number of processing steps, and machining time can be saved over a conventional cast and wrought engine part.

PM superalloy hardware is currently being produced from powder to near-net sonic shapes. These range from relatively small turbine disks to large turbine components. An Air Force-sponsored PM program is currently in progress to reduce the input weight on a large, 32-in.-(81-cm) diameter, turbine disk for Pratt & Whitney Aircraft's JT9D engine. The current input weight for this part is 800 lb (363 kg). To date, work under the program has reduced the input weight to 550 lb (249 kg). Further weight reductions of 40 to 50 lb (18 to 22.7 kg) are planned.

The PM process can also be used to produce multi-component parts. Using current technology, engine spools are normally assembled from parts made from several individual forgings. With the PM shape process, the spools can be made as a single piece. The result is a reduction in input weight and the elimination of a number of processing steps. As a typical example, a spool may be comprised of six segments: a shaft at the top, four compressor disks and a bolt flange on the bottom. The

This Williams Research compressor rotor for the F-107 demonstrates that thin, complex blades, attached to the base, can be made by the ceramic mold process. The part is currently machined from a pancake forging weighing 30 lb (13.6 kg). The as-HIPed weight is 5.5 lb (2.5 kg).

Comparison of energy requirements for conventional forgings and PM as-HIP Ti keel splice former

Process	Energy requirements (kwh)	
	Conventional forging	PM shape
Bar product	301.7	72.7
Forging	1.8	—
Atomization	—	.7
HIP	—	52.0
Machining	.5	.1
Total	304.0	125.5

Cut-up properties of as-HIP Rene 95 turbine disks[a]

Properties	Spec.	Mean	σ
R.T. tensile			
Ultimate (10^3 psi)	225.0	239.4	2.8
Yield (10^3 psi)	163.0	176.9	3.5
Elongation (%)	10	17.2	1.9
Red. of Area (%)	12	20.5	1.8
1200°F Tensile			
Ultimate (10^3 psi)	205.0	220.1	5.4
Yield (10^3 psi)	153.0	165.2	4.9
Elongation (%)	8	13.8	4.1
Red. of area (%)	10	16.5	3.3
Stress rupture[b]			
Life (hr)	50	88.8	26.3
Elongation (%)	3.0	6.5	2.1
Creep[c]			
Time (hr)	50	54.6	6.7
Elongation (%)	0.3 max	.10	.08

[a]First 75 engine sets (150 disks).
[b]1200°F/150 (10^3 psi).
[c]1100°F/150 (10^3 psi).

(Workshops on Conservation and Substitution Technology for Critical Materials, Nashville, Tenn., June, 1981.)

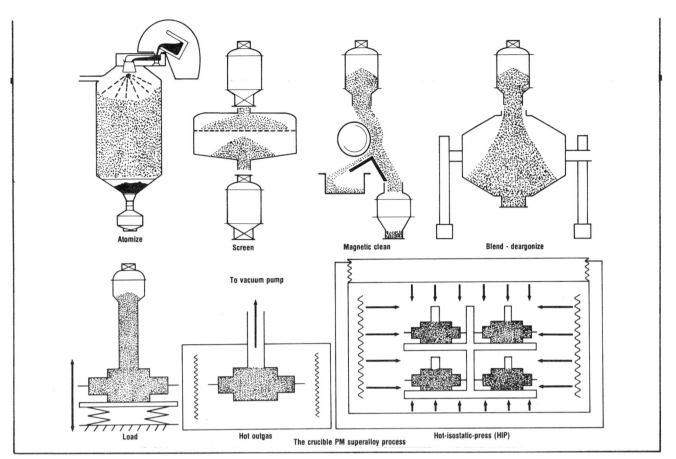

Atomize

Screen

Magnetic clean

Blend - deargonize

To vacuum pump

Load

Hot outgas

The crucible PM superalloy process

Hot-isostatic-press (HIP)

Keel splice formers — (left — forging; center — as-HIPed; right — finished part).

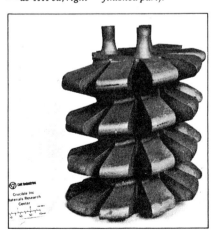

Near-net tool steel hob made by ceramic mold process.

The Northrop arrestor hook support fitting for the F-18 fighter is the largest PM part made to date. The weight of this HIP first-trial part was 60 lb (27.2 kg) compared to a forging weight of 175 lb (79.4 kg).

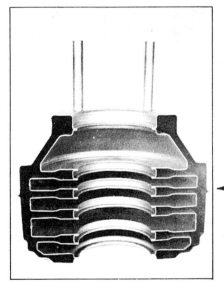

◄ *The HIP weight of a compressor spool for General Electric's TF34 engine, including an extra test ring, was 70 lb (31.7 kg). The forging for this part weighs 104 lb(47.1 kg).*

Source: Materials Engineering, November 1981, 56-61

Tensile properties of HIP-MERL-76 powder[a]

Test temperature (°F)	Yield strength (10^3 psi)	Ultimate strength (10^3 psi)	Elongation (%)	Reduction of area (%)
Test data				
RT	158	226	17	19
1000	152	203	12	19
1150	154	213	12	15
1300	152	187	22	22
Target properties				
RT	154	223	15	15
1300	150	174	12	12

[a]Heat treatment: 2125°F/1 hr/1500°F salt quench + 1400°F/16 hr/air cool

Critical element content of several superalloys used in engine disks

	Astroloy	IN-100	Rene 95	MERL 76	IN-718	Waspaloy	PA-101
Chromium	15	10	14	12	19	19	13
Cobalt	17	15	8	19	—	14	9
Titanium	3	5	2.5	4	1.0	3	4
Tantalum	—	—	—	—	—	—	4

Room temperature tensile and toughness properties of HIP Ti-6A1-4V (ELI) powder

Condition[a]	Yield strength (10^3 psi)	Ultimate strength (10^3 psi)	Elongation (%)	K_Q (10^3 psi $\sqrt{\text{in.}}$)
As-HIP	124	135	15	84
Beta anneal	130	138	10	87
Recrystallize anneal	125	134	15	86
Mill anneal	119	128	17	89
Typical wrought (mill anneal)	115	125	12	—

[a]As-HIP – No heat treatment
Beta anneal – 1300°F/12 hr (vacuum degassed) 1750°F/30 min.,
1865°F/15 min., air blast cool, 1300°F/2 hr/AC
Recrystallize anneal – 1300°F/12 hr (vacuum degassed)
1759°F/2 hr/AC, 1400°F/2 hr/AC
(to below 900°F in 45 min.)
Mill anneal – 1300°F/12 hr (vacuum degassed) 1300°F/2 hr/FC
(300°F/hr max) to 1000°F/AC

Source: G. Chanani, et al, Powder Metallurgy of Titanium Alloys, AIME, 1980.

total input weight of individual components is 490 lb (222 kg). Made as a single component, the total weight is 210 lb (95 kg), a material savings of 280 lb (127 kg) per engine. Because parts produced by the HIP process are fully dense, mechanical properties compare well with forgings. As an example, for Rene 95, cut-up test specimens for many disks show strength and ductility levels and statistical deviations in room and elevated-temperature tensile properties comparable to wrought products. The 1200 °F (648 °C) rupture and creep properties also meet specification requirements.

Titanium takes shape

PM titanium is newer and not as far advanced as tool steels and superalloys. An effort is in progress to develop a ceramic mold process for producing near-net engine and airframe hardware from titanium alloy powder. In addition to materials savings, there is an energy savings using the PM process for near-net titanium alloy parts. Rapid progress is being made in a current Air Force/Crucible manufacturing technology project aimed at producing large titanium airframe parts and complex engine parts. The properties of PM as-HIPed Ti-6Al-4V meet many current specifications. Also, tensile properties meet specification requirements in different test directions. Good fracture toughness is obtained. Properties of as-HIPed Ti-6Al-4V made from extra-low interstitial (ELI) powder compare favorably with typical wrought-mill, annealed material. Excellent fracture toughness values, K_Q, have also been observed.

From another approach, HIP near-net parts have been produced from Ti-6Al-6V-2Sn alloy powder. This process was developed at Crucible Research Center. It is currently being tested for various specific applications under Navy sponsorship at Grumman. One of these involved a fuselage brace for the F-14A aircraft. Work is currently in progress to produce a large, 48-in. (122-cm) nacelle frame using HIP-PM Ti-6Al-6V-2Sn shapes. ∎

Isostatic Hot Compaction – a Review**

H. FISCHMEISTER*

Abstract

This paper reviews the development of Hot Isostatic Pressing (HIP) through the last 20 years. Typical pressure vessel and furnace designs as well as operating procedures are described, and special attention is given to powder treatment, canning techniques, and temperature-pressure conditions. Some alternatives to HIP are discussed briefly, and recent developments of HIP processes for tool steels, superalloys and cemented carbides are reviewed in detail.

Isostatisches Heißpressen – eine Übersicht

Diese Arbeit gibt eine Übersicht über die Entwicklung des heißisostatischen Pressens (HIP) während der letzten 20 Jahre. Typische Druckbehälter- und Ofenkonstruktionen sowie Betriebsbedingungen werden beschrieben. Besondere Beachtung wird der Pulver-vorbereitung, der Kapseltechnologie, sowie den Temperatur-Druck-Bedingungen geschenkt. Einige Alternativen zu HIP werden kurz diskutiert und jüngere Entwicklungen der HIP-Verfahren für Werkzeugstähle, Superlegierungen und Hartmetalle besprochen.

Revue sur la compression isostatique à chaud

Cet article passe en revue le développement du pressage isostatique à chaud (HIP) au cours de ces 20 dernières années. On y décrit des enceintes à pression typiques, des dessins de four, ainsi que des procédés opérationnels. Une attention spéciale est accordée au traitement des poudres, aux techniques d'encapsulage et au choix des températures et des pressions. Quelques techniques concurrentes du pressage isostatique à chaud sont brièvement discutées. Les récents développements de procédés basés sur le HIP pour les aciers à outils, les superalliages et les métaux dures sont décrits en détail.

Introduction

Isostatic hot pressing resulted from the marriage of two old concepts in powder technology. Cold isostatic compaction of powders seems to have been suggested first by Madden [1] as a way of obtaining uniform and defect-free compacts of tungsten. Because of its advantage in uniform density, which was documented systematically by Borok [2], and because of its great shape flexibility, isostatic compaction has gained an important position in powder technology. During the fifties and sixties, it was used mostly for "difficult" materials like beryllium, refractory metals and ceramics [3]; recently it has also been adopted in ferrous powder metallurgy for the production of preforms for hot forging [4,5].

Hot pressing is by far older than isotatic compaction [6,7]; in modern technology it seems to have been advocated first by Gay in 1883 for making metal-bonded diamond tools[8]. The possibility of achieving density levels far beyond those of cold compaction was documented by Sauerwald, Trzebiatowski and other in the thirties [9-12]; the following two decades saw the routine application of the new technique to cemented carbides [13-16], heavy metal components [17], and refractory or hard compounds [18,19]. – Commercial hot pressing was frought with difficulties concerning die life (abrasion, reaction, oxidation, creep deformation) and contamination of the compacts by the die material or the atmospere, as well as frictional restriction of powder flow – although to a lesser degree than in uniaxial cold pressing. Nevertheless, hot pressing of large cemented carbide pieces (e. g., rolls and drawing dies) and of alumium oxide tool bits was carried out on a fairly large scale in the sixties.

The densification mechanism in hot pressing is pressure-enhanced sintering, in which diffusion creep is generally thought to predominate [20-22]; however, the very rapid densification observed under frictionless, isostatic conditions (see below) indicates strong contributions from dislocation creep.

The actual marriage of the two techniques took place in the late fifties at the Battelle Memorial Institute in Columbus, Ohio, where isostatic gas pressure was used to accomplish complicated, often multiple, bonding operations in assembling reactor components in a tight enclosure of thin sheet – e. g. "picture frame" type fuel elements [23-25]. It was observed [25] that precompacted uranium dioxide powder inside the frame compartments was further densified during the bonding operation. Soon the technique was transferred to the consolidation of powders or granules in a thin sheet container by pressurised gas.

The new technique of hot isostatic pressing (HIP) eliminated die friction problems, allowing lower densification temperatures, and did away with the cumbersome reactions between work and die material as well as atmospheric contamination.

An important incentive for HIP developments was the availability of a new fluoride process for tungsten which yielded highly pure metallic tungsten in the form of spherical granules which were too hard and smooth to be cold compacted normally, although pressures up to 10 t/cm² were tried [26]. One tenth of that pressure, applied isostatically at 1590°C, produced > 99 % density. Since the HIP temperature was far below normal sintering temperatures, the extremely fine grained structure of the granules was retained, giving a material with extraordinary ductility and strength [26,27].

Probably for economic reasons, the tungsten HIP-process did not come to commercial fruition, but it had demonstrated the technical potential of HIP. In fact, the early work at Battelle [26-28] and other laboratories [29-32] had already shown that HIP could be applied to ordinary powders of tungsten, niobium, rhenium, uranium dioxide, beryllium and metal-oxide cermets. Particular interest attached to Beryllium: consolidation in an evacuated and fully sealed envelope allowed very pure and isotropic material to be produced, and removed many of the problems connected with the toxicity of the metal.

From the beginning, ceramics and cermets formed another attractive application of the HIP technique [26,28,30,33,34]. Of particular interest are cermets with "idealized" microstructures [35-38] which can be produced by hot isostatic compaction of metal-coated ceramic granules. A highly regular distribution of the ceramic phase results, and the metal phase can be made to form a continuous network down to contents of only 2 volume per cent. In ractor fuel elements, such structures have the advantages of greatly improved heat conductivity, thermal shock resistance, and fission product retention.

Towards the beginning of the seventies, tool steels and superalloys moved into the center of interest for HIP technology. This was largely the result of the perfection, during the sixties, of inert-gas atomization techniques [39,40] required for the low oxygen levels permitted especially in the super-alloys [41], and of the parallel development of commercial large scale HIP apparatus [42-44,58]. Titanium soon joined the group of materials processed from spherical powders by HIP.

Finally, HIP has come into use for closing residual porosity in cemented carbides, and for defect healing in porous castings.

These developments will be reviewed below.

HIP Apparatus

The early work at Battelle was performed with externally heated autoclaves resembling those used in chemical process industry [46]. However, creep of the heated vessel severely limits the temperatures and pressures allowed (e. g., 700°C at 1 kbar[3]). Application of a supporting pressure, by enclosing the autoclave with its heater in an outer vessel maintained at a lower pressure, has been discussed [49] for temperatures up to 1000°C at 1 kbar.

It was realized, however, that it was better to place the heater inside the pressure vessel. This design, termed "cold wall autoclave", was pioneered at Battelle [25,26,45-48] and similar experimental installations have been constructed in other laboratories [49,50,35,36]. The large pressure vessel is expensive, but the cold wall design combines high temperature and high pressure capability with fast loading and unloading, and has therefore become the universal tool for HIP pressing.

The main problem points are the end closures of the vessel and convection currents in the hot gas volume.

The end closures have to withstand a force which is proportional to the square of the vessel diamenter. For large modern installations, with D_i = 1200 mm, P = 1 kbar, this force would amount to 14.400 tons.

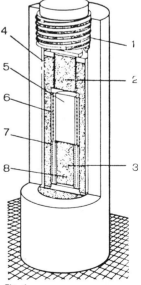

Fig. 1
HIP installation with threaded closure. 1 – resilient thread, 2 and 3 – ceramic plugs, 4 – ceramic lining, 5 – HIP zone, 6 – heater, 7 – current and thermocouple leads, 8 – heater support (Ref. [51]).

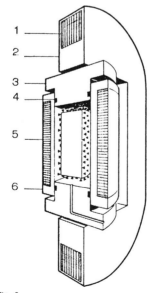

Fig. 2
HIP installation with yoke closure. 1 – wire found frame, 2 – yoke segment, 3 – end closure, 4 – seal, 5 – wire winding on HIP chamber, 6 – steel core of pressure vessel (Ref. [43, 53]).

---

* Chalmers University of Technology, Gothenburg (Sweden) and Montanuniversität Leoben (Austria)
** Invited review at P. G. Sobolevsky Memorial Symposium on Powder Metallurgy, Kiev 1977.

The closure is either held in place by a thread in the pressure cylinder (Fig. 1), or by an external yoke (Fig. 2).

For moderate loads, a breech-type closure with an interrupted thread can be used; this required only one-eight of a turn to be opened. A continuous thread on a conical plug can take a higher load and is still quick-opening.

At very high loads, continuous threads create problems. The load must be evenly distributed over the whole length of the thread to avoid galling. The thread groove in the wall of the pressure vessel constitutes a stress concentrator. In some early designs, low-cycle fatigue cracks originating from the thread groove have caused failures. One elegant way of stress spreading is the resilient thread insert (consisting of a long helical spring wound on a steel core rod) shown in Fig. 1.

In another design [52] the ordinary symmetrical thread profile has been replaced by a buttress or saw-tooth profile which offers a larger contact area to take up the load (reducing galling), and a greater moment of bending resistance. Also the thread is moved away from the pressurized part of the vessel, reducing the contribution of hoop stresses in the critical region near the bottom thread. The usual concentration of stresses at the beginning of the thread can be further reduced by slightly tapering the male thread in the plug, but not the female thread in the wall [51].

A way of avoiding threads altogether is the yoke design shown in Fig. 2. Yoke frames [55,56,42] to contain the thrust of the end closures can be made, e. g., from four bundles of heavy steel strip bolted together at the corner [52], or in the form of pillars with cross-plates at the top and bottom. Fig. 2 shows a particularly compact and safe design, with a wire-wound frame. Such wire-wound press frames were first made on a large scale by the ASEA company of Sweden [53] for diamonds synthesis, and are now standard for their "Quintus" series of heavy metalworking presses. Similar presses are now made also by a second Swedish manufacturer [54]. The wire-winding design exploits the high strength of steel that can be developed in cold drawn wire, without the risk of catastrophic crack propagation which is unavoidable in monolithic parts of high strength steel. Thus the frame for a given press capacity can be designed much lighter and smaller. In Fig. 2, the pressure vessel is wire-wound as well. The prestressed winding keeps the core cylinder under compression even at top operating load, essentially reducing fatigue risks; they are further reduced by the absence of stress concentrations on the entirely straight, unthreaded inner wall of the vessel. If the vessel should nevertheless burst, the wire winding still acts as a safety cage. This is important because in contrast to an incompressible liquid pressure medium, the expansion of a gas will accelerate burst fragments like projectiles. Long and Snowden [50] mention that a thermocouple ejected from a pressure vessel pierced the roof of their installation.

From data presented by Johansson [58], the potential energy of an isostatic pressure vessel containing about 2 m³ of gas at 1 kbar corresponds to that of 60 kg TNT.

Wire-wound HIP installations may be installed in ordinary factory premises, while for other designs the safety regulations of many countries prescribe installation in a pit, sometimes in a separate building with heavy walls and a light roof. – However, mature designs with threaded closures have proved their safety in long periods of continuous production operation.

Seals are made by O-rings with metal-to-metal seating [49] or as modified Bridgeman seals. Yoke-type designs require radially expanding seals. Some design ideas (for cold isostatic pressure vessels) are illustrated in the review by Morgan and Sands,[3] and by Dymov et al [55]. Thermocouple feed-throughs have been described by Bumm, Thümmler and Weimar [35].

Heat convection [46,50] is made dangerous by the high density of the pressurized gas. For example, the density of argon at 1000 bar and 20°C is equal to that of magnesium (1,78); at 1000°C, it is still 0.40. Its heat content is accordingly high. It also has the typical large thermal expansion coefficient of a gas, which transforms all temperature fluctuations into much greater buoyancy differences than could occur in a liquid. – In early designs, the convection generated at the cooled wall and at the cold workpiece made it impossible to attain thermal equilibrium inside the furnace. Even if a stationary state was achieved, temperature differences between furnace and charge could amount to several hundred degrees, with most of the energy input going directly to the cooling water instead of the charge [35,50]. Temperature gradients as high as 80°C/cm were reported in another work [50]. Rapid burnout of furnace heaters, uncontrollable local overheating of the charge and other difficulties occured.

Convection was restricted by vertical [48] or horizontal [50] compartmentalization of the space adjacent to the heater, and by filling the space between heater and wall with tightly packed isulating material in powder or fiber form [26,35,46,47,48,49,50]. Initially, also the space between the heater and the work was filled, e. g. with hollow alumina spheres. In present day production facilities, the work space of the furnace is sealed with ceramic plugs at the top and bottom (cf. Fig. 1,2) preventing hot gas from reaching the cold walls, and the work is often charged in a continuously heated furnace.

Helium, which gives less convection troubles, was used as pressure medium in most of the Battelle work, and experimentally in Europe [35,50]. Its density is only one-tenth of that of argon, and it is know for its excellent heat conductivity.

Bumm et coll. have demonstrated the effect of varying Ar/He mixture ratios on the temperature distribution in the furnace [35]. Cost is firmly in favour of argon, however, and with convection-proof furnace designs, argon is now used universally. The gas is recycled, but some is lost in each cycle (the residual content of the autoclave is discharged at unloading; some gas may be used for purging at the start of a new cycle). Nitrogen and air have been used [57] but since recycling losses are small and argon is readily available, argon is likely to remain the favoured pressure medium.

Heaters are usually made of nichrome or Kanthal wire (< 1250°C), molybdenum (< 1650°C) or graphite [46]. The first two can be exposed to air when recharging a hot furnace. Their temperature capability is sufficient for superalloys, tool steels and titanium. Heating wires can be wound on ceramic tubes, or mounted with ceramic insulators on a sheet cylinder [48]. Also interlocked, self-bearing wire cage constructions are used [53]. In contrast to Fig. 2, most modern HIP chambers have external water cooling. Temperatures on the inner surface of the steel vessel may amount to over 200°C; in the region of the o-ring seals, one tries to stay below 170 to 180°C.

Operation

Powders: Spherical powders, which are usually rather coarse, have been found to give the most uniform and reproducible deformation [3,24]. Typical fill densities in vibrated containers are 62–65 %. Irregular and fine powders must be precompacted to about 70 % density (preferably higher) to avoid warpage during hot densification [24,35]. Precompaction can be performed by tamping successive layers of powder into a cylindrical container[35], by ordinary cold isostatic pressing in a separate mold [24,30,35,60], or by explosive compaction [3,24,27,35,60]. In the latter two cases, the compact must be machined to fit the can. Complicated shapes have been prepared for HIP by plasma spraying of powders onto a mandrel [61].

When spherical powders are poured, their coarse and fine fractions segregate easily. Variations in packing density can result from different mixtures of coarse and fine particles; when such fluctuations occur within one container, nonuniform deformation results. A sophisticated filling system to avoid segregation has been described in connection with the ASEA-STORA tool steel process [60,58,59]. Powder is withdrawn from different points at the bottom of the storage bin into a common feed tube. The container is moved downwards at the rate of filling so that the feed orifice stays at a constant distance above the powder bed. – In the production of superalloy parts for aircraft engines, where tolerances are critical for machining costs, the powder is homogenized by double-cone blending before canning.

Some degree of precompaction may be applied even to spherical powders. In the ASEA-STORA process, the filled containers are isostatically cold pressed at 4000 bar. This raises the density from 65 to 74 % [60]; the flattening of the particle contact areas increases the thermal conductivity of the powder mass from a little over 1 % of that of the solid in the uncompacted state, to 27 % [43,59].

Containers:

Important requirements for container materials are
   weldability
   compatibility with powder and furnace ceramics
   deformability at HIP temperature
   oxidation resistance if hot loading is used
   easy removal

Compatibility is not too difficult to satisfy because normal times and temperatures of HIP give scope for little solid diffusion; superficial contamination is removed by machining. Mainly one has to look out for eutectic melt formation. Neither is deformability much of a problem. If the container is filled with powder of sufficient density, buckling will not occur, although geometrical stiffening effects at corners of sheet containers may sometimes be troublesome.

This makes weldability the main consideration. Mild steel or stainless steel are often used: They are welded by eldectron beam or TIG methods. For high temperature applications, tantalum recommends itself by its good formability and weldability. – Some typical container/powder combinations are listed in table I.

After filling with powder or with a prepressed compact, a lid with an evacuation tube is welded on to the container. Hot outgassing under continuous evacuation may be necessary where very low oxygen contents are required (superalloy powders are outgassed separately prior to canning). It is also suggested for very fine powders to avoid build-up of an internal counter-pressure when adsorbed gases are liberated at HIP temperatures [35]. An oxygen getter, e. g., titanium powder within the container, could be used instead [62]. Thanks to its chromium content, a stainless steel container attracts oxygen during heating, functioning as a self-gettering can for less reactive powders. – An oxide film on the inside of the container will also facilitate removal of the container. Bonding of the container to the powder may be desirable, if the container can serve as a protective sheat in service. Bauer and Weimar [37] state that the thermal expansion of the container material must be greater than that of the powder material, to avoid separation by cooling contraction. – Bonding can be prevented by coating the inside of the container with an oxide slurry.

Material	Temp. °C	Pressure bar	Time h	Container	Ref.	Year	Remarks
Beryllium	913	690	3	mild steel	57	1972	
	700...915	690...1030	1		61	1964	
	760...870	690...1030	1...2		32	1965	
	750...1100	35(!)...826	0,5...3		30	1962	
Titanium	955...1065	1370	—	glass	66	1973	
Tantalum	1370...1600	690...1380	0,5...5		26, 32	1964, 65	
Tungsten	1370...1700	700...1030	1...3		34, 32	1963, 65	
Tool steels	1100	1000		mild steel	60	1970	
Ni-Superalloys (general)	980...1200	690...1380	1...3		32	1965	
Astroloy	1150	1200	—		42	1970	
Astroloy	1246	52 (!)	2		97	1973	
IN100	1287	1700	1	low alloy steel	88	1972	
René 80	1260	35...52 (!)	—		86	1972	
713 LC	1204...1271	1000	—	mild steel	65	1973	
Cemented carbide	1320	1000	—		104	1974	
(WC-6Co, -11 Co)		up to 2315	—		103	1973	
WC, TaC	1595...1760	690...1030	1...3		26, 32	1964, 65	
Al$_2$O$_3$	1150...1370	690...1380	0,5...3		3	1969	
	1410	1030	—		32	1965	Translucent
MgO	800	1000	0,5		32, 107	1965, 71	Transparent
UO$_2$	1150...1480	690	1...3		26, 3, 32, 107	1964,65, 69, 71	
UO$_2$+ Mo or stainless steel	1270	690	3		32, 35	1965, 68	
UO$_2$-W	1595	690	3		26	1964	
UN	1370...1480	3450	3	stainless steel, Nb, Mo	108	1961	
	1480	690...1030			35	1968	
Graphite	1650...2315	690...2070	—		26	1964	
Ferrites (Mn-Zn, Ni-Zn); BaTiO, Pb-(Zr, Ti)O$_3$ etc.	100°C below normal sintering (1150...1300)	50... 200	0,5...22		106	1975	

Table 1   Typical HIP conditions for selected materials

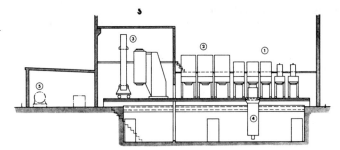

Fig. 3   HIP production line. 1, 2 – first and second stage preheating furnaces; 3 – HIP chamber and yoke frame; 4 – underfloor manipulator for loading and unloading; 5 – compressor (Ref. [43, 53]).

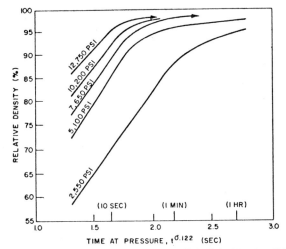

Fig. 4   Densification kinetics of fine nickel powder at 538°C (10 000 psi = 689 bar) Ref. [107].

Before sealing, each container must be carefully leak tested, e. g. by passing a jet of indicator gas (e. g., He) along all weld seams while monitoring the output of the evacuation pump. This is done even in high-volume production. Large leaks will, of course, equalize the pressure inside and outside of the container, preventing densification. If they are partially closed by sintering at HIP temperature, the container may expand on depressurization, damaging the furnace. Small leaks are treacherous because they will sinter completely after admitting high-pressure gas which is entrapped in invisible pores upon densification. Argon contents as low as 0,4 ppm have been shown to have a strong deleterious effect on the toughness of tool steels [63]. High temperature heat treatment or service will expand these pores, impairing especially the creep properties. Such "thermally induced porosity" (TIP) has been one of the main problems in HIP superalloys. (A second source are hollow particles filled with argon during atomization). TIP is now virtually overcome as a technical problem but the countermeasures continue to be an economical one.

An elegant solution of the container problem is practiced by Kelsey-Hayes of USA, using a glass whose softening range coincides with the HIP temperature [66–69]. Vycor glass is a suitable container material for superalloys. Containers of a few millimeters wall thickness are made by casting a slurry of finely ground glass scrap in plaster-of-paris forms. They are dried and fired; their outer surfaces are sealed by flame polishing, and small residual leaks can be repaired by a drop of glass. The containers are evacuated, filled with degassed powder, sealed, and preheated in a separate furnace. After HIP, they are set to cool outside the chamber. Because of the difference in thermal contraction, the container will spall from the metal part during cooling, saving a separate stripping operation. Highly complex shapes can be made at low cost by slip casting, especially if one were to borrow the shell molding technique of metal casting, where a precision molded wax core is coated with slurry, whereupon the shell is emptied by melting out the wax.

HIP Cycles

The effect of pressure on densification is shown in Fig. 4. Initial densification is extremely rapid; then a limit is reached which – for practical purposes – can be raised only by increasing the temperature. Therefore, time is not a very important parameter in HIP – although for a coarse, spherical powder, densification would be slower than for the fine powder in Fig. 4. (cf ref. [30, 61, 64]).

The pressure used in HIP is usually dictated by the capability of the installation, and the temperature is chosen to fit the available pressure rather than vice versa. Pressures around 1000 bar are usual in present-day installations. As indicated by Fig. 4, densification is not very critical to pressure variations at this level, and this is verified by production data [64] and experiments [30,61].

The effect of temperatures has been studied by many investigators. Bumm, Thümmler and Weimar [35] conclude from a collection of published data that the temperature needed to reach full density (> 99 %) in 3 hrs at 700 to 1000 bar is of the order of 0,5 times the melting temperature (T_m) for refractory metals, and 0,7 T_m for ceramics and metals with persistent oxide films and for cermets. Inclusion of other published data (Table I) indicates that these figures should be raised to 0,6 and 0,8 T_m resp. Current industrial practice varies from 1200°C (0,86 T_m) [64, 65] to 980°C = 0,73 T_m for superalloys (where low temperatures are desirable to exclude carbide precipitation during HIP); from 1050 to 1150°C (0,73...0,79 T_m) for tool steels [60], and 955 to 1065°C (0,64...0,69 T_m) for Titanium [66]. These, of course, are temperatures chosen for reliable operation, not minimum values.

Plastic deformation of the powder during precompaction can strongly reduce the temperature required for HIP densification. The lower one of the figures given above for superalloys is for mechanically deformed powder. Spherical tungsten powder normally requires temperatures around 1540°C for full densification around 1000 bar; after explosive precompaction, 1150°C was found sufficient [24]. Obviously, the low temperatures and the high initial densification rate in HIP give considerable scope for non-equilibrium defects to contribute to the densification mechanism; in ordinary sintering, such defects are normally removed by recrystallization before noticeable densification has occurred.

Where HIP is used for volume production, the containers are preheated in separate furnaces and the HIP chamber is kept hot between pressings. This is possible for temperatures up to ca. 1250°C, for which oxidation-resistant heaters are available. Both top and bottom loading is practiced in industrial production. Fig. 4 shows a line of preheating furnaces with an underfloor manipulator which moves the containers from alternate preheating furnaces into the HIP chamber. An advantage in bottom loading is that the hot argon stays captive in the chamber.

Typical cycle times in hot-loading equipment are of the order of three hours [44], limited mostly by the capacity of the pressurization equipment, and by the times for closure operation. Times at maximum temperature are usually of the order of one hour. – In high temperature operations, where cold loading is unavoidable, the thermal inertia of the furnace and of the powder mass inside the container determine the cycle time, which can be of the order of 10 to 20 hours [44,36,25]. References 25 and 36 show complete pressure-temperature-time recordings. Four hours at top temperature were used there. During heat-up, the pressure is increased gradually together with the temperature [25], or pressurization may be delayed until the furnace is hot [36]. The pressure is held for part of the cooling period, or at least decreased more slowly than the temperature.

Related Hot Consolidation Processes

Other pressure media have been tried for isostatic hot pressing, such as liquid lead [29,30] or ceramic powders [70,71]. In the latter case, compaction is not truly isostatic because friction and sintering restrict the lateral flow of the pressure-transferring powder. A very narrow pressure-temperature-time corridor must be followed to avoid premature densification of the transfer powder. The method is attractive since it needs neither pressurised gas nor containers, but so far it has not shown convincing success even in the laboratory.

Semi-isostatic hot compaction can be achieved in may ways. Lowe and Knight [72] describe an analogon to the dry bag method of cold isostatic pressing, in which a pressurized tubular steel membrane compacts beryllium powder against a cylindrical mandrel to form a tube. In 1958, Williams [29] discussed various ways of compacting metal powders in sheet containers by forging with lateral constraint. More recently, hot swaging [73] and forging [62,74] of powder-filled tubes has been used with good succes to consolidate tool steel powders. A compaction force vertical to the action of the forging tools is generated by friction within the powder itself. Densities greater than 99 % can be reached without buckling of the tubes, and a moderate further reduction by rolling produces properties comparable to those of wrought ingot material [62].

Application Trends
Tool Steels

Presently, high speed steels are made by two companies from gasatomized powders by HIP followed by conventional hot rolling [60,75]. The PM route eliminates the carbide network which develops by solidification of the residual melt between the primary crystals in an ingot, giving a virtually homogeneous distribution of very fine carbide particles in the HIP billet – finer than can be achieved by severe reduction of ingots [42,59,60,75–77]. This specific carbide distribution gives the material a fine grain size and, as a result, high toughness [76–80]. Since the carbides are not arranged in stringers, as in ingot-produced stock, the material has very little distortion on hardening. This allows large tools to be machined close to final dimensions, eliminating much of the costly finish grinding. The fine carbide distribution makes powder metallurgy steels sensitive to overheating at austenitization [75,78]; consequently, heat treatment schedules differ somwhat from those for ingot material. – The smaller size of the hard MC carbides gives a considerable improvement in grindability [81,82]. Cutting performance is claimed to be substantially improved in difficult operations where toughness is important [76–78]; in easy operations, such as continuous cutting at constant speed, the superior properties of the powder metallurgy material do not make themselves felt. Although there is still considerable debate about the true extent of the superiority of powder metallurgy high speed steels in cutting [83,84], at least the advantages of better grindability and dimensional stability are universally agreed. Special powder metallurgy grades with increased alloy and carbon content are being marketed both in the United States and in Europe. – The powder metallurgy fabrication of tool steels has recently been reviewed by the present author [4], and in a round table conference at Grenoble [84].

Superalloys

In the superalloy field [41], the development of HIP applications has been at least as promising as in tool steels. Only a few years ago, most experts felt that because of impurities or carbonitride precipitation at the particle boundaries [85,87], some kind of hot deformation would have to follow HIP forming in order to obtain satisfactory mechanical properties [88,90]. Consequently, many of the attempts at powder metallurgy processing of superalloys were based on extrusion [88,91,92]. Among the great achievements of this line of development must be counted the superplastic forming of superalloys [93–95], for which the fine grain size obtainable in extrusion is a prerequisite. The process has become know under Pratt & Whitney's trade name "gatorizing" [93,96].

In 1973, hot isostatically pressed billets of Astroloy, when deformed by forging, were shown to give properties superior to conventionally produced (cast and wrought) material [97]. In the same year, Wallace et coll. [65] showed that by proper heat treatment of undeformed HIP material, tensile strength and ductility (at room temperature) could be made superior to those of cast material. This was for the casting alloy 713 LC, which is unforgeable when produced from ingot; the tensile properties of the HIP material approached those of forged Udimet 700. Also other experiments indicated that isostatic hot pressing could produce properties which, while usually somewhat inferior to those of extruded material, were comparable or sometimes even superior to ingot produced material [64,65,67,88,98,99]. The work of Wallace at coll. [65], Larson [87] and Fox [86] had shown the importance of avoiding precipitation of MC and $M_{23}C_6$ carbides at the particle boundaries during the HIP operation. Today, the problem of carbide precipitation at particle boundaries during HIP seems to have been largely overcome by proprietory processes [100,101].

Reichmann [102] has recently demonstrated that powder metallurgy superalloys in the HIP state may have properties good enough to allow their use in aircraft engines without subsequent forging. Extruded or HIP-and-forged material has finer grain size, which gives it slightly higher strength and ductility and substantially better creep properties at low and intermediate temperatures, while unde-

formed HIP material has better creep properties at high temperatures. However, in undeformed HIP material, the grain size is still finer than in ingot-produced material. At the operating conditions of turbine disks, creep properties are not the decisive design criterium, and undeformed HIP parts may have considerable potential [102].

Several refinements of the HIP process have been proposed to allow consolidation in close-to-final shape. This is important because of the high cost of machining superalloys. Sheet containers may be combined with solid reusable tool inserts to produce accurate dimensions [26,31,102]. An interesting alternative is containerless HIP of vacuum-presintered compacts. Spherical powders can be sintered without previous compaction in re-usable ceramic molds, until a density above 95 % is reached. At that level, the pores are no longer connected to the surface, and the compact can be loaded in a HIP chamber without an envelope [102].

Also for titanium alloys, it appears possible that HIP may not be limited to the manufacture of forging preforms but may permit manufacturing some final configurations directly [66].

Cemented Carbides

Cemented carbides probably account for the largest production volume of HIP treatments to-day [103]. After ordinary liquid-phase sintering, good quality carbides still contain 0,001 to 0,01 volume per cent of pores. HIP at 1 000 bar and at 50–100°C above the solidus of the binder phase (i. e., around 1 300°C) will reduce the porosity to about 10^{-4} volume per cent [104]. Since the pores act as crack initiation points, this reduction of residual porosity drastically lowers the probability of fracture at a given stress [103,105]. Fig. 5 shows that both the mean level and the variance of the distribution of transverse rupture strength (the traditional measure of toughness in cemented carbides) are significantly improved by HIP. This is true if the material is free from other defects such as eta phase, free graphite or regions of faulty structure. – If the HIP temperature is below the solidus, the pores will be closed, but only by the binder phase [104]. The resulting "lakes" of binder phase may again act as stress concentrators (even though they would be less severe than pores).

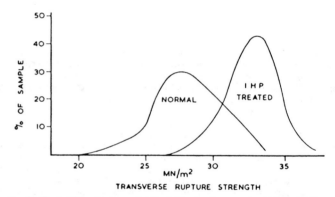

Fig. 5 Effect of HIP on fracture probability distribution of cemented carbide (rock drilling grade, WC-Co). Ref. [105].

The importance of pore removal is illustrated by the fact that a single pore which is revealed during final polishing at the surface of a drawing die or of a Sendzimir roll may cause the rejection of an expensive piece of hard metal plus a much greater amount of finishing costs already invested in the piece. For tools that are reconditioned by grinding and polishing, it is estimated that pores necessitate as much material removal as wear [105]. HIP treatment has no direct effect on hardness or wear resistance, but the improvement in strength – and especially the depression of the low-strength tail of the fracture probability curve in Fig. 5 – allows a harder, and more wear resistant, carbide grade to be used at a given level of toughness requirements. By HIP treatment, hard metal of 0,7 μm carbide grain size with only 3 % cobalt, having a hardness of over 2 000 Vickers units, can be given enough toughness to be used in drawing dies [103].

The principle of pore closure by HIP also has great potential for the healing of casting defects or of cavities formed by creep in high-temperature machine parts. HIP temperatures for such materials are probably low enough to find ways of avoiding irreversible structural degradation during the treatment. Also pores in sintered ferroelectric and magnetic ceramics have been closed by containerless HIP [106].

Conclusion

Hot isostatic pressing is a young technology. It has only recently entered the phase of instrumental maturity, but already it has made possible completely novel processes for tool steels, superalloys, cemented carbides and other materials, with substantial property improvements in all these fields, and with economical savings in some. At the present time, HIP is the most versatile and reliable compaction technique for "difficult" powders.

References

1. H. D. Madden, U.S. Patent 1 081 618 (1913)
2. B. A. Borok, Poroshkovaya Metallurgia Jaroslavl 1956 187
3. W. R. Morgan, R. L. Sands, Metallurgical Reviews 14:134 (1969) 85
4. H. F. Fischmeister, Annual Reviews Materials Sci. 5 (1975) 151
5. G. T. Brown, Communication at 6 th Intern. Powder Metall. Conf. Chicago 1976
6. W. D. Jones, "Fundamental Principles of Powder Metallurgy", E. Arnold, London 1960
7. C. G. Goetzel, "Treatise on Powder Metallurgy", Interscience, New York 1949
8. P. Gay, Brit. Patent 5255 (1883) (quoted by (6))
9. F. Sauerwald, J. Hunczek, Z. Metallkunde 21 (1929) 22
10. W. Trzebiatowski, Z.Physikal. Chemie A 169 (1934) 91
11. W. D. Jones, Metal Ind. 56 (1940) 225
12. P. Schwarzkopf, C. G. Goetzel, Iron Age 148:10 (1941) 37
13. S. L. Hoyt, Trans. AIME 89 (1930) 9
14. L. Molkov, A. V. Chochlova, Redkie Metally 4:1 (1935) 10
15. O. Meyer, W. Eilender, Arch. Eisenhüttenw. 11 (1938) 545
16. G. A. Meerson, V. I. Schabalin, Cvetnyie Metally No. 3 (1941) 77
17. G. J. Comstock, Iron Age 156:9 (1956) 36A
18. G. V. Samsonov, M. S. Kovalchenko, "Hot Pressing", Kiev 1962
19. R. Kieffer, F. Benesovsky, "Hartstoffe", Springer, Wien 1963
20. M. S. Kovalchenko, G. V. Samsonov, Poroshkov. Metall 1:2 (1961) 3
21. R. L. Coble, J. S. Ellis, J. Amer. Ceram. Soc. 46 (1963) 438
22. L. Ramqvist, Powder Metall. 9 (1966) 1
23. S. J. Paprocki, E. S. Hodge, P. J. Gripshover "Gas Pressure Bonding", Defence Metals Information Center Rep. 159 Bettelle Memorial Inst. (1961), also Materials in Design Eng. 55:3 (1962) 14
24. E. S. Hodge, Metals Eng. Quart. 1:4 (1961) 3
25. E. S. Hodge, C. Boyer, F. D. Orcutt, Industrial Eng. Chem. 54 (1962) 31
26. E. S. Hodge, Powder Metall. 7 (1964) 168
27. "Allied Chemical Tungsten" (Allied Chemical Corp., Morristown, N.J., USA, 1965)
28. G. W. Cunningham, D. E. Kizer, S. J. Paprocki, Proc. 4 th Intern. Plansee Seminar Powder Met., Reutte/Austria (1961) 483, also Reactor Materials 6 (1963):1
29. J. Williams, Powder Metallurgy 1/2 (1958) 94
30. J. L. Zambrow, Progr. Powder Metall. 18 (1962) 26
31. A. Blainey, Metal Progress 74:3 (1958) 95
32. R. R. Irving, Iron Age 1963:10 17
33. E. S. Hodge, Mat. Design Eng. 61:5 (1965) 92
34. S. W. Porembka, Ceram. Age 79:11 (1963) 68
35. H. Bumm, F. Thuemmler, P. Weimar, Ber.Deutsch.Keram.Ges. 45 (1968) 406
36. P. Weimar, F. Thuemmler, H. Bumm, Nucl.Mat. 31 (1969) 215
37. F. Bauer, P. Weimar, Zeitschr. Werkstofftechnik 1 (1970) 182
38. C. S. Swamy, P. Weimar, Powder Metall. Intern. 2 (1970) 134
39. U. P. Gummeson, Powder Metall. 15 (1972) 67
40. E. Klar, W. M. Shafer, "Powder Metallurgy for High Performance Applications" (Proc. 18th Sagamore Army Mater. Res. Conf.), Ed. J. J. Burke, V. Weiss, Syracuse Univ. Press. (1972) p. 57
41. G. H. Gessinger, M. J. Bomford, Internat. Metall. Reviews 19 (1974) 51
42. R. Johansson, S. E. Isaksson, Powder Metall. Intern. 2 (1970) 49
43. K. Zander, Powder Metall. Intern. 2 (1970) 129
44. C. W. Smith, Powder Metall. Intern. 5 (1973) 180
45. D. C. Carmichael, P. D. Ownby, E. S. Hodge, US Atomic Energy Comm. Rep. BMI-1746 (1965) (quoted from ref. 3)
46. C. B. Boyer, F. D. Orcutt, R. M. Conway, Chem. Eng. Progr. 62:5 (1966)
47. D. C. Carmichael, "Gas Pressure Bonding Techniques", Chapter 41 in "Techniques of Metals Research", R. F. Bunshah, Ed., Interscience, New York, 1968
48. C. B. Boyer, F. D. Orcutt, J. E. Hatfield, Industr. Heating 37:1 (1970) 50
49. R. L. Sands, E. P. Herbert, W. R. Morgan Powder Metall. 8 (1965) 129
50. W. M. Long, P. Snowden, Powder Metall. 12 (1969) 209
51. Autoclave Engineers, St. Erie, Pennsylvania, USA
52. National Forge Comp., Irvine, Pennsylvania, USA
53. ASEA, Västeras, Sweden
54. Carbox AB, Ystad, Sweden
55. V. V. Dymov, B. P. Lobashev, V. V. Markelov, P. G. Sabinin, "Special Design Features of the Hydrostatic Compaction Installations at the Central Scientific Research Institute of Ferrous Metallurgy", in "Researches in Powder Metallurgy", Ed.B.A. Borok, Metallurgia Press Moscow 1965 (quoted from translation: Consultants Bureau, New York 1966)
56. C. E. Muzzall, "Compendium of Gas Autoclave Engineering Studies" Union Carbide Corp. Rep. Y-1478 (1965) (quoted from ref. 3)
57. H. D. Hanes, "Powder Metallurgy for High Performance Applications (Proc. 18th Sagamore Army Mater. Res. Conf.), Ed. J. J. Burke, V. Weiss, Syracuse Univ. Press (1972), p. 211
58. R. Johansson, Proc. 7th Intern. Plansee Seminar Powder Met. Reutte/Austria (1971), paper 8
59. K. Zander, I. Strömblad, ASEA Tidning 63:4 (1971) 75
60. P. Hellmann, H. Larker, J. P. Pfeffer, I. Strömblad, Mod. Developm. Powder Met. (Proc. 3d Intern. PM Conf., New York) 4 (1970) 573
61. E. S. Hodge, P. J. Gripshover, H. D. Hanes, "Properties of Gas-Pressure Consolidated Beryllium Powder", US Atomic Energy Comm. Rept. CONF-727-2 (1964)
62. L. R. Olsson, V. Lampe, H. Fischmeister, Powder Metall. 17 (1974) 347
63. P. Hellman, E. Söderström, "Argon in Steel". Proc. Conference on Gases in Metals, Swedish Natl. Defense Res. Inst. Rept. C 20059-F9 (1975)
64. P. E. Price, R. Widmer, "Effects of Hot Isostatic Pressing Variables on Properties of Consolidated Superalloy Powders". Paper at 5 th Intern. Powder Metall. Conf., Chicago 1976.
65. W. Wallace, W. Wiebe, E. P. Whelan, R. V. Dainty, T. Terada, Powder Metall. 16 (1973) 416
66. R. H. Witt, O. Paul, M. T. Ziobro, F. D. Barberio, Mod. Developm. Powder Metall. (Proc. 4th Intern. PM Conf. Toronto) 8 (1974) 315
67. C. J. Havel, "Hot Isostatic Pressing with Vitreous Tools", Automotive Engineering Congress Detroit Mich. 1972, paper no. 720183 (Soc. Automotive Engineers, New York)
68. C. J. Havel, "Glass Bag Hot Isostatic Pressing of Superalloys" 3d Intern. Sympos. Electroslag Special Melting Tech., Ed. G. K. Bhat, A. Simkovich, Mellon Institute, Pittsburgh (1972) p. 43.
69. Brit. Pat. 1 190 123 (1970)
70. R. P. Levey, "Isostatic Hot Pressing", Union Carbide Corp. Nuclear Division, Oak Ridge, Tenn., Rep. Y-1487 (1965); quoted from ref. 3 and 71
71. F. F. Lange, G. R. Terwilliger, Ceramic Bull. 52 (1973) 563
72. J. N. Lowe, R. A. Knight, "A Simple Hot Isostat for Beryllium", 7th Intern. Plansee Seminar, Reutte/Austria (1971) paper 9
73. F. G. Wilson, P. W. Jackson, Powder Metall. 16 (1973) 257
74. H. Fischmeister, B. Aren, M. Dahlen, A. Kannappan, L. E. Larsson, L. R. Olsson, G. Sjöberg, B. Sundström, "Hot Consolidation of Metal Powders – Selected Problems and Solutions". Proc. 4th Internat. Conf. Powder Metal. CSSR, High Tatra (1974) Vol. I:293
75. E. J. Dulis, T. A. Neumeyer, "Particle Metallurgy High-Speed Tool Steel". Iron and Steel Inst. Publ. 126, "Materials for Metal Cutting" (1970) p. 112
76. A. Kasak, G. Steven, T. A. Neumeyer, "High-Speed Tool Steels by Particle Metallurgy". Automotive Engineering Congress Detroit, Mich. 1972, paper 720 182 (Soc. Automotive Engrs., New York)
77. P. Hellman, "ASEA-STORA Schnellarbeitsstahl-Eigenschaften und Anwendung" 4th Internat. Powder Metall. Conf. Dresden (1973) paper no. 44
78. P. Hellman, Werkstatt und Betrieb 108 (1975) 277
79. P. Hellman, Jernkontorets Annaler 156 (1972) 84
80. T. A. Neumeyer, A. Kasak, Metall. Trans. 3 (1972) 2281
81. P. Hellman, H. Wisell, "Effects of Structure on Toughness and Grindability of High Speed Steels", Internat. Coll. High Speed Steels, St. Etienne (1975)
82. E. J. Dulis, T. A. Neumeyer, Progress Powder Met. 28 (1972) 129
83. L. Rademacher, H. W. Müller-Stock, Werkstatt und Betrieb 107 (1974) 607
84. G. Cizeron, Materiaux et techniques, special issue 1975 "Metallurgie des poudres" (proc. 4th Europ. Symp. Powder Metallurgy Grenoble 1974) p. 182
85. K. H. Moyer, Mod. Dev. Powder Met. (Proc. 1970 Intern. PM Conf., New York) 5 (1970) 85
86. H. M. Fox, Mod. Dev. Powder Met. (Proc. Intern. PM Conf., Toronto) 8 (1974) 491
87. J. M. Larson, Mod. Dev. Powder Met. (Proc. Intern. PM Conf. Toronto 1973) 8 (1974) 537
88. L. N. Moskovitz, R. M. Pelloux, N. J. Grant, Proc., 2nd Intern. Conf. Superalloy Processing, MCIC Report (1972), paper Z
89. M. M. Allen, R. L. Athey, J. B. Moore, Met. Eng. Quart. 10:1 (1970)
90. B. Triffleman, F. C. Wagner, K. K. Irani, Mod. Dev. Powder Met. (Proc. 1970 Intern. PM Conf. New York) 5 (1971) 37
91. G. Friedman, E. Kosinski, Progr. Powder Met. 25 (1969) 3 also Met. Eng. Quart. 11:1 (1971) 48
92. B. Ewing, F. Rizzo, C. zur Lippe, in "Proceedings 2nd Intern. Conf. Superalloy Processing" MCIC Rep. (1972)
93. R. L. Athey, J. B. Moore, "Development of IN-100 Powder Metallurgy Disks for Advanced Jet Engine Applications" in Powder Metallurgy for High Performance Applications", Ed. J. J. Burke, V. Weiss, Syracuse Univ. Press (1972) p. 281
94. S. H. Reichman, J. W. Smythe, "Superplasticity in Powder Metallurgy IN-100". Proc. 1969 Fall Powder Met. Conf. (Am. Powder Met. Inst., 1970), p. 27; also Int. J. Powder Met. 6:1 (1970) 65
95. S. H. Reichman, J. W. Smythe, "High Density P/M Components for Elevated Temperature Applications", Proc. 7th Plansee Seminar Powder Met., Reutte/Austria (1971), paper 35
96. US Patent 3 519 503
97. D. J. Evans, D. N. Duhl, R. B. Slack, Mod. Dev. Powder Met. (Proc. 1973 Intern. PM Conf., Toronto) 8 (1974) 473
98. G. Raisson, Y. Honnorat, J. Morlet, "Perspectives d'application des poudres préalliées aux materiaux super-réfractaires des moteurs aeronautiques", Preprints of the 4th Europ. Symp. Powder Met., Grenoble (1975), paper 4-3
99. A. Walder, M. Marty, A. Hivert, "Elaboration par métallurgie des poudres de superalliages réfractaires", Preprints of the 4th Europ. Symp. Powder Met., Grenoble (1975) paper 4-4
100. Kelsey-Hayes Powder Consolidation Division, Birmingham, Mich., USA
101. Henry Wiggin & Co. Ltd., Hereford, England
102. S. H. Reichman, Int. J. Powder Met. 11:4 (1975) 277
103. E. Lardner, D. J. Bettle, Metals and Materials (1973) 540
104. S. Amberg, E. A. Nylander, B. Uhrenius, Powder Met. Intern. 6 (1974) 178
105. E. Lardner, Powder Met. 18 (1975) 47
106. K. H. Härdtl, Amer. Ceram. Soc. Bull. 54 (1975) 201
107. H. W. Blakeslee, "Powder Metallurgy in Aerospace Research", NASA SP-5098, Washington 1971
108. D. L. Keller, J. M. Fackelmann, E. O. Speidel, S. J. Paprocki Proc. 4th Intern. Plansee Seminar, Reutte/Austria (1961) p. 304

J. L. Bartos[1]

Review of Superalloy Powder Metallurgy Processing for Aircraft Gas Turbine Applications

REFERENCE: Bartos, J. L., **"Review of Superalloy Powder Metallurgy Processing for Aircraft Gas Turbine Applications,"** *MiCon 78: Optimization of Processing, Properties, and Service Performance Through Microstructural Control, ASTM STP 672,* Halle Abrams, G. N. Maniar, D. A. Nail, and H. D. Solomon, Eds., American Society for Testing and Materials, 1979, pp. 564–577.

ABSTRACT: Substantial increases in raw material and labor costs have provided the impetus for many diverse investigations aimed at reducing the manufacturing costs of aircraft gas turbine hardware. One of the most significant programs conducted at General Electric during the past decade involves application of powder metallurgy (P/M) processing technology to the fabrication of nickel-base superalloy rotating components. The evolution of P/M technology and its application to a high-strength superalloy, René 95, in a number of diverse aircraft engine components is described. Initial development concentrated on fabrication of P/M hardware by hot isostatic pressing (HIP) cylindrical powder preforms and forging in a conventional die system to an oversize shape. Subsequent refinement of this *HIP plus forge* process included HIP for shaped preforms followed by forging in an isothermal die system to a near net shape.

Emergence of a second P/M processing technology, designated *as-HIP,* in which the powder is simply HIP and heat treated, thus eliminating all secondary metalworking operations, is also reviewed. The technical advantages and constraints, along with the projected economic benefits of this new, highly promising P/M process, are analyzed and examples of its application to engine hardware presented. The future of P/M processing in the aircraft engine business, particularly as related to development of advanced superalloy materials, is also discussed.

KEY WORDS: steels, microstructure, nickel-base superalloy, powder metallurgy processing, hot isostatic pressing, cost reduction, isothermal forging

General Electric has had many years experience in the definition and fulfillment of goals for gas turbine materials. Achievement of these goals has occurred by either utilizing materials available in the marketplace or developing new alloys when the vendor offerings were not adequate. The demand for improved engine efficiency and thrust/weight ratio has

---

[1] Manager, Alloy Development, Aircraft Engine Group, General Electric Company, Cincinnati, Ohio 45215.

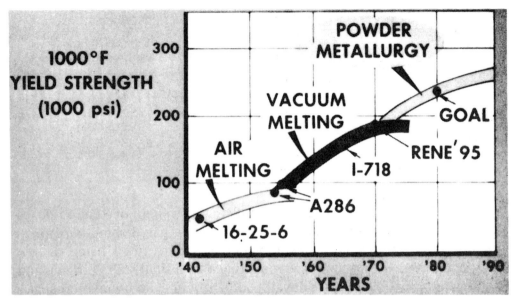

FIG. 1—*Progress in disk materials.*

resulted in a continuing development of higher strength alloys. Included in this group of alloys are aircraft engine rotor materials that have evolved from the high chromium steels and Timken alloys, through A286 and V57, René 41, Waspaloy, and up to the currently utilized Inconel 718, Astroloy, and René 95. A key to the development of higher strength alloys has been improvement in alloy processing technology. Figure 1 graphically depicts the strength benefits that accompanied new processing developments. The first gas turbine alloys were air melted, which placed a restriction on hardener content and thus on the strength level that could be attained. The advent of vacuum melting permitted significant increases in hardener content (principally gamma prime forming elements), leading to substantial improvements in mechanical properties. Table 1 illustrates the increase in hardener content and strength of turbine alloys from the early air melted compositions (A286) to the current high-strength vacuum melted superalloys.

René 95 Development

Of the wrought nickel-base superalloys used for aircraft turbine and compressor disks, shafts, rotating seals, and related parts, René 95 is by far the strongest of the commercially available alloys at temperatures up to 650°C (1200°F) as illustrated in Fig. 2. The alloy was developed in the late 1960's by General Electric under United States Air Force sponsorship [1].[2] To date, approximately 363 000 kg (800 000 lb) of double vacuum melted René 95 have been used by General Electric for application in various development and demonstrator engines.

[2] The italic numbers in brackets refer to the list of references appended to this paper.

TABLE 1—*Hardener content and strength of turbine alloys.*

Alloy	Hardener Content (Al+Ti+Cb), %	650°C (1200°F) 0.2% Yield Strength, MPa (ksi)
A286	2.4	689.4 (100)
V57	3.2	799.7 (116)
René 41	4.6	806.6 (117)
Inconel 718	6.7	930.7 (135)
Astroloy	7.5	930.7 (135)
René 95	9.5	1206.5 (175)

As would be expected, the exceptionally high strength of René 95 poses challenges to conventional methods of producing gas turbine hardware. The original René 95 alloy development was performed using double vacuum melting with vacuum arc remelting as the final step. Forging and heat treating procedures were developed to produce a unique microstructure, often referred to as the *necklace* structure. This structure consists of large (ASTM 3–7) warm worked grains surrounded by a necklace of fine (ASTM 8–12) recrystallized grains. Attainment of this microstructure is required in cast plus wrought René 95 to achieve the desired balance of strength and ductility in the alloy.

As a consequence of the extremely high strength of René 95, components made by conventional process technology are more expensive than similar parts produced from the weaker Inconel 718 alloy. The higher cost of conventional cast plus wrought René 95 components is a direct reflection of the large quantity of input material and the complex forging procedure required to fabricate engine parts. Figure 3, for instance,

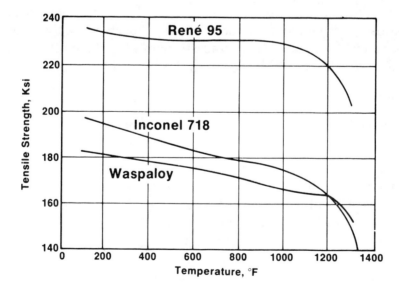

FIG. 2—*Tensile strength: René 95 versus other disk alloys.*

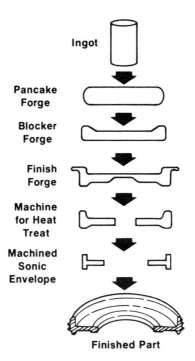

Ingot

Pancake
Forge

Blocker
Forge

Finish
Forge

Machine
for Heat
Treat

Machined
Sonic
Envelope

Finished Part

FIG. 3—*Typical processing sequence for cast and wrought René 95 disk; input/output weight ratio = 19:1.*

illustrates a typical example of the multiple forging steps, machining operations, and the rectilinear ultrasonic inspection envelope required to produce a compressor disk. Note that the finish part weight amounts to only about 5 percent of the input weight.

Powder Metallurgy Processing

The high cost of processing today's high-strength alloys such as René 95, coupled with the spiraling cost of raw materials, have shifted the emphasis trends in materials technology at General Electric. In the past 15 years, emphasis has transitioned from technologies to maximize high performance at minimun engine weight to concentrated efforts on cost improvement and increased reliability. This trend, illustrated in Fig. 4, is exemplified by developments since 1970 in René 95 technology.

The inherent forging difficulties encountered with conventional cast plus wrought René 95, along with the poor raw material conversion ratios (ratio of input to finished part weights) provided the impetus for initiation of cost reduction programs. The most significant breakthrough in reducing the costs of René 95 rotating parts was the development and application of advanced powder metallurgy (P/M) technology. The basic thrust of the P/M approach was aimed at decreasing the number of working operations and improving material utilization to ultimately reduce finish part costs.

Source: MiCon 78: Optimization of Processing, Properties and Service Performance Through Microstructural Control, ASTM STP672, 1978, 564-577

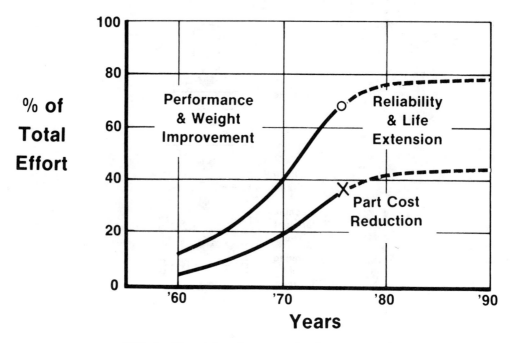

FIG. 4—*Material and process development trends.*

The René 95 composition shown in Table 2 is uniquely suited to P/M processing. A number of other commercially available superalloys have been processed utilizing P/M technology. However, in many cases an undesirable product resulted due to formation of a brittle carbide network on the surfaces of powder particles. This problem is avoided in René 95 [2] through the balance between carbon content, M_6C carbide formers (tungsten and molybdenum), and MC carbide formers (columbium and titanium). Suppression of this carbide network formation is considered a prerequisite for a viable P/M alloy.

The development of industrial capabilities to produce high purity superalloy powders by inert gas atomization and consolidate them to full

TABLE 2—*Chemical composition of P/M René 95.*

Element	Percent	Element	Percent
Carbon	0.04–0.09	Columbium	3.30–3.70
Manganese	0.15 max	Zirconium	0.03–0.07
Silicon	0.20 max	Titanium	2.30–2.70
Sulfur	0.015 max	Aluminum	3.30–3.70
Phosphorus	0.015 max	Boron	0.006–0.015
Chromium	12.00–14.00	Tungsten	3.30–3.70
Cobalt	7.00–9.00	Oxygen	0.015 max
Molybdenum	3.30–3.70	Nitrogen	0.005 max
Iron	0.50 max	Hydrogen	0.001 max
Tantalum	0.20 max	Nickel	remainder

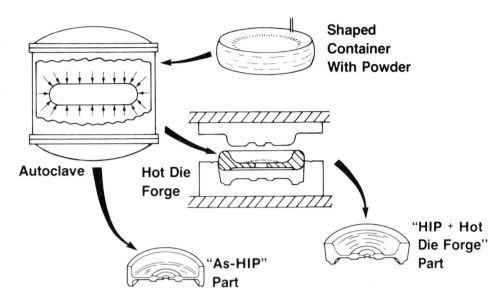

FIG. 5—*Powder metallurgy René 95 processing.*

density by hot isostatic pressing (HIP) provided the basis for initiation of P/M cost reduction programs at General Electric in 1971. Early efforts consisted of consolidating P/M preforms by HIP and applying conventional or *hot die* forging techniques to produce a HIP plus forge P/M product. More recent efforts have concentrated on development of the as-HIP P/M process, which completely eliminates all metalworking operations after HIP consolidation. A schematic description of the two processes is presented in Fig. 5.

HIP plus Forge P/M René 95

The first P/M René 95 components produced by General Electric were fabricated by forging a fully dense HIP preform on standard *cold* dies using techniques developed for conventional cast plus wrought parts [3]. The improved microstructural homogeneity of the P/M preforms relative to conventional ingot product, as shown in Fig. 6, permitted a substantial reduction in the number of forging steps required to produce the desired forge shape. This advantage, combined with the configurational flexibility offered by P/M preforms, resulted in a higher strength, more uniform forging that could be produced at somewhat lower cost than the cast plus wrought product. The success of the HIP plus cold die forge process, exemplified by the turbine disks shown in Fig. 7, led to accumulation of considerable engine test time and the first production engine commitments for P/M hardware at General Electric.

The outstanding accomplishments of the P/M René 95 HIP plus cold die forge development programs, coupled with rapidly rising raw material costs, led to refinement and further cost reduction efforts. The develop-

Source: MiCon 78: Optimization of Processing, Properties and Service Performance Through Microstructural Control, ASTM STP672, 1978, 564-577

257

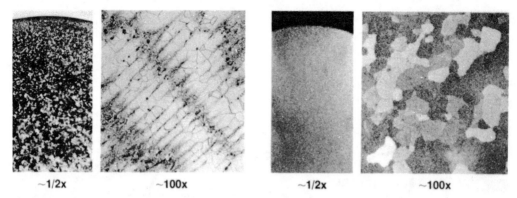

~1/2x   ~100x   ~1/2x   ~100x

FIG. 6—*René 95 homogeneity; fusion metallurgy* (left) *versus powder metallurgy* (right).

ment of hot die forging capabilities at Wyman Gordon Company and Ladish Company [4] permitted further improvements in material utilization ratios by minimizing the surface chilling effect inherent in the conventional cold die system. This innovation allowed a reduction in the protective material envelope around the forging, thus reducing input material requirements significantly. In addition, advances in shaped P/M preform technology eliminated large quantities of unnecessary input material. These improvements are graphically illustrated in Fig. 8 on a typical engine disk component produced by cold and hot die techniques. A typical HIP plus hot die forged component, shown in Fig. 9, depicts the configurational detailing and minimization of forging flash material waste possible with the process. Studies to further reduce preform input weights and forging envelope thickness are continuing in an effort to extract the maximum cost reduction from the HIP plus hot die forge process

FIG. 7—*HIP and cold die forged P/M René 95 disk components.*

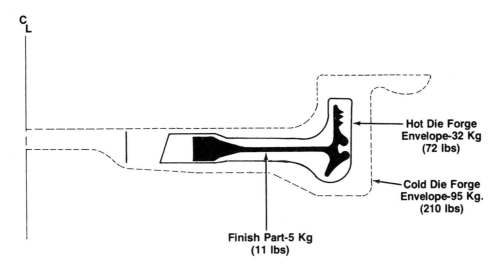

FIG. 8—*Forging envelopes for typical disk component.*

As-HIP P/M René 95

Development of the P/M forging preform technology, including the ability to fabricate components to 100 percent density, paved the way for further cost reductions from P/M processing. Initial studies of P/M preforms indicated that mechanical properties competitive with those achieved in forged products could be attained in HIP plus heat treated (as-HIP) René 95 components. Elimination of the forging operation in the P/M René 95 manufacturing process was projected to provide a substantial cost reduction relative to the HIP plus forge process. As an example, the cost reduction potential of a selected group of René 95 parts [5] is

FIG. 9—*Hot die forged P/M René 95 disk component.*

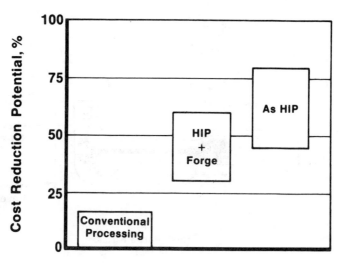

FIG. 10—*Cost reduction potential for a selected group of René 95 parts.*

shown in Fig. 10. Additional cost reductions of up to 50 percent are forecast through substitution of as-HIP processing for HIP plus forge techniques.

This discovery led to initiation of a development program at General Electric in 1973 to define processing parameters required to produce the desired mechanical property levels in as-HIP material. Basic HIP and heat treat parameters were developed [6] and the first as-HIP René 95 hardware, as shown in Fig. 11, was produced in 1975. Following extensive mechanical property evaluation, component testing, and actual engine testing, the process was accepted for production of rotating engine hardware in 1977.

One of the unique features of the as-HIP process is the mechanical property flexibility relative to conventional forged or rolled René 95 products. Strength levels in wrought products produced from either cast ingot or powder preforms are dictated primarily by the thickness of the component, that is, strength decreases as section thickness increases. Very little flexibility is possible, since it is difficult to control the amount of warm work retained in complex shaped forgings. On the other hand, properties in as-HIP products are determined primarily by the heat treatment parameters, with cooling rate from the second phase (γ') solvus being most influential. By altering this cooling rate in different section thicknesses using various quench media, a wide variety of strength levels can be achieved. This is illustrated in Fig. 12, which depicts experimentally determined cooling rates attained in various section size parts as a function of quench media. Resultant properties are dictated by the cooling rates. Thus, strength levels can be specified, within limits, in as-HIP components merely by selecting the proper quench medium.

FIG. 11—*As-HIP René 95 disk components.*

Further development of the shape-making technology required to fabricate complex, near net shape engine parts is continuing in an effort to provide additional cost reduction potential to the as-HIP process. The most striking examples of the raw materials and cost reduction capability of the as-HIP process are associated with the manufacture of complex engine shaft components. A typical shaft processing sequence comparing the conventional cast plus wrought and as-HIP technologies is presented in Fig. 13. The configurational flexibility of P/M processing permits an input material savings of approximately 156 kg (approximately 350 lb). The savings, combined with the elimination of many fabrication steps, is projected to produce an 80 percent cost reduction over conventional fabrication methods. An excellent example of the near net shape capability to produce complex shaft components is shown in Fig. 14. These full-scale parts were produced by Crucible Materials Research Center for General Electric. Mechanical properties of shafts processed by both

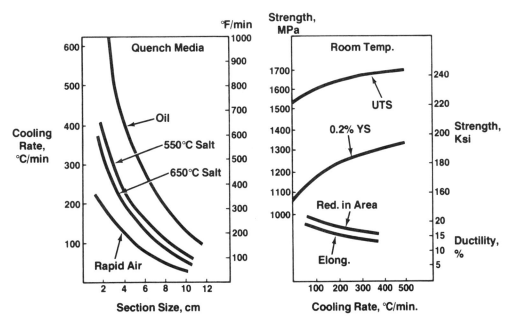

FIG. 12—*Heat treatment flexibility of as-HIP process.*

Source: MiCon 78: Optimization of Processing, Properties and Service Performance Through Microstructural Control, ASTM STP672, 1978, 564-577

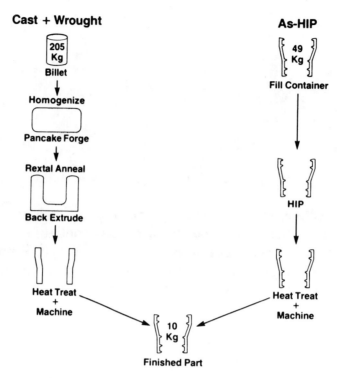

FIG. 13—*Fabrication processes for typical shaft component.*

as-HIP and conventional cast plus forging methods, presented in Fig. 15, indicate the fully comparable or superior strength levels obtained in the as-HIP product.

Powder Metallurgy Concerns

The spectacular progress achieved in P/M technology during the past ten years has also uncovered a number of potential problem areas that could have a detrimental effect on future applications in the aircraft gas turbine industry. A prime concern in all P/M products, especially when highly stressed components are involved, is the size and frequency of defects present due to powder production and handling techniques. In inert gas atomized powders, these defects are characterized into two broad categories: porosity and foreign particles.

Porosity is formed in P/M products as a result of entrapped argon gas. Argon may be entrapped between particles or inside hollow particles. Since the gas is inert, subsequent heating causes bubbles to form, producing pores up to 50 μm (0.002 in.) in diameter. Foreign particles included in the compacted P/M product are primarily nonmetallic oxides originating from either the atomization process or subsequent handling. These oxides, identified using several inspection techniques, are generally very small (less than 250 μm (0.01 in.) diameter). The potential degrading

FIG. 14—*Near net shape as-HIP René 95 shaft component.*

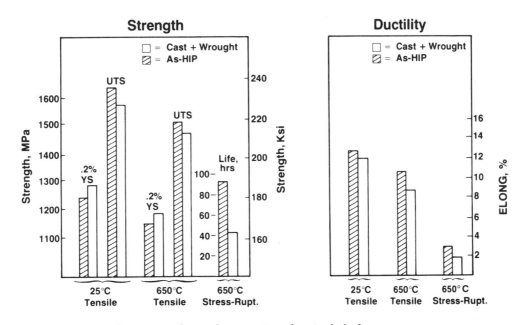

FIG. 15—*Mechanical properties of typical shaft component.*

Source: MiCon 78: Optimization of Processing, Properties and Service Performance Through Microstructural Control, ASTM STP672, 1978, 564-577

effect of these defects on mechanical properties of critical rotating hardware is a source of continuing concern in the aircraft engine industry.

Achievement of substantial raw materials and cost savings in P/M parts is dependent to a large extent on the production of near net shape components. Information on the reliability, reproducibility, and product yields associated with processes capable of making complex shapes is just now being accumulated. Results in this area must be positive to justify continued investment in P/M technology.

Powder products present new quality control challenges that must be successfully met and solved. Inspection of complex, contoured near net shape parts must be completed economically, reliably, and at higher sensitivity levels than heretofore attempted in production environments. Advances in ultrasonic inspection and surface inspection techniques are required to assure even higher quality levels in P/M components than currently attained in conventional cast plus wrought parts.

Substantial efforts in all the aforementioned problem areas are underway, with significant progress already accomplished. Through the combined efforts of powder producers and users, these and future undefined technical problems must be solved before high-strength P/M components can be applied with the same confidence associated with conventional cast plus wrought products.

Future Applications

The significant technical difficulties associated with processing highly alloyed nickel-base superalloy compositions by conventional cast plus wrought techniques essentially dictates the application of P/M technology to future advanced aircraft engine alloys. This fact is exemplified by the development of AF115 [7], an advanced P/M disk alloy designed to operate at temperatures up to 760°C (1400°F).

AF115 was developed by General Electric under Air Force sponsorship to take advantage of the compositional freedoms afforded by P/M processing. The alloy chemistry was designed to avoid the carbide segregation problem inherent in other superalloys such as IN100. Although HIP and heat treatment parameters do not resemble those specified for as-HIP René 95, AF115 will benefit from the powder processing and near net shape advances of the past ten years.

Beyond alloys such as AF115 lie a myriad of new developments fostered by the compositional and configurational flexibility of P/M processing. The achievement of unique mechanical property levels in specified component locations appears feasible through the use of two or more powder compositions or the application of different processing techniques in each area. Blending of powders or mixing with selected forms of other alloys to make *composite* P/M parts provides an opportunity to develop unique mechanical properties beyond the capability of homogeneous materials.

Concluding Remarks

The past ten years have witnessed the evolution of powder metallurgy processing from a technology suitable for making small, simple, low-strength, sintered parts to a point where large, complex, ultra-high-strength critical rotating aircraft engine hardware can be produced with mechanical properties equivalent to those of conventional forged components. Emergence of the HIP plus hot die forging process, followed by development of the as-HIP process in the last four years have presented new opportunities for substantial raw material and component cost reductions. The future of P/M technology in the gas turbine industry is limited only by currently unresolved concerns over quality and cost. Based on the progress made to date, there is every indication that the industries involved will overcome all remaining technical challenges and bring superalloy P/M processing to full production status.

References

[1] Wukusick, C. S. and Smashey, R. W., "Ultra High Strength Superalloys," Technical Report AFML-TR-68-214, Air Force Materials Laboratory, Dayton, Ohio, 1968.
[2] Allen, R. E., Bartos, J. L., and Aldred, P., U.S. Patent Number 3,890,816, 24 June 1975.
[3] Bartos, J. L., Allen, R. E., Moll, J. H., Thompson, V. R., and Morris, C. A., "Development of Hot Isostatically Pressed and Forged Powder Metallurgy René 95 for Turbine Disk Application," SAE paper 740862 presented at National Aerospace Engineering and Manufacturing Meeting, San Diego, Calif., Oct. 1974.
[4] Deridder, A. J., Hayes, A. F., Koch, R. W., and Radovich, J. F., "Deformation Processing of Superalloy Hot Isostatic Press Preforms," paper presented at 106th AIME Annual Meeting, American Institute of Mining, Metallurgical, and Petroleum Engineers, Atlanta, Ga., March 1977.
[5] Arnold, D. B., "René 95 Powder Metallurgy Opportunities for Gas Turbine Applications," AGARD Conference Proceedings No. 200 on Advanced Fabrication Techniques in Powder Metallurgy and Their Economic Implications, Hanford House, London, 1976.
[6] Bartos, J. L. and Mathur, P. S., "Development of Hot Isostatically Pressed (As-HIP) Powder Metallurgy René 95 Turbine Hardware," *Superalloys: Metallurgy and Manufacture, Proceedings,* Third International Symposium, B. H. Kear, D. R. Muzyka, J. K. Tien, and S. T. Wlodek, Eds., Claitor, Baton Rouge, 1976, pp. 495–509.
[7] Bartos, J. L., "Development of a Very High Strength Disk Alloy for 1400°F Service," Technical Report AFML-TR-74-187, Air Force Materials Laboratory, Dayton, Ohio, 1974.

EXPERIENCE WITH NET-SHAPE PROCESSES FOR
TITANIUM ALLOYS

R. H. Witt
Grumman Aerospace Corporation
Bethpage, N. Y.
and
W. T. Highberger
Naval Air Systems Command,
Washington, D. C.

Abstract

The feasibility of producing complex, deep-pocketed titanium alloy
shapes typical of airframe structural components by hot isostatic
pressing (HIP) prealloyed powders to near-net-shapes in one operation was demonstrated. Results of studies on powders, net-shape-
making capabilities, parameter control, NDI acceptance, flight
qualification and cost evaluations are presented. As a result of
this program, an HIP'd F-14A fuselage brace has been installed for
flight.

1. INTRODUCTION

Since the early 70's several efforts have
been directed toward producing powder
metal (P/M) titanium airframe and engine
parts which are either forged to shape or
compacted to near net shapes. Most efforts have been concerned with reduction
or minimization of the acquisition cost of
titanium alloy parts such as those produced from machined forgings or plate.
Titanium parts have been intrinsically
costly because of processing losses and
scrap generation in titanium production
and fabrication (see Table 1). It has
been estimated[1] that scrap generation
in primary fabrication is around 1.03 lb
per lb of mill product and secondary

fabrication around 0.75 lb per lb of mill
product. Further extrapolation suggests
only 12% material utilization or 17% if
recycled material is used.

As a result of the above considerations,
many efforts in the airframe industry in
recent years have been directed toward
reducing the cost of complex titanium
parts, especially those which are manufactured by methods that require extensive machining on all part surfaces to
final dimensions. P/M technologies that
have been investigated fairly extensively
include:

- Forging processes - isothermal,
  high energy rate, etc.

Table 1 Processing Losses and Scrap Generation for Titanium Production and Processing*

PROCESS	MAJOR INPUT	OTHER INPUT	LOSSES
• Ti SPONGE PRODUCTION (REDUCTIONS, SEPARA-TIONS, AND SIZING)	PURIFIED TiCL$_4$ (25% Ti)	• HCl • ARGON • KWH	• OFF-GRADE SPONGE FINES (-20 MESH)** • METAL LOSS (LEACH SOLUTIONS)***
• MELTING (COMPACTING, CONSUMABLE ELEC-TRODE, ARCTYPE)	Ti SPONGE (-1/2 + 20 MESH)	• HOME SCRAP & SOME OPEN MARKET SCRAP • KWH • ALLOYING ELEMENTS	• HOME SCRAP – MASSIVE (CROPPINGS, INGOT ENDS, COLLARS) • METAL LOSSES – GRINDINGS, VOLATILES, SPATTERS)
• PRIMARY FABRICATION (FORGING, ROLLING, ETC.)	Ti ALLOY INGOT (90-99% Ti)	• PICKLING BATHS • KWH FOR PRESSES & ROLLING MILLS	• HOME SCRAP – TURNINGS (INCL. CHIPS) SOLIDS (BILLET OR BAR TRIMMINGS, ETC.), SHEET CLIPPINGS • METAL LOSSES – GRINDINGS, GRIT BLASTING, PICKLING SOLUTIONS
• SECONDARY FABRI-CATION TO FINISHED HARDWARE	Ti MILL PRODUCTS (90-90% Ti)	• KWH FOR PRESSES • LUBRICANTS • TOOLS	• OPEN MARKET SCRAP – TURNINGS, SOLIDS, SHEET CLIPPINGS • METAL LOSSES – GRINDINGS, BLASTING GRIT, PICKLING SOLUTIONS, CHEM. MILL. SOLUTIONS, ECM BATHS, BURNED PRODUCTS

*ADAPTED FROM REF. 1
**FOR VACUUM DISTILLATION PROCESS
***FOR LEACHING PROCESS

1900-018V

- Powder metal processes - cold isostatic pressing/sinter (CIP-sinter), hot isostatic pressing (HIP) and hot pressing (HP)

- Castings - repaired and HIP

(Note: These P/M processes have been used for producing forging preforms or near net-shapes for use.)

Results of these efforts have been reviewed extensively elsewhere[2]. This paper is primarily concerned with experience gained on the HIP process to produce near-net deep-pocketed shapes using an approach which promises earliest implementation of the process on future aircraft without requiring major changes in powder prices. At the end of the paper, a discussion is presented on HIP net shapes summarizing their potential for use in non-structural parts and scale-up for fatigue-critical aircraft components now in progress.

2. BACKGROUND

The high cost of manufacturing titanium aircraft hardware is due primarily to industrial limitations on producing close-to-final contour shapes. This inability has made the ratio of the material bought to that which ends up in the finished article unacceptably high, especially in the aerospace industry. This buy/fly ratio problem manifested itself on the Navy F-14 and Air Force F-15 and B-1 programs. In addition to buy/fly ratios, forging lead times and machining requirements contribute to slow production rates and/or increased costs.

At the present time, many titanium alloy forgings for airframe applications must be purchased oversized with intensive machining required to yield final configurations. It is not uncommon for the material's utilization ratio (raw weight/machined weight) in many of these components to be in the range of 6/1 to 10/1.

Source: Materials Synergisms Process Conference, October 1978, 282-305

Most titanium alloy airframe parts with large potential savings have deep pockets (e.g., bulkheads and frames) and must be purchased as oversized conventional forgings which require extensive machining on all surfaces of the finished part. Most of the alloy is machined into scrap. Table 2 shows that the buy-to-fly ratio is most important for parts weighing up to about 50 lb, but when the part weight is over 100 lb even buy-to-fly ratios as low as 2.8:1 have significant machining scrap losses.

Manufacturing of parts to near-net configurations by conventional forging would require a series of intermediate (blocker) dies and several reheatings due to the inherently high flow resistance of titanium alloys in the forging dies (see Figure 1). The number of required operations would result in prohibitively high costs.

Table 2 Machined Scrap Losses For Various B/F Ratios Of Typical Airframe Parts

PART TYPE	B/F	FINAL WT., LB.	MACH. SCRAP, LB	SCRAP LOSS, %
• HORIZONTAL STABILIZER SUPPORT	6.7	9.6	54.7	85.0
• DRAG BRACE	4.45	19.1	65.9	76.4
• NACELLE FRAME	5.9	54.0	264.6	83.0
• WING PIVOT FITTING	2.8	114.0	205.2	64.3

1900-019V

Isothermal forging involves heating of forging dies to the temperature range of the forging stock thus enhancing its flowability. Although parts can be produced to final or near-final configuration, the applicability of this process is limited at the present time due to the high cost of superalloy dies, the complexity and cost of EDM machining, temperature control requirements, preform design, lubrication application, handling cycle, die life, hot inspection and cold inspection.

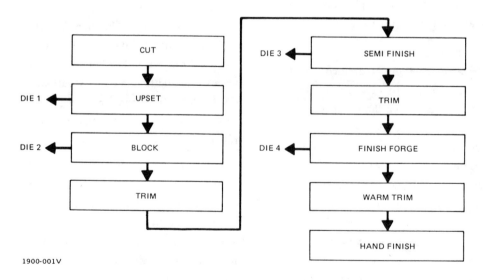

1900-001V

Figure 1 Forging Sequence for Conventional Forging Starting from Billet

In view of the difficulties involved in manufacturing airframe components to near-net configurations by forging of conventional preforms, the utilization of contoured powder preforms appears to warrant consideration to reduce the requirements for excess material and extensive machining.

In 1971, a program[3] conducted under a Naval Air Systems Command contract demonstrated the feasibility of manufacturing close-tolerance, low-draft-angle forgings from contoured sintered Ti-6Al-4V titanium alloy powder preforms. These preforms were consolidated to 97-98% of the theoretical density by cold isostatic pressing (CIP) of elemental titanium hydride/dehydride powders that were blended with aluminum-vanadium master alloy to yield the Ti-6Al-4V composition and sintered above 2500°F in vacuum for several hours. Although the feasibility of manufacturing close-tolerance forgings from contoured powder preforms was demonstrated in the course of these studies, metallographic examinations of forgings detected isolated microporosity in ribbed sections which were produced by an extrusion-type flow of metal in the die cavity rather than lateral flow under compressive stresses. This finding indicated that the utilization of powder preforms for forging applications could be greatly enhanced if these preforms could be consolidated to 100% density prior to forging. The identical CIP/sintering cycle produced only 85-90% density for prealloyed Ti-6Al-4V hydride/dehydride powders, and experimental forgings produced from these preforms exhibited unsatisfactory mechanical properties which were believed to be due to an intergranular contamination network detected in these forgings. With the advent of the commercial utilization of the hot isostatic pressing (HIP) process, the possibilities for consolidation of forging preforms to 100% dense parts became realistic. Initial trials showed that the process could not only produce 100% dense forging preforms, but potentially the final configuration as well.

In addition, the range of powder utilization could be widened to include spherical prealloyed powder produced by the Rotating Electrode Process (REP). The utilization of this powder offered the following advantages: the homogeneity of the composition could be guaranteed and the spherical configuration of particles would lead to a more uniform packing and thus facilitate dimensional control.

To obtain initial data on the process parameters and properties of HIP-processed materials, several preliminary studies were initiated in cooperation with major HIP processing concerns, including the Crucible Materials Research Center (CMRC) and the Kelsey-Hayes Company. The results of these studies indicated the following: consolidation of prealloyed spherical powders to full density is feasible by HIP; tensile, yield and elongation properties of some experimental Ti-6Al-4V and Ti-6Al-6V-2Sn configurations were comparable to those of conventional wrought materials; and CMRC demonstrated advanced capabilities of producing complex ribbed configurations in a one-step operation by HIP.

As a result of these early findings, the Naval Air Systems Command funded a program at Grumman to determine the feasibility of producing a titanium F-14A airframe component directly in a one-step process by HIP of prealloyed powders.[4] The total program scope is basically a three-phase effort on HIP of Ti-6Al-6V-2Sn titanium alloy primarily, including:

- Process Verification (Phase I)
- Reproducibility and Flight Qualification (Phase II)
- Scale-Up and Structural Verification (Phase III)

The first two phases have been completed and the scale-up phase is in its early stages. Figure 2 shows the F-14A fuselage brace which is the part selected for Phases I and II. This part was chosen since it typifies the deep-pocketed forgings used on aircraft and its size and weight (1.7 lb) allowed an economic utilization of material in experimental studies coupled with ease of handling.

3. THE HIP NET-SHAPE-MAKING PROCESS

The HIP process consists of encapsulating metallic prealloyed powders in a suitably shaped mold, evacuating and sealing the mold assembly, and isostatically pressing at a high temperature and suitable pressure in an autoclave that contains gaseous media (usually argon). Figure 3 shows schematically the principal steps in the process employed by Crucible. A typical HIP cycle is depicted in Figure 4. In considering the effects of HIP parameters, it is essential to realize that soaking at the HIP temperature for at least 8 hours is required to bring the heavily insulated mold assembly

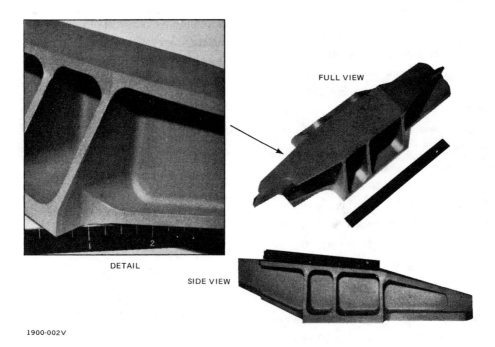

FULL VIEW

DETAIL

SIDE VIEW

1900-002V

Figure 2 Fuselage Fitting as Produced by HIP

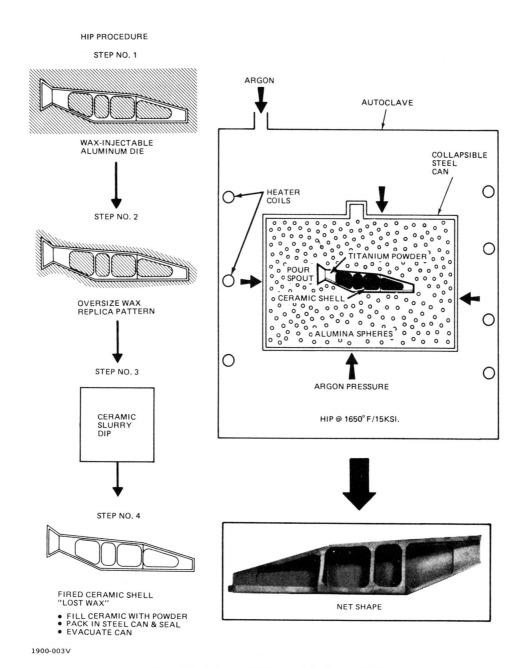

HIP PROCEDURE

STEP NO. 1

WAX-INJECTABLE
ALUMINUM DIE

STEP NO. 2

OVERSIZE WAX
REPLICA PATTERN

STEP NO. 3

CERAMIC
SLURRY
DIP

STEP NO. 4

FIRED CERAMIC SHELL
"LOST WAX"

• FILL CERAMIC WITH POWDER
• PACK IN STEEL CAN & SEAL
• EVACUATE CAN

1900-003V

ARGON

AUTOCLAVE

COLLAPSIBLE
STEEL
CAN

HEATER
COILS

TITANIUM POWDER

POUR
SPOUT

CERAMIC SHELL

ALUMINA SPHERES

ARGON PRESSURE

HIP @ 1650° F/15KSI.

NET SHAPE

Figure 3 Basic Steps in HIP Process Using Ceramic Molds

to the required temperature and the true HIP cycle, therefore, begins 10 hours after the starting time. Pressure is applied after the autoclave has been at temperature for 4 hours, as shown by the broken line in Figure 4.

4. PROCESS VERIFICATION

The Phase I effort was essentially a feasibility effort, consisting of tooling and parametric developments for the part selected and generation and evaluation of data on physical and mechanical properties of Ti-6Al-6V-2Sn titanium alloy. The F-14A Fuselage Fitting that was selected for the feasibility studies (Figure 2) was found to be readily producible with an excellent as-HIP'd surface finish in the range of 100 RMS. Figure 5 summarizes the results of dimensional analyses conducted on experimental components; it is evident that, despite limited efforts within the scope of the feasibility study, the majority of

dimensions came remarkably close to the print requirements.

The effects of HIP temperature on tensile and yield strengths, and elongation indicated that specification requirements can be met by consolidation at 1650°F. The effect of the oxygen content on the yield strength of Ti-6Al-6V-2Sn material consolidated in the range of 1650°F showed that oxygen contents above 1600 ppm were desirable to attain best strength with acceptable ductility.

Fatigue data for experimental HIP runs yielded an S/N curve comparable to that for annealed Ti-6Al-6V-2Sn plate. The data indicated that specimens consolidated at 1650°F/2 hours/15 ksi are in the immediate vicinity of the baseline curve for annealed plate.

HIP materials also exhibited good fracture toughness characteristics. The data indicated K_{ic} values equivalent to recrystallized annealed Ti-6Al-4V.

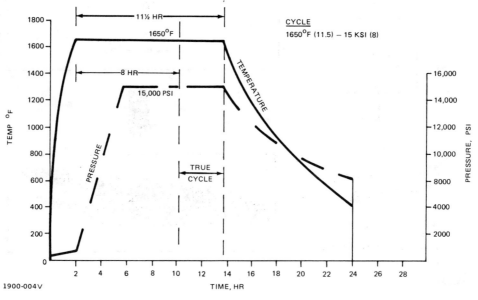

1900-004V

Figure 4. Typical HIP Cycle

5. REPRODUCIBILITY AND FLIGHT QUALIFICATION

Based on Phase I results, it was concluded that mechanical properties of hot isostatically pressed configurations depend on time at pressure and temperature, and oxygen level of the material. The initial task of the Phase 2[5] effort was designed to evaluate effects of increased time at temperature and pressure on mechanical properties and to improve dimensional tolerances of the selected part. Special efforts were required, for instance, to eliminate beveling at the lower base of the brace.

The objectives of the Phase 2 effort were:

- Verify mechanical properties of HIP materials

- Evaluate dimensional reproducibility of a pilot lot of ten parts

- Identify NDI criteria

- Verify the flight-worthiness of HIP-produced F-14A fuselage braces by performing spectrum fatigue tests simulating four design life cycles

The powder utilized in these experimental studies was produced by REP techniques,

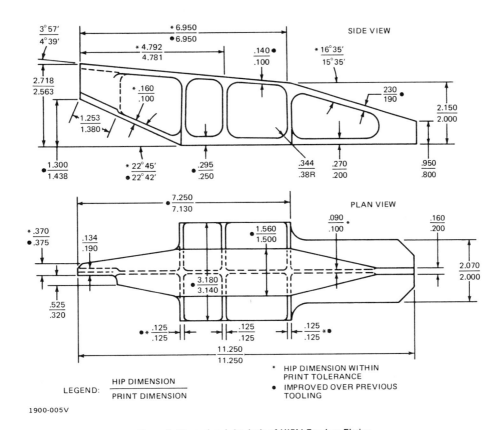

Figure 5 Dimensional Analysis of HIP'd Fuselage Fitting

identical to those used in the previous program. Powder remaining from the first phase was utilized in the first HIP run to expedite the refinement of dimensional tolerances.

In view of significant volume changes (approximately 35 percent) which occur on HIP processing, uniform packing of powder particles is of paramount importance in consolidation of net or near-net configurations. Spherical powders have good flow characteristics that are essential to obtain uniform packing of powder throughout the mold cavity. At the time this program was initiated, the Rotating Electrode Process (REP) was

Table 3 Chemical Analysis of REP Powder

POWDER	ELEMENT WT, %									
	Al	V	Sn	Fe	O	Cu	C	N	H	Ti
LOT 1	5.6	5.4	1.8	0.62	0.19	0.48	0.04	0.01	0.001	BALANCE
LOT 2	5.7	5.5	2.0	0.74	0.18	0.69	0.02	0.01	0.009	BALANCE

1900-020V

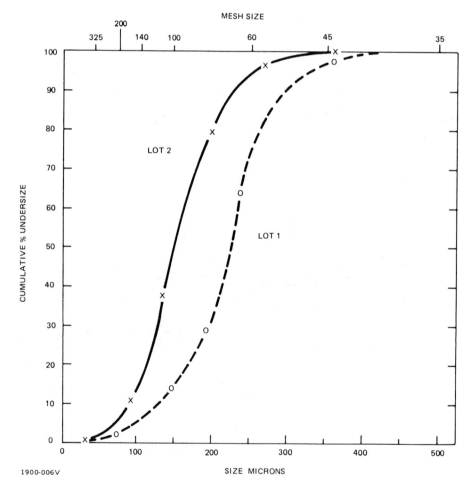

1900-006V

Figure 6. Particle Size Distribution of Ti-6Al-6V-2Sn Titanium Alloy Powder

the only commercially available method to produce Ti-6Al-6V-2Sn spherical powders. Mechanical and chemical properties of two REP powder lots used are presented in Table 3 and Figure 6. Lot 1 was used in the feasibility studies (Phase 1) and initial experiments of Phase 2 involving improvement of dimensional tolerances. Lot 2 was used in the remainder of this phase of the program.

Radiography detected the presence of isolated fine, high-density particles which were subsequently identified as tungsten and thus were apparently associated with the use of tungsten electrodes in the REP process. Attempts to remove tungsten by AVCO's ferromagnetic separation process were partially successful, since only larger (0.050 in.) tungsten particles could be effectively removed. Effects of fine tungsten inclusions on room-temperature tensile and fatigue appear to be minor, however, since fractographic studies never detected failure initiation sites at tungsten particles. Modification of the REP process, currently in progress at Nuclear Metals, Inc., will replace

tungsten with consummable titanium alloy electrodes thus eliminating the source of the problem.

Of far greater concern, were isolated occurrences of Si- and Al-rich inclusions (probably oxides) which served as failure initiation sites and caused low fatigue endurance. Some of these particles may escape radiographic detection, since their density is much closer to that of the base-metal. Thus, extensive metallographic evaluations of representative cross-sections is required to verify freedom from these inclusions.

It has been reported that the source of these inclusions has been identified and corrective measures are currently being incorporated in powder manufacturing operations so that future powder lots will be free of foreign inclusions.

5.1 MECHANICAL PROPERTIES

Based on the data generated in the previous phase, parameters for HIP processing were 1650°F/15 ksi/3 hr. Post-vacuum heat-treat times at 1300°F were 2 to 24 hours. The vacuum level was 10^{-5} microns Hg at 1300°F.

Table 4 Effect of Vacuum Annealing on Tensile Properties of HIP'd Ti-6Al-6V-2Sn Configurations

	AS-MACHINED HIP'd PROPERTIES		HIP'd PROPERTIES AFTER ANNEALING AT 1300°F			
POWDER LOT	1	2	1	2	2	GRUMMAN
PPM O_2	1900	1800	1900	1800	1800	FORGING SPECIFICATION
HRS AT 1300°F	NONE	NONE	2 HR	24 HR	24 HR	GM 3117
F_{tu}, HR	149.1	144.7	148.0	142.5	144.2	150.0
F_{ty}, HR	142.2	134.6	141.0	135.1	136.4	140.0
ELONG, %	17.7	14.7	18.0	22.0	20.0	8.0
RA, %	35.1	43.9	43.0	47.8	42.2	20.0

1900-021V

5.1.1 Tensile Properties

Representative tensile properties of Ti-6Al-6V-2Sn HIP materials are given in Table 4. Higher tensile ultimate and yield strength values obtained from Lot 1 powders appear to be associated with the higher oxygen content of 1900 ppm. Elongation and reduction-of-area characteristics easily meet forging requirements. Vacuum annealing at $1300^\circ F$ did not significantly affect tensile strength, but elongation and reduction-of-area were improved by the 24-hr treatment.

Although ultimate and yield tensile strength values for configurations consolidated from Lot 2 (1800-ppm oxygen) powders were 3-4 percent below the specified values for wrought materials in the longitudinal grain direction, the isotropic characteristics of HIP materials will permit use without significant changes in design criteria. Isotropy of microstructure and mechanical properties were established by metallography

(Figure 7) and comparable test data for randomly selected transverse test specimens.

5.1.2 Fracture Toughness

The previous results presented in Table 5 demonstrated that HIP'd Ti-6Al-6V-2Sn exhibits excellent toughness (70 and 60 K_{ic}, and 1400 and 1900 ppm oxygen, respectively). These values exceed the normal range for this alloy of 35-60 K_{ic} and are equivalent to that of Ti-6Al-4V in the RA condition. Those tested in this phase showed K_q values of 85 ksi $\sqrt{in.}$ and indicated that these parts can surpass the normal forging range consistently.

Table 5 HIP Ti-6-6-2 and Ti-6-4 Fracture Toughness Properties (1400 ppm Oxygen)

ALLOY	HIP TEMP, °F, POWDER MESH	FRACTURE TOUGHNESS (Kq), KSI INCH ½			AVG
Ti-6Al-6V-2Sn	1650, -35	67.5	68.8	69.5	68.6
		70.5*	70.9*		70.7
	1650, -100	53.0	54.0	63.3	56.6
	1800, -35	71.8	73.2	76.9	74.0
	1800, -100	60.4	63.8	65.3	63.2
Ti-6Al-4V	1550, -35	62.3	66.1	68.5	65.6

*K_{ic}

1900-022V

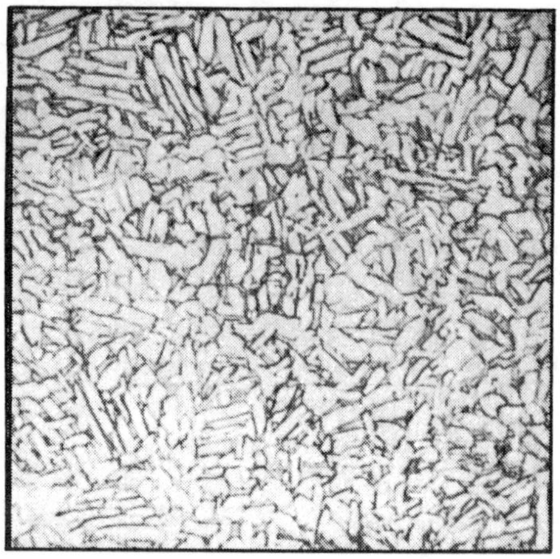

1900-007V

Figure 7. HIP Grain Structure (250X Mag)

1900-008V

HIP   WROUGHT

Figure 8 Topography of Fracture Surfaces in K_{IC} [Fracture Toughness] Specimens

Topography of fracture surfaces in HIP materials was different from that exhibited by forgings (Figure 8). The "pebbly" contours of HIP fracture surfaces indicated the possibility that the path of crack propagation in a fine isotropic grain structure undergoes frequent diversions which lead to a significant improvement in fracture toughness characteristics. Previous work on crack growth (da/dn) has shown that HIP materials also exhibit good crack propagation resistance.

5.1.3 Fatigue Strength

Figure 9 shows that most data points obtained in this phase fell within the representative data band for annealed forgings. In the specimen designated with an "M" subscript, low fatigue endurance was apparently associated with failure initiation at an inclusion as discussed previously. This shows that a

clean powder is urgently required for parts that are fatigue-critical and must operate with the equivalent of fully forged properties.

5.2 REPRODUCIBILITY (PILOT LOT)

The secondary objective of this phase was to evaluate dimensional tolerances of the brace parts produced in a pilot run (see Figure 10). Table 6 lists the critical target dimensions selected and variations from allowable tolerances determined by the dimensional analysis. Variations were attributed to an excessive load in the autoclave. The total weight of configurations processed in the reproducibility run was 113 lb - far in excess of previous autoclave loadings which ranged from 8 to 27 lb.

Although the dimensions were generally acceptable, the No. 31 dimension (web) was of some concern. As a result, another brace was produced under more

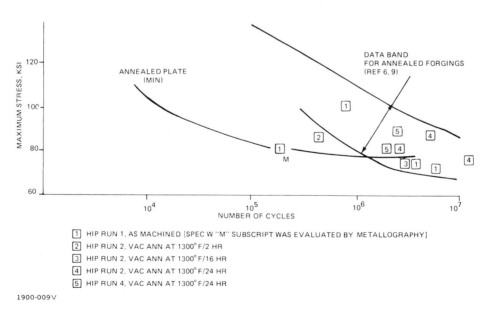

1. HIP RUN 1, AS MACHINED [SPEC W "M" SUBSCRIPT WAS EVALUATED BY METALLOGRAPHY]
2. HIP RUN 2, VAC ANN AT 1300°F/2 HR
3. HIP RUN 2, VAC ANN AT 1300°F/16 HR
4. HIP RUN 2, VAC ANN AT 1300°F/24 HR
5. HIP RUN 4, VAC ANN AT 1300°F/24 HR

1900-009V

Figure 9. Fatigue Endurance of Ti-6Al-6V-2Sn Titanium Alloy Power Consolidated by HIP at 1650°F

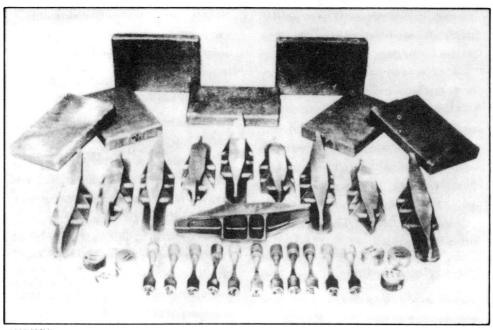

1900-010V

Figure 10. Pilot-Run Parts (HIP)

controlled conditions and with a heavy autoclave loading similar to the pilot run. Apparent temperature gradients detected in the course of the pilot run indicated that the temperature of each individual mold assembly must be controlled to ascertain that the required temperature is reached and maintained during the HIP cycle, particularly when heavy autoclave loads are processed. One approach is to weld a thermocouple to the metallic enclosure of the mold assembly. The soundness of this approach was established by production of a high-quality brace with essentially no defects that was well within all critical dimensional tolerances including the No. 31 dimension. Temperature data automatically recorded

Table 6 Dimensional Reproducibility Studies

DIMENSIONS[1]	TARGET	VARIATIONS FROM ALLOWABLE TOLERANCES[2] [3]								
		SM-506	SM-507	SM-508	SM-509	SM-510	SM-511	SM-513	SM-514	SM-515
4	4.781	- 0.011	WT	WT	WT	WT	- 0.005	WT	WT	WT
5	3.438	WT	WT	WT	WT	WT	WT	WT	WT	WT
8	0.125	WT	WT	WT	WT	WT	WT	WT	WT	WT
27	1.515	WT	WT	- 0.002	- 0.005	- 0.005	WT	WT	WT	- 0.003
28	3.140	- 0.025	- 0.010	+ 0.025	- 0.027	- 0.025	- 0.010	- 0.030	- 0.020	- 0.025
30	1.500	WT	WT	WT	WT	WT	WT	- 0.010	WT	WT
31	0.100	(4) WT TO - 0.020	(4) WT TO - 0.020	(4) WT TO - 0.010	(4) WT TO - 0.015	(4) - 0.006 TO - 0.019	(4) - 0.003 TO - 0.016	(4) WT TO - 0.020	(4) WT TO - 0.020	(4) WT TO - 0.020
35	2.000	WT	WT	WT	WT	WT	WT	WT	WT	WT

NOTES:
(1) SEE FIG. A-1 FOR IDENTIFICATION OF DIMENSIONS TAKEN
(2) ALLOWABLE TOLERANCES: +0.030, -0.010 IN.
(3) WT = WITHIN TOLERANCES
(4) RANGE

1900-023V

indicated the temperature to be within $-0/+8°F$ of the required setting.

Representative test data on specimens machined from rectangular test blocks and braces produced in the pilot run are shown in Table 7. All ultimate and yield tensile strength values were within 5 percent of the current requirements for annealed forgings. Typical elongation properties exceed applicable specification minimum requirements by a factor of two. Fatigue data showed that saturation shot peening to 0.1 Almen intensity gave acceptable results for these parts and fracture toughness values were 72-74 ksi $\sqrt{in}$.

5.3 FLIGHT QUALIFICATION

To verify the flight-worthiness of HIP braces, a truncated 12-level spectrum fatigue test was designed based on an actual 212-level load spectrum experienced by the F-14A wing-box in flight. The acceptance criteria were based on four life cycles, i.e., 24,000 equivalent flight hours (EFH).

The test assembly designed was an actual replica of a 20-inch-long section of the F-14A wing center-section as shown in Figure 11. The completed test assembly shown in Figure 12 simulated accurately the stiffness of the wing center-section at the location where the braces were to be installed. Strain gages were positioned on the test braces in locations of maximum anticipated stresses as shown in Figure 13.

One HIP brace withstood the 24,000 EFH test at 100 percent operational stress level and a second HIP brace withstood the identical test at 150 percent operational stress level. The functional spectrum fatigue tests in the spectrum fatigue testing fixture (Fig. 14) thus verified the flight-worthiness of the component.

5.4 QUALITY CONTROL PROVISIONS

Some defects encountered in HIP titanium powder materials differ significantly from those encountered in wrought and cost materials, and consequently their detection requires modification and/or refinement of existing NDI methods or development of entirely new techniques. For sections up to 0.300-in.-thick, detection of porosity of 0.005 in. in diameter and larger by standard radiographic techniques presented no problems.

Table 7 Tensile Properties of Configuration Processed in the Course of Reproducibility Studies

PARENT CONFIGURATION	CONDITION	$F_{tu'}$ KSi	$F_{ty'}$ KSi	ELONG %
TEST BLOCK	AS HIP + Vac Ann.	145.5	137.4	17.0
TEST BLOCK	AS HIP + Vac Ann.	145.5	137.6	16.5
TEST BLOCK	AS HIP + Vac Ann.	150.2	141.2	16.0
TEST BLOCK	AS HIP + Vac Ann.	150.2	140.2	17.0
BRACE 513	AS HIP + Vac Ann.	147.4	137.6	18.0
BRACE 513	AS HIP + Vac Ann.	149.2	139.1	18.0
TEST BLOCK	AS RE-HIP + Vac Ann.	145.5	141.9	15.0
TEST BLOCK	AS RE-HIP + Vac Ann.	144.9	141.8	15.5
TEST BLOCK	AS RE-HIP + Vac Ann.	146.7	139.4	16.0
TEST BLOCK	AS RE-HIP + Vac Ann.	148.1	141.4	15.0
BRACE 509	AS RE-HIP + Vac Ann.	145.3	135.5	18.0
BRACE 509	AS RE-HIP + Vac Ann.	142.0	133.2	18.0

1900-024V

1900-011V

Figure 11. Simulated Portion of F-14A Wing Center Section (Looking Aft)

280

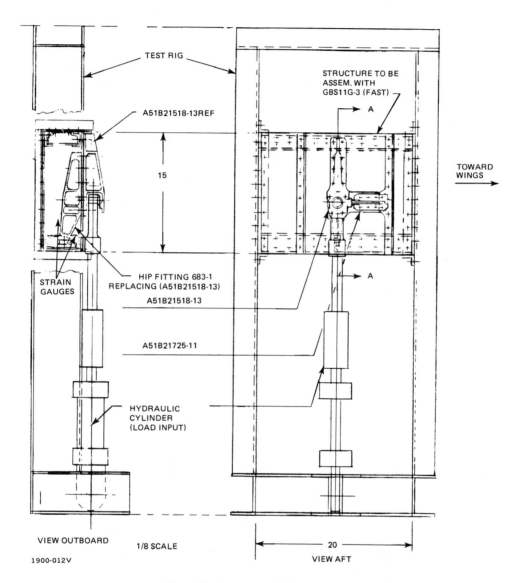

TEST RIG

A51B21518-13REF

15

STRAIN
GAUGES

HIP FITTING 683-1
REPLACING (A51B21518-13)

A51B21518-13

A51B21725-11

HYDRAULIC
CYLINDER
(LOAD INPUT)

VIEW OUTBOARD   1/8 SCALE

1900-012V

STRUCTURE TO BE
ASSEM. WITH
GBS11G-3 (FAST)

A

TOWARD
WINGS

A

20

VIEW AFT

Figure 12. Arrangement for Testing HIP Fitting

By using finer grain films, increasing the exposure time and decreasing the intensity of radiation, the limit of detectability could be widened to include 0.003-in.-diameter pores. By inspecting "suspect" areas of the film by an image-enhancement system utilizing edge-refinement and magnification techniques, porosity with diameters as low as 0.001 in. could be detected.

Contrast between radiographic images of high-density tungsten particles and the matrix permitted a reliable detection of such particles down to 0.002 in. in size using fine-grain films and optimized exposures. Detectability of fine tungsten particles could be further improved by the image-enhancement approach. It is expected that with the elimination of the tungsten electrode in powder production, REP or other types of spherical prealloyed powder produced by electron-beam melting will be available in time for critical structural part implementation. In all of the program work accomplished to date, no failures were attributable to tungsten inclusions; as a result, this

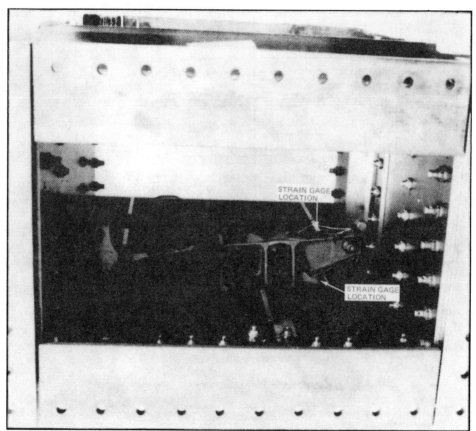

1900-013V

Figure 13. Test Brace Looking Inboard Along Front Beam of Simulated Wing Box

powder should be acceptable for secondary structural applications.

As indicated previously, silicon- and aluminum-containing inclusions (probably oxides) were detected by metallographic techniques. Although microprobe analysis confirmed the presence of the metals, it could not establish whether oxide was present. These contaminants were apparently introduced during the powder manufacturing process, since they were detected in configurations consolidated in both ceramic and metal molds. The contrast in sensitivity between radiographic techniques is not fool-proof, including the image-enhancement system. Metallographic observations of prolongations can be used to ascertain such quality. Recent reports[6] indicate that Nuclear Metals has determined the source of these in-

1900-014V

Figure 14. Spectrum Fatigue Testing Fixture

clusions and are making provisions to produce a clean powder product.

In the initial stage of applying HIP processes to production parts for aircraft use, it will be essential to utilize stringent (100%) inspection provisions to assure quality. These requirements may be relaxed as more experience is gained on part reproducibility and reliability. In addition to radiographic, dimensional and microstructural analyses, it will be necessary to perform fluorescent penetrant inspections for surface defects. It is also necessary to destructively evaluate at least one part per autoclave run for tensile, soundness and chemistry determinations. Table 8 lists the requirements for these quality provisions.

6. COST AND IMPLEMENTATION

The impact on machining requirements for the fuselage brace after HIP manufacture is illustrated in Figure 15 which shows the costs for the raw forging and the net HIP part. The as-HIP'd fuselage brace weighs 2.2 lb and the raw forging 6.2 lb, a savings of 65% in scrap chips. Figure 15 indicates the original estimate for 176 pieces comparing forging and HIP (including machining costs), indicating a potential savings of 40% with an assumed powder

Table 8 Inspection Methods For Acceptance of Parts

METHOD	INSPECT FOR
• RADIOGRAPHY (RT)	POROSITY, HEAVY & LIGHT INCLUSIONS
• FLUORESCENT PENETRANT (FPI)	SURFACE POROSITY, CRACKS, ETC.
• METALLOGRAPHY (PROLONGATION, POUR SPOUT)	MICROSTRUCTURE, POROSITY INCLUSIONS
• CUT-UP PART	TENSILE PROPERTIES (F_{t_u}, F_{t_y}, e, E) MICRO- STRUCTURE (ALPHA/ BETA) SOUNDNESS CHEMISTRY
• DIMENSIONAL	TO DRAWING TOLERANCES

1900-025V

cost of $35/lb. This estimate, however, did not consider complete inspection requirements that were added later to assure quality of initial purchases.

To gain experience with HIP parts, the fuselage brace is being flown in an F-14A aircraft and an order has been placed for 37 parts for later flight and complete evaluation for quality and reproducibility. Even though this is a reduced buy in number of parts compared to the original estimate and 100% inspection is required, an estimated savings of 17% is contemplated. This approach will provide the necessary experience with the process on a secondary structural part that will be valuable in assessing the usefulness of the process for application in future aircraft.

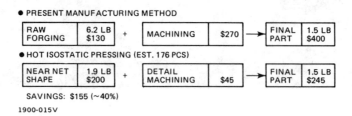

● PRESENT MANUFACTURING METHOD

RAW FORGING 6.2 LB $130 + MACHINING $270 → FINAL PART 1.5 LB $400

● HOT ISOSTATIC PRESSING (EST. 176 PCS)

NEAR NET SHAPE 1.9 LB $200 + DETAIL MACHINING $45 → FINAL PART 1.5 LB $245

SAVINGS: $155 (~40%)

1900-015V

Figure 15 Cost Savings by HIP of Fuselage Brace

7. SCALE-UP EVALUATION

To scale-up the process from the 20-in.^{2} plan area of the fuselage brace fitting to a 500-in.^{2} plan area component, a program is planned by Naval Air Systems Command this winter. The large HIP autoclave has a 40-in. working diameter and could be used to process two full-size nacelle frames (Figure 16) per cycle. This part was selected for the program because extensive test data on conventional forgings are available. In order to increase the autoclave packing density up to the equivalent of ten parts per cycle and, thereby, increase throughput and decrease cost per part, Grumman proposed to utilize electron-beam (EB) welding by joining HIP sections as illustrated in Figure 17. This hybrid (HIP/EBW) approach will increase the part range open to HIP processing so that many types of large structures can be built-up. Preliminary EB welding parameter work performed indicated that excellent quality and mechanical properties were attainable on transverse butt welds. Tensile ultimate strength, yield strength and elongation for the EB butt welds were 148.5 ksi, 158.3 ksi and 16%, respectively. Figure 17 illustrates the capability of EB welding to join thin web sections up to 2 inches thick in a single-pass weld.

The scale-up program will consist of the following tasks:

- Powder Evaluation, Selection and Specifications

- Determination of Additional Design Properties

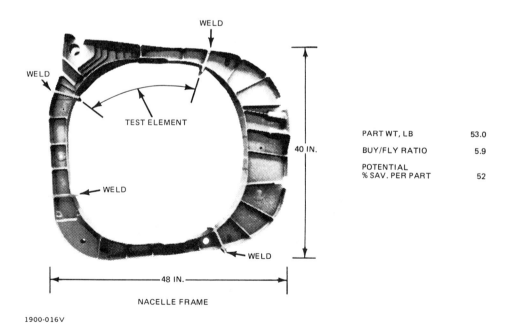

PART WT, LB	53.0
BUY/FLY RATIO	5.9
POTENTIAL % SAV. PER PART	52

NACELLE FRAME

1900-016V

Figure 16 Part Selected for Hybrid Manufacture

- Manufacture and Qualification of Test Element

- Manufacture of One Prototype Hybrid Nacelle Frame

- Hybrid Frame Design and Preliminary HIP Tooling

Based on the results of this work, further efforts will be proposed to confirm reproducibility and flight qualification.

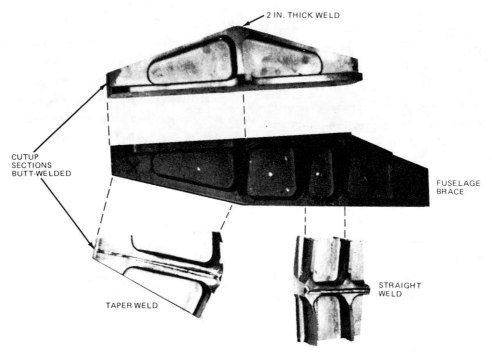

a. TYPES OF EB WELDS IN POWDER METAL PARTS

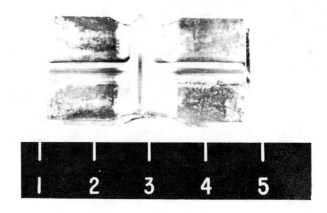

b. EB WELD IN HIP Ti-6-6-2 POWDER PART
- RADIOGRAPHIC & ULTRASONIC ACCEPTABILITY
- THIN WEB/HIGH DEPTH-TO-WIDTH WELD
- NO DISTORTION & LOW RESIDUAL STRESS

1900-017V

Figure 17 EB Weldability of HIP Parts

8. CONCLUSIONS

- Fuselage braces manufactured by the hot isostatic pressing (HIP) process withstood F-14A wing-box spectrum tests simulating 24,000 equivalent flight hours (four design lives) without failure and have been qualified for flight and purchase.

- Projected savings related to the utilization of the HIP process in manufacturing these parts are in the range of 30 to 40 percent for a 176-part buy and 15 to 20 percent for 35 to 50 parts.

- Microstructure of HIP configurations exhibited completely isotropic characteristics.

- Vacuum annealing at 1300°F is essential to assure acceptable mechanical properties and electron-beam weldability of HIP-processed Ti-6A1-6V-2Sn titanium alloy.

- Fracture toughness (K_{ic}) values were in the range of 69 to 74 ksi $\sqrt{in.}$ and significantly exceeded corresponding values for annealed forgings, typically 57 ksi $\sqrt{in.}$ Improvement of fracture toughness characteristics is believed to be associated with the diversion of the crack tip propagation at former grain boundaries.

- Elongation was in the range of 15 to 16 percent and thus exceeded the specified value for forgings (8 percent) by a factor of two.

- At the present oxygen level (1800 ppm), the values for $F_{tu'}$ and $F_{ty'}$ were approximately 3 percent below the design specifications. Specified values for $F_{tu'}$ and $F_{ty'}$ (150 ksi and 140 ksi, respectively) can be obtained, however, by minor adjustment of the oxygen content within the allowable limits.

- Fatigue endurance limit of HIP'd Ti-6A1-6V-2Sn titanium alloy in tension-tension (R=0.1) is in the range of 70 to 80 ksi and is thus comparable to that of annealed plate.

- Precise control of parameters is of utmost importance in processing of Ti-6A1-6V-2Sn titanium alloy powders. Freedom from temperature gradients and proper outgassing of ceramic molds and packing materials warrant particular attention.

- Nondestructive inspection (NDI) capabilities can be greatly improved by utilization of radiographic image-enhancement techniques.

- Detection of isolated, fine, foreign ceramic particles can be ascertained at the present time only by extensive metallographic studies. Availability of certified inclusion-free powders is an essential prerequisite for utilizing the HIP process in manufacturing of highly stressed airframe components.

- Existing tight tolerances for machined
components may have to be relaxed
slightly to assure that all dimensions
of HIP parts fall within design specifications, thereby ensuring that the
significant cost saving potential is being fully realized.

10. REFERENCES

1. Sutcliffe, P.W. and Marden, P.G.,
Titanium Powder Metallurgy,
AGARD Report No. 627 on Advanced
Manufacturing Methods and Their
Economic Implications, 1975.

2. Advanced Fabrication Techniques in
Powder Metallurgy and Their Economic Implications, AGARD Conference Proceedings No. 200, April
1976.

3. Witt, R.H. and Paul, O., Feasibil-
ity of Producing Close-Tolerance
F-14 Titanium Forgings from
Sintered Powder Preforms, Final
Report, Contract N00019-70-C-0598,
December, 1971.

4. Witt, R.H. and Magnuson, J., Manufacturing of Titanium Airframe
Components by Hot Isostatic Pressing, Contract No. N00019-74-C-0301,
Contract No. 5018-002-4, April 1975.

5. Witt, R.H. and Magnuson, J., Flight
Qualification of Titanium F-14A Air-
frame Components Manufactured by
Hot Isostatic Pressing (HIP), Final
Report, Contract No. N00019-76-C-
0143, June 1977.

6. Private communication with Nuclear
Metals Inc. and AFML.

11. ACKNOWLEDGMENT

The work reported herein was performed under Naval Air Systems Command contracts listed in References 3,
4, 5 and 7. The latter two contracts
were administered under the technical
direction of Mr. W.T. Highberger of
NASC.

The authors also wish to acknowledge
the contributions of Mr. Olev Paul and
Mr. Joel Magnuson to the work
reported.

12. BIOGRAPHIES

Mr. R.H. Witt is Group Head of
Advanced Metallurgy and Welding Development at Grumman Aerospace Corporation, where he has been employed for
the last 15 years. He is responsible for
manufacturing and process development
in the metallics regime. Mr. Witt's
diversified experience includes fabrication of aluminum, titanium, steel, high-
temperature and refractory alloys by
brazing, welding, powder metallurgy,
arc welding, rolling and diffusion bonding. In addition, Mr. Witt was the
Project Engineer for the EB welding
program leading to the introduction of
this process on the F-14. While at
Grumman, Mr. Witt has been associated with the following projects: Isothermal Forging of Titanium Near-Net

shapes, EB Welding of Ti-6Al-4V Titanium Wing Box Structure, Pulsed-Arc Tube Welding of Stainless Steel and Titanium Alloys for Permanent Hydraulic Connections, Hot Isostatic Pressing (HIP) of Pre-Alloyed Powders, Weld Quality Definition, Sliding-Seal Electron-Beam Welding of Aerospace Structures, Laser Welding, and Plasma-Arc Repair Welding of Discrepant Machined Parts. Prior to joining Grumman, Mr. Witt was previously employed by Republic Aviation Corporation and Sylvania Electric Products Company on metallurgy, welding and brazing of high-strength, high-temperature materials. He is a member of SAMPE, ASM, AWS and Sigma Xi, and has presented more than forty technical papers in various fields of metallurgy and welding. He holds the B. Met. E and M. Met. E. degrees from Polytechnic Institute of New York.

Mr. W. T. (Ted) Highberger is a metallurgist in the Materials and Processing Branch, Engineering Division, Naval Air Systems Command, Washington, DC. He is reponsible for titanium applications, metal matrix composites and manufacturing technology for metals. Prior to this assignment, he was an investigator and project engineer in various ordnance programs at the Naval Surface Weapons Center, Dahlgren, Va. He graduated from the College of Wooster in 1938 with a BA in mathematics and worked in mill metallurgy for Republic Steel Corporation, Massillon, Ohio, from 1940-1953, with time out for naval service in World War II. Mr. Highberger is currently a member of AWS and AIME.

ISOSTATIC PRESSING OF COMPLEX SHAPES
FROM TITANIUM AND TITANIUM ALLOYS

Stanley Abkowitz

Dynamet Technology Inc., Burlington, Massachusetts, U. S. A.

Introduction

Cold isostatic pressing (CIP) of metal powders differs from conventional mechanical pressing in several ways that offer specific advantages.

(1) Since hydrostatic (water-oil) pressure is applied uniformly on all surfaces of the part to be consolidated, the density achieved is of significantly greater uniformity.

(2) With this uniform pressure the size of the part to be compacted is limited only by the dimensions of the pressure chamber.

(3) Since the mold material is elastomeric (as opposed to punch and die hard tooling) no lubricant need be added to the powder to minimize galling and die wear.

(4) With the elimination of lubricant no pre-sinter for lube removal is necessary and lubricant contamination of the reactive metal powders need not be contended with.

(5) With the use of elastomeric tooling, complex shapes beyond the capabilities of conventional pressing are feasible.

In the past, isostatic pressing was employed to compact ceramic materials at relatively low pressures and to form simple billets of difficult to melt matals such as tungsten and beryllium. More recently the technology has progressed to the manufacture of complex shapes from a variety of metals. The greatest economic advantages of isostatic pressing lie in its application to metals that are relatively expensive and which are rather difficult to machine (since the process requires less input material and reduces the extent of machining required). Certainly one of the metals that fulfills these characteristics is titanium and its alloys.

As-Sintered Properties and Forging Preforms

Intensive developmental efforts of recent years [1] [2] have shown that titanium and its alloys could be cold compacted and vacuum sintered to achieve satisfactory density and useful mechanical properties. This was accomplished using inexpensive elemental powders in combination with master alloy powders as opposed to the more expensive pre-alloyed powders which in addition to their high cost are not readily cold compactable.

The tensile properties achieved with commercially pure titanium and Ti-6Al-4V alloy are reviewed in Table 1 below.

TYPICAL PROPERTIES

	C.P. Titanium (.12% oxygen)			Ti-6Al -4V (.12% oxygen)	
	Sintered (94% dense)	Forged Preform (100% dense)		Sintered (94% dense)	Forged Preform (100% dense)
UTS	62,000 psi	66,000 psi	UTS	120,000 psi	133,500 psi
YS	49,000 psi	53,000 psi	YS	107,000 psi	122,000 psi
Elong.	15%	23%	Elong.	5.0%	11.5%
R.A.	23%	30%	R.A.	8.0%	24.7%

TABLE 1

The as-sintered properties represent a 94% of theoretical density product produced by isostatic pressing at 60,000 psi followed by vacuum sintering at 2250°F for 3 hours. The further improvement in properties achieved by forging from a 94% dense preform is also apparent from this data. The as-sintered tensile properties compare well with those properties achieved by conventional casting and offer an economic process for like applications. Where higher density and further improvement in mechanical properties is required, forging the pre-shaped P/M preform may be employed. Other metal working techniques which have been proven to fully densify the as-sintered preform are rolling and extrusion. Typical forging preforms and a finished forged part are shown in Fig. 1 and Fig. 2.

By designing the elastomeric tooling in conjunction with a mandrel, cored shapes can be produced with configurations pressed and sintered close to size. The large titanium valve ball of Fig. 3 and the titanium alloy liner of Fig. 4 were produced by isostatic pressing around a mandrel which was removed from the "green compact" prior to vacuum sintering. The rotor cannister shown in Fig. 5 demonstrates the producibility of very thin walls (.035"). These parts are subsequently finish machined to final dimensions.

In the above applications the components possess the desirable mechanical properties and corrosion resistance in the as-sintered condition. It is interesting to point out that in addition to the considerable material and machining time saved by the preformed shape, further savings are offered in the finish machining operations which still must be performed. This is due to the free machining aspects of the P/M material. This ease of machinability is indicated in Fig. 6. At 100 hole "tool life" the cutting speed for annealed Ti-6Al-4V improves from 25 feet per minute for the wrought product to 60 feet per minute for the P/M material [3].

Subsequent Hot Pressing - "CHIP"

It has been pointed out that for certain applications where full density is required, the hot working operations of forging, extrusion, or rolling can be performed on P/M preforms. More recently, Dynamet has demonstrated that cold isostatic pressing and sintering can be followed by hot isostatic pressing (HIP) to produce fully dense shapes [6]. The microstructures of Figs. 7A and 7B demonstrate the closing of porosity for the Ti-6Al-4V transformed beta structure achieved by "HIPing" the as-sintered P/M shape.

The chemical analysis and mechanical properties achieved after HIP
are shown in Table 2 below and compare favorably with conventional 100%
dense wrought product:

TENSILE TEST AT ROOM TEMP.:		QUANTITATIVE ANALYSIS (%):	
Tensile P.S.I.	132,797	Iron	.085
Yield .2% offs	120,024	Vanadium	4.07
Elongation in 4D	13.0%	Aluminum	5.97
Reduction Area	26.0%	Carbon	.035
		Hydrogen-Core	.0018
		Nitrogen	.0109
		Oxygen	.1978

TABLE 2
MECHANICAL PROPERTIES & CHEMICAL ANALYSIS OF
SINTERED AND HIP PREFORM (CHIP) OF Ti-6Al-4V

This combination of CHIP (CIP plus HIP) is a highly economical
technique for achieving the maximum density, particularly for titanium
alloys. The prior cold isostatic pressing and sintering permits use
of the elemental-master alloy powder as opposed to the more expensive
pre-alloyed powder and in addition, only the elastomeric tooling need
be employed in the consolidation since the hot pressing step can be
accomplished on the preform sintered shape without the need for any
expensive and expendable hard tooling. In other words, the sintered
preform can be HIPed as inexpensively as many commercial titanium
castings are currently densified. Fatigue endurance limits at 10^7 cycles
of 52,000 psi have been obtained with CHIP Ti-6Al-4V.

Complex Shapes

The technology of both CIP and CHIP processing permits significant
advantages in the manufacture of complex shapes such as impellers. Fig. 8
illustrates the detail that can be achieved in the complex configuration of
a radial impeller. Cost savings in the manufacture of such shapes can be
as high as 70%, even where subsequent machining might be required. Fig. 9
indicates the potential (with recent developments in tooling) for production
of multi-holed structures and perforations in both thin and thick sections.
The titanium alloy disc shown has over 1000 holes and is 3/16 inch thick.
Heavier sections with square, triangular, or any shaped hole could readily
have been incorporated into the tooling.

Cost and Energy Savings

Cost saving applications have emerged for isostatic pressing from
industries as diverse as instant photography and missile manufacture.
Fig. 10 illustrates a Polaroid titanium nozzle section where very
expensive machining operations were previously employed. The use of
isostatic pressing with a detailed hard mandrel produced a net shape
for the internal configuration and thus eliminated the major costs of
machining these components.

292

In the missile wing configuration shown in Fig. 11 which is currently under development, similar cost savings are forseen with the advantageous application of elastomeric tooling in combination with mandrel technology.

One component currently in production at the rate of 300 parts per month is a titanium alloy "Dome Housing" for the Sidewinder Missile (See Fig. 12). This item had been machined from a billet blank weighing over 5 lbs. The final shape is machined to close tolerances and very thin wall. Application of isostatic pressing and vacuum sintering to this component resulted in a preform blank weighing 1-1/4 lbs. and significantly reduced the amount of finish machining necessary. In addition, the final machining required was on a free machining material, thus enabling the close tolerances to be more readily achieved.

In many cases the cost of the P/M preform might well exceed the cost of the billet stock, with significant eventual cost savings still achieved in reduced finish machining. In this specific case however, the cost of the preform is only 60% the cost of billet, with additional still better savings in finish machining costs.

Today, cost savings should not be considered our only goal. Energy savings take equal billing. Previous manufacture of the Dome Housing from "mill product" required the billet forging (ENERGY) of 5 lbs. of ingot, which required the ingot melting (ENERGY) of at least 5 lbs. of sponge, which required the sponge reduction and refining (ENERGY) of over 5 lbs. of ilmenite or rutile, which required the extraction (ENERGY) of well over 5 lbs. of ore.

Since the P/M process starts with powder produced in the manner of sponge, and since the preform blank uses 1-1/4 lbs. of material, the energy requirements for extraction, reduction and refining are reduced to 25% energy consumption. In contrast to the significant energy required to vacuum double arc melt and forge to billet 5 lbs. of material is the much reduced energy to press and vacuum sinter only 1-1/4 lbs. to preform shape. Finally, even finish machining to final dimensions from the P/M preform results in still additional energy savings due to the free machining aspect of the material.

Based on the above qualitative analysis it has been concluded that the percent energy conservation by this process is significantly greater than even the high dollar conservation which is achieved, although the use of scrap titanium in melting would reduce this advantage to some degree.

Other Materials

The technology above has been applied to materials other than titanium [4] [5]. Stainless steels, tool steels, aluminum alloys, copper, nickel, Hastelloy C and the refractory metals among others have all shown technical promise and economic advantage with these processes. Increasing commercial acceptance and significant utilization of this technology is anticipated in the immediate future.

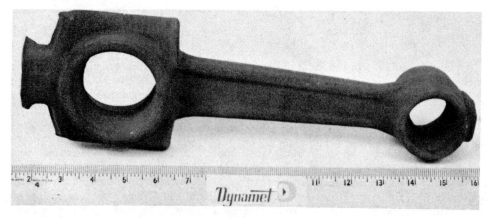

FIG. 1 - TITANIUM ALLOY CONNECTING ROD AS
PRESSED AND SINTERED FORGING PREFORM

FIG. 2 - TITANIUM COMPRESSOR BLADE
FORGING P/M Ti-6Al-4V
PREFORM AND FINISHED FULL
DENSITY FORGING

FIG. 3 - TITANIUM VALVE BALLS AS PRESSED AND SINTERED.

FIG. 4 - Ti-6Al-4V LINER (8" Dia.) AS PRESSED AND SINTERED

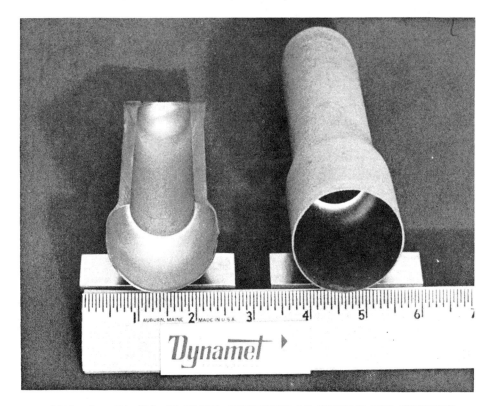

FIG. 5 - Ti-6Al-4V ROTOR CANNISTER AS PRESSED AND SINTERED

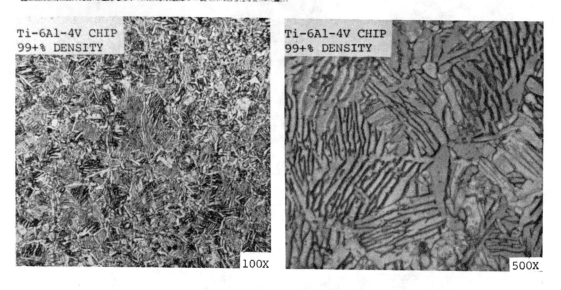

FIG. 6

MACHINABILITY STUDY ON
P/M TITANIUM 6Al-4V ALLOY

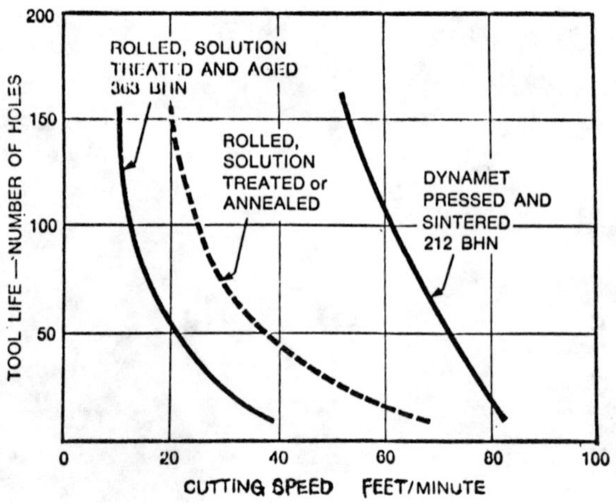

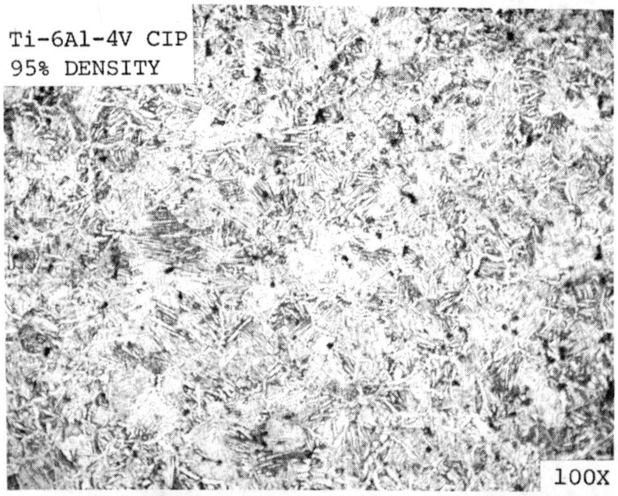

Ti-6Al-4V CIP
95% DENSITY

100X

FIG. 7A

MICROSTRUCTURE Ti-6Al-4V
AS PRESSED AND SINTERED

Ti-6Al-4V CHIP
99+% DENSITY

100X

Ti-6Al-4V CHIP
99+% DENSITY

500X

FIG. 7B - MICROSTRUCTURE Ti-6Al-4V COLD ISOSTATIC
PRESSED AND SINTERED FOLLOWED BY HIP

FIG. 8

Ti-6A1-4V
RADIAL ROTOR

FIG. 9

Ti-6A1-4V PERFORATED DISC
3-1/2" DIAMETER BY 3/16"
THICK WITH OVER 1000 HOLES

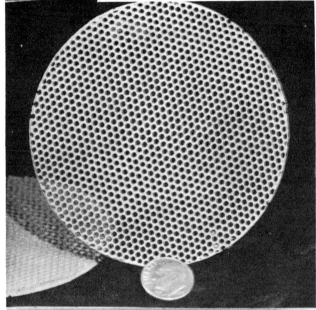

FIG. 10 - Ti-6A1-4V NOZZLE PREFORM AS ISOPRESSED AND AFTER VACUUM SINTERING

FIG. 11 - Ti-6Al-4V MISSILE WING ISOSTATICALLY PRESSED AND VACUUM SINTERED

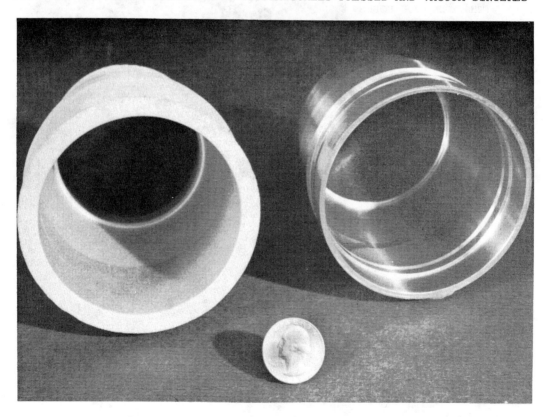

FIG. 12 - Ti-6Al-4V ALLOY DOME HOUSING PREFORM 94%
DENSITY AND FINISHED MACHINED COMPONENT

References

1. S. Abkowitz, "Titanium P/M Preforms, Parts and Composites". Titanium Science and Technology - Proceedings of the Second International Titanium Conference (AIME) held at M.I.T., Cambridge, Massachusetts, May 2 - 5, 1972, Vol. 1, Pages 381-398

2. S. Abkowitz "Cost Savings in the Application of P/M Titanium and P/M Aluminum Alloys". 1974 National Powder Metallurgy Conference Proceedings, Vol. 30, Pages 85-101.

3. Metcut Report to Dynamet Technology "Drilling Pressed and Sintered Titanium 6Al-4V". Metcut Report No. 1023-12894-1, February 1969.

4. "Developments in Titanium and Aluminum P/M Technology", Precision Metal Magazine, September 1974, Pages 44-48.

5. "Cold Pressing Opens the Door to Complex Shapes", Iron Age Magazine, May 9, 1977, Pages 39-41.

6. "Isostatic Pressing of Complex Shapes from Titanium and Other Materials" 10th National SAMPE TECHNICAL CONFERENCE, Kiamesha Lake, New York, Oct. 17-19th, 1978.

MANUFACTURE OF TITANIUM COMPONENTS
BY HOT ISOSTATIC PRESSING

W. Theodore Highberger
Naval Air Systems Command
Washington, D. C.   20361

Introduction

Structural titanium parts for military aircraft now being forged by conventional methods require extensive machining to yield final configurations. A 10/1 "buy-to-fly ratio" is quite common. To reduce this prohibitive cost alternative systems were considered. Such a system is the hot isostatic pressing (HIP) process, which consists of heating powder under pressure in an autoclave to conform to the net or near net shape of a finished part. Titanium alloys are particularly adapted to this process since they diffuse readily and dissolve their own oxides.

Program Objectives

In 1974 the Naval Air Systems Command awarded a contract to the Grumman Aerospace Corporation teamed with Crucible Division of Colt Industries to develop a hot isostatically pressed powder part for the F-14 aircraft. The first year program was designed (1) to apply HIP technology to the manufacture of an F-14 fuselage fitting, made from a Ti-6Al-6V-2Sn alloy and (2) to generate data on physical and mechanical properties of Ti-6Al-6V-2Sn and Ti-6Al-4V alloys. The latter alloy was studied through a supplemental project with Kelsey Hayes, using a similar process.

Manufacture by HIP

Figure 1 is a schematic of the hot isostatic pressing operation as performed by Crucible, Inc. Essentially, it consists of enclosing metallic powders in a suitably shaped mold of a special ceramic composition, evacuating and sealing the mold assembly, and placing it in an autoclave under a high temperature and pressure with an argon atmosphere. The applied pressure is transmitted to the powder charge through the walls of the mold, softened by the high temperature. This high temperature further promotes bonding by self-diffusion of the titanium particles. This powder was spherical in shape, supplied by Nuclear Metals.

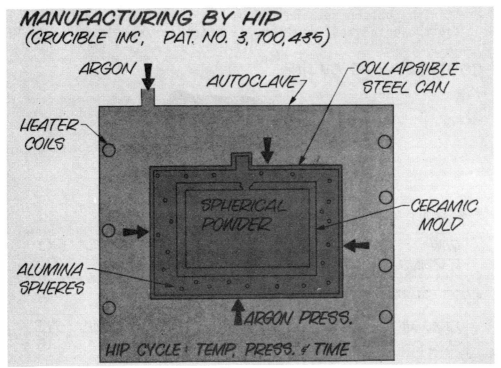

Figure 1

Schematic of Hot Isostatic Pressing Operation, as Performed by Crucible

<u>Cost Savings</u>

A detail and full view of the F-14 fuselage brace produced by HIP as a pilot part is shown in Figure 2.

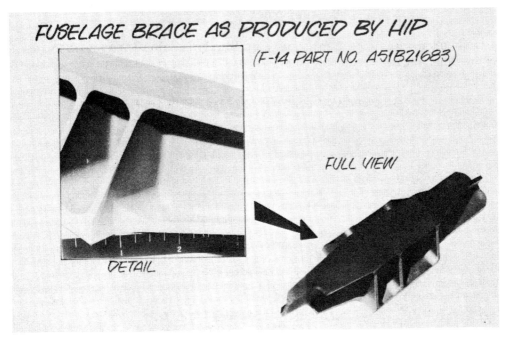

Figure 2

Fuselage Brace as Produced by HIP

Source: Powder Metallurgy in Defense Technology, Vol. 3, 1977, 145-150

Only a light finish pass on top and bottom are needed for a net shape to drawing tolerance. The projected cost savings for this part are depicted in Figure 3.

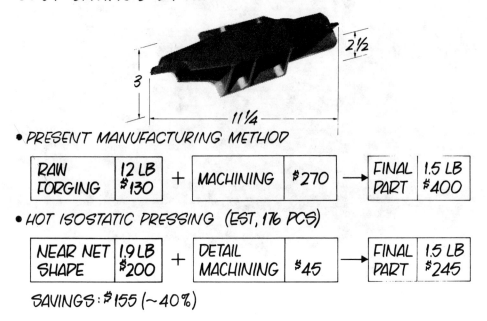

COST SAVINGS BY HIP

2½

3

11¼

• PRESENT MANUFACTURING METHOD

| RAW FORGING | 12 LB $130 | + | MACHINING | $270 | → | FINAL PART | 1.5 LB $400 |

• HOT ISOSTATIC PRESSING (EST, 176 PCS)

| NEAR NET SHAPE | 1.9 LB $200 | + | DETAIL MACHINING | $45 | → | FINAL PART | 1.5 LB $245 |

SAVINGS: $155 (~40%)

Figure 3

Cost Saving by HIP

$155 per part is equivalent to about $15,500. per 50 unit lot. When applied to all titanium pocketed airframe configurations for the F-14, up to 100 lb. sizes, a savings of $40,000. per plane is projected.

Non-Destructive Testing

Figure 4 shows an excellent correlation between ultrasonic velocity, as indicated by the abcissa, and density of a Ti-6Al-4V test piece; density is shown on the ordinate.

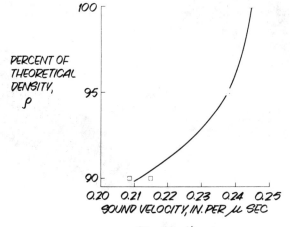

Ti-6Aℓ-4V NDI DENSITY DETERM. BY SOUND VEL.

PERCENT OF THEORETICAL DENSITY, ρ

SOUND VELOCITY, IN. PER μ SEC

Figure 4

Ti-6Al-4V, Density Determined by Ultrasonic Velocity

This would give a high level of quality assurance for the parts.

Mechanical Properties

Fracture toughness, yield strength and some representative microstructures are shown in Figure 5.

FRACTURE TOUGHNESS: HIP Ti-6Al-6V-2Sn

HIP TEMP. °F	MESH SIZE	F_{ty} KSI	TOUGHNESS, K_Q KSI INCH 1/2			MICROGRAPH x500
1650	-35	132	67.5	68.8	69.5	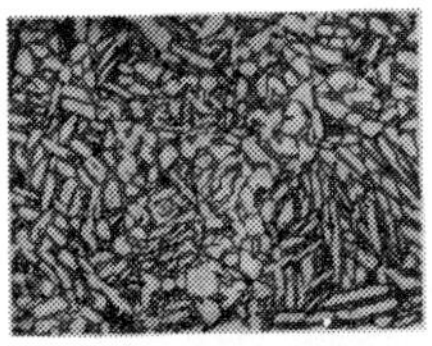
			70.5*	70.9*	*KIc	
	-100	135	53.0	54.0	63.5	
1800	-35	128	71.8	73.2	76.9	
	-100	129	60.4	63.8	65.3	
BARSTOCK	-	143	59.5	60.9	62.6	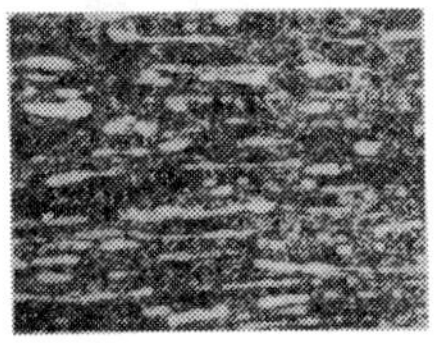

Figure 5

Fracture Toughness, HIP Ti-6Al-6V-2Sn

K_{Ic} values of 70 ksi $\sqrt{in}$ for the 35 mesh size are excellent and could be traded off for increased yield strength by varying processing parameters. Figure 6 shows tensile and yield strength properties equivalent to minimum plate requirements in the 35 mesh powder size.

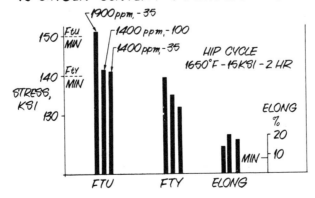

TENSILE RESULTS OF HIP Ti-6Al-6V-2Sn vs OXYGEN CONTENT & POWDER MESH

Figure 6

Tensile Results for HIP Ti-6Al-6V-2Sn

Figure 7 depicts the fatigue results which gave values equivalent to annealed plate under two processing conditions.

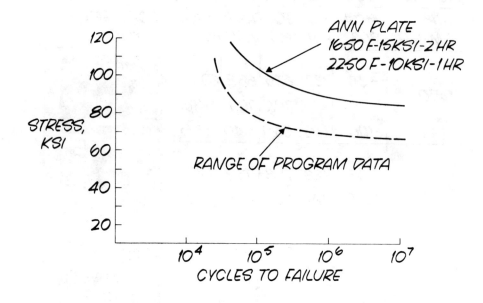

Figure 7

Fatigue Results for HIP Ti-6Al-6V-2Sn

The tensile and fatigue requirements for Ti-6Al-4V are within the required range for plate properties from MIL-T-9046, as shown in Figures 8 and 9.

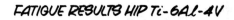

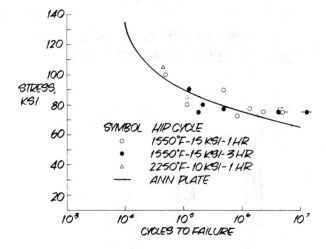

Figure 8

Effect of HIP Cycle on Fatigue Life of Ti-6Al-4V

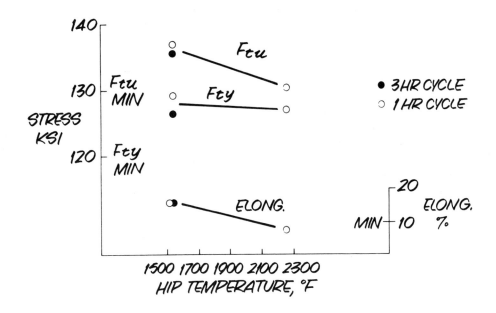

Figure 9

Tensile Properties of HIP Ti-6Al-4V

Conclusions

Feasibility of manufacturing titanium parts to near net shape by hot isostatic pressure has been established. In addition, preliminary mechanical
property values were in the range of annealed plate minimums for both
Ti-6Al-6V-2Sn and Ti-6Al-4V.

Current Work

A program is currently underway with Grumman and Crucible to determine
reproducibility and to verify these property values. In addition, qualification
and flight testing of the fuselage brace is scheduled and up-scaling to larger
parts is projected.

TITANIUM NET SHAPES BY A NEW TECHNOLOGY

G.R. Chanani[1], R.H. Witt[2], W.T. Highberger[3], and C.A. Kelto[4]

Abstract

Increased costs and lead times for current aluminum and titanium parts on high-performance aircraft have led to development of ways to compensate for cost increases and delivery delays; one promising approach for titanium alloys employs hot isostatic pressing (HIP) of spherical titanium alloy powders to produce complex, near-net shapes having high materials utilization factors (MUF). This paper is concerned with the status of the Crucible Inc. Ceramic Mold HIP process as it is being developed on Navy and Air Force programs for near-net parts which could make significant use of this technology possible in the future. Part I deals with F-14A Parts Evaluation and Part II with an F-18A Part Evaluation.

PART I — F-14A PARTS EVALUATION

by R.H. Witt, and W.T. Highberger

Synopsis

An F-14A fuselage fitting (1.7 pounds) demonstrated MUF could be increased from 28 to 81%, costs reduced 25 to 40% (depends on quantity) and delivery times improved by 50%. A production lot of this part will be flown in 1980. This success has led to a scale-up effort on an F-14 nacelle frame (53 pounds). Progress for these programs will be reported on mechanical properties, shape making capability, dimensional reproducibility, flight worthiness and quality.

Introduction

The high cost of manufacturing titanium aircraft and engine components is due primarily to industrial limitations on producing close-to-final dimension shapes. This inability has made the ratio of the material bought to that which flies unacceptably high. In addition to high buy-to-fly ratios, forging delivery lead times, now up to as high as 100 to 150 weeks, and overall machining requirements contribute to delays, slow production rates and/or increased costs.

As a result of these considerations, many efforts in recent years have been directed toward reducing the cost of complex titanium parts, especially those which are manufactured by methods that require extensive machining on all surfaces to final dimensions. The work that has been accomplished recently on HIP of titanium and superalloy shapes has clearly shown the cost-reduction possibilities by producing HIP net or near-net shapes. The major work that remains to gain acceptance of this process in the industry relates to scale-up of the process for large parts and development of a design data base.

---

1. Northrop Corporation, Hawthorne, CA
2. Grumman Aerospace, Bethpage, NY
3. Naval Air Systems Command, Washington, D.C.
4. AFML/LTM, WPAFB, OH.

Reprinted with permission from Titanium '80 Science and Technology, Vol. 3, 1980, 2309-2319, © 1980 American Institute of Mining, Metallurgical and Petroleum Engineers, Inc.

This paper discusses the status of Navy programs that have been directed toward attaining this goal. Table 1 summarizes the parts and alloy selection that have been employed for the F-14A parts and shows the buy-to-fly ratios and materials utilization factors that are being sought relative to machined forgings which are presently in use. In the following, after a brief description of the HIP and powder processes utilized, a discussion is given of results obtained on studies related to F-14A aircraft.

Process Description

Hip Process — The Crucible Ceramic Mold HIP process consists of encapsulating metallic prealloyed titanium powders in a suitably shaped mold, inserting the mold in a special can surrounded by a compressible medium, evacuating, and sealing the can. The assembly is then isostatically pressed at a high temperature (1600 to 1750°F) at a suitable pressure (10 to 15 ksi) in an autoclave that contains gaseous media (usually argon). Figure 1 shows schematically the principal steps in the process employed by Crucible. Recent work on scale-up of the process has indicated that is more efficient to machine wax patterns from large blocks in the early iterations, thereby eliminating the complex die. In considering the effects of HIP parameters, it is essential to realize that soaking at the HIP temperature is required to bring the heavily insulated mold assembly to the required temperature; the true HIP cycle, therefore, begins 8 to 10 hours after start of the cycle. Work is presently being concentrated on reducing this precycle time by reducing the amount of compressible ceramic powder in the can. Consideration is also being given to preheating cans prior to the HIP cycle.

TABLE 1 — PART SELECTION, ALLOY, B/F & M.U.F.

PART	F-14A FUSELAGE BRACE SUPPORT FITTING	F-14A NACELLE FRAME
ALLOY	Ti-6Al-6V-2Sn	Ti-6Al-6V-2Sn
FLYING WT, LB	1.7	53.0
B/F RATIO [1]		
• FORGING	3.6	6.6
• HIP	1.2	1.3
M.U.F. – % [2]		
• FORGING	28	14
• HIP	83	77

NOTES:

(1) BUY-TO-FLY RATIO $= \dfrac{\text{RAW WT}}{\text{MACHINED WT}}$

(2) M.U.F. = MATERIALS UTILIZATION FACTOR $= \dfrac{\text{MACHINED WT}}{\text{RAW WT}} \times 100$

Powder Process — In all of the studies discussed, the powder used has been produced by Nuclear Metals Inc. using the patented Rotating Electrode Process (REP).

Chemical analysis of the alloy is virtually unchanged by this process. The titanium alloy powders produced to date for the parts discussed and related studies include regular grade Ti-6Al-4V, ELI grade Ti-6Al-4V, Ti-6Al-6V-2Sn, Ti-6Al-2Sn-4Zr-6Mo and Corona 5 (Ti-5Mo-4.5Al-1.5Cr).

F-14A Aircraft Parts — In the early 1970s, a program conducted under Navair funding demonstrated the feasibility of manufacturing close-tolerance, low draft-angle titanium forgings from contoured sintered preforms using elemental powders. Microporosity in ribbedsection tips remained, however; this was attributed to preform design and density. Utilization of 100% density preforms as produced by HIP, rather than cold isostatic pressing and sintering, was proposed as the next step. Initial trials showed that the HIP process not only could produce 100% density, but that properties were good enough to consider the HIP process for producing near-net shapes in an essentially one-step operation.

Fuselage Brace — As a result of these early findings, NAVAIR funded a program to determine the feasibility of producing an F-14A titanium airframe component of prealloyed powders by HIP. The success of this process verification study resulted in a continuation program to further evaluate the process.

This program established the feasibility of manufacturing F-14A small components (1.7 lb) by establishing flight-worthiness, reproducibility, economics and NDI criteria. The part, an F-14A fuselage fitting (Figure 1), showed that the MUF could be increased from 28% to 81% and cost savings of 25 to 40% could be realized depending on the quantity purchased. A production lot of this part has been ordered and is scheduled for flight in 1980.

An initial effort on feasibility, consisting of tooling and parametric investigations for the shape selected, proved-out shape-making capability with surface finishes around 100 rms and showed that mechanical properties equivalent to annealed plate were attainable for Ti-6Al-6V-2Sn REP powders. Based on these results, it was concluded that mechanical properties depend on time at pressure and temperature, and oxygen content of the material. Table 2 shows the tensile properties obtained for samples with and without post-vacuum anneals. It was found that vacuum anneals benefited the fatigue properties. A 2-hour vacuum anneal, therefore, has been retained as a required post-HIP treatment.

A reproducibility pilot-lot run was then made which confirmed properties and showed that dimensional tolerances were attainable. The total weight of the configurations processed was 113 lb — far in excess of previous loading which ranged from 8 to 27 lb. All ultimate and yield strength values were within 5% of the current requirements for annealed forgings. Typical elongation properties exceed applicable specification minimum requirements by a factor of two. Fatigue data showed that saturation shot peening to 0.1 Almen intensity gave acceptable results for these parts and fracture toughness values were 72-74 si $\sqrt{\text{in}}$.

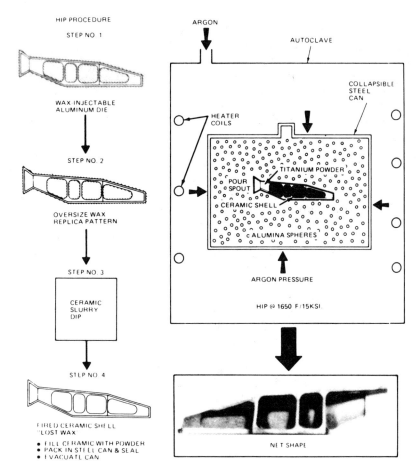

FIGURE 1. BASIC STEPS IN CRUCIBLE HIP PROCESS USING CERAMIC MOLDS

TABLE 2 — TENSILE PROPERTIES OF HIPed Ti–6AI–6V–2Sn CONFIGURATIONS

	AS MACHINED HIPed PROPERTIES			HIPed PROPERTIES AFTER ANNEALING AT 1300°F			
POWDER LOT	1	2	1	2	2	GRUMMAN	
PPM O_2	1900	1800	1900	1800	1800	FORGING SPECIFICATION	
HR AT 1330°F	NONE	NONE	2 HR	24 HR	24 HR	GM 3117	
F_{tu}, H	149.1	144.7	148.0	142.5	144.2	150.0	
F_{ty}, H	142.2	136.6	141.0	135.1	136.4	140.0	
ELONG, %	17.7	14.7	18.0	22.0	20.0	8.0	
RA, %	35.1	43.9	43.0	47.8	42.2	20.0	

To verify flight-worthiness of HIP braces, a truncated 12-level spectrum fatigue test was designed based on the actual 212-level load spectrum experienced by the F-14A wing-box in flight. The acceptance criteria were based on four life cycles, i.e., 24,000 equivalent flight hours (EFH). One HIP brace withstood the test at 100% operational stress level; a second brace withstood the test at 150% operational stress level. The functional spectrum fatigue tests in the simulated box fixture thus verified the flight-worthiness of the HIP component.

Nacelle Frame — The other F-14A aircraft part is a nacelle frame weighing 53 lb (as-machined for flight) which will be fabricated by electron-beam (EB) welding four HIP'd powder parts into an oval-shaped F-14A frame section. Scale-up to parts ranging from 20 to 50 lb promises significant MUFs (around 80-95%), cost savings over 50% and potential for reducing delivery times by about 50% or more. In this case, an attempt is being made to increase the MUF from 14% for machined forgings to over 90% for the HIP component. Qualification of this type of part should increase autoclave throughput and result in reduced autoclave charges per part. This will help improve delivery schedules.

Preliminary EB welding parameter work performed to date indicates that acceptable quality and mechanical properties are attainable on transverse butt welds. Tensile ultimate strength, yield strength and elongation for EB butt welds were 158.3 ksi, 148.5 ksi and 16% respectively.

The program, which is just starting, consists of three phases: powder certification (Phase I); design and processing of EB welded hybrid nacelle frames (Phase II); and reproducibility study and structural component tests (Phase III).

TITANIUM NET SHAPES BY A NEW TECHNOLOGY
PART II - F-18A PARTS EVALUATION*

G.R. Chanani, W.T. Highberger, and C.A. Kelto

Synopsis

The F-18A arrestor hook support fitting (AHSF) was selected to demonstrate the viability of hot isostatic pressing of prealloyed titanium powder in the manufacture of large, complex, and structurally critical airframe parts. Three iterations are planned to obtain a near-net shape AHSF. The first iteration of this complex part proved the process to be feasible. Even though some of the areas on the P/M AHSF were not completely filled, overall dimensions and shape of the part were very good. Two 3x5x5-inch P/M Ti-6Al-4V ELI witness blocks with different annealing treatments were evaluated by Northrop. Tensile properties of the heat treated ELI P/M product were within the acceptable range for annealed wrought Ti-6Al-4V ELI alloy. Process verification of the final iteration support fittings will include dimensional analysis, NDI, mechanical testing, and microstructural examination. In addition, a P/M AHSF will be structurally tested to evaluate the durability and damage tolerance of the component. Successful completion of these tests and coupon testing will demonstrate the structural integrity and applicability of the P/M manufacturing processes developed during the program.

Introduction

The overall objective of this program is to apply low cost titanium powder metallurgy technology to advanced Naval aircraft structures. Specifically, the objective is to reduce the cost of manufacturing selected titanium components for the F-18. This will be accomplished by hot isostatic pressing of titanium powder to make a near-net shape of the F-18A Arrestor Hook Support Fitting (AHSF). In this paper, results obtained on witness blocks and the first iteration HIP part will be described. Three different heat treatment conditions as well as the as-HIPed condition were evaluated for use on the actual component. The three heat treatment conditions were beta anneal (BA), recrystallize anneal (RA), and mill anneal (MA).

Process verification of the first iteration part included dimensional analysis, NDI, and microstructural examination.

The F-18A AHSF serves as the fuselage attachment of the arrestor hook used for cable braking in carrier landing. This selection was made for the cost reduction potential, complexity, use potential, generic potential, process validation, and accessibility of the fitting.

---

*This work was performed under joint AFML/NAVAIR sponsorship, Contract No. F33615-77-C-5005.

Currently the AHSF is machined from a hand-forged billet with a buy-to-fly weight ratio of 15/1. However, a die forging with a buy-to-fly weight ratio of 6.2/1 will be used for producing this part. This compares with a ratio of 2/1 when produced from a near-net HIP shape. The machining of the P/M part will be limited to facing and drilling the lug and the final machining of the interior of the fitting along the upper flanges which mate with the keel structure.

Program and Process Description

The Crucible Ceramic Mold HIP process described in Part I of this paper is being used for this program. Since it is difficult to predict consolidation characteristics of complex shapes like the AHSF, three iterative changes in mold geometry are anticipated to establish the final wax and mold shapes to produce an accurate near-net AHSF.

Due to the critical nature of the part, an ELI (Extra Low Interstitial) grade of Ti-6Al-4V powder is being used. The powder will be produced by the rotating electrode process (REP) at Nuclear Metals, Inc. (NMI). The powders will be fully characterized for chemistry, cleanliness, size distribution, flow rate, density, and microstructure.

The HIPed components produced from each iteration will undergo dimensional analysis, NDI, mechanical testing, and microstructural examination.

For the final heat treatment selection, different annealing treatments on witness blocks were evaluated. This evaluation included tensile, fatigue, fracture toughness, and fatigue-crack growth tests on the heat treated as well as the as-HIPed material.

In addition, a titanium P/M AHSF obtained from the pilot production run of the final shape parts will be structurally tested to evaluate the durability and damage tolerance of the component.

Successful completion of these tests and coupon testing will demonstrate the structural integrity and applicability of the P/M manufacturing processes developed during the program.

Results and Discussion

The Ti-6Al-4V ELI powder used for the first iteration AHSF and witness blocks was produced by NMI using the rotating electrode process. As expected, the REP powder was uniform and predominantly spherical. The powder exhibited an acicular microstructure, typical of material rapidly cooled from the beta phase field.

The powder was examined for foreign particles both metallographically and magnetically. Polished and etched cross sections of the powder and the powder compacts were carefully scanned for irregular or foreign particles. This examination revealed a large number of tungsten particles.

Figure 1 shows a machined wax pattern for the P/M AHSF. The target drawing for this wax pattern was prepared by modifying the engineering drawings to include tooling check and machining index points to assure that the P/M part will machine to final dimensions of the AHSF at the lowest cost. The ceramic shell mold made from this wax pattern is shown in Figure 2.

The first iteration part was HIPed at 1750F using 15,000 psi for 10 hours. Figure 3 shows a comparison of the first iteration P/M AHSF with the actual machined part.

A dimensional analysis of the first iteration AHSF was performed by both Northrop and Crucible. Even though some of the areas on the P/M AHSF were not completely filled, overall dimensions and the shape of the part were considered very good. Radiographic and penetrant examination revealed some surface voids as well as several areas which were not completely filled. Indications of dense particles (probably tungsten) were found throughout the part by radiography. These areas averaged approximately 0.035 inch in size with the largest particles being on the order of 0.085 inch. No internal cracks were found by radiographic inspection (limit of detection: 1.5 percent of thickness of the part, the thickness of the part varied from 0.032 to 1.5 inch).

A study using witness blocks was performed to select the annealing treatment which would provide satisfactory mechanical properties with the minimum amount of distortion during the heat treatment cycle. For this work two 3 x 5 x 5-inch Ti-6Al-4V ELI blocks were HIPed at 14.7 ksi for 10 hours. These blocks were cut into four groups of specimen blanks. Since the HIPed blocks had a high hydrogen content*, three of the four groups of specimen blanks were vacuum degassed at 1300F for 12 hours while the fourth one was used in the as-HIPed condition. Table 1 lists the heat treatment parameters for the three anneal treatments (BA, RA, and MA).

Tensile test results for the above annealing treatments are shown in Table 2. These results meet the specification requirements presently used for wrought Ti-6Al-4V ELI parts, which are also shown in Table 2. The strain-life fatigue results on smooth specimens are shown in Table 3. The as-HIPed and mill annealed samples showed somewhat longer smooth fatigue lives than the other treatments. The low fatigue life of one of the recrystallized annealed specimens was due to a large nonmetallic inclusion. The fatigue crack-growth rate testing showed essentially similar crack-growth rate behavior for all the specimens except for the beta-annealed material, which showed a slightly superior fatigue crack growth behavior. Fracture toughness values were essentially similar for all the specimens and exceeded specification requirements for wrought ELI Ti-6Al-4V material.

*This high hydrogen content was attributed to the decomposition of moisture in the secondary pressing medium during hot isostatic pressing and could also be the cause of the substantial (0.012 to 0.016 inch thick) alpha case. This problem is being rectified in future shape trials.

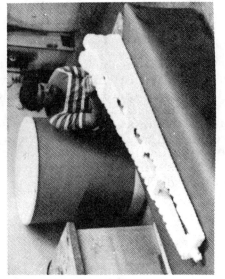

FIGURE 2. CERAMIC SHELL MOLD MADE
FROM THE WAX PATTERN IN FIGURE 1

FIGURE 3. COMPARISON OF MACHINED AND FIRST ITERATION HIPED ARRESTOR HOOK SUPPORT FITTING

FIGURE 1. MACHINED WAX PATTERN OF
THE ARRESTOR HOOK SUPPORT FITTING

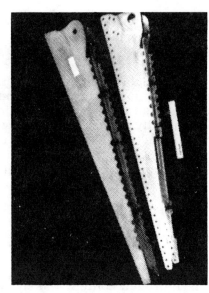

TABLE 1. HEAT TREATMENT

BETA ANNEAL (BLOCK 725B – 1750/30 MIN, HEATED TO 1865F/15 MIN, AIR BLAST COOLED, 1300F/2 HR, AIR COOLED

RECRYSTALLIZE ANNEAL (BLOCK 726A) – 1750F/2 HR, AIR COOLED, 1400F/2 HR, AIR COOLED (TO BELOW 900F WITHIN 45 MIN)

MILL ANNEAL (BLOCK 726B) – 1300F/2 HR, FURNACE COOLED (300F/HR MAX) TO 1000F, AIR COOLED

TABLE 2. TENSILE PROPERTIES FROM HIPED Ti-6Al-4V (ELI) MATERIAL

HEAT TREATMENT (HYDROGEN CONTENT)	YS (KSI)	UTS (KSI)	%E	R.A.	MODULUS (10^6 PSI)
AS HIP (236 PPM)	124 124	135 135	15 16	41 41	17.1 16.8
BETA ANNEAL (64 PPM)	129 130	138 138	10 10	23 24	16.8 16.9
RECRYSTALLIZE ANNEAL (82 PPM)	125 124	134 133	14 15	40 40	16.5 16.9
MILL ANNEAL (40 PPM)	119 119	128 127	17 16	41 40	16.4 16.7
MIN SPEC MMS-1225 FOR ANNEALED Ti-6Al-4V (ELI) MAX HYDROGEN 125 PPM	110	120	10	20	—

TABLE 3. STRAIN-LIFE FATIGUE RESULTS ON SMOOTH SPECIMENS

CONSTANT AMPLITUDE, STRAIN CONTROL
MAX STRAIN = 0.010, R = 0.0, F = 1 HZ,
TRIANGULAR FUNCTION, ROOM TEMPERATURE

HEAT TREATMENT (HYDROGEN CONTENT)	CYCLES TO FAILURE
AS HIP (236 PPM)	12,195 11,243
BETA ANNEAL (64 PPM)	7,258 8,684
RECRYSTALLIZE ANNEAL** (82 PPM)	13,501 7,552*
MILL ANNEAL (40 PPM)	11,039 11,857

*FATIGUE-CRACK INITIATED AT A VOID IN THE VICINITY OF A 0.014-IN. SILICON-RICH PARTICLE.
**WROUGHT Ti-6Al-4V (ELI)
Nf – 6000 CYCLES
MAX STRAIN = 0.010, R = 0.0 F = 1Hz

The results of the heat treatment study indicated that all three annealing heat treatments investigated had acceptable tensile properties. The as-HIP or mill anneal condition is preferred over the other conditions due to the low warpage and distortion expected to result from these lower temperature treatments (No heat treatment for as-HIP, 1300F for mill anneal vs 1865F for beta anneal, and 1750F for recrystallize anneal).

Fractographic examination was performed on the failed fatigue specimens obtained from the heat treatment study. None of the primary or secondary crack initiation sites were found to be associated with the presence of tungsten. In most cases, nonmetallics such as silicon or calcium compounds were detected surrounding voids associated with initiation sites. These inclusions could have been introduced in the powder during production or handling.

Based on the heat treatment investigation results, the first iteration AHSF was vacuum annealed at 10^{-4} torr pressure at 1300F for four hours. Vacuum anneal was necessary to remove hydrogen. The heat treated AHSF was dimensionally analyzed and compared with the as-HIP AHSF to determine whether any warpage occurred due to heat treatment. The results showed that the dimensional changes were generally minimal and could be taken into account in design and final machining of the part. The maximum dimensional change over the total length of the HIPed AHSF was a 0.14 inch decrease, probably due to some warpage.

Samples were removed from various locations of the heat treated AHSF for microstructural analysis. The microstructural examination revealed that removal of an 0.020-inch surface layer by chem milling will eliminate the alpha case formed during HIPing. Except for presence of tungsten particles, the micrographs were basically similar to those for wrought material. Dimensional analysis performed after chem milling indicated that the surface removal was uniform.

Residual stress measurements at various locations of the AHSF using a Rigaku X-ray Analyzer indicated the heat treated AHSF was essentially in the stress-relieved condition.

Summary and Conclusions

The overall dimensions and shape of the first iteration AHSF were very good considering the complexity of the part. The heat treated part did not have any significant residual surface stresses. Fractographic investigation together with the mechanical test results did not indicate any harmful effect of tungsten particles. However, in all the low cycle fatigue tests the fatigue-crack initiation was associated with nonmetallic inclusions. The heat treatment investigation on HIPed witness blocks of Ti-6Al-4V ELI powder indicated that the tensile properties of the annealed P/M material met the specification requirements of wrought titanium material and the heat treatment response of the P/M material is similar to that for the wrought material.

CHARACTERIZATION AND PROCESSIBILITY OF
HIGH SPEED TOOL STEEL POWDERS

Dr. Kishor M. Kulkarni

Glidden Metals Group of SCM Corp.
Cleveland, Ohio

ABSTRACT

New developments in near net shape fabrication of high speed tool steel components via cold press and sinter route can in many cases lead to substantial cost reduction. The recent commercial availability of high quality water atomized powders is an important element in this development. This paper describes characteristics and processibility of water atomized powders and to a limited extent compares them with similar properties of impact pulverized and gas atomized powders. The powder characteristics studied include size distribution, shape, tap and apparent densities, flow rate, hardness and oxygen content. Then, compressibility and green strength of different types of powders are compared. An attempt is made to show the interrelationship of different powder characteristics and material processibility. Some background information is provided about tool steel metallurgy, P/M mill forms, full dense sintering and powder production.

Introduction

In the past few years water atomized tool steel powders have been commercially available. These high quality powders can be compacted using existing equipment and then the compacts can be sintered to full density without the need for any post sintering operations for further densification. This paper is primarily about the characteristics of such water atomized tool steel powders. Through out the paper M-2 high speed steel (HSS) is used for illustration although most comments are practically equally applicable to other HSS powders and to a lesser extent to other non-HSS tool steels. Details of powder production and sintering are beyond the scope of this paper but these items are referred to briefly in view of the close interrelationship between powder production, characteristics and sinterability. To a limited extent the water atomized powders are compared to gas atomized powders and to those produced with an additional step of impact pulverization.

In this paper by "full dense" sintering we mean densification in excess of 98 percent of the density of wrought material or similar composition. This is of a lower limit as in most practical applications the density is significantly higher. Often density measurement cannot be relied on as an inspection method for porosity and only microscopic examination must be employed to detect if there is any porosity.

Background

Traditionally the "as sintered" density of ferrous or copper based powder metallurgy (P/M) materials rarely exceeds 90 percent of the wrought density. In some applications such as in bronze bearings and filters the applications require porosity. In others the target density is selected on the basis of a combination of product requirements, commercial feasibility and cost. Further increase in density can be obtained at added cost by techniques such as infiltration, P/M forging or other type of mechanical working and hot isostatic pressing (HIP). For various reasons so far P/M forming has not found wide applications; but the investigation effort has clearly demonstrated the sensitivity of important properties like fatigue life and impact strength to even a few percent of residual porosity. In shaped components HIP has been used primarily for nickel base superalloys for aerospace industry and for carbides as tooling components or cutting tools. By implication HIP is still too expensive for other applications. Property improvements obtained by infiltration are modest in comparison with those possible by P/M forging or HIP and infiltration is not relevent in the context of processing of tool steels.

At present most of the HSS are used in cutting tools in operations involving interrupted cuts. In such demanding applications full density is required. Yet, till recently cost effective P/M methods were not available for processing of tool steels and therefore few tool steel parts were made via P/M. Progress in technology and increasing cost for key alloying elements like Cobalt, Tungsten and Molybdenum are now providing impetus to P/M processing of tool steels.

Alloy	Wt. %					
	C	W	Cr	V	Mo	Co
M-2	1	6	4	2	5	-
T-15	1.5	12	4	5	-	5

PRINCIPAL FUNCTIONS

W - HARD CARBIDES, RED HARDNESS

Cr - HARDENABILITY, HARDNESS WITH HIGH CARBON

V - VERY HARD CARBIDES, FINE GRAINS

Mo - HARD CARBIDES, RED HARDNESS, HARDENABILITY

Co - HIGH RED HARDNESS

Fig. 1   Compositions of Two Common High Speed Tool
Steels and Functions of Key Alloying Elements

Particular attention is attracted to near net shape production methods in view of the increasing labor and material costs and material shortages.

HSS Metallurgy

High speed tool steels are probably the most complex amongst the common ferrous materials and there are several excellent books[1-3] and numerous technical papers devoted to their metallurgy. We shall only review some basic facts important for the remainder of this paper.

Fig. 1 shows nominal compositions of M-2 - the most common grade in U.S.A. - and of T-15 - one of the more highly alloyed grades. Functions of important alloying elements are also summarized in this figure. High hardness, its retention at high temperatures and adequate toughness are major attributes of HSS. The room temperature hardness derives from the usual martensitic reaction and from hard carbides of Cr, Mo, W and V. The high temperature or hot hardness comes because of the stability of carbides of Mo, W and V. The classic paper by Kayser and Cohen[4] clearly shows how a large portion of MC type carbides of V and much of M_6C type carbides of W and Mo remain undissolved even at the high solution treating temperature employed for HSS. Overall these alloying elements greatly reduce the kinetics of most reactions, make HSS air hardening, necessitate multiple tempering and give HSS a very high resistance to grain growth. Most HSS are solution treated at temperatures close to start of melting and for optimum properties close process control is essential during their heat treatment.

P/M Mill Forms

There is general agreement that it is desirable to have fine, uniformly distributed carbides. Unfortunately, the carbides in tool steels are prone to severe center line segregation at the ingot stage. Extensive degree of hot working is essential to break up and distribute the carbides and even then some banding and non-uniformity of carbide size persists. The carbides also cause poor hot workability. As a result of these factors conventionally produced mill forms i.e. round or rectangular bars and other products of uniform cross section, are expensive to produce and of less than optimum properties. Therefore much work has been carried out on production of P/M mill forms and Crucible Inc.[5-8] division of Colt Industries and Uddeholms (formerly Stora Kopparberg) have commercially produced P/M mill forms since early 1970's.

The P/M method in its most common form comprises production of gas atomized powder, preparation of cold isostatically pressed (CIP) compact, canning the compact and then subjecting it to HIP. Subsequent hot mechanical working may or may not be employed. The main advantages claimed of the P/M mill form are reduced distortion during heat treatment, improved grindability, finer grain size, improved toughness and improved tool performance. The products appear to be receiving growing market acceptance specially for grades such as T-15 and M-4 which are more difficult to produce by the conventional cast and hot work method. In addition, both Crucible and Uddeholms have developed[6-8] special more highly alloyed grades of tool steels which would be impossible to produce by conventional methods.

There are many applications for which the P/M mill forms will be preferable because of their above mentioned advantages. However, in a majority of the cases involving shapes they suffer from the same disadvantages as the conventional mill forms, namely, poor material usage and high machining cost. Although the basic process of gas atomization, CIP and HIP is technically feasible for near net shape making, the author believes that any method incorporating HIP is likely to be prohobitively expensive for most HSS components. Further the whole approach is aimed at end users or machine shops and not at the P/M industry. As described later, the gas atomized powder is not suitable for the usual P/M method of compaction and sintering and as far as is known such gas atomized powder is not commercially sold.

Full Dense Sintering

The cold press and full dense sinter process is a relatively recent development. The method uses water atomized powder which is compacted by either CIP or die pressing. Lubricant is unnecessary in CIP and only die wall lubrication is preferred in die pressing. (Subsequent sections will give more details of pressing and green properties). Good quality of powder and closely controlled high temperature sintering are essential.

The principal advantage of the process is its near net shape capability. This means that complex shapes can often be made with only finish grinding so that material utilization is high and machining cost can be reduced substantially. In addition, other advantages of P/M processing such as fine uniformly distributed carbides are also obtained. Successful use of full dense sintered cutting tools attests to their good toughness. Most of the work thus far has been on full dense sintering of high speed tool steels. Availability of high quality tool steel powders that can be pressed on existing equipment is an important element in the development of the technology. However, the furnace for this process has more exacting demands than normal P/M furnaces and vacuum furnaces are preferred. Also, as with any near net shape process careful attention is needed to selection of candidate components, manufacturing tolerances and tooling design.

Published literature indicates[9-13] that there are many companies engaged in studying full dense sintering of HSS and several patents have been issued relating to the technology. Two companies active in commercially using the process are Consolidated Metallurgical Industries, Inc. (CMI) in which SCM Corporation has a minority interest and Powdrex Inc. in England. We shall later comment more about the applications and future prospects; but, suffice to mention that we expect rapid growth in full dense technology both by this method and by other competing techniques.

Powder Production

Basically water atomization consists of melting the work material, allowing the melt to fall through a tundish and impacting the stream by water jet to form powder particles. Typically the water atomized particles are irregular in shape and consequently low in apparent density and high in green strength. As atomized HSS powder is very hard (with microhardness on Vicker's scale of around 850 HV) and has a high oxygen content of about 1,500 - 3,500 ppm. It must therefore be annealed and deoxidized to make it suitable for further processing. As for the particle shape and size distribution, some variation is possible through control of atomization conditions. Powdrex Inc. has developed a novel modification which gives powder of even greater irregularity of shape than other methods of water atomization. (SCM has licensed this powder making technology.) Unless otherwise mentioned, the characteristics of water atomized powders discussed below are those of prealloyed HSS powder produced by Powdrex Inc.

Water Atomized Powders and Their Processing

There are several powder characteristics which are customarily provided by a manufacturers along with the material. In most cases they are easy to measure and yet provide important clues about the processibility of the powder and its suitability for the intended application. Fig. 2 gives such data about a parti-

APPARENT DENSITY   2.2 G/CM^3

TAP DENSITY   3.0 G/CM^3

FLOW RATE   40.0 s/50 G.

SIEVE ANALYSIS

+ 60 MESH	0
−60 + 80	10
−80 + 100	8
−100 + 150	16
−150 + 200	17
−200 + 325	21
−325	28

CHEMISTRY

0.85 C, 0.2 Mn, 0.2 Si, 6.1 W, 4.0 Cr,
5.3 Mo, 1.8 V, 800 ppm O_2, 150 ppm N_2

GREEN DENSITY   6.6 G/CM^3 ⎫
TRANSVERSE RUPTURE ⎱   7600 psi   ⎬ 60 TSI (827 N/mm^2) COMPACT,
GREEN STRENGTH   ⎰   (52 N/mm^2) ⎭ DIE WALL LUBE ONLY

Fig. 2

Example of Data For One Lot of Water Atomized
and Annealed Powdrex Type M−2 Powder

cular lot of M-2 powder produced by Powdrex Inc. In this section we shall discuss the various powder properties in Fig. 2 with regard to their importance and compare them with similar data for other types of powders. Much of this section is about the powder or about green compacts but some information is provided about full dense sintered items. The data were obtained with procedures in general conformity with applicable MPIF standards

Particle Shape

The as atomized particle shape is primarily a function of the cooling rate during atomization with faster cooling leading to a more irregular shape. A highly irregular shape is typical of water atomized powder and is obvious in Fig.3. Several methods have been tried to define the irregularity of the shape quantitatively; but, from a more practical view point irregularity of shape can be better judged by apparent density (A.D.) and green strength which it affects greatly as discussed later. Another technique called the Coldstream process (14) can be employed to produce fine powders from a variety of feed stock. The process depends on size reduction through a high speed impingement of the feed stock against a fixed target and would naturally lead to a far greater uniformity of particle shape than the irregularly shaped water atomized powder in Fig. 3. Also, it is well known that gas atomized powder as used for example by Crucible and Uddeholms is quite spherical and thus very regular in shape.

Sieve Analysis

The sieve analysis data in Fig. 2 is presented graphically in Fig. 4. Sieve analysis is a representation of the particle size distribution and can be used to check for lot to lot variation of the distribution. It has a direct effect on other properties like the apparent density. But, it must be noted that the sieve analysis is influenced by the particle shape so that even for the same equivalent sphere (mean volume) diameter the irregular water atomized powder may appear coarser than the more regular powders obtained with other processes. In any case, the size distribution can be varied as desired within practical limits by manipulating atomization conditions. Currently the water atomized powder is supplied to -60 mesh size and can be full dense sintered as such; although in general, finer powder gives a smaller as sintered grain size.

Apparent Density, Tap Density and Flow Rate

Apparent density is important from a practical view point because it must be considered in designing tooling to get the target as sintered dimensions. For the same reason, any given tooling can be employed satisfactorily only with powder having a particular or very similar apparent density. While all the above properties can be controlled during powder manufacture, in general, irregular powder has low A.D. and proportionately larger difference between A.D. and T.D. and a longer flow time than a more regular powder. However, for a given particle shape, finer the powder the more difficult its flow with the resulting longer flow time. Very fine powder may not flow at all when checked according to MPIF standard No. 03. As for consistency of A.D., note that A.D. follows the law of mixtures so that blending of different lots can be resorted to if unusually close uniformity is required.

130X

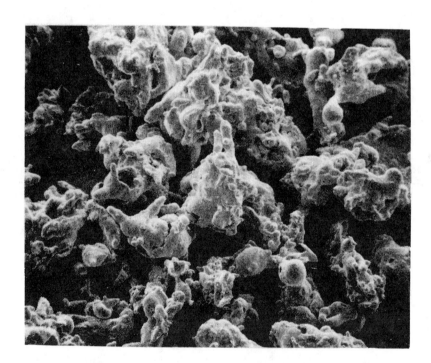

260X

Fig. 3   Appearance of Water Atomized Powdrex Type M-2 Powder
(Notice the highly irregular particle shape)

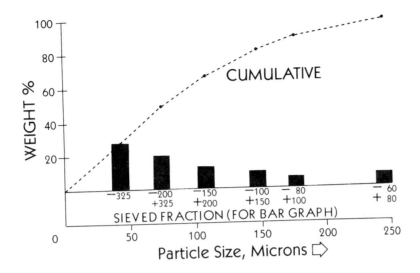

Fig. 4   Particle Size Distribution for Powdrex M-2 powder.

Compressibility and Hardness

The microhardness of annealed water atomized HSS is approximately 250 - 300 HV with some variation depending on the grade of HSS and the exact cycle used for annealing. Microhardness measurement on powder particles is tricky. A more direct indication of hardness is the pressed or green density at a given pressure, that is, compresibility. Fig. 5 relates green density to compacting pressure for samples pressed with die wall lubrication alone. Slightly higher densities result with a small quantity of admixed lubricant. It is important to point out that density values in Fig. 5 are similar to those for commonly used 410L SS powder [15] thus showing the excellent compressibility of the water atomized powders and the feasibility of compacting them in existing equipment.

The HSS powders can be compacted either in hard tooling or by CIP. Larger production quantities and need for closer tolerances favor die pressing where as smaller quantities but more complex shape and/or larger size suggest isostatic pressing. Elimination of lubricant and greater uniformity of density are other advantages of isostatic pressing. The upper limit for most CIP facilities is 30 tsi or 414 N/mm^2 (which is adequate for most applications) where as die pressing is often carried out at twice that pressure.

Comparison of differnt types of powders shows some interesting trends illustrated in Fig. 6. The water atomized powder has low A.D. and T.D. but the pressed density increases rapidly with compaction pressure. The increase in density continues at a high pressure of 60 tsi or 827 N/mm^2 and probably beyond. For any particular alloy the density of annealed powders at high pressure of 30 tsi (414 N/mm^2) or more is primarily dependent on the material properties. The densification at high pressure occurs by a large amount of plastic flow which is nearly independent of the initial particle shape or size distribution. The gas

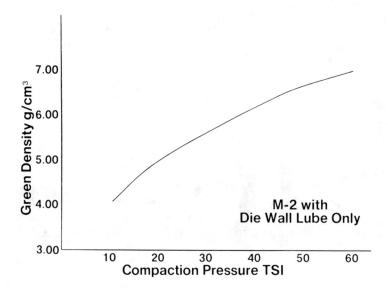

Fig. 5 Effect of Compaction Pressure on Green Density of Annealed
Powdrex M-2. (1 TSI = 13.8 N/mm^2)

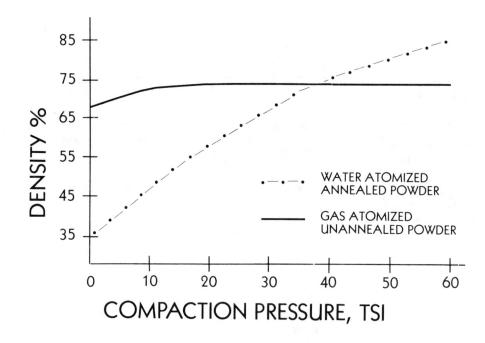

Fig. 6 Comparison of Compressibility of Water/Atomized and gas Atomized
Powders (The former is the same as in Fig. 5. The curve for gas
atomized powder is based on Ref. 7)

atomized spherical powders have high A.D. and T.D. but are normally compacted without annealing. Fig. 6 suggests that densification occurs at low pressures by some rearrangement of the spherical particles and some local plastic deformation at what must be initially point contacts. However, the hard particles cannot be deformed much and density becomes insensitive to higher pressures in excess of about 30 tsi or 414 N/mm^2.

Green Strength

Water atomized irregular powders have a higher green strength than spherical powders. A reasonable amount of green strength is essential in P/M industry to permit normal in process handling of the green compacts. Tests on annealed M-2 powder made with conventional (not Powdrex type) water atomization show its green strength to be similar to that of 304 L SS powder. More significantly as shown in Fig. 7 the water atomized powder made by the Powdrex process had a high green strength which is nearly three times that of powders made by other processes. This high green strength of the Powdrex type powder will be valuable in making parts with thin sections or complex details often needed on tooling components. In some cases green compacts can be machined to incorporate features that would be impossible or very expensive to achieve during compaction.

The gas atomized spherical powders normally compacted without annealing have no green strength to speak of. This factor coupled with their high hardness makes them totally unsuitable for the usual compact and sinter route used in P/M industry. The green strength depends on many different factors as shown in some careful studies by Klar and Shafer.[16] Considering the pertinent factors it is speculated that gas atomized powders, if it was annealed, would show some increase in green strength but it would still be very low compared to Powdrex

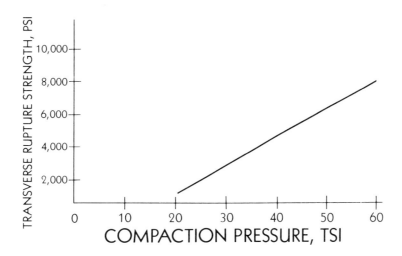

1 PSI = 0.007 N/mm^2 1 TSI = 13.8 N/mm^2

Fig. 7   Effect of Compaction Pressure on Transverse Rupture Green
         Strength of Annealed Powdrex M-2.

type of powder. In any case, such annealing of gas atomized powder would make it much more expensive than water atomized powder without any compensating advantages.

Chemical Analysis

The highly alloyed nature of HSS make it imperative to control chemical composition closely. Usually X-ray spectroscopy is utilized to measure most of the elements and care must be taken to follow proper procedures and use suitable standards (when available, to be secured from National Bureau of Standards). From a sintering point of view carbon and oxygen contents are the most important and to measure these, we have satisfactorily used commercially available carbon and oxygen analysers. The carbon content must be adequate for all the carbide formers like V, Mo, W and Cr and for matrix hardness but not so excessive that undesirable quantity of austenite stabilizes. Of course, carbon content can be adjusted by mixing carbon in HSS powders the same way as in the case of other ferrous powders.

Oxygen content of annealed water atomized powders is less than 1,000 ppm and is indicated for each shipment as shown in Fig. 2. The powder can be satisfactorily full dense sintered with such oxygen content which in fact decreases during sintering in the necessary non-oxidizing atmosphere so that oxygen content in sintered HSS may be similar to that in gas atomized and HIPped HSS. It is essential that the oxygen content be low to promote densification as well as to minimize the possibility of forming oxide inclusions and to avoid imparing the balanced carbon content during sintering. The caveat here is only about the oxygen content of the as received HSS powder. Our tests have shown that the HSS powder can be stored in closed containers in which they are supplied, for months without much oxygen pick up. Naturally it is desirable to use the powder quickly and as far as possible keep the storage time short. The importance of the interrelationship of C and O_2 contents is well documented.[17-19]

Dimensional Changes During Sintering

Sintered part precision depends on predictable changes in dimensions of the compact during sintering. In P/M industry consistent and low dimensional change during sintering is a highly desirable attribute of the powder. In this regard, processing of water atomized HSS powders differs greatly from sintering of other common powders. Full dense sintering of HSS powder may show linear shrinkage of the order ten percent because of the large difference between green and sintered densities. If the green density is uniform and sintering technique controlled, it would be reasonable to expect similar linear shrinkage in different directions. But, this clearly demands a good deal of expertise from the parts producer.

Sintered Microstructure

How does one judge if sintering is of good quality? One easy method is to inspect the microstructure in as sintered condition. Provided cooling is not deliberately slow the structure would be similar to but more uniform than austenitized and as quenched wrought HSS. Fig. 8 depicts schematically what to look for. The carbides should be small in size and well distributed and must not form a continuous network around the grain boundary. There should be no or very little porosity and grain size should be small. Considering the variety of grades, requirement of different applications and the heat treatments that the

Microstructure of As Sintered P/M Tool Steels

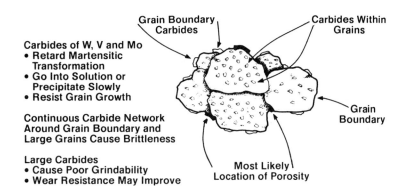

Grain Boundary Carbides

Carbides Within Grains

Carbides of W, V and Mo
• Retard Martensitic Transformation
• Go Into Solution or Precipitate Slowly
• Resist Grain Growth

Continuous Carbide Network Around Grain Boundary and Large Grains Cause Brittleness

Large Carbides
• Cause Poor Grindability
• Wear Resistance May Improve

Grain Boundary

Most Likely Location of Porosity

Fig. 8   Schematic of Microstructure of As Sintered P/M Tool Steels

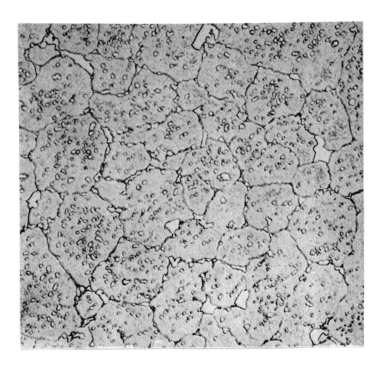

Fig. 9   Microstructure of As Full Dense Sintered M-2 Sample (Prepared from -60 mesh annealed Powdrex type powder. Magnification 500X. 5% Nital Etchant)

samples will subsequently undergo, quantitative specifications are difficult. But, these general guidelines may suffice in most cases. An example of good sintered microstructure is given in Fig. 9. Subsequent heat treatment will normally render the grain size much finer. Satisfactory performance in actual application is of course, the real acid test for the quality of the part.

Response to Conventional Processing

Let us comment briefly on how the water atomized HSS powder may respond to conventional processing. As mentioned earlier the preferred furnace is of vacuum type and sintering in the more common nitrogen bearing atmospheres may not lead to full density specially if the dew point is high. The nitrogen pick up will be considerably higher than allowed in wrought or vacuum sintered tool steels causing higher as sintered hardness as has been observed for stainless steel.[15] (It is interesting to mention that such nitrogen pick up has been studied[20] as a source of increased hardness.) Erratic results in densification and quality of microstructure are also possible because the controls on the existing furnaces may be far inferior to those built for full dense sintering. Having said all that, a conventional parts producer may still get good enough results to get a feel for the new opportunities that could develop with water atomized tool steel powders.

Applications and Future Prospects

Current Applications

Hervey[10] has discussed at length several current and potential applications of full dense sintered HSS parts. Examples of some components sintered by CMI are illustrated in Fig. 10. By and large the current applications cover metal cutting tools with one major application (not shown in Fig. 10) in a different category of wear parts viz. a cam for a diesel engine fuel injection pump. The part geometries that can be full dense sintered cover a broad spectrum and interestingly the new technology was cost effective in making relatively simple shaped T-15 indexable inserts produced by CMI since late 1978.[21] Powdrex Inc. has also reported[11] many applications of full dense sintered HSS including many of the threading dies uses in U.K. which are made from M-35 grade.

Expected Developments

P/M processing of tool steels is a rapidly developing technology and progress is expected both in work materials and in processing techniques.

A major advantage of P/M processing is the ease with which composition can be changed. Minor modifications in response to customer needs are possible right now and undoubtedly new special grades will be introduced. In fact, Powdrex has already introduced a special grade[11,12] for full dense sintering just as Crucible and Uddeholms have done[6,8] for P/M mill forms.

New applications will come not only as replacements for wrought HSS components but also for other materials. On one hand, new high hardness P/M grades may compete with carbides. At the other extreme, common grades such as M-2 processed by P/M route may replace lower alloy components either because of lower life time cost or because the increased performance demands make the lower

Approx. 3/4 Full Scale

Fig. 10 Examples of Full Dense Sintered HSS Components (Courtesy of Consolidated Metallurgical Industries Inc., Farmington Hills, MI.)

alloy material unsatisfactory. The full dense sintering technology may very likely be extended to other tool steels and to non-tool steel materials. Finally, even though this paper emphasizes full dense applications, it is recognized that not all properties are equally sensitive to residual porosity. Indeed, it is conceivable that somewhat lower density applications will develop or there will be technological break-through which will make it possible for a wide segment of conventional P/M parts producers to process HSS powders.

Acknowledgements

Thanks are due to Dr. C. I. Whitman, Director of Research, Mr. M. F. Marchitto, Manager of New Business Development, and other SCM Management for their support of this program and the permission to publish this paper. I am also indebted to many of my co-workers who have contributed to the experimental work. My special thanks to Ms. M. Cornelius for her careful typing.

References

1. G. A. Roberts, et_x al., "Tool Steels", Third Edition, American Society for Metals, Metals Park, Ohio, 1962.
2. P. Payson, "The Metallurgy of Tool Steels", John Wiley & Sons, Inc., New York, N.Y., 1962.
3. R. Wilson, "Metallurgy and Heat Treatment of Tool Steels", McGraw-Hill, New York, N.Y., 1975.
4. F. Kayser and M. Cohen, "Carbides in High Speed Steel--Their Nature and Quantity", Metal Progress, Vol. 61, June 1952, pp. 79-85.
5. E. J. Dulis and T. A. Neumeyer, "New and Improved High-Speed Tool Steels by Particle Metallurgy", in Progress in Powder Metallurgy 1972, pp. 129-142, Metal Powder Industries Federation (MPIF), Princeton, N.J., 1972.
6. A. Kasak and E. J. Dulis, "Powder-Metallurgy Tool Steels", Powder Metallurgy, Vol. 21, No. 2, 1978, pp. 114-121.
7. P. Hellman, et_x al., "The ASEA/STORA Process", Modern Developments in Powder Metallurgy, Vol. 4, pp. 573-582. MPIF, 1971
8. P. Hellman, "Powder Metallurgy High-Speed Tool Steels for Broaches", Technical Paper MR 78-287, Society of Manufacturing Engineers, Dearborn, Michigan, 1978.
9. R. P. Hervey, "Performed P/M Tool Steel for Cutting Tools", Technical Paper MR 78-315, Society of Manufacturing Engineers, Dearborn, Michigan, 1978.

10. R. P. Hervey, "Automotive Applications for Full Density Powder Metallurgy". Presented at the Society of Automotive Engineers, 1979 Congress & Exposition, February 27, 1979.
11. J. Smart and P. Brewin, "Complex Shapes in High Alloy Steels", Metal Powder Report, Vol. 33, No. 5, May 1978, pp. 207, 210-211.
12. "Powdrex Advances P/M High-Speed Steel Powder Production", announcement from Metal Powder Report, Vol. 33, No. 11, November, 1978, pp. 523.
13. F. L. Jagger, et_x al., "Heat-Treatment Response and Cutting Properties of Sintered Type M2 High-Speed Steel", Powder Metallurgy, Vol. 20, No. 3, 1977, pp. 151-157.
14. J. Walraedt, "The Coldstream Process: A New Powder Production Equipment", Powder Metallurgy International, Vol. 2, No. 3, August, 1970, pp. 77-80.
15. H. D. Ambs, and A. Stosuy, "The Powder Metallurgy of Stainless Steels", from Handbook of Stainless Steels by Peckner and Bernstein, Chapter 29, McGraw-Hill, New York, 1977.
16. E. Klar and W. M. Schafer, "On Green Strength and Compressibility in Metal Powder Compaction", from Modern Developments in Powder Metallurgy, Vol. No. 9, pp. 91-113. MPIF, 1977

References (cont.)

17. P. Lindskog and S. Grek, "Reduction of Oxide Inclusions in Powder Preforms Prior to Hot Forming", from Modern Developments in Powder Metallurgy, Vol. 7, pp. 285-301. MPIF, 1974

18. A. R. Kieffer, et al., "P/M High Speed Steels Modified by Addition of Monocarbides", from Modern Developments in Powder Metallurgy, Vol. 8, pp. 519-536. MPIF, 1974

19. G. T. Brown, "Development of Alloy Systems for Powder Forming", Metals Technology, May-June 1976.

20. J. J. Dunkley and R. J. Causton, "Powder Metallurgy Extrusion of High Speed Steels and Other Alloys", as reviewed in Metals and Materials, Feb. 1977, p. 36.

21. K. Gettelman, "Powdered T-15 Inserts Fill the Carbide Gap", Modern Machine Shop, Vol. 51, No. 6, November, 1978, pp. 86-89.

Production of components by hot isostatic pressing of nickel-base superalloy powders

Turbine disks for gas turbine engines are made from nickel-base superalloys but can be fabricated by several processing methods. The choice of a specific processing route depends on a variety of interrelated factors which include design property requirements, materials selected, and processing costs. It is shown that disk fabrication by direct hot isostatic consolidation of nickel-base superalloy powders, is the most cost-effective method of production; however, such a manufacturing route must yield a product compatible with the property requirements of the component. Some of the metallurgical factors that influence such properties in as-hot isostatically pressed (HIP) material are examined. It is shown that by control of the HIP condition, manipulation of composition and heat treatment can be used to control properties. Discussion centres on the balance between strength and stress rupture properties, and results for low-carbon Astroloy and a modified IN 100 alloy, identified in a recent alloy development programme, are used to illustrate the property balances that can be achieved in HIP materials. MT/432

M. J. Blackburn
R. A. Sprague

© 1977 The Metals Society. The authors are with the Materials Engineering and Research Laboratory, Pratt and Whitney Aircraft Division, East Hartford, Conn., USA.

The overall process sequence used to fabricate a component can be complex, especially for high-technology systems such as the gas turbine engine. All stages of such a sequence can have an impact on the final properties of the component, and thus process selection and control are important. With increasing emphasis on acquisition costs, maintainability, and efficiency, the search for low-cost fabrication techniques compatible with the component property requirements will continue. In this paper the production of turbine disk components by the consolidation of metal powders by hot isostatic pressing (HIP) is considered. Emphasis is placed on some of the metallurgical factors which influence the properties of HIP products. In order to put these factors in perspective, however, it is first necessary to consider the materials and processing methods used and the specific properties required. In several cases, these are contrasted with the methods, materials, etc. presently used for manufacturing turbine disks.

Materials

Nickel-base superalloys, because of their unique high-temperature properties, are used extensively in the turbine section of gas turbine engines. As the performance requirements of such powerplants become more demanding, the material properties must be improved to compensate for the more severe operating conditions. However, the specific property improvements required in an alloy depend upon the component for which it is to be used. For example, for the airfoil section of a blade or vane, high-temperature creep and rupture properties coupled with good thermal fatigue resistance are critical properties.

Turbine disk alloys which operate at lower temperatures have a different set of property requirements, which are enumerated below. Improved engine performance can be achieved by raising the turbine inlet temperature and increasing the rotational speed in the turbine. For a turbine disk, these changes in operating conditions lead to requirements for higher strength materials with superior stress rupture characteristics. Any development programme aimed at producing such properties cannot ignore other factors such as low density for lightweight structure and compatibility with low-cost manufacturing methods.

It is outside the scope of this paper to trace the efforts that have led to the development of the various nickel-base superalloys in use today. It will suffice to point out that many alloys were initially developed to be used in the cast form, and therefore for use in blade and vane applications. In several cases, slight modifications were made to such a cast alloy so that it could be hot worked, and thus a material/process for disk applications evolved. More recently, with the advent of powder metallurgical materials, further modifications have resulted in alloys that can be fabricated by this technique.

The status of powder metallurgy of nickel-base superalloys was reviewed in 1974 by Gessinger and Bomford.[1] In this review the production methods of powder making were described, and they have not changed markedly in the past two years, although several other system types are under development.[2] The alloys and properties discussed in the review were mechanically worked at some stage of the consolidation or processing cycle and thus are not directly relevant to this paper. Therefore, a better basis for comparison of properties is the alloys currently used for turbine disk applications within Pratt and Whitney Aircraft, which can be classified by increasing strength and temperature capability (other producers would cite a different series of alloys for a similar classification). This

This paper was presented at the conference on 'Forging and properties of aerospace materials' held jointly by The Metals Society, the Leeds Metallurgical Society, and the Bradford Metallurgical Society at the University of Leeds on 5–7 January 1977. The purpose of the conference was not only to survey current forging practice – primarily in titanium- and nickel-base alloys – but also to discuss future trends and possibilities including powder-forming methods and isothermal and superplastic forging.

1a Modern turbine disk

series is as follows:

INCO 901→Waspaloy*→Astroloy→IN 100

To give only one property point for this alloy series, the yield strength at 704°C increases from 700 to 1050 N/mm² from INCO 901 to IN 100. This alloy series is also arranged in order of increasing solute element content, for it is the precipitation of the gamma prime (γ') phase that is primarily responsible for the strength increases.

Turbine disk technology

Figure 1a shows a turbine disk and Fig.1b its cross-section with a tabulation of critical mechanical properties at various locations. The specific property requirements are location sensitive and reflect the temperature and stress gradients present in such a component under service conditions. The bore section of the disk which experiences the highest stresses demands high yield and ultimate tensile strengths in order to minimize displacements in service and to provide adequate burst margin. The rim of the disk operates under lower stresses but at higher temperatures; in such locations the creep rupture properties are critical.

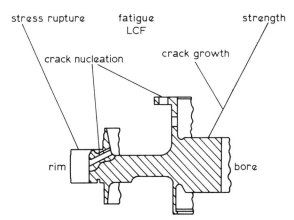

1b Cross-section of disk in Fig.1a, showing critical property requirements at various locations

*Registered name, United Technologies Corporation.

2 Manufacturing sequences for production of turbine disks

It should also be emphasized that as the rim slots are serrated to hold the blades, notch stress rupture properties must thus also be adequate. Discontinuities in a disk, such as rim slot cooling holes or bolt circle holes, are critical from the standpoint of low-cycle fatigue. The bore section could be fatigue limited by the growth of a fatigue crack from a small embedded processing defect flaw. Although the body stresses are the highest in this location, the situation can be dealt with by fracture mechanics, since these stresses are elastic. It can be seen, therefore, that a balanced set of properties is required for any material to be used in a disk component. In this paper two properties, strength and stress rupture characteristics, are emphasized. Fatigue capability is such a complex subject that it cannot be treated adequately here.

Manufacturing methods

There are essentially two methods of manufacturing disks; the various stages in both processes are shown in Fig.2. The conventional method begins with a cast ingot, usually produced by VIM and subsequent VAR melting and casting. This material is drawn down to billet by mechanical working from which forging multiples are sectioned. A series of forging operations reduce such a section to the general shape of the required disk. This forging is then machined to the final shape, as shown in Fig.1a. The

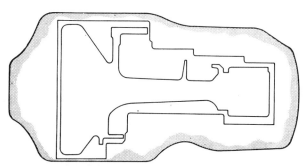

3 Cross-sections of a typical winged turbine disk at various stages of fabrication: typical weights for forging, 170 kg; sonic shape, 90 kg; finished shape 50 kg

second production route begins by reduction of the alloy to powder form, either by remelting ingot or producing a (virgin) melt directly and then fragmenting a stream of metal into small droplets. Reconsolidation of the powdered alloy to the desired preform shape can be accomplished by several methods, also listed in Fig.2. Subsequent processing can range from forging, of a type similar to the operations described in the conventional processing cycle, to no additional shaping in the case of a directly hot isostatically pressed net shape (direct HIP).

There are various input parameters to the selection of a manufacturing route to produce a disk. Obviously, the operational requirements, and hence the properties, will be set in the design stage, when the alloy to be used is determined. In an advanced engine turbine disk, this will almost certainly be a high-strength alloy with good (rim) creep rupture properties in addition to fatigue (life) goals. High-strength nickel-base alloys contain large concentrations of hardening elements, e.g. Al, Ti, Hf, Nb, in order to produce a high volume fraction of hardening phase $Ni_3(Al, Ti, \ldots)$, denoted γ'. Such high hardener content alloys cannot be processed (with an adequate component yield) by the conventional cast plus wrought method. Ingot segregation occurs on such an extensive scale as to preclude homogenization by thermal treatments, and local areas of low melting point constituents render the material unforgeable without such homogenization. Thus, high-strength alloys have to be processed by powder metallurgical methods, which eliminate the segregation problem.

As shown in Fig.2, several methods are available for producing consolidated shapes starting with powdered raw material. Of these, direct hot isostatic pressing is the most cost effective. Such processing minimizes the input material and thus subsequent machining and eliminates any subsequent working operations. Further details of the cost benefits of this approach may be found in Refs.3 and 4. The various sequence of shapes that occur during the fabrication of a disk are shown in Fig. 3, in which the relative material weights for each shape are also listed. The square-cut envelope is produced from

the forging shape for sonic inspection purposes; this is the type of shape which has been produced in the JT8D turbine disk programme described below. It could be thought that the nearer the shape produced approached the finished shape, the greater the cost savings. However, there are barriers at the present time to producing components in configurations very close to the finished shape. These include component surface condition and lack of adequate inspection procedures for complex shapes. In addition, techniques for producing reproducible shapes and for maintaining such shapes through subsequent heat treatment and machining are presently in the development stage.

The basic method for producing HIP shapes is quite simple, as illustrated in Fig.4. A container is filled with outgassed powder, sealed, and placed in an autoclave, where it is subjected to pressure at a high temperature. After consolidation, the container is removed and the shape proceeds through the normal inspection, heat-treatment, and machining procedures. In practice, the individual stages of the process have to be considered in some detail. For example, the selection of the container shape and material will have a major impact on cost and the dimensions of the product. Sheet metal containers have been widely used for large simple shapes, while quite complex small-scale parts have been produced in ceramic containers.

The shapes illustrated in Fig.3 were made in sheet metal containers from low-carbon Astroloy. They form part of an internal Pratt and Whitney Aircraft programme which has the objective of replacing the present Waspaloy disks in certain models of the JT8D engine with a directly hot isostatically pressed product. In this programme over 20 disks have been fabricated, and an extensive test programme has culminated recently in the first successful engine test. One of the most encouraging aspects of this effort has been the excellent and reproducible properties that have been obtained on large consolidation sections of material. Fears expressed in the early 1970s about the intrinsic properties of direct HIP products have proved

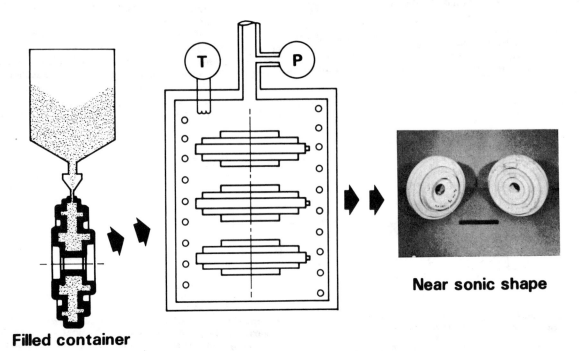

Filled container

Near sonic shape

4 Schematic representation of hot isostatic pressing method of disk manufacture

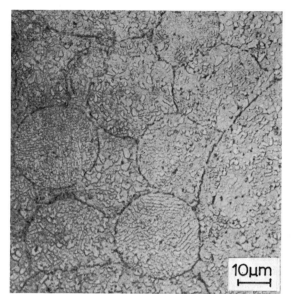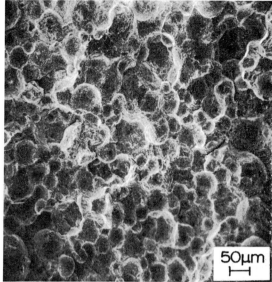

a   b

5 *a* microstructure of as-hot isostatically pressed IN 100 showing carbide outlining of prior powder particle boundaries; *b* stress rupture surface showing particle boundary separation

to be unfounded. One obvious question from the above summary is why was low-carbon Astroloy selected to replace an alloy with lower strength capability? The reason rests with the metallurgy and properties of HIP products, which is the subject of the remainder of the paper.

Metallurgy of HIP products

In this section the similarities and differences between wrought alloys, whether they be fabricated from ingot or powder, and material consolidated by direct HIP are examined. Some of the basic factors in the comparison are first discussed and then two specific examples are given to illustrate and expand upon some of these general points.

The composition of powder metallurgy superalloys is similar to that of their basic wrought counterparts. Therefore, the basic principles behind the attractive high-temperature properties, the formation of a controlled dispersion of the γ'-phase, and the methods used to accomplish useful dispersion remain the same for both classes of alloys.

Minor phases such as carbides or borides are in many cases different for HIP products, the differences deriving from the powder production process. During rapid solidification, carbides are preferentially nucleated on the surface of powder particles.[5] These are usually of the MC type, and although the specific composition is dependent upon base-alloy composition they are often titanium rich. The chemistry and structure of these carbides are often changed during subsequent HIP processing and heat treatment, but in certain cases their location remains unchanged: for example, Fig.5a shows an IN 100 alloy in the hot isostatically pressed condition, showing that the carbides still reside at the prior particle boundaries. Obviously, the presence of a continuous film of an intrinsically brittle phase can lead to premature failure under certain conditions. This is illustrated in the properties listed in Table 1, which show that the hot isostatically pressed material is notch brittle under stress rupture conditions. Figure 5b shows that under these test conditions separation occurs along prior particle boundaries. In a wrought alloy of the same composition the working operation will break up carbide films and a redistribution of carbides will occur. In this condition the alloy exhibits improved stress rupture properties.

Table 1 also shows that a direct HIP product usually has a lower strength than its wrought counterpart after the same heat treatment. In the latter product type some element of thermomechanical processing is inherent in the final product. This takes the form, in the example cited, of a very fine grain size introduced through the Gatorizing* process applied to IN 100. In other cases such as Astroloy processed below the γ' solvus residual work can be present in the material, which contributes to the

*Registered trade mark, United Technologies Corporation.

Table 1   Properties comparison of as-hot isostatically pressed and Gatorized IN 100

	0.2% yield strength, N/mm²	Ultimate tensile strength, N/mm²	Ductility, %	Life, h	Elongation, %
21 °C tensile					
Gatorized	1 120	1 610	26		
As-hot isostatically pressed	1 048	1 491	10		
704 °C tensile					
Gatorized	1 085	1 225	28		
As-hot isostatically pressed	1 008	1 176	9		
732 °C, 748 N/mm² stress rupture					
Gatorized				35	12
As-hot isostatically pressed				10	Notch failure

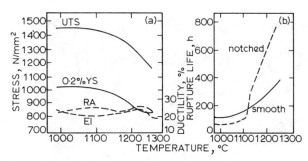

6 **Variation of** *a* **538°C tensile and** *b* **732°C, 560 N/mm²** **stress rupture properties of low-carbon Astroloy** **with pressing temperature**

Table 2 Phase I alloy screening: alloy compositions

Base alloy	Modifications
Astroloy	W, Ta, Hf, Nb, 0·03–0·07C
IN 100	No V, Nb, Hf, 0·02–0·07C
MAR–M–432	No W, high Al : Ti, Hf, 0·03C
PA 101	Low W, Nb, 0·05C
AF2–1DA	Nb, no Ta, 0·10C
René 95	No modifications
AF115	No modifications

strength of the alloy.[6] Methods of controlling grain size in HIP products are examined below, but it is obvious that, using current technology, residual work effects will not be present in such material. Thus, the only available method for changing the strength of direct HIP materials is through heat treatment and compositional modification.

In both wrought alloys and HIP material there is often a trade of specific mechanical properties. Material development efforts in the 1960s showed that a compromise between strength and fracture toughness occurs in many high-strength alloys. In the specific case of nickel-base super-alloys this balance is between strength and creep rupture properties. The balancing of these two properties is also a dominant theme in the following discussion.

LOW-CARBON ASTROLOY
This material is a simple modification of the conventional Astroloy composition produced by lowering the carbon content from 0·08 to ≤0·05%.

As noted above, this alloy was selected as the material for the JT8D programmes as a substitute for Waspaloy. Previous programmes had shown that the properties of direct HIP powder Waspaloy showed poor mechanical properties, including lower strength capability. Concurrent work on low-carbon Astroloy showed that after certain heat treatments properties equivalent to those of

Waspaloy could be obtained. During the course of the present programme the strength capability of low-carbon Astroloy was determined in some detail, and the results serve to illustrate some of the basic relationships in HIP alloys. Some details of the material compositions and processing used in this study can be found in papers by Podob[7] and Evans and Judd.[8] Two groups of variables can be used to manipulate the properties of this alloy: HIP consolidation temperature and heat treatment.

The variation in strength and ductility with HIP temperature, after a constant heat treatment, is shown in Fig.6. This shows that yield strength level can be changed by 70–100 N/mm² with little change in tensile ductility (Fig.6a). However, the stress rupture characteristics change dramatically, as shown in Fig.6b, a discontinuity occurring near the γ' solvus temperature (of ~1 120°C in this alloy). Material hot isostatically pressed below this temperature is notch brittle and shows poor rupture life and ductility. If processing is performed above this temperature, both life and ductility increase and the material is notch ductile. Space precludes any detailed description of the reasons, as far as they are known, for these changes. Grain size increases with HIP temperature, as may be seen from Fig.7, and contributes to the increased stress rupture capability. The low carbon content of the alloy reduces the prior particle boundary decoration; however, in material hot isostatically pressed below the γ' solvus, some outlining of prior particles is present. Extensive boundary migration and grain growth occur during hot isostatic pressing above the solvus, and the carbides are distributed uniformly at these newly created boundaries. The improved stress rupture characteristics can be traced to both an

a   b   c

7 **Microstructures of low-carbon Astroloy after hot isostatic pressing at** *a* **1 100°,** *b* **1170°, and** *c* **1 215°C**

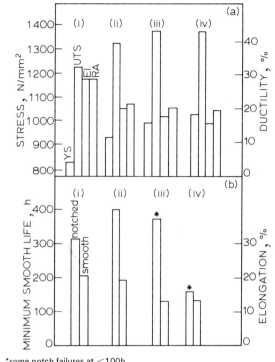

*some notch failures at <100h

8 Influence of heat treatment on a 620°C tensile and b 732°C, 560 N/mm² stress rupture properties of low-carbon Astroloy:
   (i) 1100°C, air cool, +4 h at 982°C, +8 h at 760°C
   (ii) 1100°C, air cool, +8 h at 760°C
   (iii) 1100°C, 482°C salt quench, +8 h at 760°C
   (iv) 1100°C, 315°C salt quench, +8 h at 760°C

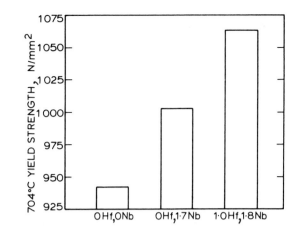

9 Yield strength levels measured at 704°C for IN100 and modifications after heat treatment of γ'−18°C, air cool +8 h at 760°C, air cool

increased grain size and the modification of the carbide distribution. The results indicate that for this specific alloy to produce an adequate property balance, hot isostatic pressing above the γ' solvus is mandatory.

The discussion of heat-treatment techniques for changing properties is restricted to material hot isostatically pressed above the solvus. Standard heat treatments for most wrought alloys consist essentially of three stages: solution treatment, carbide stabilization anneal, and aging treatment. The specific heat treatment used for low-carbon Astroloy to yield properties approximately equal to those of Waspaloy is given in Ref.7. This treatment is modified with respect to that used for conventional Astroloy in that a slower cooling rate from the solution-treatment temperature is used and single-stage carbide stabilization and aging treatments are substituted. Increasing the cooling rate from the solution-treatment temperature and/or eliminating the carbide stabilization stage of the treatment sequence are methods of raising the strength capability. Figure 8 illustrates this point, showing that elimination of the intermediate annealing temperature raises the yield strength by about 100 N/mm², while using

quenching treatments from the solution-treatment temperature are effective in raising the yield strength up to 175 N/mm² over conventionally heat-treated material. However, using more rapid quenching treatments can lead to problems with quench cracking. Such treatments also lead to more scatter in the stress rupture properties and a tendency to notch brittleness in such testing. It can also be noted from Fig.8 that elimination of the carbide stabilization treatment does not reduce the stress rupture properties. This latter point is perhaps not surprising, for, as shown by Podob,[7] the carbide distribution and structure set during the HIP cycle and remain essentially unchanged by the stabilization anneal. This stage of the treatment apparently causes additional growth or precipitation of the γ'-phase, reducing the amount of the phase available for precipitation during aging. Wrought alloys typically contain a higher carbon content with different carbide size and distribution characteristics. Elimination of this heat-treatment stage is thus not possible in such materials without compromising the elevated-temperature properties.

HIGH-STRENGTH ALLOYS

It can be seen from the above discussion that although

Table 4   Phase II alloy selection: compositions

Base	Modifications
Astroloy	1·75 Nb, 0·03C
Astroloy	0·75 Hf, 0·03C
IN100	1·5 Nb, 0·75 Hf, 0·015C
IN100	1·5 Nb, 0·03C
IN100	1·75 Nb, 1·0 Hf, 0·035C
IN100	0·0125C
PA101	1·75 Nb, 0·75 Hf, 0·03C, no Ta
AF2−1DA	1·5 Nb, 0·03C, no Ta

Table 3   Phase I alloy screening: HIP and processing variables

Hot isostatic pressing
  18 kg compacts
  hot isostatic pressing at 9°−18°C above and below γ' solvus (γ's)
Heat treatment
  γ's−14°C for 2 h, air cool/982°C for 8 h, air cool/732°C for 8 h, air cool

Table 5   Phase II alloy selection: HIP and processing

Hot isostatic pressing
  22 kg compacts
  hot isostatic pressing 9°−18°C above and below γ' solvus (γ's)
Heat treatments
  Thick (75×1 125×1 250 mm) sections
    γ's−35°C for 2 h, air cool/732°C for 8 h, air cool
    1 135°C for 2 h, 350°C salt quench/732°C for 8 h, air cool
    1 135°C for 2 h, 482°C salt quench/732°C for 8 h, air cool
  Thin (19×75×1 250 mm) sections
    γ's−14°C for 2h, air cool/732°C for 8h, air cool

Source: Metals Technology, August 1977, 388-395

339

Table 6 Phase II alloy selection: typical properties

Alloy	Section size	Cooling rate	Tensile properties at 21°C 0·2YS, N/mm²	UTS, N/mm²	El, %	RA, %	704°C 0·2YS, N/mm²	UTS, N/mm²	El, %	RA, %	Stress rupture properties at 732°C, 648 N/mm² Life, h	El, %
Astroloy +1·75 Nb	Thick	AC*	1088	1592	18·8	19·2	1014	1334	18·6	28·3	110·9	10·6
Astroloy +0·75 Hf	Thin	AC	1119	1533	14·7	17·7	1038	1257	14·8	22·4	59·8	9·6
IN 100 +1·5 Nb. 0·75 Hf	Thick	AC	1100	1579	19·1	21·8	1057	1334	21·1	28·1	95·8	8·2
IN 100 +1·5 Nb, 0·75 Hf	Thin	AC	1096	1592	18·3	21·8	1067	1330	19·3	28·8	88·3	7·5
IN 100 +1·75 Nb, 1·0 Hf	Thick	AC	1117	1563	16·1	18·7	1047	1327	12·5	15·7	67·4	5·9
IN 100 +1·75 Nb, 1·0 Hf	Thin	AC	1113	1617	18·5	19·5	1094	1330	16·2	22·6	64·4	6·0
PA 101 +1·75 Nb, 0·75 Hf, no Ta, 0·03C	Thick	AC	1121	1621	14·5	14·8	1047	1347	7·9	13·6	111·5	6·6
AF2–1DA +1·5 Nb,	Thick	AC	1143	1434	7·7	9·0	1056	1388	14·4	21·7	15·4	4·3
Goal			1047	1505	15	15	1029	1190	12	12	23	5

*Air cool

high-strength versions of the low-carbon Astroloy can be produced, problems with stress rupture properties and quench cracking can be encountered. In order to produce higher strength materials to be used in direct HIP products, a new approach is needed. Therefore, in 1974 a programme was initiated within Pratt and Whitney Aircraft to identify a high-strength alloy (with properties equivalent to those of current high-strength alloys) produced by consolidation of powder and subsequent Gatorizing. The approach was to take existing base-alloy compositions with demonstrated high-strength capability; to make appropriate alloy modifications of these compositions and to evaluate using the criterion of achieving the tensile and stress rupture properties of IN 100. The increased strength capability arises from the larger γ' volume fraction in these alloys: Astroloy contains about 40 vol.-% of the phase, and the selected alloys

contain more than 60%. Increasing the hardening solute content also increases the solvus temperature in most of the alloys. A two-phase programme was conducted evaluating over 70 alloy and process modifications. The base alloys selected for investigation were Astroloy, MAR–M–432, IN100, PA101, AF2–1DA, René 95, and AF115. Phase I of the programme contained 34 modifications of these alloys, summarized in Table 2. Processing conditions for these alloys are given in Table 3, from which it can be seen that two HIP conditions were studied, but the heat-treatment conditions were held constant. Property evaluation of these initial conditions showed the following results:

(i) all of the alloys showed yield strength deficiencies at elevated temperatures: the material hot isostatically pressed at temperatures below the γ' solvus

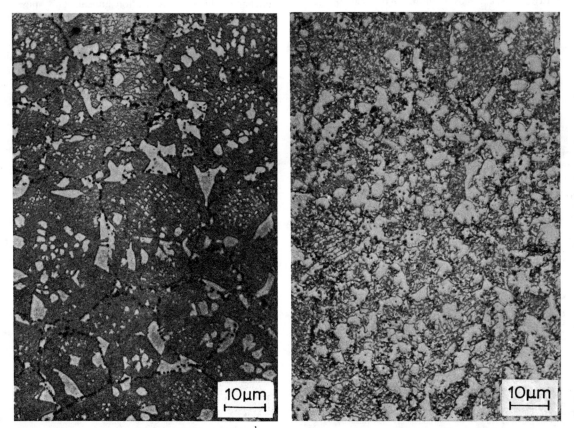

10 Microstructures of a unmodified and b hafnium–niobium modified IN100, showing change in carbide distribution

(fine grained) was stronger than material hot iso-statically pressed above the solvus

(ii) PA 101, MAR–M–432, René 95, and AF115 did not meet the tensile ductility requirements

(iii) only alloys from the Astroloy and IN 100 series and A F 115 met the stress rupture goals.

In order to correct the strength deficiencies, an extensive heat-treatment study was carried out. It was found that, as in the Astroloy programme, using a more rapid cooling rate from the solution-treatment temperature and/or elimination of the carbide stabilization age were effective in raising the strength level. The higher strength capability of material hot isostatically pressed below the γ' solvus was confirmed, and thus the bulk of the work was subsequently performed on such material. However, it was also found, not unexpectedly, that an inverse relationship between strength and both stress rupture and tensile ductility properties existed in all alloys. Thus, if any alloy exhibited inadequate ductility or stress rupture properties after the initial Phase I heat treatment, these properties became even lower at an increased strength level. Two positive factors were identified in this section of the study: first, the lower carbon levels improved tensile ductility and stress rupture properties, and secondly, an apparently synergistic hafnium and niobium effect on strength level was identified in IN 100 base alloys, as illustrated in Fig. 9. These two elements also produced very uniform carbide distributions, with little or no sign of prior particle boundary decoration after consolidation. This no doubt contributed to the good stress rupture and ductility properties observed.

In Phase II work was concentrated on Astroloy and IN 100 compositions but low-carbon modifications of PA 101 and AF2–1DA were also included. The eight alloys studied are summarized in Table 4. The process conditions evaluated (Table 5) were extended to include a wider range of heat treatments. The results observed on selected alloys, after a constant heat treatment, are shown in Table 6, in which data from alloys which came closest to, or actually met, the property goals is presented. It can be seen that the Hf and Nb modifications of IN 100 meet all the property requirements. These compositions are the basis for the alloy formulation, designated MERL 76, which has been selected for current and future high-strength disk applications.

This programme has shown that there is no intrinsic barrier to hot isostatic pressing below the γ' solvus of an alloy and retaining an adequate balance between stress rupture and tensile properties. It should also be noted that the selected alloy contained hafnium and niobium, which modify both the type and distribution of carbides present in the material; the latter point may be seen from Fig. 10. The γ' solvus of MERL 76 is 1199°C, and thus hot isostatic pressing below the solvus can be performed at considerably higher temperatures than in (unmodified) low-carbon Astroloy. Thus, it is not clear at present whether the increased interdiffusion produced by the higher temperatures or the carbide modification effect is responsible for the property balance in this alloy; perhaps both factors contribute. Another important factor is the relatively simple heat treatments that can be used to achieve the required properties. This is not only an overall benefit but could be a special advantage in the production cycle for complex, near-net shapes.

Summary

In the first part of this paper materials and processing used in the fabrication of turbine disks for gas turbine engines were discussed in relatively general terms. Direct hot isostatic pressing of nickel-base superalloy powders to intermediate shapes has been identified as a cost-effective approach, as long as the properties of the product can be tailored to the component requirements. Some of the metallurgical factors that must be considered in obtaining a balance between strength and stress rupture properties have been discussed. Earlier problems experienced in hot isostatically pressed material which were traced to poor carbide distributions can be eliminated by using lower carbon contents and/or using additions of elements which form more stable carbides. If these factors are coupled with heat treatments specific to this type of alloy, in contrast to a conventional forging alloy type, attractive property balances can be achieved. Such treatments are, in many cases much simpler than those used for the conventional alloy counterparts. The authors have not discussed the limits of strength achievable in hot isostatically pressed material nor, except in relatively general terms, the mechanics of how specific properties are obtained. Although some preliminary work has been performed in both areas, it is clear that much more is required to produce definitive answers.

References

1. G. H. GESSINGER and M. J. BOMFORD: *Int. Metall. Rev.*, 1974, **19**, 51.
2. 'Advanced fabrication techniques in powder metallurgy and their economic implications'; 1976, Neuilly sur Seine, AGARD.
3. 'Summary of Air Force/industry manufacturing cost reduction study', AFML–TM–LT–73–1, Jan. 1973.
4. H. A. HAUSER: *SAMPE Q.*, April 1975, 1.
5. J. M. OBLAK: unpublished work, Pratt and Whitney Aircraft, *see also* Ref. 7.
6. e.g. R. F. DECKER: 'High-temperature materials for gas turbines', (ed. P. R. Sakin and M. O. Speidel); 1974, New York, Elsevier.
7. M. T. PODOB: 'Modern developments in powder metallurgy' (ed. H. H. Hausner and P. W. Taubenblat), Vol. 11, 25; 1977, Princeton, NJ, Metal Powder Industries Federation.
8. D. J. EVANS and G. M. JUDD: *ibid.*, Vol. 10, 199.

Microstructural Instabilities During Superplastic Forging of a Nickel-Base Superalloy Compact

J-P. A. IMMARIGEON AND P. H. FLOYD

The high temperature flow behavior of a nickel-base superalloy powder compact, prepared by hot isostatic pressing has been examined by means of uniaxial compression testing in terms of the microstructures developed during plastic flow. The tests were done isothermally at 1050 and 1100 °C and at constant true strain rates between 10^{-5} s^{-1} and 1 s^{-1}. The fine grained compact exhibits some degree of superplasticity which always increases with compressive flow as the grain structure is refined. The faster the rate of deformation, the finer is the grain size produced at high strains, when steady state conditions of flow appear to develop. By deforming to different strains at a given strain rate or into the steady state regimes at various strain rates, grain sizes in the range 1 to 5 μm were produced. By unloading and restraining the test pieces in situ, the effect of grain size on the onset of plastic flow has been examined and the yield stress observed to increase with grain size. It is shown that, in this material, hardening or softening occurs during flow depending on the size of the initial grains. The changes in microstructure and flow stress observed during deformation are analyzed and the potential offered for control of the microstructure during isothermal forging is discussed.

RECENTLY, opportunities for improvements in the properties of materials for aerospace applications have arisen from developments in new processing technologies, particularly in the area of powder metallurgy and in the applications of superplastic forming.[1,2] These developments concern the high temperature materials for propulsion gas turbines, mainly the nickel-base superalloys, as well as the structural alloys based on titanium. Traditionally, improvements in the properties of these materials have derived from advances in alloy composition and heat treatment[2-4] In many instances the limits of this approach have been reached.[3] It is becoming increasingly evident, however, that further progress can be achieved by control of the microstructure during thermomechanical processing.[3,5,6]

The combination of powder metallurgy processing and isothermal forging offers design freedoms and opportunities for microstructural control not possible with more conventional forming operations. Control of the microstructure can be exercised both during consolidation of prealloyed powders[7-9] and during the forging of compacts.[3,8-12] The factors which influence control of the microstructue during compaction have been extensively examined and are well understood.[7,8] However, property control during isothermal forging has not been explored to the same extent, in spite of the potential offered by this approach.[13,14] By appropriate control of the forming variables, it should be possible, in principle, to produce forgings with predictable variations in microstructure across the part. This tailoring of structures and properties to design requirements remains, however, a relatively unexplored area. This can be attributed to a lack of understanding of the hot working behavior of these types of alloys, which stems in turn from the inherently complex relationship between forming variables and structures developed during working.

The success of this development will therefore require a detailed understanding of the physical phenomena involved during hot working of these materials. In this work, the high temperature flow behavior of a nickel-base superalloy compact is examined. The compressive flow curves generated under conditions of constant true strain rate and constant temperature are examined in terms of the microstructures developed during working. The role of grain size during superplastic flow is analyzed, and the behavior of the alloy rationalized in terms of microstructural instabilities.

EXPERIMENTAL MATERIAL

The compact examined had been prepared previously[15] by hot isostatic pressing from a commercial powder of Inco 713LC, a high strength casting type nickel-base superalloy. In producing this material, attempts had been made to maximize the hot workability of the consolidated product by appropriate choice of powder type, particle surface treatments and pressing conditions.[15] The powder selected had been argon atomized. Its chemistry and mesh size distribution are given in Table I. The powder had been washed in a boric acid solution prior to canning in evacuated and sealed mild steel containers. The hot isostatic pressing conditions used are given in Table II.

The alloy contains approximately 50 vol pct of the γ' intermetallic phase, Ni$_3$(Al, Ti). Since consolidation was

J-P. A. IMMARIGEON, Research Officer, and P. H. FLOYD, Technical Officer, are with the National Aeronautical Establishment, National Research Council Canada, Ottawa.
Manuscript submitted October 1, 1979.

Table I. Powder Chemistry and Mesh Size Disribution of Argon Atomized Powder*

Composition, Wt Pct		Sieve Analysis		
Cr	12.10		+ 80	5.1
Al	5.96	− 80	+ 100	9.6
Ti	0.78	− 100	+ 150	16.5
Mo	4.61	− 150	+ 200	21.4
Zr	0.13	− 200	+ 270	14.2
C	0.05	− 270	+ 325	7.7
B	0.008	− 300	+ 400	7.4
Nb	†	− 400	+ 500	9.3
Ni	Balance	− 500		8.6

*Data Source: NRC NAE LTR ST 586.
†Not analyzed—specified at 1.5 to 2.5.

Table II. Hot Isostatic Pressing Conditions

Temperature °C	Time h	Pressure MPa
980	4	105
+ 1175	2	105

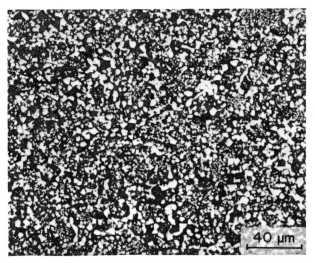

Fig. 1—Microstructure of the hot isostatically pressed 713LC compacts after preheating for 30 min at the test temperature of 1050 °C.

performed below the γ′ solvus (1210 ± 10 °C), a two phase γ + γ′ condition was retained during hipping. Under these conditions, the presence of the second phase prevents coarsening of the fine as-atomized grains and the compact can be expected to exhibit some degree of superplasticity and have good workability.[8,12]

The microstructure present in the compacts immediately before testing is shown in Fig. 1. There is no evidence of original powder particle boundaries due to minimal carbide decoration during consolidation.[8] The structure is uniform, with a grain size in the range 4 to 7 μm. The γ′ phase is in a coarse overaged condition, mostly in the form of large angular particles (∼1 to 5 μm) located along grain boundaries and also within the grains. A finer spheroidal γ′ (<0.5 μm) can be detected within some of the larger grains.

EXPERIMENTAL PROCEDURE

The flow behavior of the compact was examined by means of uniaxial compression testing. The tests were carried out on subscale blanks, 9.65 mm long and 6.35 mm in diam, in a high temperature compression train built to operate with a 100KN MTS hydraulic testing machine[14,16] which was modified for constant true strain rate deformation. Temperatures and strain rates used were typical of isothermal forging operations for this type of material.[6] The temperature was kept constant to within ± 2 °C and the strain rate did not deviate by more than 2 pct from the selected values. The specimens were preheated for 30 min and then deformed between flat silicon nitride dies, under a flowing argon atmosphere. Frictional effects at the die-specimen interfaces were minimized by the use of a molten glass lubricant. Annular grooves machined into the end faces of the specimens helped maintain proper lubrication throughout the test. Under these conditions, the coefficient of friction was measured at less than 0.04.[17] Finally, when required, the specimens were quenched in oil within 2 s of the end of the test.

In the course of the test, the developed load and specimen height were continuously monitored, the latter by means of a high temperature displacement transducer designed to follow the relative displacement of the compression dies.[16] In deriving the true stress-true strain curves from the digitized load-height data, it was assumed that no volume change takes place during deformation.

Previous work[12] had demonstrated that the microstructure in nickel-base superalloy compacts can change substantially during hot working and that the high temperature flow strength of this type of material is strongly grain size dependent. In the present work experiments were intended to further evaluate these effects.

Firstly, to characterize the flow behavior of the as-hipped material, and the microstructures produced during forming, a series of specimens was tested under various conditions of constant true strain rate and temperature to a strain of 0.8 and then quenched for metallographic examinations. This was done at 1050 and 1100 °C and at strain rates between 10^{-5} s^{-1} and 1 s^{-1}.

Secondly, to evaluate the interrelationships between microstructure and flow properties during forming, and in particular the effects of grain size, a series of specimens was tested using an interrupted test technique that involved prestraining, unloading and immediately restraining the test piece. By varying the prestrain conditions, different microstructures were produced and their effects on flow behavior directly evaluated from the flow curves generated during retesting. In these tests, the prestraining variables examined were strain and strain rate. These interrupted tests were performed at 1050 °C only. At this temperature, well below the solvus, the high volume fraction of γ′ tends to slow down microstructural changes induced by working, which makes it easier to follow these changes.

The quenched specimens were examined by means of optical microscopy, following standard metallographic procedures and were electroetched in a chromic acid solution (6 g chromium trioxide in 100 ml distilled water) at 16 10³ A·m⁻² for 25 s.

RESULTS

The flow curves for the as-hipped material are shown in Fig. 2 for the two test temperatures. Under all testing conditions, the as-hipped material exhibits flow softening. The degree of softening decreases with decreasing strain rate at both temperatures. It is also less pronounced at the higher temperature. At high strains, steady states of deformation are approached during which the flow stress appears to remain constant with increasing strain. The strain for the onset of steady state flow increases with strain rate and decreases with temperature. It should be noted that, the strains achieved at the higher strain rates are not sufficient for a steady state regime of flow to be fully established. However, the shape of the flow curve at high strains suggests that such a regime is approached asymptotically. Steady state stresses and peak stresses are stongly dependent on strain rate and temperature. Both increase with strain rate and decrease with temperature.

The as-quenched microstructures corresponding to steady state stresses or highest strains achieved at 1050 °C are shown in the micrographs of Fig. 3. They are typical of thermomechanically processed nickel-base superalloy compacts[12,19] and consist of a micro-duplex arrangement of γ and γ' grains. Under all testing conditions, the grains remain equiaxed in spite of the large deformations involved.

The grain sizes produced at the various strain rates at 1050 °C were measured following Heyne's procedure[20] and the mean lineal intercepts obtained are shown in Table III. The data indicate that the microstructure is refined during plastic flow at all strain rates investigated, and that the greater the strain rate, the finer the grain size produced. Since the final flow strength increases and the final grain size decreases with an increase in strain rate, the final flow stresses appear to be inversely correlated to the final grain sizes as shown in Fig. 4. The trend observed for the range of strain rates examined may be represented by an expression of the form

$$\sigma_{ss} \propto \bar{\lambda}_{ss}^{-n} \tag{1}$$

where σ_{ss} is the steady state flow stress, $\bar{\lambda}_{ss}$ is the average microduplex steady state grain size and n is of the order of 2 in Fig. 4. It must be pointed out, however, that the value of n would be somewhat lower if steady state regimes of flow had been fully established at the fastest strain rates.

Results from the interrupted tests are presented in Fig. 5 to 8. The effects of the amount of prior deformation at a fast strain rate on flow behavior at a slower strain rate are demonstrated in Fig. 5. With increasing prestrain, a transition in flow behavior can be observed from one where the compacts flow soften during straining to one where the flow strength rises continuously instead. In all cases, however, the stresses developed at high strains all converge to the same level.

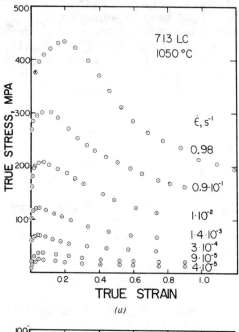

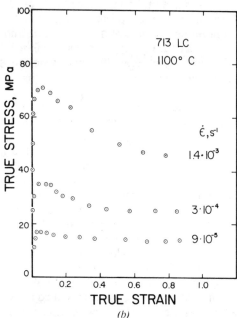

Fig. 2—Compressive true stress-true strain curves for 713LC compacts deformed at different strain rates at (a) 1050 and (b) 1100 °C.

Table III. Average Grain Sizes (Measured as Mean Lineal Intercepts) Produced After Different Deformation Histories at 1050 °C*

Effect of Strain Rate†		Effect of True Strain‡	
Strain Rate s⁻¹	Grain Size μm	True Strain	Grain Size μm
9.8 10⁻¹	1.4	0.1	3.0
8.8 10⁻²	1.7	0.3	2.4
1.4 10⁻³	2.7	0.6	2.0
8.8 10⁻⁵	3.6	0.8	1.6

* Average grain size of compact before deformation, 4 to 7 μm.
† After deformation corresponding to strains of Fig. 2.
‡ At a constant true strain rate of 10^{-1} s⁻¹

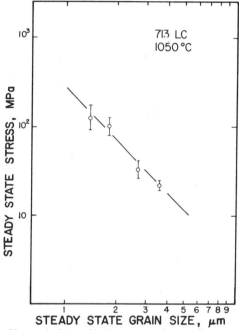

Fig. 4—Observed relationship between steady state flow stresses and steady state grain sizes.

This steady state stress appears to be independent of prior history and to depend only on the imposed strain rate. For small and large prestrains, the convergence is not complete in spite of the large strains involved. However, for an intermediate prestrain of 0.6 steady state flow is rapidly established within a true strain of about 0.2.

The microstructures corresponding to the interruption strains of Fig. 5 are shown in Fig. 6. Some degree of structural inhomogeneity is evident in all cases, although this is greatly diminished after a strain of 0.8. The overall effect of straining appears not only to

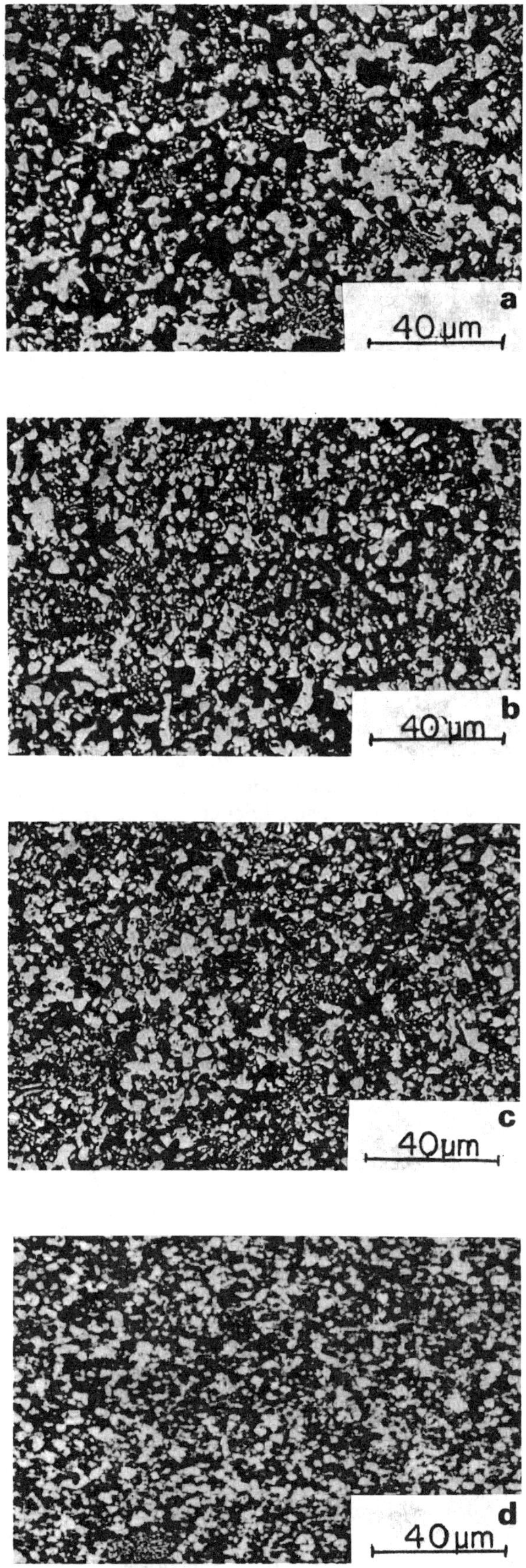

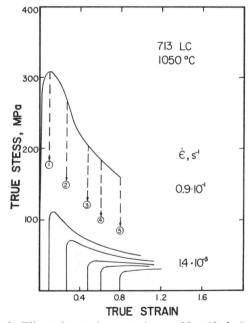

Fig. 5—Effects of prestrain, at a strain rate of $9 \times 10^{-2} s^{-1}$, on the flow curve at a lower strain rate of $1.4 \times 10^{-3} s^{-1}$ (1050 °C).

Fig. 3—Microstructures developed after forging at 1050 °C at the constant true strain rates of (a) $9 \times 10^{-5} s^{-1}$, (b) $1.4 \times 10^{-3} s^{-1}$, (c) $9 \times 10^{-2} s^{-1}$, and (d) $9.8 \times 10^{-1} s^{-1}$. Amounts of deformation corresponding to maximum strains of Fig. 2.

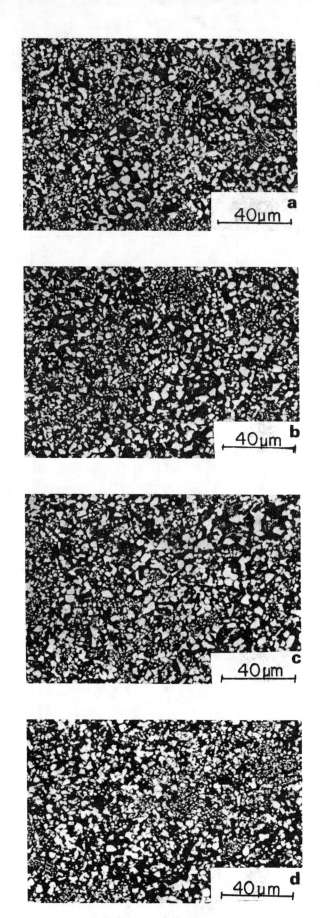

Fig. 6—Microstructures developed during forging at 1050 °C and at a strain rate of 9×10^{-2} s^{-1} for true strains of (a) 0.1, (b) 0.3, (c) 0.6 and (d) 0.8.

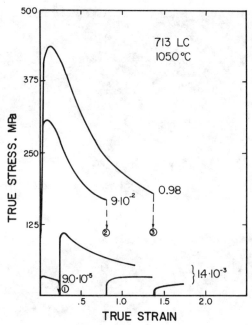

Fig. 7—Effects of prestrain at three different strain rates on the flow curve at 1050 °C and a constant true strain rate of 1.4×10^{-3} s^{-1}.

distribute the phases more uniformly but also to refine the microduplex structure of γ and γ' grains. In separate studies on similar materials[12,19] it was shown that deformation induced recrystallization is partly responsible for these structural modifications. From the present work, there is also evidence that the larger γ' particles are broken up during flow by intruding γ phase while the finer γ' particles tend to coalesce and grow into larger units. Since these morphological transformations occur at constant volume fraction of the second phase, this produces changes in interphase boundary area as well as in the mean free path between second phase particles. The grain boundary area also appears to increase with deformation, since the grains which form within the γ phase are considerably finer than the original as-hipped grains. Their size appears to be related to the mean free path between the γ' particles and is, on the average, close to that of the particles.

Average grain sizes were measured for all interruption strains. The results are presented in Table III. By matching the grain size and the corresponding flow curve in Fig. 5 it is clear that the finer grain sizes initially offer the least resistance to deformation. The gradual elimination of flow softening at 1.4×10^{-3} s^{-1}, as the starting grain sizes decreases should also be noted. For grain sizes below 2 μm, the compacts flow harden at this strain rate.

Results from another series of interrupted tests are shown in Fig. 7. Here the flow behavior of the as-hipped material is compared to that of compacts previously exposed to different strain and strain rate histories, that is, having different starting microstructures. When initially deformed at a slow strain rate, the compacts show little difference in flow behavior with the unprocessed material. Flow softening is still evident, and a slight decrease in the peak stress can be observed. In contrast, when the compacts are first processed at intermediate or fast strain rates, their initial resistance

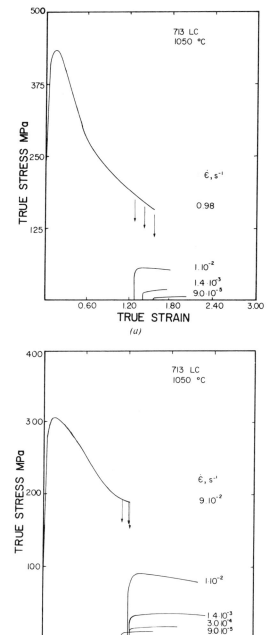

Fig. 8—Effects of prestrain at (*a*) 0.98 s^{-1} and (*b*) 9×10^{-2} s^{-1} on the flow curves at different strain rates and 1050 °C.

strain rate at constant temperature.

Other interrupted test results are shown in Fig. 8. The two finest grain sizes examined in this work were produced by prestraining at the two fastest strain rates and their effect on flow strength was compared by retesting at various strain rates as shown in the figure. In line with previous observations, the finer the grain size after prestraining, the lower the yield stress upon retesting at a given strain rate. It can also be seen that either softening or hardening occurs during restraining depending on the test strain rate and on the grain size present upon resumption of the deformation.

While it was shown, before, that softening during flow could be associated with grain refinement, hardening appears to be associated with coarsening of the grains as suggested in the micrograph of Fig. 9. This shows that after retesting at the slowest strain rate (9.0 10^{-5} s^{-1} in Fig. 8(*a*)) the grain size is coarser (2 μm) than that present at the time of the test interruption (1.4 μm). The results from these interrupted tests infer that flow behavior prior to the onset of steady state flow depends on the starting and finishing (steady state) grain sizes as discussed later.

The effects of starting grain size on yield strength are summarized in Fig. 10. Yield stresses (0.002 offset) at 1050 °C are plotted as a function of strain rate for all the different starting microstructures. The as-hipped material has the highest resistance to flow. Its strain rate sensitivity, $m = (\partial \ln \sigma_y / \partial \ln \dot{\epsilon})_{\text{structure}}$, increases with decreasing strain rate until m is approximately 0.5. This value is characteristic of superplastic flow properties and is not unexpected from a superalloy compact which has been consolidated at a temperature below the γ' solvus.[12] This material does in fact exhibit abnormally high ductilities when stress-ruptured tested. The strain rate sensitivities for all the microstructures tested are of the same magnitude. They remain of the order of 0.5 to increasingly higher strain rates as the starting grain size decreases. This is, again, characteristic of superplastic behavior.

The grain size dependence of the yield stress, at

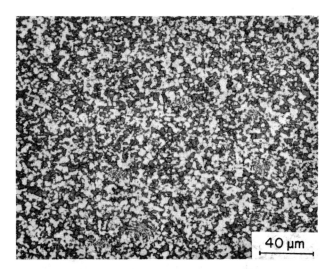

Fig. 9—Microstructure developed after testing at a strain rate of 0.98 s^{-1} and then retested *in situ* at a strain rate of 9.0×10^{-5} s^{-1} as shown in Fig. 8(*a*).

to deformation is considerably reduced. No flow softening occurs and instead the flow strength increases continuously upon restraining. However, irrespective of the prior deformation histories, a steady state regime is approached at high strains. The material initially processed at the intermediate strain rate has reached this condition after a true strain of 0.5. After the same amount of deformation, the compacts initially processed at the low and fast strain rates are still softening and hardening respectively. However, their flow curves appear to converge toward that of the materials prestrained at the intermedite strain rate which again suggests that steady state stress is only a function of

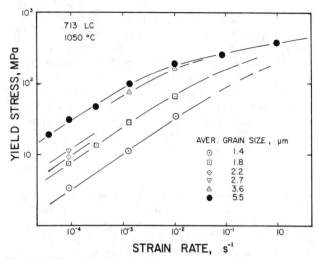

Fig. 10—Effects of strain rate and average initial grain size on the high temperature flow strength of 713LC compacts at 1050 °C.

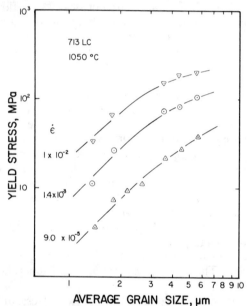

Fig. 11—Effect of average microduplex grain size on the high temperature compressive yield strength of 713LC compacts at 1050 °C and three strain rates.

constant strain rate and temperature, is shown in Fig. 11 for three different strain rates. These data were derived from those of Fig. 10. This dependency may be represented by an expression of the form:

$$(\sigma_y)_{\dot{\epsilon},T} \propto \bar{\lambda}^a \qquad [2]$$

where $\bar{\lambda}$ is the average microduplex grain size at the *start* of the test and the magnitude of the strain rate sensitivity index, $a = (\partial \ln \sigma_y / \partial \ln \lambda)_{\dot{\epsilon},T}$, varies with strain rate and grain size. For the finer grain sizes, this index is of the order of 2, but it decreases rapidly with either an increase in strain rate or grain size. These trends are to be expected under superplastic conditions. It may be noted that the magnitude of a falls in the range normally quoted for fine grained superplastic materials,[21,22] although it is somewhat higher for the finer grain sizes examined.

DISCUSSION

It has been shown that the flow strength of 713LC compacts is highly strain rate and structure sensitive at hot working temperatures. Under these conditions, the microstructure is unstable and evolves during deformation at constant strain rate and temperature towards a steady state regime of flow. At this stage, both the grain size and flow stress remain constant and independent of prior history as might be expected in metals that recrystallize during hot working.[23] It has also been shown that differences in flow behavior prior to steady state deformation arise from differences in initial grain size and differences in direction of change of grain size. If a proper framework is to be established for the prediction and control of microstructure in forged components of this kind of material, it is most important that these observations be fully rationalized.

The trend observed between steady state flow stresses and steady state grain sizes, Eq. [1], is not unexpected under hot working conditions.[23-25] Similar correlations have been reported between dynamically recrystallized grain sizes and steady state flow stresses for many other alloy systems, including nickel alloys.[23-25] In these cases, the size of the grains is known to be dictated by the defect densities associated with the stored energy of working.[23] In the present material, the nucleation of new grains and the grain refinement during deformation can also be attributed to the stored energy of working.[12,19] Therefore it is not surprising that similar trends are obtained. It must be emphasized that the correlation given by Eq. [1] does not imply that steady state stresses are causally related to steady state grain sizes (see discussion to that effect in Ref. 23, p. 450). The data for Eq. [1] were collected at different strain rates and the observed trend only reflects the fact that both quantities are related to strain rate, as shown schematically in Fig. 12. This figure indicates that the flow stress and grain size under steady state conditions are only related to strain rate at constant temperature and are independent of the starting grain size. The latter only affects

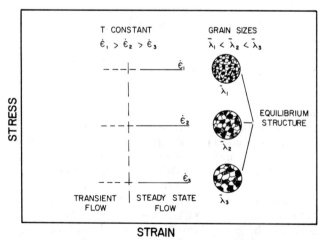

Fig. 12—Schematic representation of the steady state portions of the flow curves and the corresponding average microduplex grains produced at three strain rates in 713LC compacts deformed in compression at 1050 °C.

transient flow, that is deformation before steady state conditions are achieved.

The trends observed prior to the onset of steady state flow are not unexpected either. Under hot working conditions, the flow strength in metals depends *only* on strain rate, temperature and structure.[23] Under these conditions, the total derivative of the stress $d\sigma$ can be written as:

$$d\sigma = \left(\frac{\partial \sigma}{\partial \dot{\epsilon}}\right)_{T,S,} \cdot d\dot{\epsilon} + \left(\frac{\partial \sigma}{\partial T}\right)_{\dot{\epsilon},S} \cdot dT$$
$$+ \left(\frac{\partial \sigma}{\partial S}\right)_{\dot{\epsilon},T} \cdot dS \qquad [3]$$

where $\dot{\epsilon}$ is the strain rate, T the temperature and S a parameter that accounts in the most general case for the effects of microstructural features such as dislocation densities, size and shape of grains, precipitate morphologies or texture.[23] In the present case Eq. [3] reduces to

$$\sigma = \left(\frac{\partial \sigma}{\partial S}\right)_{\dot{\epsilon},T} \cdot dS \qquad [4]$$

since the tests were carried out at constant strain rate and constant temperature (*i.e.* $d\dot{\epsilon}/dt = dT/dt = 0$). Also since the most important microstructural modification during deformation of the compact appears to be a change in microduplex grain size, it can be assumed, as a first approximation, that changes in the parameter S reflect mainly those in grain size. In this special case, Eq. [4] can be rewritten as:

$$\frac{d\sigma}{dt} = \left(\frac{\partial \sigma}{\partial \bar{\lambda}}\right)_{\dot{\epsilon},T} \cdot \frac{d\bar{\lambda}}{dt} \qquad [5]$$

where $\bar{\lambda}$ is the average microduplex grain size. It is recognized that other features of the microstructure may affect the magnitude of the parameter S and also that the temperature may not remain constant during testing because of adiabatic heating. Nevertheless Eq. [5] indicates that changes in flow strength during testing at constant strain rate and temperature are dictated by the form of the function relationship between flow strength and grain size and by the change in grain size induced by the deformation.

At sufficiently high temperatures, the influence of grain size on flow strength in metals becomes opposite that at room temperature and takes the form:[26]

$$\sigma_y \propto \bar{\lambda}^{a>0} \qquad [6]$$

where the exponent, a, may be a function of temperature, strain rate and grain size. It is generally agreed that this behavior can be attributed to the increasing contribution from grain boundary sliding to deformation. Consequently, such dependencies are to be expected in superplastic materials for which grain boundary sliding is the dominant mode of deformation. This is indeed the case in the superplastic compacts as shown previously, Eq. [2]. Because of this grain size dependency, and because of Eq. [5], the direction of change in flow stress at constant strain rate and temperature can be expected to be the same as the direction of change in

grain size. For a given strain rate and temperature, the latter is governed by the initial grain size relative to the steady state grain size. If the initial grain size λ_i is larger than the steady grain size $\bar{\lambda}_{ss}$, then $d\bar{\lambda}/dt < 0$ and the flow stress decreases to its steady state level. Conversely, if $\lambda_i < \lambda_{ss}$, the $d\bar{\lambda}/dt > 0$ and the flow stress increases during plastic flow. Thus, depending on the magnitude of the ratio $\bar{\lambda}_i/\bar{\lambda}_{ss}$, the superplastic compacts either flow soften or flow harden initially. When $\bar{\lambda}_i \sim \bar{\lambda}_{ss}$ no transient behavior is expected and the compacts enter a steady state of flow upon straining. These three possible cases are presented schematically in Fig. 13.

A simple rationalization of transient deformation in the 713LC compacts now becomes possible. The data on the steady state flow stresses and steady state grain sizes, Fig. 4, are replotted schematically in Fig. 14

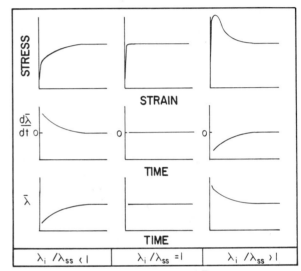

Fig. 13—Effect of the ratio of initial grain size $\bar{\lambda}_i$ to steady state grain size $\bar{\lambda}_{ss}$ on the shape of the flow curve of 713LC compacts.

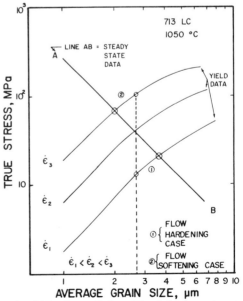

Fig. 14—Model for the prediction of flow behavior and control of grain size during superplastic forging of nickel-base superalloy compacts at 1050 °C.

together with the grain size dependence of the yield stress measured at different strain rates, derived from Fig. 11. It can be seen that for a given starting grain size, the material flow hardens after yielding if the imposed strain rate is such that a coarser grain size is produced at steady state (Case 1, Fig. 14). Conversely, flow softening would be expected from the same initial microstructure for any rate of deformation larger than $\dot{\epsilon}_2$ (Case 2, Fig. 14). In this case, the steady state grain size produced is finer than the starting grain size.

The mechanisms by which flow propagates in these two cases are not identical. Whenever the conditions of flow are such that the yield stress falls below line AB in Fig. 14 (*i.e.* case 1), the compacts can be regarded as fully superplastic and their behavior characteristic of stage II superplasticity. In this case, the grains tend to grow during flow and since the dominant mode of deformation is grain boundary sliding, the flow stress rises[21,22] towards its steady state value. In contrast, in the region immediately above line AB (*i.e.* case 2, Fig. 14), the conditions of flow are initially only partially superplastic. They are characteristic of the transition between stage II and stage III superplasticity, meaning that both intragranular deformation and grain boundary sliding occur simultaneously during flow.[19,27,28] In this case, the grains are refined during compression which produces an increase in the degree of superplasticity and reversion to stage II behavior,[12,19] and as a result the flow stress decreases towards its steady state value. This suggests that the steady states observed in the compacts deformed in compression might be regarded as states during which superplastic properties are simultaneously lost through growth and regenerated through refinement of the average microduplex grain size.

The presentation of the data in the form of Fig. 14 is very useful because it provides not only the grain size and strain rate dependence of flow strength independently, but also the direction of changes in grain size and flow strength that may be expected during flow as a function of grain size and strain rate. However, it must be emphasized that the figure describes only the *magnitude* of changes in flow strength and grain size. It does not indicate the *rate* at which these changes may occur. The results indicate that this varies with strain rate and grain size and probably as well with temperature. For instance, grain growth occurs slowly at a low strain rate of $9 \ 10^{-5} \ \mathrm{s}^{-1}$, Fig. 9, and the *rate* at which the material hardens remains low accordingly, Fig. 8(*a*). Nevertheless, the *degree* of hardening involved in this case is not inconsistent with the model suggested, Fig. 14, since the final grain size of 2 μm, Fig. 9, is far from having reached its steady state value of 3.6 μm, Fig. 3(*a*).

This implies that to obtain a complete description of deformation under hot working conditions in this type of material it is also necessary to take into considertion the rate of change of grain size as a function of strain rate, temperature and instantaneous value of the grain size. This information may be extracted from data similar to that contained in Table III.[29]

The phenomenological model presented provides an interpretation of microstructural changes and flow strength variations during plastic flow of this type of material. For instance, during compaction of the same

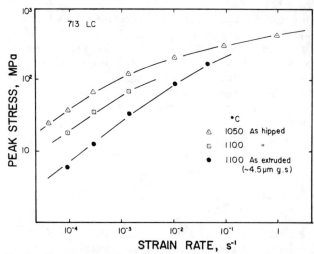

Fig. 15—Comparsion of the effects of strain rate on the peak flow stresses of 713LC compacts consolidated by hot isostatic pressing and by hot extrusion.[18]

powder by extrusion,[18] it is the higher strain rate and strain achieved during compaction which lead to a grain size finer than that produced after hipping. The as-extruded grain size is approximately 4.5 μm,[18] and the peak flow stresses for the extruded material are accordingly reduced, as shown in Fig. 15.

Perhaps more important, the observations made in this work imply that the microstructure in forged parts can be predicted and therefore controlled. The degree of control might at first appear somewhat limited. However, an increase in forging temperature an be expected to widen the range of grain sizes that can be produced during isothermal forging. Furthermore, additional grain size control should be feasible during subsequent heat treatment. By combining the information contained in Fig. 14 and the rate of change in grain size expressed as a funciton of strain rate and grain size with analytical techniques for plasticity modelling, it might be possible to produce forgings with different structures and therefore properties in different parts of the component. By specifying the shape of the preform, prepared by powder processing, as well as the necessary die shapes, appropriate amounts of strain and strain rate could be applied in order to develop optimum grain structures in specific locations of the finished product.[29]

CONCLUSIONS

Flow stress and microstructural changes during isothermal hot working of 713LC powder compacts have been examined and the following general conclusions were reached:

1) The as-hipped compact behaves superplastically at 1050 °C and exhibits stage II/stage III transition behavior over the entire experimental range.

2) In this case the degree of superplasticity increases with compressive deformation as a result of grain refinement.

3) Under isothermal and isostrain-rate conditions, steady state regimes of flow develop at high strains. The

fastest strain rates produce the highest flow stresses and finest grain sizes. As a result, steady state flow stresses and steady state grain sizes are inversely related.

The flow strength of this material increases with grain size. Deformation before the onset of steady state flow is therefore governed by the size of the initial grains relative to that of the steady state grains. Refinement of the average grain size during flow produces softening and growth produces hardening.

5) Grain sizes can be predicted as a function of deformation history and therefore grain size control during isothermal forging should be possible. To extend the range of grain sizes achievable, the effect of working temperature needs to be examined.

ACKNOWLEDGMENTS

The authors are grateful to Dr. W. Wallace of the National Aeronautical Establishment for the supply of the test material.

REFERENCES

1. G. H. Gessinger and M. J. Bomford: *Int. Met. Rev.*, 1974, vol. 19, pp. 51–76.
2. C. Hammond and J. Nutting: *Met. Sci.*, vol. 11, pp. 474–90.
3. N. A. Wilkinson: *Met. Technol.*, 1977, vol. 4, pp. 346–59.
4. R. F. Decker and C. T. Sims: *The Superalloys*, pp. 33–77, John Wiley and Sons, New York, 1972.
5. W. H. Couts and J. E. Coyne: *Proc. 2nd Int. Conf. Superalloys —Processing, Seven Springs*, 1972, Book MCIC-72-10, pp. K1–K20, Nat. Tech. Info. Serv., Springfield, VA.
6. J. E. Coyne: *Met. Technol.*, 1977, vol. 4, pp. 337–45.
7. M. J. Blackburn and R. A. Sprague: *Met. Technol.*, 1977, vol. 4, pp.488–95.
8. W. Wallace, J-P. A. Immarigeon, J. Trenouth, and B. D. Powell: *Proc. AGARD Conf. on Advanced Fabrication Techniques in Powder Metallurgy and Their Economic Implications*, Ottawa, 1976, AGARD CP-200, paper 9, pp. 1–13.
9. C. A. Morris and J. W. Smythe: *Proc. 2nd Int. Conf. Superalloys —Processing, Seven Springs*, book MCIC-72-10, pp. Y.1–26, Nat. Tech. Info. Serv., Springfield, VA.
10. J. F. Barker and E. H. VanDermolen: *ibid*, pp.AA1–23.
11. B. Ewing, F. Rizzo, and C. ZurLippe: *ibid*, pp.BB1–20.
12. J-P. A. Immarigeon, G. Van Drunen, and W. Wallace: *Proc. 3rd Int. Conf. Superalloys—Metallurgy and Manufacture, Seven Springs*, 1976, pp. 464–72, Claitor's Pub. Div., Baton Rouge.
13. G. W. Greenwood, W. E. Seeds, and S. Yue: *Proc. Conf. on Forging and Properties of Aerospace Materials*, pp. 255–65, Book 188, The Metals Society, London, 1978.
14. R. L. Hewitt, J-P. A. Immarigeon, W. Wallace, A. Kandeil, and M. C. de Malherbe: *DME/NAE Quarterly Bulletin*, 1978, vol. 4, National Research Council Canada, Ottawa, pp. 1–23.
15. W. Wallace: unpublished results, NAE, National Research Council Canada, Ottawa, 1974.
16. J-P. A. Immarigeon, A. Y. Kandeil, W. Wallace, and M. C. de Malherbe: *J. Test. Eval.*, Nov. 1980, vol. 8, no. 6, pp. 273–81.
17. J-P. A. Immarigeon and R.L. Hewitt: unpublished research, National Research Council Canada, 1980.
18. R. T. Holt: unpublished results, NAE, National Research Council, Canada, 1978.
19. A. Y. Kandeil, J-P. A. Immarigeon, W. Wallace, and M. C. de Malherbe: *Met. Sci.*, Oct. 1980, vol. 14, no. 10, pp. 493–99.
20. ASTM Standard Designation E 112-77.
21. J. W. Edington, K. N. Melton, and C. P. Cutler: *Prog. Mater. Sci.*, 1976, vol. 21, pp. 61–159.
22. M. Suery and B. Baudelet: *Rev. Phys. Appl.*, 1978, vol. 13, pp. 53–66.
23. H. J. McQueen and J. J. Jonas: *Treatise on Materials Science and Technology*, vol. 6, R. J. Arsenault, ed., pp. 393–493, Academic Press, New York, 1975.
24. C. M. Sellars: *Philos. Trans. R. Soc. London*, 1978, pp. A-288, pp. 147–58.
25. J. P. Sah, G. J. Richardson, and C. M. Sellars: *Met. Sci.*, 1974, vol. 8, pp. 325–31.
26. D. McLean: *Proc. 4th Int. Conf. on Strength of Metals and Alloys, Nancy, France*, 1976, vol. 3, pp. 958–75, ed. Lab. Phys. Solid —ENSMIM, Nancy, France.
27. T. H. Alden: *Treatise on Material Science and Technology*, vol. 6, R. J. Arsenault, ed., pp. 225–66, Academic Press, New York, 1975.
28. H. J. McQueen and B. Baudelet: *Proc. 5th Int. Conf. Strength of Metals and Alloys, Aachen*, August 1979, Haasen *et al*, eds., pp. 329–36, Pergamon Press, 1979.
29. R. L. Hewitt, J-P. A. Immarigeon, and P. H. Floyd: *Proc. 8th North Amer. Res. Conf., Rolla, MO*, May 18–21, 1980, pp. 129–34, Society of Manufacturing Engineers, Dearborn, MI, 1980.

SECTION VIII
Other Net Shape Processes

New Aluminum Hot Press Method Provides Heat Treated Condition

ENGINEERS WITH the Illinois Institute of Technology Research Institute (IITRI), Chicago, have recently developed an isothermic hot pressing technique for the production of precision aluminum parts at net shape and of preforms for forging and extruding.

A feature of the process is that compacted aluminum parts, susceptible to age hardening, naturally attain a T4 condition after quenching without subsequent heat treatment.

The process begins by converting aluminum scrap into free-flowing uniform die-filling particulates which are then heated and pressed using special lubricating methods in a hot die.

In isothermal hot pressing both the particulates and the die are kept at the same temperature.

To form the particulates, engineers at IITRI melt in air EC aluminum clippings or other alloys such as 7075 or 2014, and pour them into a rotating perforated cup. The molten metal is centrifugally forced through orifices in the cup, and discrete, uniform particulates are snapped off the liquid streams of the emerging metal.

These are generally needle shaped, about .25 of an inch in length and .010 of an inch in diameter. The particulates are smooth, lustrous and clean.

Once these needles are dropped into the die cavity, a feedbox retracts and levels the particulates off. At this point, the top and bottom punches of the press approach each other, and the needles between them are then hot consolidated and welded to each other to form a compact.

The method applies 12 tsi at a temperature of 950°F held for a pressing time of two seconds or less to form acceptable parts. The short dwell period makes the process economically attractive, according to IITRI engineers, and the speed of production enables commercial compacting presses to be used.

Process Sequence

Once the part has been pressed, the top punch retracts from the die orifice and the bottom punch ejects the compacted piece. As the powder feedbox goes forward to replenish the die cavity, the previously ejected part, still at 950°F, is moved across the platen of the press by the powder feedbox and is quenched in water.

Because the solution annealing temperature range for the particles is around 900°F, and because the part has been maintained at a temperature within this range (950°F), parts made of aluminum alloys susceptible to age hardening quench from the solution-annealed condition and age harden at room temperature without further heat treatment, thus providing the T4 condition at virtually no extra cost.

Quenching contributes other basic advantages. One is that tolerance control is improved over the fine control already obtained by hot pressing (± .0005 of an inch per inch). Also,

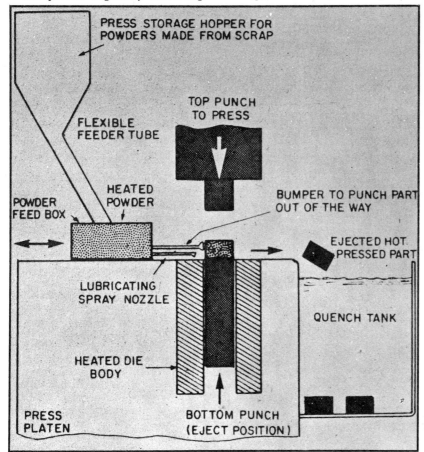

Figure 1: Schematic above details work flow of hot pressed aluminum parts from needle-shaped powders which quench out from press in T4 condition.

oxidation of the compact is limited so that better surfaces and properties can be obtained.

The advantage of using the needle particulates is that they are free flowing at either room or elevated temperatures—almost to their melting point. This differs from conventional aluminum powders made by atomization techniques which do not flow freely hot or cold, according to IITRI.

One problem encountered in hot pressing is inadequate lubrication of the dies. Compaction of the particulates must not cause seizing or galling of the pressed part to the die which can weld shut if lubrication is inadequate.

In the IITRI approach, the die is sprayed with graphite immediately prior to the powder being fed into the die. This occurs with each compaction to assure a fresh coating of lubricant on the die walls.

By means of the proprietary lubrication approach, IITRI engineers believe there is almost no limit to the thickness of a part that can be hot pressed other than that imposed by the press and die.

A large compact, such as the one shown in Figure 2 (2 inches in diameter by 2 inches in length at 300 grams, single-action pressed), cannot be readily made by conventional powder metallurgy methods. Organic lubricating waxes mixed with aluminum powder will allow for full compaction of such a preform, but normal sintering methods will not be able to remove all the wax from the center portion of the compact, making it unusable.

If the die wall is lubricated and unwaxed powder is pressed in such a die, only compacts of a maximum of about half an inch in thickness can be produced without galling or seizing.

At IITRI, engineers have overcome this problem and can compact 4-inch hot needle lengths via single-action pressing to full-density finished parts with uniform properties—even at the bottom end.

When the parts come out of the die they are coated uniformly with the black graphite that was originally on the die wall. Such a coating can be used in subsequent forming operations as with preforms that can be forged or extruded either hot or cold. This eliminates the conventionally applied black coating to forging preforms normally needed for rapid heat transfer when heating prior to forging.

Cleaning

For precision parts that are to be used directly because they are net shape, the black coating may be undesirable. If this is the case, the parts may undergo a vibratory cleaning operation which not only removes the graphite, but also deburrs the components giving them a frosty finish. If a high-gloss finish is desired, the part can then be ball burnished.

This is basically a peening operation which not only gives the part a good surface appearance, but also imparts compressive stresses into the surface to enhance the properties of the finished component.

The hot pressing technique produces two kinds of parts—those ready for use as finished components and those which are at a preform state. Net shapes are produced in just one major step—the hot compaction cycle. A modest clean-up plus minimum quality control requirements compare favorably with other processes involving multiphase manufacture and test procedures. Savings of 15 percent to 75 percent are possible for such parts that can be pressed in a die, according to IITRI.

The preforms are unusual in that they are claimed to offer the forging fabricator a component approximately 30 percent less expensive than a cast

Figure 2: Newly-developed aluminum hot pressing process permits production of outsize compacts such as that shown above, not feasible with conventional techniques.

Figure 3: Aluminum scrap, shown at lower left, is transformed into needles, lower right, whose properties permit hot forming of automotive shapes above left and right.

356

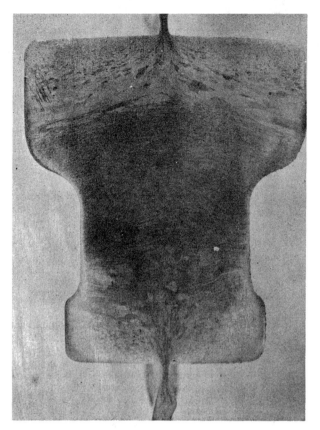

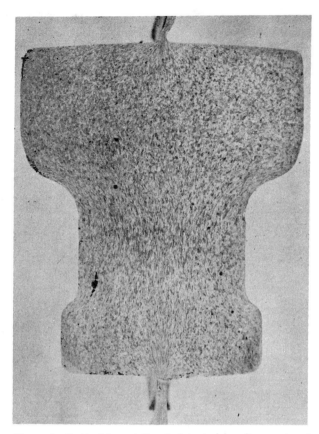

Figure 4: Microstructure of forged aluminum 6061 cast-wrought alloy has range of grain sizes and heavy deformation or flow lines of the grains.

Figure 5: Particulate metallurgical part shows fine equiaxed grains throughout. Dark spots are graphite inclusions, preventable by changing molten metal cup.

or wrought part, with microstructures as shown in Figures 4 and 5. The cast-wrought part (Figure 4) has a range of grain sizes from extremely coarse to very fine and shows heavy deformation or flow lines of the grains. In contrast, the particulate metallurgical part (Figure 5) shows uniform, fine equiaxed grains throughout the structure and minimal distortion as evidenced by a lack of heavy flow line patterns.

Powder metallurgy and particulate metallurgy parts offer isotropic properties whereas cast-wrought parts, because of the unidirectional working of ingot down to bar or plate stock, show definite directional properties and such products are anisotropic. Because of this, the critical limiting factor for current design work is the short-transverse direction, which has the lowest mechanical properties of the three directions. In particulate metallurgy the properties in all three directions are almost the same regarding ultimate tensile strength (5 percent variation) and close regarding elongation.

Aside from isotropy, these parts have other unusual properties (Table 1). They are virtually 100 percent dense and therefore leakproof to a minimum of 2500 psi hydraulic oil pressure and 400 psi helium gas pressure (Figure 6).

The particulate metallurgy parts have the properties of neither castings nor powder metallurgy parts, but of wrought type products such as forgings. Parts eject from the die in conditions ranging from cast-wrought annealed to cast-wrought cold worked. The hot pressing method stress-strains the part with rapidity, so that less time is allowed for recrystallization and grain growth.

Strengthening has been found also to be a function of the surface-area to volume relationship of the particles. The lower this value is, the greater the strength developed on ejection from the die. Engineers at IITRI feel that

Figure 6: Hot pressing method produces parts virtually 100% dense and virtually leakproof to minimum of 2,500 psi hydraulic oil pressure and 400 psi helium.

the higher the volume of material to be worked, the more "cold working" can be induced into the particulate so that it comes out stronger and harder from the die.

Aside from the mechanical properties noted, parts are capable of being welded and of taking virtually any conventional aluminum finish including mechanical, chemical and electrochemical finishing.

Pricing of hot-pressed parts has shown that equivalent parts made by aluminum die casting can be undersold by 15 percent; those fabricated by conventional powder metallurgy by 25 percent to 50 percent; and forgings or machined bar stock by 50 percent to 75 percent.

Savings are derived because raw material used is about half the cost of primary material and because the process is essentially a single-stage operation. Aluminum powder metallurgy, by comparison, requires cold pressing, sintering, coining and then heat treating.

Capital investment is lower than conventional powder metallurgical processes. A new hot pressing line needs but one modified compacting press, whereas the usual powder metallurgical line needs a press, a two hot-zone belt furnace with pollution exhaust, a protective atmosphere generator and distribution system, a coining press plus extra die and a heat treating furnace.

Another factor is that the hot-pressing process is virtually pollutionless and requires less energy.

This process is presently being made available to industry by IITRI through its licensing program. As of now, a main generic patent application has been applied for with the U. S. Patent Office and additional applications are being processed in some foreign countries. ∎

			Table 1		
		TYPICAL MECHANICAL PROPERTIES OF HOT PRESSED ALUMINUM PARTS, AIR COOLED FROM 950°F/12 tsi/2 sec			
Material	Density, % of Theoret.	UTS, psi	Elong- ation, %	Rockwell Hardness, H scale	SA/Vol x 10^{-1} in.
1100 Al, needles	>99	14,600	-	70.3	500
Same, cold rolled 40%	∿100	19,900	-	88.7	
1100 Al -100 mesh powder	>99	13,585	31.8	44.5	1,500
Scrap Al + 3% Cu	∿100	22,770	2.4	97.6	180
Scrap Al + 3% Cu	∿100	21,875	4.2	84.2	415
Scrap Al + 3% Cu	∿100	18,995	5.0	90.7	760
7075 Swarf	>99	52,655	-	97 R/E	200
Mg Ribbon 950°F/24 tsi/2 sec	>99	27,200	5.2	28 R/H	360
Iron Carbonyl Powder 950°F/50 tsi/2 sec (10 micron diameter)	>99	96,200	2	12 R/C	50,000

Appliancemakers Evaluate Net-Shape Process

By Harry E. Chandler

MAKERS OF HOME APPLIANCES have been unusually successful in holding their cost line over the years by diligently seeking out materials and processes that are more economical than those they are currently using. At this time, there is interest, for example, in possibly replacing enameling iron with vacuum-degassed, cold-rolled steel. In partmaking, injection molding of a powder metal-plastic mixture (a near-net shape process) appears to have potential as an economical alternative. Plastic in pores is melted away in sintering. Pores can then be infiltrated with a lubricant.

Another net-shape process attracting serious consideration involves the conversion of aluminum scrap into needles that are subsequently hot pressed into finished parts or preforms for forging or extruding. Virtually any small aluminum appliance part now made by die casting, conventional P/M, forging, or by machining from bar stock is a candidate, claims IIT Research Institute (IITRI), Chicago, licensor of the process.

Other potential applications, including gears and connecting rods, in autos and business machines are also cited. Information about parts being evaluated by the U.S. appliance industry is regarded as proprietary. However, appliancemakers in foreign countries are looking at such parts as aluminum pistons and brass valve components.

"Without qualification, the most attractive part of this process is the economics," states Samuel Storchheim, manager of metalworking and foundry technology at IITRI.

"Costing of hot-pressed parts by the new IITRI process," he continues, "has shown that equivalent parts made via aluminum die casting can be undersold by 15%, those via conventional P/M by 25 to 50%, and forgings or machined bar stock by 50 to 75%.

"Such savings come from the fact that the raw material used is about half the cost of primary material and . . . the entire process is really a single-stage operation."

Hot-pressed parts, it is reported, require only a modest cleanup (vibratory cleaning) and minimum quality control. Forging and extrusion preforms are said to have better microstructures (Fig. 1) than those of their cast-wrought counterparts. The latter show a range of grain sizes — from extremely coarse to very fine — and heavy deformation or flow lines of the grains. By comparison, the hot-pressed preform has uniform, fine equiaxed grains throughout the structure, and distortion is minimal, as evidenced by a lack of heavy flow line patterns.

Other Improvements: mechanical properties are said to be isotropic; parts are fully dense and leakproof; and microstructures are comparable to those of wrought, annealed products.

Parts are reported to be leakproof to a minimum of 2500 psi (17 MPa) hydraulic oil pressure and 400 psi (2.8 MPa) helium gas pressure. They can be used as-formed in vacuum, hydraulic, liquid transfer, and other types of fluid flow applications. Leakproofing of conventional P/M parts and die castings is an extra step.

Wrought-type properties are obtained, it is explained, because hot pressing stress-strains parts with extreme rapidity. There isn't sufficient time for recrystallization and

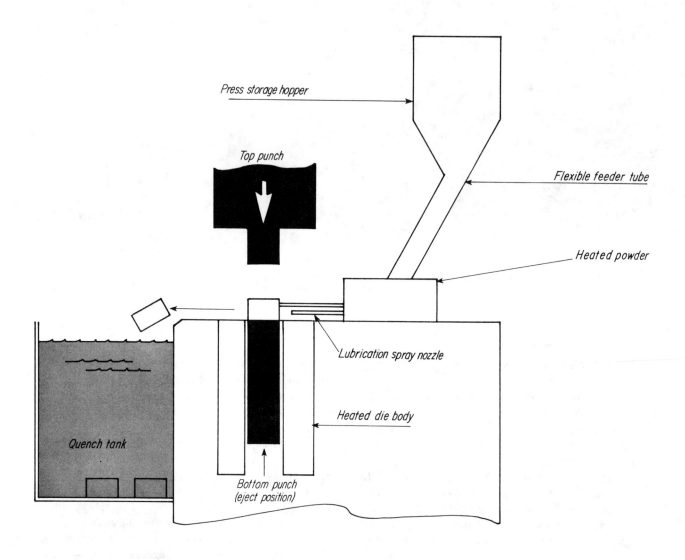

Press storage hopper

Top punch

Flexible feeder tube

Heated powder

Lubrication spray nozzle

Heated die body

Quench tank

Bottom punch
(eject position)

grain growth (conventional annealing or softening of the component).

Still speaking about the partsmaking process, Mr. Storchheim adds, "the parts are at an elevated temperature where they are plastic and flow beautifully under pressure to not only level out very uniformly (within 0.001 in. [0.025 mm] or less on thickness) but also move completely and properly into such volumes as gear teeth, splines, and narrow openings. The response is like that in cold-working." Microstructures are shown in Fig. 2.

Strengthening is also a function of the surface-area-to-volume relationship of the particles (see Table I). The lower the value, the greater the strength developed. It is felt, Mr. Storchheim explains, that the higher the volume of material to be worked, the more cold working can be induced into

the particulate, so that it comes out stronger and harder from the die.

In addition, parts can be welded, and they will take practically any conventional aluminum finish — mechanical, chemical, or electrochemical.

The Process: From Raw Material to Partsmaking

IITRI starts with EC aluminum scrap clippings which are basically more than 99.5% pure. However, there is a tramp pickup of about 3% copper. Alloys such as 7075 and 2014 can be used as scrap, including machine shop turnings and borings.

Scrap is melted in air and poured into a rotating perforated cup. Molten metal is centrifugally forced through

Fig. 1—Left, macrostructure of forged aluminum 6061 cast-wrought alloy. Right, macrostructure of forged hot pressed preform. Dark spots are graphite inclusions picked up during molten metal spinning in a graphite cup. Inclusions can be eliminated by changing to another cup material.

orifices in the cup; discrete particulates are snapped off the liquid streams of emerging metal. The particulates are generally needle shaped (acicular) and run about ¼ in. (6 mm) long by 0.010 in. (0.25 mm) in diameter. Surfaces are smooth, lustrous (there is a thin oxide coating), and clean.

The needles or particles are free flowing at room or elevated temperatures. Conventional atomized aluminum powders, by comparison, are not free flowing, hot or cold. This is a key to the process.

Lubrication is an important part of the process. Immediately before particles are fed into the die, the die is sprayed with graphite. This is done before each compac-

Table I—Typical Mechanical Properties of Hot Pressed Aluminum Parts[1]

Material	Density, %Theoretical	UTS, psi (MPa)	Elongation, %	Hardness, R h	Surface Area/ Volume x 10⁻¹ in.
1100 Al needles	>99	14 600 (103)	—	70.3	500
1100 Al cold rolled 40%	~100	19 900 (138)	—	88.7	
1100 Al 100 mesh powder	>99	13 585 (97)	31.8	44.5	1500
Scrap Al + 3% Cu	~100	22 700 (159)	2.4	97.6	180
Scrap Al + 3% Cu	~100	21 875 (152)	4.2	84.2	415
Scrap Al + 3% Cu	~100	18 995 (131)	5.0	90.7	760
7075 Al swarf	>99	52 665 (365)	—	97[2]	200
Mg ribbon[3]	>99	27 200 (186)	5.2	28	360
Iron carbonyl[4] powder	>99	96 200 (662)	2	12[5]	50 000

1. Air cooled from 950 F (510 C) / 12 tsi (165 MPa) / 2 s; 2. Rockwell E scale; 3. 950 F (510 C) / 24 tsi (7 kPa) / 2 s; 4. 950 F (510 C) / 50 tsi (690 MPa); 5. Rockwell C scale.

tion. Action of the top and bottom punches hot consolidates the particulates, which are welded to each other to form a compact.

Size capability is greater than that of the conventional P/M process. For example, aluminum needles can be formed into a 300 g slug measuring 2 in. (51 mm) long by 2 in. (51 mm) in diameter with a single-action press. An aluminum compact of that size can be obtained with conventional P/M, but organic lubricant waxes are required; and wax in the center portion of the compact could not be removed with normal sintering, making such a part ususable.

Mr. Storchheim says if the die wall is lubricated in conventional P/M and unwaxed powder is pressed, maximum compact thickness is ½ in. (13mm). Galling or seizing is the problem.

Isothermal: IITRI uses what it describes as a slightly modified (for hot pressing) commercial compacting press (see lead figure). Both the needles and the die are heated to 950 F (510 C). Pressure of 24 000 psi (166 MPa) is applied. Maximum pressing time is approximately 2 s. This speed, points out IITRI, makes the process attractive to the appliance industry.

After compaction, parts are water quenched (see lead figure). This means an aluminum part quenches from the solution-annealed condition and naturally age hardens at room temperatures. The T4 condition is obtained at virtually no extra cost.

Quenching also improves tolerance control (it's better than ±0.0005 in./in. [or mm/mm]); and oxidation is limited so that better surfaces and properties can be obtained.

Preform or Part: The preform or part comes out of the die with a black graphite coating (transferred from the die during pressing). The coating can be used as a lubricant in subsequent forging or extrusion, eliminating the conventional black coating normally used for rapid heat transfer when furnace heating prior to forging.

The coating may not be desirable if the product is a ready-to-use part. Vibratory cleaning is used to remove the graphite and also deburr the part, giving it a frosty finish.

If a high gloss or mirror finish is wanted, the part can be ball burnished. This is essentially a peening operation which also impacts compressive stresses into the surface, enhancing properties.

Another Facet: The IITRI patent on the process is not limited to aluminum. Work has also been done with magnesium, copper, and iron. Some results are summarized in Table I.

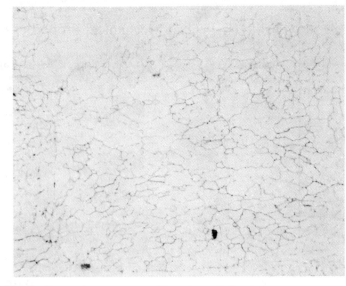

Fig. 2—Top, microstructure of hot pressed aluminum, transverse view; Kellers etch; 50 X; middle, longitudinal view; Kellers etch; 50 X; bottom, needle interior showing fine grain structure. Uniform line in structure is solid-state weld joint between two needles; Kellers etch; 200X.

Squeeze forming of aluminium-alloy components

The squeeze-forming process, involving the pressurized solidification of liquid metal in re-usable dies is described, and the importance of various process parameters is discussed. A range of aluminium-alloy components made by GKN Technology Ltd using this technique is shown, and details are given of the types of component geometry amenable to the process. Mechanical-property data determined in several of these components are presented for a number of aluminium alloys. It is observed that both 'forging' and 'casting' alloys can be successfully squeeze formed, and that the mechanical properties achieved compare favourably with conventional properties. It is concluded that the work performed at GKN Technology Ltd has confirmed squeeze forming to be a potentially powerful technique for the manufacture of a wide range of aluminium components in competition with more conventional production methods. MT/800

G. Williams

K. M. Fisher

© 1981 The Metals Society. The authors are at GKN Group Technological Centre, Wolverhampton.

The squeeze-forming process allows close-tolerance high-integrity components to be produced from liquid-metal feedstock. The process is not a new one, and it has several synonyms in the literature, for example: squeeze casting,[1-4] extrusion casting,[5] liquid pressing,[6,7] pressure crystallization,[8,9] and Cothias Casting.[10] There are claims that the process may be applied to several different feedstock metals[2,3,5,6,8] but work at GKN has concentrated upon aluminium alloys. GKN Technology Ltd have been developing this process for several aluminium-alloy components since 1973. This paper presents a description of the basic process, outlines the important process parameters, and provides mechanical-property data for several squeeze-formed aluminium alloys.

Process outline

Squeeze forming is a generic term to describe a range of tooling concepts which promote the pressurized solidification of a component in re-usable dies. The process involves the following steps.

1. An accurately metered amount of molten metal is poured into a preheated die cavity, located on the bed of a hydraulic press.

2. The press is activated to bring down a top die or punch to close off the bottom die cavity, and pressurize the liquid metal. This is done as rapidly as possible after pouring so that only a small amount of solidification can occur without pressure.

3. The pressure is held on the metal until solidification is completed. This forces the metal into intimate contact with the die faces thereby significantly increasing the heat flowrate. Most importantly the pressure encourages any micro/macro-shrinkage cavities to be filled up.

4. Finally, the punch is withdrawn and the component is ejected.

Two basic forms of the process may be distinguished depending on whether the movement of the punch causes any liquid metal movement. These are shown diagrammatically in Figs.1 and 2. Figure 1 illustrates the case where there is no metal movement. This type of squeeze forming is suitable for ingot-type components or those with an essentially planar shape. Figure 2 illustrates the case where there is an appreciable amount of metal movement involved: in this case backward extrusion. The profile of the punch causes the liquid metal to be displaced to fill the die cavity. Once this has happened the metal is pressurized and the press is then stalled. This type of squeeze forming is more versatile and can be used to make a wide range of shaped components.

The types of tooling may be further elaborated. For example, movable cores may be incorporated into the design, while forward and/or horizontal displacement of the metal can be achieved in order to generate a range of possible component shapes. Furthermore, it is not always necessary to have the hydraulic press acting through the complete area of a top die or its equivalent, but it can operate upon a reduced area, as illustrated in Fig.3. This type of tooling variation can increase the freedom of component shape, but usually leads to the need for higher pressures.

Example components

Several of the components that have been made by GKN Technology Ltd are illustrated in Fig.4. This range of examples shows the wide variety of shape capability of the squeeze-forming process, both in terms of configuration and size.

The smallest component that has been made was a mere 125 g, and the largest to date weighed 35 kg. Currently, components of about 18 kg are being made on a regular basis.

Most of the components have been round in shape, but Fig.4 shows that rectangular shapes are also technically and commercially viable. The car wheel shows that undercut shapes can be made by use of split-die techniques, and the transmission yoke component is an example of a component made with retractable side cores.

It has proved possible to manufacture components with a wide range of section thicknesses. The thinnest tackled to

This paper was presented at the international conference 'Solidification technology in the foundry and casthouse' held by The Metals Society at the University of Warwick, Coventry, on 15–17 September 1980.

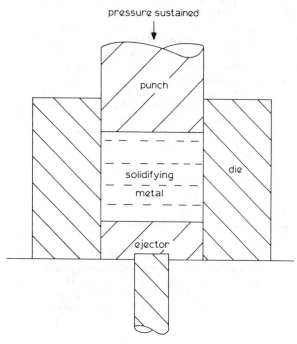

1 Representation of squeeze-forming process with no metal movement

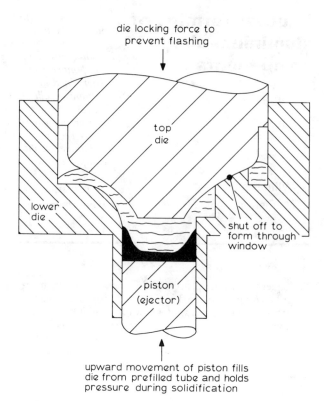

3 Variation of squeeze forming analogous to vertical die casting

date has been 2 mm, and the thickest has been 50 mm. Obviously, exaggerated sectional changes within one component are not to be recommended, but in certain circumstances it is possible to combine heavy and light sections without the metallurgical problems which would result in most other foundry techniques.

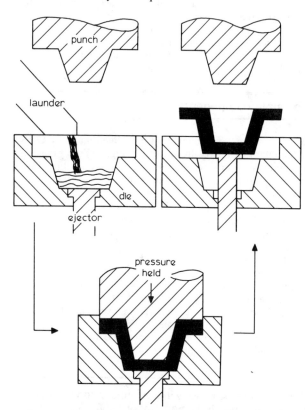

2 Representation of squeeze-forming process with backward extrusion of liquid metal

The components illustrated in Fig.4 have been made in a very wide range of aluminium alloys. Several have been made in conventional casting alloys, but several have also been produced in alloys which would normally be restricted to forging processes. This additional freedom of alloy choice results from the pressing action which causes the metal to flow and fill the die cavity. There is not such a great need for a reasonable fluidity in order to reproduce the die features faithfully. The final solidification pressure overcomes problems of feeding shrinkage and often suppresses tendencies for hot tearing. The range of alloys that has been investigated is discussed below.

The components represented in Fig.4 compete both technically and economically with alternatives made by low-pressure and gravity die casting, high-pressure die casting, forging, and fabrication.

General process features

The squeeze-forming process uses rigid metal dies, which are kept at a controlled temperature. In this respect it is very similar to high-pressure die casting. The die materials used are the same as for high-pressure die casting: usually the hot-working die steels such as H13. Before each press cycle the dies are sprayed with a thin coating of parting agent, the usual one being an aqueous dispersion of graphite.

The high quality of the re-usable dies, and the very thin die coating, leads to a process which has an inherently good dimensional reproducibility. The dimensional tolerances are in general at least as good as those obtained by high-pressure die casting. Figure 5 shows the reproducibility in the diameter of a typical component over a batch of 1000.

4 A range of squeeze-formed components made by GKN Technology Ltd

The most variable dimensions are those governed by the horizontal partline of the tools. This is because the final rest position of the tools is mostly governed by the quantity of metal which has been put into the dies. A variety of metal metering techniques may be used to control this weight of metal within acceptable limits, e.g. automatic ladling or several commercially available metal metering pumps. If particularly close control of across-the-partline dimensions is required, then it is possible to build overflow systems into the toolsets.

In most cases, the use of overflow systems can be avoided, and the material yield in these circumstances is excellent. It is a feature of squeeze forming that runner systems are not usually required, and the need for feeder methods is completely eliminated. Therefore, apart from the possible need for limited machining operations (bolt holes, etc.) or removal of sharp edges, the metal placed in the dies is used in its entirety.

The pressures used depend upon the component configuration, but generally fall within the range 31–108 MPa. Components which are comparatively tall and slender will require higher pressures than those which tend to be more squat.

It is interesting to note that sound components of forging quality can be made from wrought-specification alloys by the use of press loads significantly lower than those required for solid-phase forging. A single die only is required for squeeze forming, but several may be required for different stages of a forging process.

The pressurized solidification of squeeze forming ensures a particularly faithful reproduction of the surface of the dieset. This reflects the intimate contact between the die and solidifying metal at all stages of the solidification. The full contact area, and the suppression of any airgap, which often forms at the component/mould interface in other foundry methods, causes a heat flowrate which is at least one order of magnitude greater than in gravity die casting.[9,11] Solidification times are therefore much shorter than in any process other than high-pressure die casting, and fine-grained structures with small dendrite cell sizes result.

Mechanical properties of components

It has been mentioned that the process is amenable to the use of a very wide range of alloys. Alloy specifications normally restricted to forging applications may be used, as well as the alloys usually associated with casting operations. Alloys that respond to heat treatment can quite successfully be put through the necessary solution treatment and age hardening.

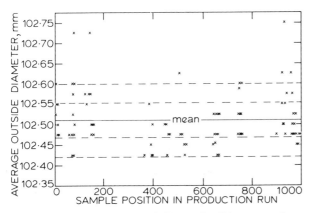

5 Dimensional reproducibility of diameter of a squeeze-formed component: batch size 1000

Table 1 Tensile data for squeeze-formed material (casting-type alloys) compared with conventional properties of chill castings

| Alloy designation | Approximate composition | Temper condition | Mechanical properties of squeeze-formed material | | | Comparable properties of conventionally chill-cast material | | | |
			0.2% PS MN m^{-2}	UTS, MN m^{-2}	Elong., %		0.2% PS, MN m^{-2}	UTS, MN m^{-2}	Elong., %
LM5	Al–5Mg	M	142	250	14	Minimum spec.	>90	>170	>5
						Typical prop.	90	230	10
LM18	Al–5Si	M	103	187	13	Minimum spec.	>60	>140	>4
						Typical prop.	80	150	6
LM25	Al–7Si–Mg	M	124	195	15	Minimum spec.	>80	>160	>3
						Typical prop.	90	180	5
		TE	165	235	7	Minimum spec.	>130	>190	>2
						Typical prop.	150	220	2
		TF	250	300	10	Minimum spec.	>220	>280	>2
						Typical prop.	240	310	3

Temper condition: M=as formed; TE=solution treated and stabilized; TF=solution treated and aged to maximum strength.

CASTING-ALLOY TYPES

Table 1 lists three aluminium casting alloys which have been squeeze formed into several components. The mechanical properties are compared with the specified minima for each alloy type, and also typical values of properties obtained by chill casting (either gravity or low-pressure die casting).

It will be noted that the squeeze-formed mechanical properties are superior in all respects to those given by conventional casting. The proof stresses have been improved by squeeze forming, and of particular significance is the marked improvement in the elongation to failure, which is a measure of the toughness and ductility of the material. These improvements reflect both the fine-grained structure and more importantly the elimination of microporosity in the squeeze-formed material.

The results with the alloy LM25 are quite interesting because they have been achieved with an alloy containing a significant amount of iron (about 0.5%Fe). The alloy LM25 is regularly used for gravity and low-pressure die casting of safety-critical components. When it is used in these applications it is necessary to employ primary-quality alloy (with a cost premium) to restrict the iron to less than 0.2%. This avoids the embrittling effect of iron aluminide needles within the cast structure, which renders it susceptible to impact failure. The rapid solidification of the squeeze-forming process, however, has prevented the formation of coarse iron aluminide needles. The intermetallic compound was still present, but was extremely finely divided, and so rendered harmless. It is thus possible to consider the squeeze forming of secondary-quality foundry alloys (recycled scrap) for safety-critical components, which would conventionally require the more expensive primary-quality alloy.

Fatigue tests have been carried out with LM25 samples, which were cut from a bracket component. A servohydraulically controlled fatigue machine was used to execute push–pull tests about mean zero. The results are presented in Fig.6, which includes for reference the results of similar tests carried out on conventionally cast LM25. It can be seen that a significant improvement in the fatigue performance has been achieved by squeeze forming this type of alloy.

FORGING-ALLOY TYPES

The results of mechanical-property testing of six alloys which are regarded as forging materials are presented in Table 2. These values have all been obtained from test samples cut from actual components.

Table 2 includes comparative data for each alloy type when it has been conventionally forged. It is important to be

Table 2 Tensile data for squeeze-formed material (forging-type alloys) compared with conventional properties of forged material

| Alloy designation | Approximate composition | Temper condition | Mechanical properties of squeeze-formed material | | | Comparable properties of conventionally forged materials | | | |
			0.2% PS MN m^{-2}	UTS, MN m^{-2}	Elong., %		0.2% PS, MN m^{-2}	UTS, MN m^{-2}	Elong., %
AA7075	Al–5Zn–2Mg–Cu	T6	525	565	6	Die forgings, || grain flow	>427	>503	>7
						Minimum prop ⊥ grain flow	>414	>483	>2
						Typical longitudinal properties	503	572	11
		T73	415	455	4	Die forgings, || grain flow	>379	>441	>7
						Minimum prop., ⊥ grain flow	>359	>421	>2
AA2014	Al–4·5Cu–Si–Mg–Mn	T6	455	485	2	Die forgings, || grain flow	>379	>434	>6
						Minimum prop., ⊥ grain flow	>372	>434	>2
						Typical longitudinal properties	414	483	13
AA6061	Al–1Mg–1Si–0·5Cu–Mn	T6	325	335	8	Die forgings, || grain flow	>241	>262	>7
						Minimum prop., ⊥ grain flow	>241	>262	·5
						Typical longitudinal properties	276	310	12
AA6066	Al–1Mg–1·5Si–1·5Cu–Mn	T6	385	405	5	Die forgings, || to grain flow	>310	>345	>8
						Minimum prop.	..	..	..
						Typical longitudinal properties	359	393	12
AA6082	Al–1Mg–1Si–1Cu–Mn	T6	260	285	7	Forging stock and forgings	255	295	8
						Typical longitudinal properties	..	..	..
SF1	Al–4Zn–2Mg	Non-standard T6/T73	335	385	10	Forgings, || grain flow	>330	>385	>8
						Minimum prop. ⊥ grain flow	>310	>370	3
						Typical longitudinal properties	350	400	10

Temper condition: T6=solution treated and aged to maximum strength; T73=solution treated, double aged to maximize stress-corrosion resistance; T6/T73=heat treated, modified to maximize stress-corrosion resistance in this alloy.

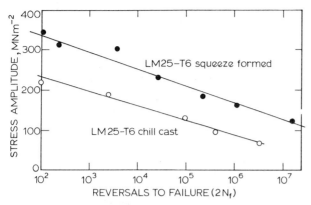

6 Fatigue data for casting alloy LM25–T6: squeeze-formed and conventionally cast conditions

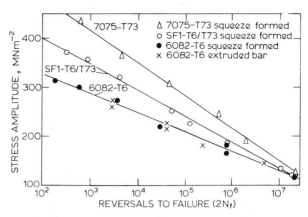

7 Fatigue data for several squeeze-formed 'forging-type' alloys with data for extruded AA6082–T6 for reference

well aware of the inhomogeneity in properties which is introduced by forging operations. In order to quantify this effect, Table 2 contains data for minimum mechanical properties in the longitudinal and transverse directions wherever possible.

Squeeze-formed components are isotropic in their properties, and not subject to the directionality which often affects a forging. It can be seen that in general the squeeze-formed components compare extremely well with forgings. In most cases, there is a good comparison between the isotropic squeeze-formed property and the longitudinal forging property.

In two cases (AA 7075 and AA 6066) the ductility values were less than those of the longitudinal forging data, but greater than the transverse forging data. The strengths in these cases were the equivalent of a longitudinally tested forging, or better.

Alloy AA 2014 exhibited quite low ductility, similar to a transversely tested forging, but had quite respectable strength values.

The major cause of the reduction of ductility in these three alloys was the presence of an appreciable copper content (1%Cu for 6066, 1·5% for 7075, and 4·5% for 2014). The copper formed grain-boundary concentrations of eutectic materials which were not readily dispersed by heat treatment. Even in these cases however, the properties may be considered as usable since they are still better than the transverse properties of forgings.

It is fair to conclude that in most cases the squeeze-formed materials compare well with forgings in all mechanical respects, and they can be superior if isotropic properties are important. Particularly noteworthy is the ability to produce components from 7000 series alloys (Al–Zn–Mg), which are the highest strength aluminium alloys currently available.

Figure 7 presents results from push–pull, about mean zero, fatigue tests which have been carried out on a servo-hydraulically controlled machine. The tests have been carried out on samples cut from actual components, not from separately made testpieces. The results from conventionally extruded AA 6082 (H30) are included for reference: in this case the data are in the longitudinal direction, it not being possible to obtain samples of sufficient size from the transverse direction.

The results show good fatigue properties for squeeze-formed material, which in one case compare favourably with conventionally extruded material. This further substantiates the claim that squeeze formings in general are comparable with forgings with respect to mechanical performance.

General conclusions

1. The work at GKN Group Technology Ltd confirms that squeeze forming is a potentially powerful technique for manufacturing a wide range of aluminium components.

2. Example components have been made which compete favourably with those made by many conventional production methods, including low-pressure and gravity die casting, forging, and fabrication.

3. The process produces sound components with excellent dimensional reproducibility and very good material utilization.

4. Casting-type alloy specifications result in significantly improved mechanical properties following squeeze forming.

5. Forging-type alloy specifications usually result in isotropic mechanical properties comparable to conventional forgings measured in the longitudinal direction. This includes the capability to process the high-strength 7000 series alloys.

Acknowledgment

The authors wish to thank the GKN organization for the facilities provided over a period of time, which have enabled the development programme to be carried out, and for permission to publish the findings. Particular thanks are due to Dr T. L. Johnston, the management of GKN Technology Ltd, and the management of GKN Sankey Ltd. Especial acknowledgment is given to the members of the development team, by whose physical and mental effort the work programme has been carried through.

References

1. Y. KANEKO, H. MURAKAMI, K. KURODA, and S. NAKAZAKI: 10th SDCE International Die Casting Congress, 1979, Society of Die Casting Engineers Inc.: also *Foundry Trade J.*, 1980, **148**, (3183), 397.
2. J. C. BENEDYK: 6th SDCE International Die Casting Congress, Nov. 1970, paper 86, Society of Die Casting Engineers Inc.
3. K. M. KULKARNI: *Foundry Manage. Technol.*, Aug. 1974, 76.
4. R. F. LYNCH, R. P. OLLEY, and P. C. J. GALLAGHER: *Trans. AFS*, 1975, **83**, 569.
5. V. M. PLYATSKII: 'Extrusion casting'; 1975, New York, Primary Sources.
6. V. P. SEVERDENKO and T. P. MALEI: *Dokl. Akad. Nauk SSSR*, 1961, **5**, 253.
7. A. F. ASTASHOV and S. I. TISHAEV: *Kuznechno-Shtampov.*, 1967, (8), 16.
8. B. B. GULYAEV *et al.*: *Liteinoe Proizvod.*, Dec. 1960, **12**, 33.
9. S. CHATTERJEE and A. A. DAS: *Br. Foundryman*, 1973, **66**, (4), 118.
10. L. W. CHAMBERS: *Foundry Trade J.*, 1980, **148**, (3187), 802.
11. Y. NISHIDA and H. MATSUBARA: *Br. Foundryman*, 1976, **69**, 274.

FEASIBILITY OF SQUEEZE CASTING FERROUS COMPONENTS

by

K. M. Kulkarni, Manager, Metalworking Research, IIT Research Institute, Chicago, Illinois 60616
D. Stawarz, Research Engineer, La Salle Steel, Hammond, Indiana 46323
R. B. Miclot, Metallurgist, U.S. Army Weapons Command, Rock Island, Illinois 61201

ABSTRACT. Squeeze casting is a hybrid of conventional forging and casting techniques and consists of pouring molten work material into a lower forging die, allowing the melt to solidify partially, closing the dies, and maintaining pressure directly on the work material until it solidifies completely. The process offers several advantages over conventional forging and casting techniques such as improved material yield, dimensional precision and quality and reduced finish machining. This paper describes recent work on the fabrication of two steel weapon components viz., a receiver base which is trough-shaped, 17-in. long and 2-1/2 in. wide and a barrel support which is 14 in. in height, much of it with thin-walled tubular geometry. Receiver base squeeze castings weighing only 12 lb each were successfully made whereas sand castings for the same component weighed 47 lb with gates and risers. The barrel support squeeze casting weighed only about 14 lb whereas the corresponding sand casting with gates and risers weighs 80 lb.

INTRODUCTION. The conventional casting and metalforming process often result in oversize and heavy weapon components, which then must be extensively machined. In sand casting, much expense is also associated with the preparation of molds and trimming of castings. Squeeze casting or liquid metal forging, which is a hybrid of conventional casting and forging techniques, has been under continuous study at IIT Research Institute. Of special interest is a recent project carried out to investigate the feasibility of squeeze casting large ferrous components such as a barrel support and a receiver base.

Currently the two components are machined from sand castings, and the sand-cast and finish-machined forms of each are shown in Figs. 1 and 2. The receiver base is a U-shaped channel-like component with 17 in. in overlength. The finish-machined part weighs only 6.2 lb and is produced from a sand casting which weighs 47 lb with risers and gates. The barrel support is 14 in. in overall length, over half of which is tubular in shape. The component weighs 7.4 lb in finish-machined condition and 80 lb as a sand casting with risers and gates. Both the components are made from 8630 type steel, the receiver base to a class of 140-110 and the barrel support to 150-125 class.

Both the components show poor material yield from sand casting to the finished components, require extensive finish machining, and incur much expense with the various foundry operations. The reject rate also tends to be high.

Because of the long length, thin sections, and the geometry, both these components--particularly the barrel support--are extremely difficult parts to produce as squeeze castings. Moreover, the difficulty is compounded by the low carbon content of the material and consequently very high melting temperature range. In this program, the dies were designed and fabricated, a series of squeeze casting tests were conducted, and the products were evaluated. This brief report summarizes the work and assesses the potential of the process for fabrication of large ferrous components to reduce the cost of their fabrication.

SQUEEZE CASTING--PROCESS PARAMETERS AND ADVANTAGES. The process is variously termed liquid metal forging, squeeze casting, or extrusion casting. It is essentially a hybrid of the forging and the casting processes. Apparently, the process has been investigated in Russia for quite some time although it is a relative newcomer in this country. There are various publications discussing the major facets pertaining to the technique.[1-2]

The four stages of squeeze casting are shown in Fig. 3. They consist of introducing the molten metal into the die cavity, allowing it to solidify partially, applying pressure to fill the die cavity and to accomplish solidification under pressure, and, finally, removing the completed casting from the die cavity. The pressure is high enough and is maintained for adequate duration to eliminate all traces of interdendritic porosity by pressure feeding of the molten metal and of shrinkage porosity by keeping any gas in solution. The main process variables are the melt temperature, the time delay before the pressure is applied to the solidifying material, the pressure level and its duration, and the tooling temperature. The variables must be optimized to obtain excellent product quality without adversely influencing the die life.

The squeeze casting process differs significantly from ferrous die casting and rheocasting. Ferrous die casting is suitable mainly for small components of around 1 lb in weight, has poor material yield, and cannot be used for producing thick components. Rheocasting is a multi-step process consisting of agitating molten metal until about half the volume fraction solidifies, transferring it into forming dies, and then subjecting the metal to the forming process. So far it has been applied only for making small components of simple shapes from nonferrous materials or cast irons.

The principal advantages of the squeeze casting process can be listed as follows:

1. Ability to produce parts with complex profile and thin sections beyond the capability of conventional casting and forging techniques.

2. Substantial improvement in the material yield in comparison with sand casting because of the elimination of risers and gates.

3. Elimination of the labor associated with sand casting such as for molding, trimming of risers and gates, and cleaning of casting.

4. Significant reduction in pressure requirements in comparison with conventional forging while at the same time increasing the degree of complexity that can be obtained on the parts.

5. Ability to utilize both cast and wrought compositions of work materials.

6. Substantially increased production rates in comparison with conventional casting techniques.

7. Possibility of substantial cost reduction in comparison with sand castings because of items 2 and 3 and in comparison with forgings because of the lower cost of the melt stock used for squeeze casting instead of the wrought materials used for forging.

8. Improvement in product quality such as the surface finish and mechanical properties in relation to sand castings. The generally harder as-squeeze cast material may also, in some cases, eliminate the need for further heat treatment.

TOOLING, EQUIPMENT, AND PROCEDURE.
Squeeze Casting Dies. The receiver base dies are shown in Fig. 4. Apart from a small insert at the point where the metal first touches the dies, the entire bottom die was made from H13 die steel with a hardness of R_C 38. The entire top die, or punch, was also made from the same material. (The first series used Anviloy 1150 tungsten-base material for the projecting part of the punch. See later discussion.) It may be noted that the squeeze casting was to be made with the length of the part in the horizontal plane. The melting and melt transfer furnace can also be seen in the photograph. Two ejection pins were provided in the bottom die. They were actuated by the press ejection system in the initial series. Subsequently, special auxiliary cylinders were provided for each of the pins.

A schematic of the die design for the barrel support is shown in Fig. 5. The die set is designed so that the punch first comes down and enters the bottom die to the desired point, and then the two move together against the large springs to the end of this stroke. In this fashion, the relationship of the inside and outside detail of the squeeze casting could be controlled precisely in spite of the minor variation of the weight of the molten metal trapped in the die cavity. The entire bottom die was again made from H13 die steel with the exception of an Anviloy plug at the bottom where the molten metal impinges. The punch support system was again made from the H13

steel and the long, cylindrical pin was made from a variety of different materials to compare the performance of different die materials. In the majority of the tests, Inconel 718 was used as the pin material and was found to be the most suitable.

Press. A 250-ton hydraulic press was utilized for the squeeze casting tests. The press can come down at a free speed of about 500 ipm, and the working speeds are up to about 80 ipm. The retraction force available to separate the dies was only 10% of the press capacity and caused some problems as is mentioned later. The ejection system consisted of a small hydraulic cylinder mounted below the press bed and had a maximum tonnage capacity of about 20 tons. In addition, auxiliary cylinders were used in the receiver base die set in the later squeeze casting series.

Melting and Melt Transfer. The work material was molten in a 20-lb capacity induction melting furnace mounted on a bracket directly on the hydraulic press. The induction coils had a power supply of 9600 cycles at 40 kva. After melting, the furnace could be tilted by means of a handle, and the melt could be transferred directly into the die cavity without any manual ladling. The stock for each experiment was weighed separately and melted in the furnace with appropriate provision for the loss due to skull remaining in the furnace. The melt stock was 8620 type steel, to which additions were made for carbon, molybdenum, and manganese. Also, aluminum was used for deoxidation just prior to transfer of the molten metal into the die. The aim composition was similar to that used by Rock Island Arsenal for sand castings and was as follows:

0.25%C, 0.75% Mn, 0.50%Si, <0.010%P, <0.010% S, 0.50% Ni, 0.55% Cr, 0.55% Mo, bal Fe.

No problem was encountered in achieving a composition similar to this.

Test Procedure. The experimental procedure consisted of opening the dies, applying the mold wash, then melting the work material and transferring it into the dies, closing the dies and building up and maintaining the pressure for the desired duration, and then releasing the load. Attempt was generally made to retract the press immediately but, because of the inadequate press capacity, often several minutes of delay elapsed before the ram could be retracted. Then the ejection force was applied, and the squeeze casting was pushed up from the die. It could then be manually lifted off the die. The pins were then retracted. The die was visually inspected and, in rare instances when there was any local welding, it was hand polished. Then the die was again precoated for the next experiment. In the case of the receiver base, the two ejection cylinders applied auxiliary load after the main press tonnage was built up. It was maintained for the same duration as the main squeeze casting pressure.

The different process parameters were varied, and it was found that the optimum press speed was about 300 ipm. The press was completely closed at this speed without changeover to the slower speed. The melt temperature was measured by a platinum-rhodium thermocouple and the temperatures were generally kept to within 25°F of the desired value. Both the melt temperature and the melt transfer time are very important, and the latter was of the order of 5 to 10 sec for both components. The die temperature was measured prior to each experiment. The duration of application of load was measured after the load built up to the desired value.

EVALUATION OF RECEIVER BASE SQUEEZE CASTINGS. The long length of the component and thin sections make this a difficult component to squeeze cast. Nevertheless, optimization of the various process parameters (Table 1) resulted in excellent quality squeeze castings; a comparison with the machined component and a sand casting is shown in Fig. 6. Whereas the sand casting weighed 47 lb with risers and gates, the squeeze casting weighed only about 12 lb, thus substantially reducing the degree of any machining required and increasing the material yield.

The successful squeeze casting of this component represents a length-to-thickness ratio of nearly 40 with the length in the horizontal plane.

Surface Quality and Dimensions. As in any squeeze casting, the surface quality is somewhat better on the areas made by the top die in relation to those made by the lower die. This is because the molten metal after its introduction into the bottom die begins to solidify immediately and this solidification of the surface prior to the application of the pressure causes a poorer surface finish. The surface quality on the exterior of the squeeze casting was comparable to the surface of the good quality sand castings. The inside of the channel that was formed by the top die had an excellent finish when produced by the Anviloy punch that was used in the first few tests. In the squeeze castings made with the die steel punch, the surface finish was initially very good but gradually showed a rough grainy appearance because of the deterioration of the surface of the punch.

The lateral dimensions of the squeeze casting are controlled mainly by the die dimensions and, to a certain extent, by the die and melt temperatures. Excellent control can therefore be maintained on the lateral dimensions, and in measurements carried out on 33 castings the tolerance spread was of the order of ± 0.002 in. per in. for the length and ± 0.008 in. per in. for the width. In the vertical direction, that is, the direction of motion of the dies, the dimensional precision depends on the quantity of melt trapped in the die cavity. This, in turn, depends on the quantity of melt introduced into the die and any loss during buildup of the pressure. In these tests, the premeasured quantity of material was individually melted in the furnace. Loss of metal occurred because of the skull remaining in the furnace and also because of some metal that was pushed out of the die during pressure buildup. The weights of the castings were of the order of ± 0.4 lb although some castings did show greater variation. Despite fairly high precision in the squeeze casting, the combination of the

surface quality and dimensional tolerance makes it unrealistic to attain machining tolerances necessary to produce any surfaces of the receiver base net.

Internal Integrity. The internal quality of the squeeze casting was checked by sectioning and etching and by radiography. Defect-free castings were made through optimization of the process parameters, Table 1. In the initial squeeze casting tests, porosity of about 1/4 in. linear dimensions was noticed in the thick end (3/4 in.) of the casting where the solidification takes much longer time than the remainder which is of much lower thickness (3/8 in.). This porosity was effectively eliminated through two auxiliary hydraulic cylinders by applying the auxiliary pressure with the help of the two ejection pins, one of which was located in the thick end. The pressure capacity of the auxiliary cylinders was about 9,000 to 10,000 psi, and this was adequate to eliminate the porosity in the thick end. Cross sections of a squeeze casting and a sand casting are compared in Fig. 7. Note that the sand casting has a significant amount of distributed porosity whereas the squeeze casting shows only a very slight trace of microporosity on the centerline of the section.

Mechanical Properties. Evaluation of the mechanical properties of the squeeze casting was difficult since the properties are generally specified for a sand casting as measured in a keel block and are rarely checked on samples cut from the casting proper. In this case, tensile test specimens were cut directly from the squeeze castings, thus necessitating usage of subscale specimens, and compared to the property specification. The heat treatment was varied to a limited extent and, in particular, the normalizing temperature was investigated with regard to its effect on the tensile properties.

Three specimens were tested from a squeeze casting after austenitizing at 1570°F, water quenching, and tempering at 1070°F for 1 hr. The specimens had a gage length of 1 in. and gage area of 0.25 x 0.25 in. The average properties were 129,700 psi yield stress, 139,000 psi ultimate strength, 14% elongation, and 31% reduction in area. A comparison of the properties of specimens cut from a sand casting against those cut from a squeeze casting is provided in Table 2. Note the excellent ductility of the squeeze castings in spite of the shorter gage length and smaller gage area. Additional tests were conducted to study the influence of normalizing temperature on properties of squeeze casting. As expected, normalizing made the yield stress values of replicate tests very close. Unfortunately, a flaw in the sample caused the ductility to be only 6%. The tests show that there should be no problem meeting property specifications under production conditions. Changes from this development program such as usage of draft on the dies and better control over melt quality will further enhance the properties that can be obtained under production conditions.

EVALUATION OF BARREL SUPPORT SQUEEZE CASTINGS. The barrel support is the most difficult component squeeze cast to date. In spite of the long length of 14 in. and a length-to-wall thickness ratio of 40, it was successfully produced as a squeeze casting and is compared with a sand casting and a finished component in Fig. 8. The sand casting with risers and gates weighs 80 lb in comparison to a squeeze casting weight of only 13 to 15 lb. Successful squeeze casting of this component shows the feasibility of producing long tube-like components with thin walls.

Surface Quality and Dimensions. The barrel support was made in a vertical orientation so that the molten work material when introduced into the bottom die formed a deep pool with a relatively small open surface. Consequently, the surface appearance of the barrel support squeeze castings was far superior to that of the receiver base. This difference was undoubtedly enhanced by the totally undamaged nature of the lower and the top dies for the barrel support whereas some surface roughness occurred in the dies for the receiver base.

Initially, the squeeze castings were made with a ceramic material as the mold wash. Subsequently, carbon soot was used as the mold-separating agent applied prior to each squeeze casting. The ceramic material was still periodically used and, in this case, the carbon soot was applied over the ceramic material. With this type of mold wash application the surface appearance of the squeeze castings was exceptionally good, as exemplified in Fig. 9. The appearance is far better than that of a sand casting or a hot forging and approaches the type of finish that can be obtained by cold forging.

The dimensional precision of the barrel support squeeze casting is influenced by such factors as discussed above for the receiver base. However, the die design used for the barrel support makes the influence of variation in the melt volume less important than in the case of the receiver base. In making the barrel support the bottom die and the punch come to a definite relative position, and then they move together against the plug at the lower end. As a result, any excess melt stock sppears as a thicker end plug which is, in any case, machined during the finish machining operation.

Once again, although the dimensional precision was good, it does not approach finish machining tolerances which are generally of the order of a few thousandths of an inch. Consequently, some finish machining should be provided on the barrel support squeeze casting also. Initially, squeeze castings were made with very sharp corner radii in attempts to produce various surfaces net. The sharp radii did cause some cracking in the corners in the rectangular portion of the squeeze casting. These cracks were eliminated on increasing the radii.

The hydraulic press utilized for the squeeze castings has a retraction force capability of only up to 25 tons to separate the dies after squeeze casting. The long length and the complex geometry of the barrel support often made this force capacity inadequate and, in many cases, 8 min or so were required after releasing the squeeze casting load before the dies could be opened. During this period, the squeeze casting shrank on the solid Inconel 718

punch. This caused tearing on the tubular portion or the squeeze casting. Such defects can be eliminated in production by retracting the punch immediately after release of the squeeze casting load. This points to the need of a hydraulic press having a higher retraction capacity than is commonly available and is a factor to be considered in obtaining a press for squeeze casting.

Internal Integrity. The squeeze castings made with optimized process parameters (Table 3) were completely free of porosity on the tubular section of the squeeze casting. In the initial tests, some squeeze castings were completely defect-free whereas large porosity occurred in some squeeze castings in the thick rectangular parts. With a single-acting press it was difficult to eliminate porosity in this area, and another practical approach was developed. This was to provide some additional machining allowance on the lugs on the rectangular portions as shown in Fig. 10. The thick portions on the lugs act like risers and eliminate the porosity in the other portions of the rectangular portions. The porosity does occur in the thick lug but, since this is subsequently machined off, it does not make the part unacceptable. This approach is useful in producing complex-shaped squeeze castings where there may be substantial variation in the section thicknesses.

The interface of the tubular section and the rectangular portions of the squeeze casting is complicated in geometry. Initially, the squeeze castings showed no porosity at the interface. However, some cracking occurred at the sharp corners near the interface and, therefore, the corner radii were increased substantially. At the same time, since it was apparent that some machining would be desirable, the geometry of the interface was simplified and various steps that were originally there were eliminated. Unfortunately, during this modification, the thickness of the interface was inadvertently increased and this caused some porosity to occur there because the interface now became thicker than the tubular portion or the rectangular portion and the pressure could not be transmitted through these thinner sections which solidify earlier. The porosity can be eliminated by keeping the steps similar to those occurring in the finish-machined component so that the section thickness at the interface is small or alternatively by using a double-acting hydraulic press.

Squeeze castings normally show a fine-grained cast structure. In the case of the barrel support, finely spaced columnar grains were seen growing from the inside as well as the outside surfaces of the component normal to the surface. The two fronts meet near the centerline of the thickness, and some equiaxed grains were seen in this area. A defect-free structure is more easily obtained if the drafts on the die components producing the inside and outside of the squeeze casting are such that there is a slight increase in thickness towards the direction from which the load is applied. Thus the wall thickness of the tubular portion should preferably increase away from the closed end. In the tests conducted, equal draft (1/4°) was used on the inside and outside, and yet good quality castings could be produced.

Source: 4th North American Metalworking Research Conference Proceedings, 1976, 58-66

Mechanical Properties. The work material used for barrel support and the receiver base is identical so that the general comments about the mechanical properties are about the same for the two components. Once again, whereas the property specification was for chill blocks made separately during sand casting, the tensile test specimens in the case of the squeeze castings were cut directly from the casting.

Four specimens were tested from a barrel support squeeze casting--two from the tube section and two from the rectangular portions. The heat treatment used was 1690°F normalizing for 1 1/2 hr, austenitizing at 1575°F, water quenching, and tempering at 1100°F for 1 hr. The gage length was 1 in., and the gage diameter 0.25 in. The average properties were 140,000 psi yield stress, 149,200 psi ultimate strength, 6% elongation, and 13% reduction in area.

There is need for additional work on heat treatment and work material composition. Under production conditions, better melt quality should also result; but on the whole, no difficulty is anticipated in meeting the specifications developed for the sand castings.

TOOLING PERFORMANCE. The tooling for squeeze casting the receiver base and the barrel support had to withstand very severe operating conditions because of the very high melting temperature of the work material, usage of common die steels for practically all the tooling components, and the attempt to produce many of the surfaces net on the squeeze casting. In spite of this, on the whole, the tooling withstood the squeeze casting trials very well. This work demonstrates that it is practical to utilize die steel tooling for producing squeeze castings from low-carbon steels in production quantities. In this section, additional details of the tooling performance are provided for the two components studied.

Receiver Base Dies. The lower die for the receiver base made from H13 die steel, and no serious damage to the die was noticed in approximately 180 squeeze casting tests. Because of the attempts to minimize the machining on the squeeze cast component, the sides of the bottom die cavity were vertical, that is, without any draft. These sides suffered some damage, especially at the level of the molten metal immediately on pouring the material into the die. Some under cut was apparent which made it difficult to remove the squeeze casting and the die set had to be cleaned to widen the sides by 0.015 in. after 53 squeeze castings. This damage can be practically eliminated by providing a small draft of the order of 1° on the sides.

When the molten metal is poured into the bottom die, it impinges at the same point. This caused some erosion and, in a few cases, local welding. Ceramic mold wash was applied to the area but, because of the washing action of the molten metal, sometimes this was not adequate. Usage of an Anviloy or similar refractory-base material would be desirable to improve tool life in this area. This need only be a small insert. (Such an insert was utilized in the bottom die but, to keep the pouring time to a minimum, the technique of pouring had to be modified which allowed the molten metal stream to impinge on the die steel instead of the Anviloy.)

The only modification incorporated into the bottom die was the addition of two auxiliary cylinders. As discussed in the section on evaluation of receiver bases, some porosity was noticed initially in the thick end of the squeeze casting. To eliminate this, hydraulic cylinders were placed below each of the two ejection pins so that additional pressure could be applied independent of the main press tonnage. This modification was for improving the squeeze casting quality and does not show any shortcoming of the tooling proper.

The projecting part of the punch was initially made from Anviloy, a tungsten-base material. This material is not wetted by molten steel and, consequently, gave excellent surface finish on the squeeze castings. However, because of the poor mechanical strength of the material, the punch broke in several pieces and some blistering also occurred on the punch. Therefore, the material seems unsuitable for long production runs. Over 120 squeeze castings were then produced by a punch made from H13 die steel which had a provision for water cooling. Minor scoring was noticed on the part because of the vertical walls without any draft, but, on the whole, H13 withstood the squeeze casting trials very well and seems to be a satisfactory candidate for production runs. For even longer tool life, the punch should be made from Inconel 718 type of material, which has proved to be so satisfactory for the barrel support as discussed below.

In summary, the entire bottom die for squeeze casting the receiver base can be made from H13 die steel with the exception of a small refractory material insert at the point of initial contact between the molten metal stream and the die. The punch can be made from H13 die steel with, perhaps, only the small projecting part of the punch made from Inconel 718. Some draft should be provided on the walls of the squeeze casting. The usage of common material and relatively much thinner section thicknesses should lead to a substantial reduction in the cost of the squeeze casting dies in relation to the cost of the conventional hot forging dies.

Barrel Support Dies. The lower die for the barrel support was made from H13 die steel and withstood the entire squeeze casting series, in which over 130 squeeze castings were made, very well. The only problem encountered arose because of the long, tubular portion of the squeeze casting, especially since initially no draft was provided on the die in this area. In one experiment, minor welding occurred between the squeeze casting and the die and, in pushing the squeeze casting out, the die was scraped off in one area. The cleanup cut on the die increased the diameter by approximately 0.050 in. The die is in excellent condition and should show very long life of several thousand squeeze castings.

The main body of the punch was again made from H13 die steel and, in most of the experimental work, the long small-diameter portion, or the punch pin, was made from Inconel 718 material. The first Inconel pin was used in 64 squeeze castings and was in excellent condition until it was damaged because of experimental difficulties. A second pin was made and was used in over 40 squeeze castings. It is still in excellent condition and should show a very long life. The only indication of damage on the pin was at the lowest corner radius where slight upsetting occurred. This merely shows that a larger corner radius should have been used instead of the 1/4 in. radius utilized in this program. Significantly, this performance was obtained in spite of the absence of any water cooling of the punch pin.

During squeeze casting, the 2 1/4 in. diameter punch pin is surrounded completely by molten metal and this, coupled with its long length and small diameter, makes it the most severely stressed component of either of the die sets. For tooling to be utilized for long production runs, this pin should be made so that, if necessary, it can be replaced quickly. Once again, some draft is necessary on both the punch pin and the lower die although a small draft of only 1/4° will suffice.

POTENTIAL FOR COST REDUCTION AND LARGE-SCALE PRODUCTION. The feasibility of producing large, complex ferrous components has been demonstrated. This was achieved in spite of the fact that a work material with low carbon content is the most difficult type to squeeze cast because of its high liquidus temperature. Stainless steel and high-carbon steel would have been much easier to squeeze cast. The barrel support--with the dimensions to which a squeeze casting can be made--is completely beyond the capability of conventional forging and sand casting techniques. The barrel support squeeze casting had a large height-to-wall thickness ratio of 40, and a nearly similar length-to-wall thickness ratio was achieved in the receiver base. In the latter case, it was shown how auxiliary pressure can be utilized to squeeze-cast components with widely differing section thicknesses.

In the development project, the squeeze casting component and the tooling designs were selected to study the very limits of the process. Several changes would facilitate the production greatly, as has been discussed earlier. This would include provision of some draft on the side walls and provision of machining allowance on all surfaces. Usage of programmable controller or some other similar automatic controls would be highly desirable in production for uniform product quality. Better melt quality control inherent in a large-scale foundry operation will also be helpful. The forging press should be designed with adequate ram retraction and ejection force capacities. The dies should be designed to facilitate quick changing of all components such as the punch pin in the barrel support that must withstand more severe service.

The production quantities of interest to the Arsenal are around 2000 components of each type per year. The capability of the squeeze casting

process is far in excess of this even with one set of tooling and one hydraulic press. Thus, the same press could be used for producing a variety of other candidate components also. However, in view of the complexity of the receiver base and, in particular, the barrel support and the lower carbon content of the work material, production of these components should be undertaken only after adequate experience has been gained with production of simpler and, preferably, nonferrous components.

The squeeze casting process shows the potential for a substantial saving in production cost for both the receiver base and the barrel support. The significant reduction in weight of these squeeze castings in relation to the sand castings will mean much lowering of the material cost and also of the machining costs. The saving in machining cost will be smaller in proportion since a finish machining cut will still be required on both the components. The largest potential for saving comes from the elimination of foundry steps such as making and breaking of the molds and trimming of risers and gates. Savings will also accrue in other associated steps such as storage and handling of sand and patterns. Finally, for the same production capability, the squeeze casting process will require a substantially smaller size of melting and melt handling facility than a conventional sand casting facility.

In general, the potential cost saving is influenced greatly by the production quantity requirements, the complexity and size of the component, and the work material. When only a few dozen components are required, sand casting and machining from solid blanks are favored. When larger production quantities of the order of a few thousand are required, squeeze casting quickly becomes a very promising production technique. Complex geometry and difficult-to-machine materials would favor squeeze casting. When quantities are very large and the component geometry is relatively simple, forging comes into the picture and should be compared to squeeze casting. Of course, a complex geometry such as the barrel support is completely beyond the capability of the forging technique regardless of the required quantities.

CONCLUDING REMARKS. Squeeze casting is a hybrid of conventional casting and forging techniques which shows a potential for cost reduction in fabricating many components. The difficulty in squeeze casting increases with the melting temperature range of the material and, thus, the aluminum alloys are the easiest to squeeze cast and the low-carbon steels are the most difficult among the common groups of materials. The work presented here has demonstrated the feasibility of squeeze casting two complex large components, namely, the barrel support and the receiver base, in ferrous materials of low-carbon content. The barrel support had height-to-wall thickness ratio of 40, and the receiver base had a similar length-to-wall thickness ratio. Both the components showed a tremendous reduction in weight compared to conventional sand castings and a significant increase in the part complexity that can be achieved. These factors would translate into substantial savings in overall production costs.

Source: 4th North American Metalworking Research Conference Proceedings, 1976, 58-66

ACKNOWLEDGMENTS. This presentation is based on Contract No. DAAF-03-73-C-0099 sponsored by the Rock Island Arsenal at IIT Research Institute in Chicago. Thanks are due to both the organizations for permission to publish this work and help in its preparation. The views expressed by the authors do not represent the opinion of either the Department of the Army or the Department of the Defense. The paper was typed by Mrs. Mary Dineen and Mrs. Mary Scroll and edited by Miss Violet Johnson, all of IIT Research Institute.

REFERENCES

(1) K. M. Kulkarni, Machine Design, Vol. 46, No. 11, p. 125, May 2, 1974.

(2) J. C. Benedyk, ASME Publication 72-DE-7 (1972).

Table 1

OPTIMIZED CONDITIONS FOR SQUEEZE CASTING THE RECEIVER BASE

Melt Weight	13-13.5 lb
Die Temperature	600° to 700°F
Melt Temperature	2900° to 2950°F
Time from Start of Pour to Application of Load	6 sec (max.)
Press Speed	300-600 ipm
Load Cycle	125 tons for 30 sec
Auxiliary Ram Load	8 tons for 30 sec

Note: Depending on the draft on the dies and the number of ejection pins, the casting may have to be cooled for a few seconds prior to ejection to minimize bending during ejection.

Table 2

COMPARISON OF TENSILE PROPERTIES OF RECEIVER BASE SAND CASTINGS AND SQUEEZE CASTINGS

Identification	Test Bar Location (see Fig.)	Treatment	Hardness (avg), BHN	YS (0.2% offset), psi	UTS, psi	Elongation, %	Reduction of Area, %
Aim Properties	--	--	--	115,000	140,000	10	25
Keel Block[a]	--	b	341	147,000	164,000	12	40
Sand Casting[a]	A	b	341	150,500	164,000	10	26
Sand Casting[a]	A	b	341	154,500	165,000	10	25
Sand Casting[c]	B	b	341	150,500	154,500	3	9
Squeeze Casting[d] B84	A	e	319	131,500	138,500	13	26
Squeeze Casting[d] B84	A	e	311	131,000	138,000	14	29

[a]0.505 gage diameter, 2.0 gage length.

[b]Normalized at 1700°F, austenitized at 1625°F, water quenched, Tempered: 2.5 hr at 1025°F, 1.0 hr at 1050°F, 1.0 hr at 1100°F.

[c]0.375 gage diameter, 1.4 gage length.

[d]0.25 x 0.25 gage area, 1.0 gage length, 0.05 ipm crosshead speed.

[e]Normalized at 1700°F, austenitized at 1620°F, water quenched. Tempered: 1.0 hr at 1050°F, 1.0 hr at 1110°F, 1.0 hr at 1150°F.

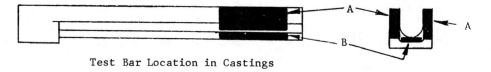

Test Bar Location in Castings

Table 3

OPTIMIZED CONDITIONS FOR SQUEEZE CASTING THE BARREL SUPPORT

Melt Weight	14.5-15 lb
Die Temperature	600° to 700°F
Melt Temperature	2950° to 3000°F
Time from Start of Pour to Application of Load	9 sec (max.)
Press Speed	300-600 ipm
Load Cycle[a]	200 tons for 5 sec then open dies immediately
Ejection Cycle	Remove stripper plate and eject immediately

[a]
Depending on the component and die designs, usage of double-acting press may improve the product quality.

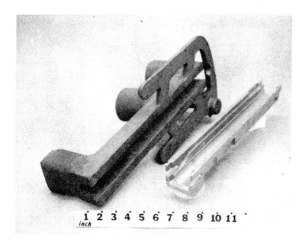

FIGURE 1. RECEIVER BASE COMPONENT WEIGHING 6.2 LB, MACHINED FROM AN UNTRIMMED SAND CASTING WEIGHING APPROXIMATELY 47 LB.

FIGURE 2. BARREL SUPPORT COMPONENT WEIGHING 7.4 LB, MACHINED FROM AN UNTRIMMED SAND CASTING WEIGHING APPROXIMATELY 80 LB.

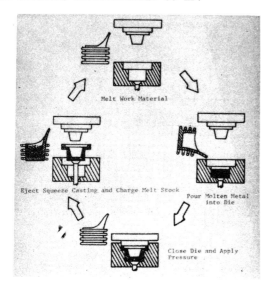

FIGURE 3. PRODUCTION SQEUENCE FOR SQUEEZE CASTING.

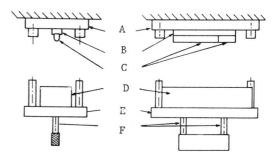

A - TOP PLATE ON DIE SET   D - H13 BOTTOM DIE
B - H13 PUNCH SUPPORT   E - BOTTOM PLATE
C - H13 (OR ANVILOY   OF DIE SET
    PUNCH INSERTS   F - EJECTION PINS

FIGURE 4. RECEIVER BASE SQUEEZE CASTING DIE ASSEMBLY.

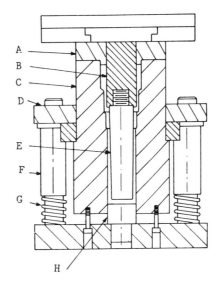

A - STRIPPER PLATE H13, R_C 38-40

B - PUNCH BODY H13, R_C 38-40

C - DIE BODY H13, R_C 38-40

D - 4 POST DIE SET

E - PUNCH

F - SPACER SLEEVE H.R.S.

G - SPRING - 1400 LB/IN.

FIGURE 5. BARREL SUPPORT DIE ASSEMBLY, FRONT VIEW.

FIGURE 6. COMPARISON OF A MACHINED COMPONENT, A SAND CASTING (CENTER), AND A SQUEEZE CASTING FOR THE RECEIVER BASE.

Source: 4th North American Metalworking Research Conference Proceedings, 1976, 58-66

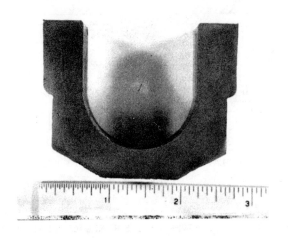

(a) SQUEEZE CASTING

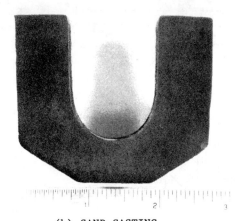

(b) SAND CASTING

FIGURE 7. COMPARISON OF THE CROSS-SECTION OF A SQUEEZE CASTING AND A SAND CASTING.

FIGURE 8. COMPARISON OF A MACHINED COMPONENT, A SAND CASTING, AND A SQUEEZE CASTING FOR THE BARREL SUPPORT.

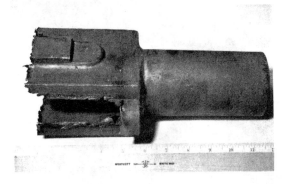

FIGURE 9. AN EXAMPLE OF THE EXCELLENT SURFACE APPEARANCE OF BARREL SUPPORT SQUEEZE CASTINGS (SHOWN AS-CAST WITHOUT ANY TREATMENT OF THE SURFACE).

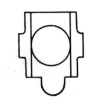

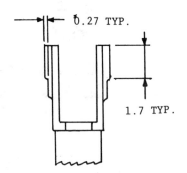

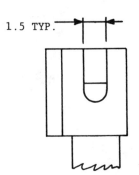

FIGURE 10. SCHEMATIC OF THE MODIFICATION FOR THE RECTANGULAR PORTION OF THE BARREL SUPPORT SQUEEZE CASTING.

M.J. Stewart

Superplastic forging of Zn-Al-Cu alloys

Abstract. Superplastic forging of non-ferrous alloys offers the advantages of very low forging loads, inexpensive die materials, and the ability to forge precision complex-shaped parts in a single step. The main disadvantages of the process are the required low forging rates and the necessity of temperature controlled dies. Upset forging and closed die forging have been performed on Zn-22% Al and Zn-20% Al - 2% Cu alloys. Cylinders 1.4 in (35.6 mm) in diameter by 2.5 in (63.5 mm) high have been upset to a reduction in height of 97% and also have been forged into a complex part with thin web sections. The optimum forging conditions are discussed for these materials.

Résumé. Le forgeage superplastique des alliages non-ferreux offre les avantages suivants: très faibles charges, matériaux de matrices peu dispendieux, possibilité de former en une opération des pièces de précision de forme complexe. Les principaux désavantages du procédé sont les faibles taux de forgeage qu'on doit utiliser et la nécessité d'avoir des matrices à température contrôlée. Les alliages Zn-22%Al et Zn-20% Al-2%Cu ont été formés par refoulement et en matrice fermée. Des cylindres de 1.4 po (35.6 mm) de diamètre et de 2.5 po (63.5 mm) de hauteur ont été refoulés jusqu'à une réduction de hauteur de 97% et forgés en une pièce complexe comportant des nervures fines. Les conditions optimales pour forger les deux alliages sont discutées.

Superplastic metals have been studied in the laboratory for many years (1-11), but only recently have these materials been used commercially. Superplasticity is characterized by very large tensile elongations at very low stresses. Elongations of over 1000% are not uncommon in zinc and nickel alloys. Superplastic materials have grain sizes of the order of one micron, and the finer the grain size the lower the stress and greater the elongation at a particular strain rate (3-6). The flow curves of these materials may be defined as:

$$\sigma = K \, \overset{\circ}{\epsilon}{}^m$$

where σ is the true flow stress, K is constant, $\overset{\circ}{\epsilon}$ is the strain rate and m is the strain rate sensitivity. An alloy is said to be superplastic if the strain rate sensitivity m is greater than 0.3. This will usually occur at temperatures greater than 0.4 of the absolute temperature melting point and at strain rates of 10^{-4} to 10 sec^{-1}. For comparison, conventional forging is at strain rates of the order of 10^3 sec^{-1} and deep drawing of the order of 10^2 sec^{-1}.

Superplastic metal forming can offer distinct advantages over conventional metal forming.
(a) Due to the large strains possible, complex shapes can be produced in a single step without the danger of tears and cracks.

(b) The loads required to produce large parts are low due to the small flow stresses.
(c) Most superplastic materials revert to normal metal behaviour outside a narrow elevated temperature range or the material can be heat-treated after forming to a non-superplastic state.
(d) Many materials are superplastic (1), such as zinc alloys, copper alloys, steels, nickel alloys, aluminum alloys and various lead-tin-cadmium alloys.

There are two disadvantages of superplastic metal forming.
(a) To produce the desired forming properties, the rates of forming are necessarily low.
(b) Due to the longer forming times and the narrow temperature range required, dies have to be maintained at the forming temperature.

The most common utilization of these materials is in forming sheet metal parts. Hollow bodies similar to deep drawing can be made with techniques similar to the plastics industry methods such as vacuum or bulge forming (12-15). Due to the low flow stresses, low pressures can be used.

Forging of solid shapes with superplastic material (16-18) has not received as much attention as sheet parts. Forging of these parts would offer an alternative to die casting, permanent mould casting, and conventional forging. The large strain rate sensitivity, m, which accounts for the very large tensile elongations is not as important in a superplastic forging process. Necking is not a problem in normal closed die forging. However, for the complex parts envisioned for superplastic materials, there could be regions

M.J. Stewart is a Research Scientist with the Metal Forming Section, Physical Metallurgy Division, Mines Branch, Department of Energy, Mines and Resources, Ottawa, Canada.

where tensile stresses are induced and the high strain rate sensitivity could be beneficial. The large strain rate sensitivity restricts the forging to low deformation rates because the forces will rapidly rise if the forging rate is increased.

Because limited quantitative data are available that can be used to establish forging rates, temperatures and proper material preparation and handling, this work was undertaken. The alloys chosen were a Zn-22 wt% Al eutectoid alloy and a Zn-20 wt% Al-2 wt% Cu alloy. Alloys similar to these are superplastic (2-11) and are being used commercially in sheet form. The addition of copper to the Zn-Al alloy increases the creep resistance without seriously affecting the superplasticity (10).

Initially, upset forging tests were conducted to analyze the material under controlled experimental forging conditions. This was followed by the totally enclosed die forging (flashless forging) of a complex part.

Material preparation

For these eutectoid alloys a simple heat treatment can be used to obtain the desired fine microstructure for superplasticity. This involves solution treating in the high-temperature single-phase region, above 275°C, and rapidly quenching the material to room temperature, forming the ultra-fine grain size two-phase material.

The alloys used in this study were melted and cast into extrusion billets 4.5 in (114 mm) in diameter by 10 in (254 mm) long. The melts were prepared from 99.99% pure zinc, aluminum and copper metal. The extrusion billets were homogenized at 375°C for 48 hr after which the billets were either ice-water quenched or allowed to air cool. Extrusion was done at 200°C to a final diameter of 1.50 in (38 mm) in a 750-ton extrusion press. The extrusion ratio was 9 to 1 and the extruded rods were allowed to air cool. The forging preforms were machined from the extruded bars to 1.40 in (35.6 mm) in diameter by 2.50 in (63.5 mm) long.

Four different heat treatments of the forging preforms were investigated:

Condition A: As-extruded material from the extrusion billets that were ice-water quenched after homogenization.
Condition B: Solution treated preforms at 375°C for $2^1/_2$ hr followed by an ice-water quench. This heat treatment removed effects of any prior thermal-mechanical treatment.
Condition C: As-extruded material from the extrusion billets that were air-cooled after homogenization.
Condition D: Solution treated preforms at 375°C for $2^1/_2$ hr followed by air-cooling. This heat treatment removed the effects of any prior thermomechanical treatment.

Upset forging apparatus and procedure

Flat steel dies 6 in (152 mm) square and 2 in (51 mm) thick with embedded resistance heaters were used for the upset forging tests. The upper and lower die temperatures were maintained within ± 1°C with a two-zone temperature controller. The control thermocouples were 0.25 in (6.4 mm) below the centre of the die faces. Forging was done in a 500-ton (4.45 MN) hydraulic forging press. A load cell was mounted above the upper die, to measure the load during forging and a water-cooled brass block protects the load cell from the hot dies. Sheet mica was used to insulate the bottom die from the press. The movement of the press crosshead was measured as the voltage output from a ten-turn potentiometer, mounted on the crosshead, with a stationary wire snubbed around the potentiometer. A small hole was drilled in the sample to accommodate a thermocouple for recording the sample temperature.

For a forging test, the cold preform was inserted between the two hot dies and allowed to heat to forging temperature. When the preform reached the desired temperature, typically twenty minutes, it was upset-forged to a predetermined height. Strip chart recorders charted the forging load, the crosshead displacement, the sample temperature and the upper and lower die temperature. The forging rate could be varied continuously from 0.5 to 110 in/min (0.2 to 47 mm/sec).

As in all upset compression tests, lubrication is of the utmost importance. The degree of barrelling that occurs is an indication of the coefficient of friction. Ideally, the test piece sides should remain vertical to give homogeneous metal flow. The lubricants tried were colloidal graphite, sodium stearate soap, various oils, grease, 0.015-in (0.4 mm) and 0.005-in (0.13 mm) thick teflon sheet, spray teflon, molybdenum disulphide powder, and molybdenum disulphide mixed with bearing grease. Only the 0.005-in (0.13 mm) thick teflon and the molybdenum disulphide plus grease eliminates barrelling completely. Teflon was used for the upset testing because of handling ease. The graphite lubricant was almost as good because it allowed only slight barrelling.

Forging results

Typical upset forgings are shown in Figure 1 with reductions of 0%, 22%, 38%, 57%, 82%, 93% and 96%. The load vs time and the displacement vs time results for each forging were converted to true stress (σ) vs sample height (H) curves for comparative interpretation. Figures 2 and 3 show typical curves and this type for the Zn Al and Zn-Al-Cu alloys respectively. On both figures the dashed lines represent the condition B material and the solid lines the condition A material. The curves are all for a forging temperature of 270°C and the forging rates are shown on each curve. For all the results presented the forging temperature is taken to be the initial uniform preform temperature prior to forging. The initial gradient throughout the specimen was less than 1°C. During forging, the adiabatic heating raised the sample temperature as much as

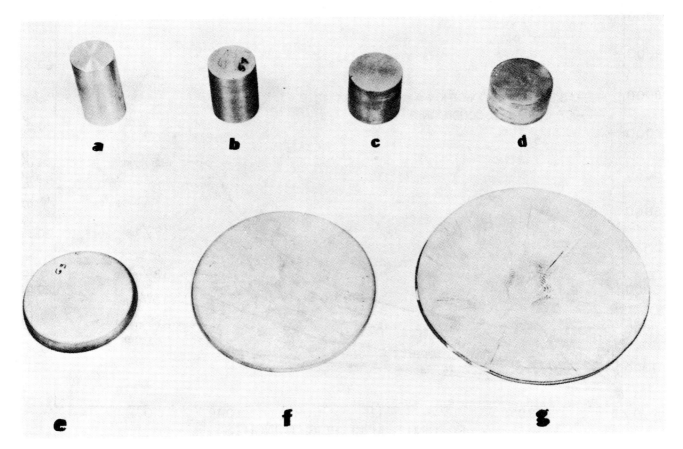

Figure 1. Upset forging specimens after a reduction in height of (a) 0%, (b) 22%, (c) 38%, (d) 57%, (e) 82%, (f) 93% and (g) 96%. Underformed specimen (a) is 1.4 in. in diameter by 2.5 in. high.

15°C depending on the initial temperature, forging rate, preform condition and lubrication.

Generally, there are two shapes of the σ-$\dot{H}$ curve. In condition A material, the stress rises rapidly to a relatively flat plateau region without any pronounced maximum. In condition B material the stress reaches a maximum in the initial forging stage and then decreases to a plateau. All the curves eventually show an increase in stress in the later stages of forging. This can be attributed to a breakdown in the teflon sheet lubrication resulting in an increase in the load. Also the true strain rate is rapidly increasing as the sample becomes thinner and there is a change in the grain size as will be discussed later.

For a comparison of the various forging conditions, the true stress at a fixed specimen deformation has been chosen. Figures 4 and 5 are plots of the forging rate ($\dot{H}$) in inches per minute vs the true stress for the Zn-Al alloy at a specimen height of 2 in (50.8 mm) and 1 in (25.4 mm), respectively, for various temperatures. Because the press speed is constant throughout the compression the true strain rate is increasing continually during forging. From the σ-$\dot{H}$ curves, as the forging rate increases, the forging load rapidly increases, as expected, due to the large strain rate sensitivity of superplastic materials. Also, as the

temperature decreases from 270°C the curves are displaced upwards with the increased stress. The complete curve is not shown for the lower forging temperatures for the condition A and B material. The curve for the condition C material is at a much higher stress level than the condition A or B material at an equivalent forging temperature. The condition D material is also substantially higher.

Figures 6 and 7 are similar types of σ-$\dot{H}$ curves for the Zn-Al-Cu alloy. The curves show the same general behaviour as the Zn-Al alloy for the same forging conditions. For both materials, at 270°C the forging stresses at rates of about 1 in/min (0.42 mm/sec) are down to only 600 psi (4.1 MN/m^2).

The effect of temperature can be seen more clearly on a true stress vs forging temperature plot, Figure 8. This is a plot for a forging rate of 3 in/min (1.3 mm/sec) at the time during forging when the sample has a height of 2.0 in (50.8 mm). A number of such plots can be drawn for various forging rates and deformation, but the trends followed will be the same. For all the curves, the true stress initially increases gradually as the temperature drops from 270°C, but beyond a certain point the change becomes much faster. Results for both alloys are shown with the Zn-Al alloy maintaining a lower stress level than the Zn-Al-Cu

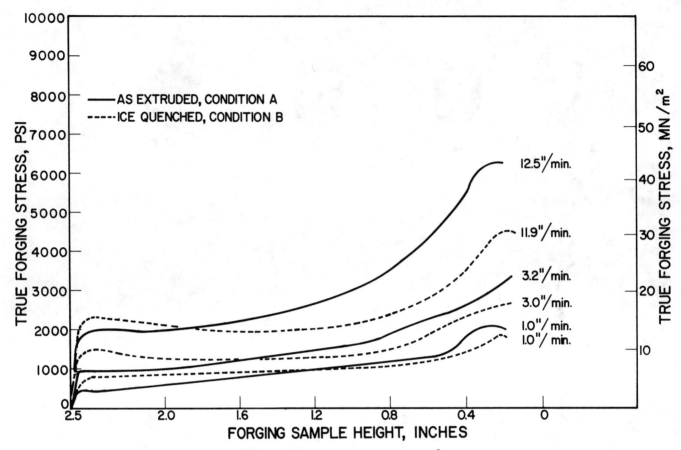

Figure 2. True stress vs sample height for the Zn-Al alloy forging at 270°C at the forging rates shown.

alloy at all temperatures. At temperatures around 270°C the difference between the two alloys is slight, whereas at the lower temperatures this difference becomes quite large. The line extending above 270°C from the condition B Zn-Al material shows the change in the forging stress when the material enters the single-phase region at 275°C.

Metallography

Specimens for metallography are mounted in plastic, coarse polished on silicon carbide grinding papers down to 600 grit and finish polished with 1.0 and 0.05 micron alumina powder. Etching is done by immersion in a 5% NaOH solution for 5 sec, rinse, and immersion in a 2% HNO_3 in ethyl alcohol for 5 sec. The structures are observed optically and on a scanning electron microscope.

The structures of the Zn-Al forging preforms in conditions A, B, C and D are shown in Figures 9a, 9c, 10, and 11, respectively. The ice-water quenched, condition B, material has the finest grain size, approximately 0.4 micron, followed by the condition A material, approximately 0.8 micron. The condition C material is more complex, consisting of a mixture of fine, equiaxed eutectoid regions and of lamellar eutectoid. The condition D material has a very coarse grain structure consisting of a large amount of the lamellar structure. The structures of the Zn-Al-Cu alloys

are equivalent to the Zn-Al alloy except for small islands of Cu-Zn ϵ phase (19) as shown in Figure 12 for the condition A material. The identification of this phase was confirmed using the electron microprobe.

The forged structures of the condition A and B material after a reduction of 93% in height at 270°C are shown in Figure 9b and 9d, respectively. In both cases the structure has remained equiaxed although the grain size has increased to 1.7 microns for the condition A material and to 1.4 microns for the condition B material.

The forged structures of the condition C and D material did not change significantly due to forging. The Zn-Al-Cu alloy forged structures paralleled the Zn-Al alloys, again, except for the islands of Cu-Zn ϵ phase.

Discussion of upset tests

Very large deformations are possible with Zn-Al and Zn-Al-Cu alloys under low stresses with superplastic forging. From a commercial point of view, the factors to consider when determining the desired forging conditions are:

(a) the simplest method of obtaining the forging preform;

(b) the simplest heat treatment prior to forging;

(c) the minimum time for heating prior to forging;

(d) the minimum forging temperature;

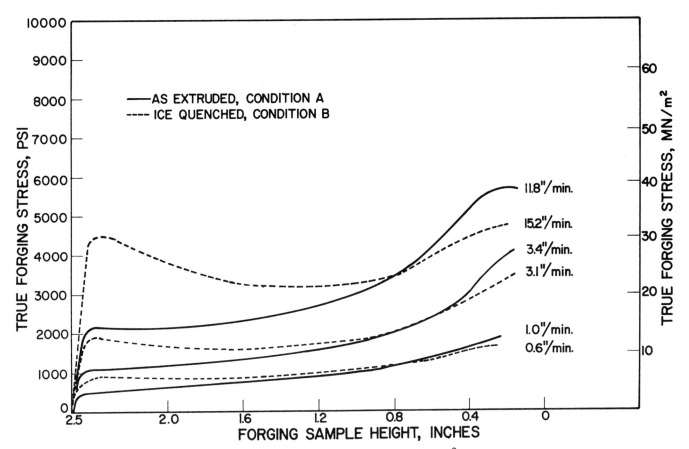

Figure 3. True stress vs sample height for the Zn-Al-Cu alloy forged at 270°C at the forging rates shown.

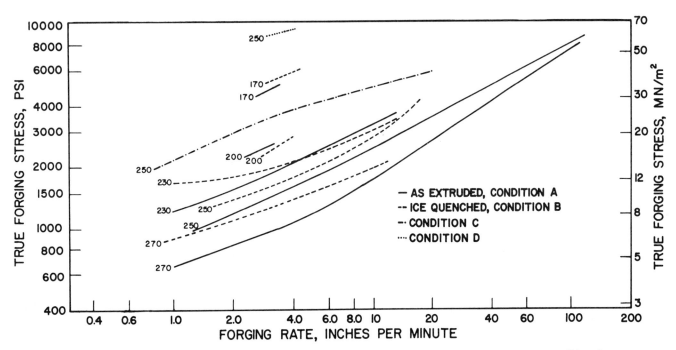

Figure 4. True stress vs forging rate for the Zn-Al alloy at a forging height of 2.0 in. at the temperature and condition shown.

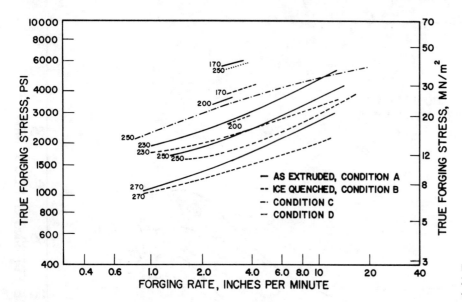

Figure 5. True stress vs forging rate for the Zn-Al alloy at a forging height of 1.0 in. at the temperature and condition shown.

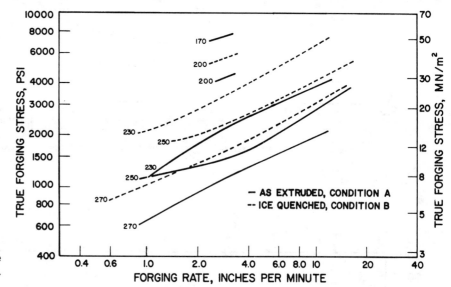

Figure 6. True stress vs forging rate for the Zn-Al-Cu alloy at a forging height of 2.0 in. at the temperature and condition shown.

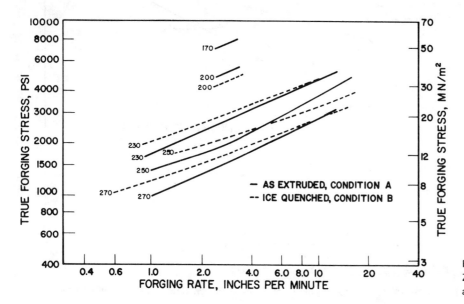

Figure 7. True stress vs forging rate for the Zn-Al-Cu alloy at a forging height of 1.0 in. at the temperature and condition shown.

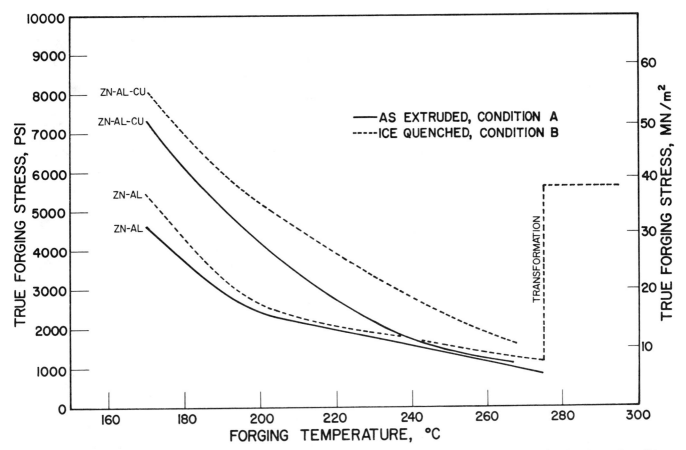

Figure 8. True stress vs forging temperature at a forging height of 2.0 in. and a forging rate of 3.0 in. per min for the alloy and condition shown.

(e) the minimum forging load;
(f) the maximum forging rate;
(g) the best lubricant;
(h) the best post-forging mechanical properties.

Obviously, from the upset forging results, these conditions cannot all be met simultaneously. If existing presses are used, the forging loads and rates may be partially fixed. Due to the low rates required, hydraulic presses are the only type suitable.

From the σ-H curves, Figures 4 and 5, the condition D material is unsuitable because of the unnecessary high loads. This large-grained material will not behave in a superplastic manner and, hence, die fill would be restricted. The air-cooled heat treatment would, however, be desirable in a finished part that is to be used at an elevated temperature because the superplastic tendencies are eliminated and the strength is retained.

The condition C material is also unsuitable because it also has a large flow stress, but not as large as the condition D material. From the microstructure, Figure 10, the fine-grained regions would behave in a superplastic manner, but the large islands of lamellar eutectoid would not, resulting in an increase in the flow stress.

The condition A and condition B material are both suitable for forging. Comparing the condition A and B material for equal forging conditions, Figure 2, the maximum in the σ-H curve that occurs in the condition B material is substantially greater than the stress in the condition A material. The two curves become close in the later stages of deformation and a cross-over is observed. The as-extruded, condition A material has the larger grain size because it was extruded at 200°C, near the forging temperature. The ice-water quench material, condition B, has an initial ultra-fine grain size which, when heated to the forging temperature, is seen to increase slightly. The grain size also increases during the forging.

The strain rate coefficient, m in the equation, is the slope of the σ-H curve. Exact values cannot be determined due to the limited number of tests and the wide range of strain rates between tests, but the value is between 0.30 and 0.55 for both the condition A and condition B material, which is within the superplastic range of 0.3 to 1.0.

Whether condition A or B material would be used commercially would be determined by the associated processes used to obtain the preforms. If extrusion is used to obtain rod for preforms, the as-extruded material would be the logical choice if the extrusion billets are heat treated.

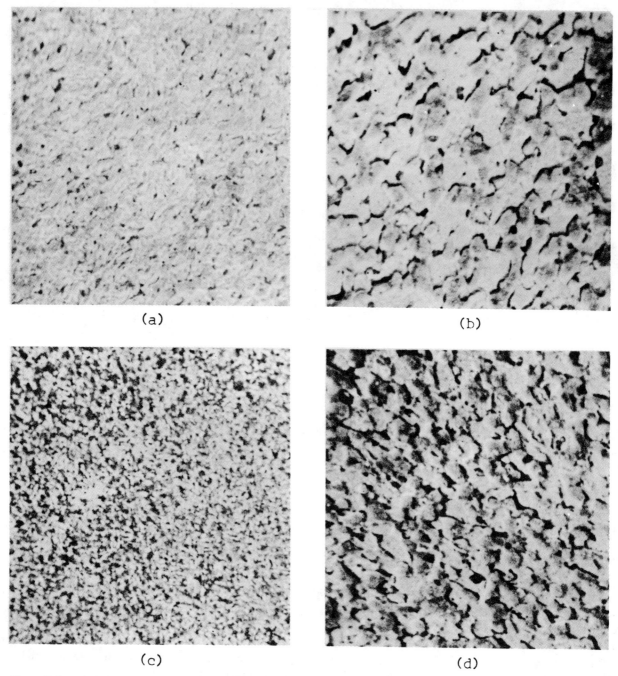

Figure 9. Scanning electron micrographs of the Zn-Al alloy: (a) in condition A, (b) in condition A followed by a 93% reduction at 270°C, (c) in condition B, and (d) in condition B followed by a 93% reduction at 270°C. (2600X)

This heat treatment would be an homogenization treatment followed by a quench. Similarly, rolling can be used in place of extrusion. It is possible that with a much larger degree of mechanical working, the heat treatment prior to working could be eliminated. If a cast preform is used, the ice-water quench, condition B, treatment must be used, because the as-cast material would have a grain structure similar to condition C or D material.

To utilize the low flow stresses of these superplastic materials, forging rates of less than approximately 5 in/min (2.1 mm/sec) would be desirable. This would keep the flow stress below 2000 psi (13.8 MN/m²). Although this is relatively slow, production can be fairly rapid because only one press stroke will be required to form even complex parts. Owing to the low stresses, parts with very large plan areas will be possible with light presses, or, alternatively, multiple cavities might be used in existing large bed presses.

From the plot of stress vs temperature, Figure 8, it is reasonable that the alloys be forged at a temperature as close to the transformation, 275°C, as possible. There is,

Figure 10. Optical micrograph of the Zn-Al alloy in condition C. (370X)

however, only a moderate increase in load down to about 210°C for the Zn-Al alloy and down to 250°C for the Zn-Al-Cu alloy. Because of the slow forging rates, the dies must be maintained at these temperatures. Die life should be long, because of the very low loads and good lubrication.

During the experiments it was noted that these materials are insensitive to cracks in the preform. If a crack was present initially due to an extrusion defect, there was no tendency for the cracks to propagate even after very large deformations.

Closed die forging

The two alloys have been forged into the shape shown in Figure 13. The die is such that no flash is produced. The forging preform, on the left in Figure 13, is the same in size and in shape as used in the upset tests. The part was forged in one press stroke with graphite for lubrication. The base of the forging is 2.5 in (63.5 mm) in diameter and the part is 1.75 in (44.5 mm) high. The webs on the part taper from 0.25 in (6.4 mm) thick at the base to 0.188 in (4.78 mm) thick at the free end or from 0.125 in (3.18 mm) thick at the base to 0.063 in (1.6 mm) thick at the free end. There are two of each type of web on the part. The thin cylindrical rod projecting from the web corner is where the material has been extruded into the 0.045-in (1.1 mm) diameter vent holes.

Completely filling the die was not difficult. With a forging rate of 0.5 in per minute (2.1 mm/sec) the maximum load was 100 tons (0.89 MN) at 260°C. Even though the material flow is complex there were no laps formed in the finished part. This shape shows that these superplastic materials can easily be formed into very complex shapes with thin sections.

Figure 11. Optical micrograph of the Zn-Al alloy in condition D. (150X)

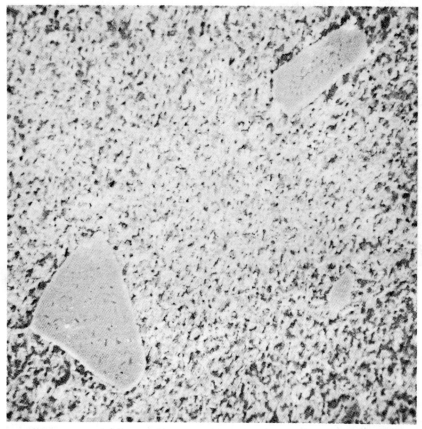

Figure 12. Scanning electron micrograph of the Zn-Al-Cu alloy in condition A. (1300X)

Conclusions

1. Large deformations at low stresses can be obtained in Zn-Al and Zn-Al-Cu alloys under actual forging conditions.

2. The optimum superplastic forging conditions are at a forging rate of less than 5 in/min (2.1 mm/sec) and at a temperature as close to the transformation as possible.

3. Good lubrication during forging is obtained with teflon sheet, a grease-molybdenum-disulphide mixture, or colloidal graphite.

4. Two satisfactory material preparation methods are:

(a) as-extruded material from billets that were ice-water quenched from above 275°C;

(b) ice-water quenching from above 275°C.

5. A complex shape can be forged with excellent die fill under relatively low loads.

References

(1) R.H. Johnson. Superplasticity. *Met. Reviews*, 15 (146): 115-134 (1970).

(2) W.A. Backofen, I.R. Turner and D.H. Avery. Superplasticity in an Al-Zn Alloy. *Trans. ASM*, 57 (4): 980-990 (1964).

(3) D.H. Avery and W.A. Backofen. A Structural Basis for Superplasticity. *Trans. ASM*, 58 (4): 551-562 (1965).

(4) K. Nuttall and R.B. Nicholson. Microstructure of Superplastic Alloys. *Phil. Mag.*, 17 (149): 1087-1091 (1968).

(5) C.M. Packer and O.D. Sherby. An Interpretation of the Superplasticity Phenomenon in Two-Phase Alloys. *Trans. ASM*, 60 (1): 21-28 (1967).

(6) D.L. Holt. The Relation Between Superplasticity and Grain Boundary Shear in Aluminum-Zinc Eutectoid Alloy. *Trans. TMS AIME*, 242 (1): 25-31 (1968).

Figure 13. The closed die forged part produced in a single forging step from the preform shown on the left.

(7) T.H. Alden and H.W. Schadler. The Influence of Structure on the Flow Stress-Strain Rate Behaviour of Zn-Al Alloys. *Trans. TMS AIME*, 242 (5): 825-832 (1968).

(8) R.H. Johnson, C.M. Packer, L. Anderson and O.D. Sherby. Microstructure of Superplastic Alloys. *Phil. Mag.*, 18 (156): 1309-1314 (1968).

(9) C.M. Packer, R.H. Johnson and O.D. Sherby. Evidence for the Importance of Crystallographic Slip During Superplastic Deformation of Eutectic Zn-Al. *Trans. TMS AIME*, 242 (12): 2485-2489 (1968).

(10) H. Naziri and R. Pearce. The Influence of Copper Additions on the Superplastic Forming Behaviour of the Zn-Al Eutectoid. *Int. J. Mech. Sci.*, 12 (6): 513-521 (1970).

(11) K. Nuttall. Superplasticity Above the Invariant in the Eutectoid Zn-Al Alloy. *J. Inst. Metals*, 99: 291-292 (Sept. 1971).

(12) G.C. Cornfield and R.H. Johnson. The Forming of Superplastic Sheet Metal. *Int. J. Mech. Sci.*, 12 (6): 479-490 (1970).

(13) T.Y.M. Al-Naib and J.L. Duncan. Superplastic Metal Forming. *Int. J. Mech. Sci.*, 12 (6): 463-477 (1970).

(14) T.H. Thomsen, D.L. Holt and W.A. Backofen. Forming Superplastic Sheet Metal in Bulging Dies. *Met. Eng. Quart.*, 10 (2): 1-7 (1970).

(15) K. Lacy. Metals that Form Like Thermoplastics. *Metalworking Prod.*, 113: 29-32 (Jan. 22, 1969).

(16) Superplastic Alloy Can Mold Prototype Parts Cheaper, Faster. *Product Eng.*, 42 (8): 36 (1971).

(17) W.A. Backofen. Superplasticity Enchants Metallurgy. *Steel*, 165 (24): 25-28 (1969).

(18) L.T. Feng. Closed-Die Forming Process Brings Use of Superplastic Alloys Closer. *Automot. Eng.*, 78 (9): 44-45 (1970).

(19) L. Taylor, ed., *Metals Handbook, 1948 Edition*, The American Society for Metals, 1948, p. 1244.

Precision Forging of a High Strength Superplastic Zinc-Aluminum Alloy

R. W. BALLIETT, J. A. FORSTER, AND J. L. DUNCAN

Precision forgings can be produced in a high strength zinc aluminum alloy exhibiting superplasticity by means of an isothermal forging cycle which combines an approach at constant speed followed by a dwell at constant load. This study shows that integral threads, sharp corners, and very thin webs can be formed directly in the forging operation. A pressure compensated time parameter is developed which permits forging operations over a range of different conditions to be correlated. This parameter may be used to predict the result of actual precision forging processes from scale model tests.

SUPERPLASTIC alloys have a characteristically low flow stress when they are deformed at low strain-rates. This flow stress increases rapidly with strain-rate and at strain rates in the range normally employed in commercial hot working, the forming loads required are similar to those which must be applied to conventional alloys. With the development of superplastic alloys which have good mechanical and physical properties at ambient temperatures, it has become desirable to evaluate economic techniques for producing precision forged parts utilizing the superplastic properties of these alloys. It has been suggested that economical slow speed forging processes could be developed for superplastic alloys in certain applications. The disadvantage of longer forming times would be outweighed by the lower cost of tooling and press equipment which is made possible by the significant reductions in forging loads.

A number of forging experiments carried out with the eutectoid zinc-aluminum alloy have been reported. This alloy can readily be produced in an ultrafine grain structure and deforms under low stresses at temperatures just below the transformation at 527°F (275°C). Processes such as those described by Fields,[1] Stewart,[2] and Feng[3] are extensions of compression molding practice carried out in the plastics industry. Other experiments in which zinc-aluminum dies for coining operations were produced by hobbing are described by Saller and Duncan.[4] The most extensive work on closed die forging is that by Stewart.[2] He performed a number of upsetting and forging operations on zinc-aluminum-copper alloys over a range of forging speeds and temperatures. Using pressures of up to 60,000 psi (414 MPa), he showed that excellent die filling could be obtained even in very thin web sections. In addition, very small draft angles could be used and sharp corners were readily produced on the parts. All of Stewart's experiments were carried out on a hydraulic press at very low strain-rates. Closed die forging of a Pb-Sn-Zn ternary eutectoid alloy was carried out by Moles[6] using dies which had been prepared for conventional hot forging of gas turbine blades. He showed that forging could be

achieved at one-tenth the usual load but that conventional design of flash gutters was not appropriate for the superplastic alloys.

The object of this present work is to efficiently utilize superplasticity in the production of precision forgings using a modified eutectoid zinc-aluminum alloy. This alloy, known commercially as HSZ*, is

*St. Joe Minerals Corp.; Monaca, PA, USA.

superplastic at 500°F (260°C) and possesses good mechanical properties at room temperature. The forging cycle investigated combines an approach at a constant and rapid ram speed followed by a dwell at constant load. The forging cycle is illustrated in Figs. 1 and 2. Figure 1 shows total forging load as a function of time while Fig. 2 shows the forging load as a function of displacement. A comparison of these two figures reveals that in this process, most of the forging occurs during the initial period of rapid ram advance. The balance of the deformation required to fill the "detail" in the part occurs at maximum load during the dwell provided at the end of the stroke. This paper describes details of a series of experiments to determine the relationship between load and time in the forging of detail in a simple part. In addition, a number of different parts were produced to illustrate the feasibility of producing features such as screw threads, thin webs, and raised lettering in precision forged parts using the technique described previously.

R. W. BALLIETT, formerly with St. Joe Mineral Corp., is now Director of Technical Development, Phelps Dodge Brass Company, Anniston, AL 36202. J. A. FORSTER and J. L. DUNCAN are Graduate Student and Professor, respectively, Department of Mechanical Engineering, McMaster University, Hamilton, Ontario L8S 4L7, Canada.
Manuscript submitted June 14, 1977.

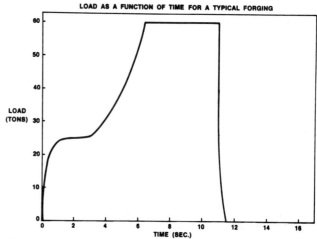

LOAD AS A FUNCTION OF TIME FOR A TYPICAL FORGING

Fig. 1—Load as a function of time for a typical precision forging.

Reprinted with permission from Metallurgical Transactions A, 9A, September 1978, 1259-1264, © 1978 American Society for Metals and The Metallurgical Society of AIME

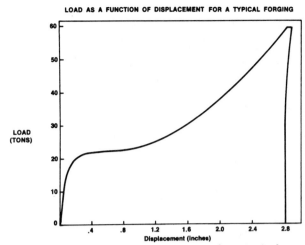

Fig. 2—Load as a function of displacement for a typical precision forging.

PROCEDURE

The composition of the alloy studied is as follows:

Al	Cu	Mg	Ca	Zn
22.1 wt pct	0.95 wt pct	0.045 wt pct	0.025 wt pct	balance

The alloy is prepared from Special High Grade Zinc (99.99 pct) and 99.9 pct aluminum ingot with alloying additions made from commercially pure materials. It is continuously cast into 5 in. (127 mm) diam extrusion billets using a conventional vertical casting technique. The cast billets are solution treated at 690°F (365°C) for 48 h, water quenched, and subsequently extruded at 500°F (260°C) into rods of the appropriate diameter. The microstructure in the extruded condition, Fig. 3, consists essentially of a two phase, finely divided granular pearlite with an apparent grain size of approximately 3 μm.

Flow stress data for this alloy, Fig. 4, has been presented by Balliett and Abramowitz[5] who performed isothermal compression tests at constant strain-rate. The usual constitutive equation

$$\sigma = k \, \dot{\epsilon}^{\,m} \qquad [1]$$

has been fitted to the stress strain rate data in the

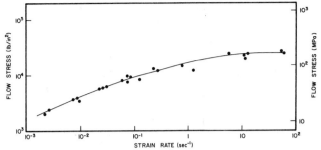

Fig. 4—Flow stress at 500°F (260°C) for the HSZ alloy as a function of strain-rate.

range $10^{-3} < \dot{\epsilon} < 10^{-1}$ per s. Values for the constants m and k in Eq. [1], derived from tests at 500°F (260°C) with total strains in the range $0.2 < \epsilon < 0.5$, are summarized in Table I. The room temperature properties of this alloy are shown in Table II.

Forging experiments were performed using a 200 T (1.78 MN) metalworking press, Fig. 5, which is fully instrumented and fitted with closed loop electro-hydraulic control permitting accurate control of ram speed and load.

In an attempt to compare the time taken to form a particular feature at different pressures, a tool was constructed to forge a lip on a billet as shown in Fig. 6. The height of the lip was measured as a function of both time and pressure. The second set of tools, Fig. 7, consists of a container into which various inserts

Table I. Stress/Strain Rate Properties at 500°F (260°C)

Strain-Rate Sensitivity, m	0.4	
Flow-Stress constant, k	25000	s⁴ lb/in.²
	172.5	s⁴ MPa

Table II. Room Temperature Mechanical Properties of HSZ

Tensile Strength	60 ksi	414 MPa
Yield Strength	48 ksi	331 MPa
Pct Elongation	19	19
Hardness	64 R_B	
Impact, CVN	8 ft-lbs	10.8 J

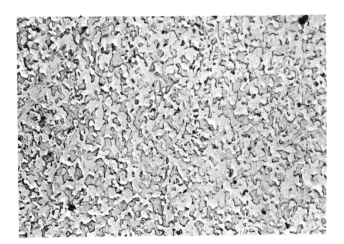

Fig. 3—Microstructure of the HSZ alloy as-extruded. Nital etch, magnification 11,700 times.

Fig. 5—200 T (1.78 MN) metalworking press at McMaster University.

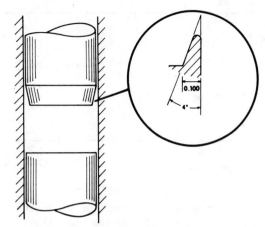

Fig. 6—Schematic of tool used to produce forged 'lip'.

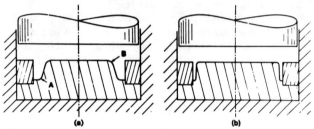

Fig. 7—Schematic of precision forging tools used to produce ribs of varying thickness.

can be placed to produce parts having various features such as thick and thin webs.

PRESSURE-TIME ANALYSIS

A series of experiments was conducted to develop an analytical parameter which could be used to evaluate the effects of changes in pressure and time on forgings produced in the tools shown in Fig. 6. The experiments were performed with a closed forging die which produces a lip on a billet. To avoid problems of billet eccentricity, machined billets were used which were a close fit in the cylinder. The die was lubricated with mineral oil and the forging temperature was 482°F (250°C). The ram speed was 0.02 in. (0.5 mm) per s and forging pressures up to 38,000 psi (262 MPa) were employed. The cross-section of a fully forged part is shown in Fig. 8. Billets of three different heights were used and after forging the average height of the lip, h_l, as shown in Fig. 6, was measured with a depth micrometer. It may be anticipated

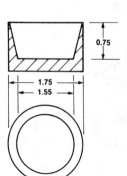

Fig. 8—Cross section of disk with forged lip.

that for thinner billets, frictional restraint across the faces of the punch and the base would restrict forging. Cups formed from billets of different thicknesses subjected to the same load and time are shown in Fig. 9. The parameter h_l/D is plotted against initial billet size in Fig. 10 and in some cases the middle thickness billet had the smallest lip.

The geometric parameter h_l/D was selected because it represents a suitable design variable. In any analysis it is essential to evaluate a parameter which is indicative of the process under consideration and also applicable to geometrically similar processes. The pressure is a function of the diameter, while the height of the lip is a function of the pressure, time and tooling geometry. It is apparent that the parameter h_l/D is a function of the major process variables and can be used in scaling the results of these tests for any geometrically similar process.

Considerations of deformation in a simple process, as shown in the Appendix, indicate that in a constant pressure process of a particular part, the time taken to achieve a given change in shape would depend on the pressure and the material properties. The analysis indicates that the geometric parameter h_l/D would be a function of a pressure compensated time parameter, namely,

$$h_l/D = f\left\{t \cdot \left(\frac{\bar{p}}{k}\right)^{1/m}\right\} \qquad [2]$$

where $\bar{p}$ is the mean forging pressure, $4P/\pi d^2$. The

Fig. 9—Cups formed from billets of various thicknesses.

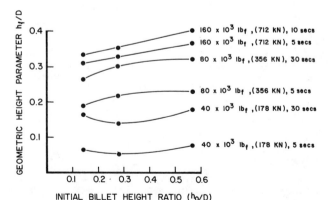

Fig. 10—Geometric height parameter (h_l/D) as a function of initial billet height ratio (h_o/D).

form of the function *f* would be dependent on the tool geometry and initial billet height. To test this hypothesis the results of measured lip heights are plotted against the time parameter in Fig. 11.

The results are for billets having an initial height to diameter ratio of 0.14 to 0.57 and although there is considerable scatter, the results include tests over a wide range of parameters; pressures from 40,000 to 160,000 lb/in.^{2} (276 to 1104 MPa) and forging times from 5 to 60 s. The functional relationship demonstrated would appear to be sufficiently accurate for process design purposes.

In all the tests only one material, HSZ, was used and further work is required to demonstrate the generality of the pressure/time parameter suggested. It would also be interesting to see whether such a parameter could be extended to take account of temperature using an Arrenhius type function as employed by others[7,8] in extrusion.

In many cases the thickness of the lip at the top was less than 0.005 in. (0.13 mm); this could not normally be produced in conventional materials.

FORGING OF PARTS

The composite closed forging die, Fig. 7, permitted forging of a number of different parts by changing inserts within the die. The diameter of these parts was 3.5 in. (89 mm) and forging pressures up to 20,000 psi (138 MPa) were used. All the forgings were in HSZ alloy and the initial billet was about 2 in. (50 mm) in diam. Various billet lengths were employed. In all these experiments the approach speed of the ram was the maximum speed the machine was capable of, about 20 in. (508 mm) per min and the temperature was 482°F (250°C). The maximum load was 100 tons (890 kN). Figure 12 shows the first of two parts that were designed to demonstrate the effect of forging thin webs adjacent to a thick section. The part with the thicker flange was readily formed in less than 5 s. The thinner flange however required almost 2 min. The thickness of the flange was 0.05 in. (1.3 mm), the draft one degree, and the height of the flange approximately 0.625 in. (16 mm).

It is interesting to note that the most difficult corner to fill in the die, Fig. 7(*a*), was at A, and that

Fig. 12—Parts showing both thin and thick forged webs.

a cavity sometimes appeared in the part at B if the billet was not exactly centered before forging.

Forming Threads

Previous experiments on the forging of a screw thread[9] indicated that whiskers of material could be shaved from the billet during the deformation. To avoid this problem of shaving, the threaded parts, as shown in Fig. 13, were produced from a preform. Complete forging of a 2.625 in. (67 mm), 8 TPI (0.31 TPMM) thread was readily obtained with a load of 100 tons (890 kN) in less than 10 s.

Raised Lettering

By replacing the flat top punch with an engraved one, raised lettering was produced on the top surface of a part with a load of 100 tons (890 kN) applied for less than 5 s. A successfully formed part is shown in Fig. 14. The lettering is 0.05 in. (1.3 mm) square in cross-section with sharp corners and no draft.

Spline

The part having a castellated head, zero draft and sharp corners shown in Fig. 15 was easily produced at a load of 100 tons (890 kN) in less than 5 s. In addition, and also shown in Fig. 15, a part with a hexagonal head, with zero draft, was also produced in less than 5 s.

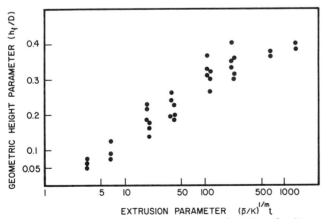

Fig. 11—Geometric height parameter as a function of a dimensionless pressure compensated time parameter for billets having an initial height/diameter ratio of 0.14 to 0.57.

Fig. 13—Part showing as-forged threads.

Source: Metallurgical Transactions A, 9A, September, 1978, 1259-1264

Fig. 14—Part showing raised lettering.

Fig. 15—Parts showing castellated and hexagonal heads forged from the HSZ alloy.

The experience in forming these parts can be summarized as follows:

1) At forging pressures of about 20,000 psi (138 MPa), complete filling was obtained in less than 5 s for most parts.

2) Sharp corners and raised features can be produced readily.

3) Minimum web thicknesses and draft angles can be significantly less than in conventional forging practice.

4) Threads can be forged completely and accurately.

5) Both machined billets and ones sawn from extruded bar were used. There was no difference in the surface quality of the forged part.

6) The surface of the die is reproduced in the part and mirror finishes can be obtained by using highly polished dies.

7) Poor die filling in some parts was obtained when the billet was not concentric in the die.

8) Very little flash was produced at the pressures employed in these tests.

CONCLUSIONS

These forging experiments indicate that by using a constant load dwell period of a few seconds at the end of a high strain rate forging stroke, precision forgings of a high strength zinc alloy can be produced at very modest forging pressures. A pressure-com-

pensated time parameter has been developed which, for the HSZ alloy, can be used to evaluate the effects of additional pressure and time on actual precision forgings. The demonstration parts produced indicate that precise detail such as screw threads, raised lettering, and very thin webs can be forged. Using this technique it is not unreasonable to believe that precision forgings could be produced on a commercial basis at a rate of 200 to 500 parts per h having all the characteristics described.

APPENDIX A

We consider the simple frictionless compression of a cylinder, Eq. [A1], which obeys the constitutive law:

$$\sigma = k \, \dot{\epsilon}^{\,m}. \tag{A1}$$

The governing equation for the process is:

$$\frac{P}{A} = \bar{p} = k \cdot \left[\frac{1}{h} \frac{dh}{dt} \right]^m$$

which can be written:

$$\left[\frac{\bar{p}}{k} \right]^{1/m} dt = -\frac{dh}{h}. \tag{A2}$$

The time taken for a given in shape from h_o to h is given by:

$$\int_0^t \left[\frac{\bar{p}}{k} \right]^{1/m} dt = \int_{h_o}^{h} -\frac{1}{h} \, dh. \tag{A3}$$

For a constant pressure process:

$$t \cdot \left[\frac{\bar{p}}{k} \right]^{1/m} = \ln \left[\frac{h_o}{h} \right]. \tag{A4}$$

We postulate that for any similar process:

$$h = f \left\{ t \left[\frac{\bar{p}}{k} \right]^{1/m} \right\} \tag{A5}$$

where h is an equivalent change of dimension associated with the process and f is a function of the tool geometry. It should be noted that h is related to the total strain by

$$\epsilon = \ln \left[\frac{h_o}{h} \right]. \tag{A6}$$

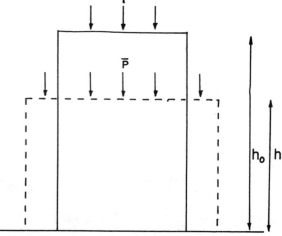

Fig. A1: Uniform frictionless compression of a cylinder.

REFERENCES

1. D. S. Fields and T. J. Stewart: *Int. J. Mech. Sci.,* 1970, vol. 13, p. 63.
2. M. J. Stewart: *Can. Met. Quart.,* 1973, vol. 12, no. 2, pp. 159–69.
3. L. T. Feng: *SAE Booklet 700133,* January 1970.
4. R. A. Saller and J. L. Duncan: *J. Inst. Metals,* 1971, vol. 99, p. 173.
5. R. W. Balliett and P. H. Abramowitz: *Proceedings Third North American Metal-working Research Conference,* pp. 100-14, 1975.
6. M. D. C. Moles: *D. Phil. Thesis,* University of Cambridge, 1972.
7. W. Wong and J. Jones: *Trans. TMS-AIME,* 1968, vol. 242, p. 2271.
8. M. M. Farag and C. M. Sellars: *J. Inst. Metals,* 1973, vol. 101, p. 137.
9. J. A. Forster and J. L. Duncan: Metalworking Research Report No. 55, McMaster University, 1975.

INDEX